T0205765

Lecture Notes in Computer Science 14429

The series Lecture Notes in Computer Science (LNCS), including its subseries Lecture Notes in Artificial Intelligence (LNAI) and Lecture Notes in Bioinformatics (LNBI), has established itself as a medium for the publication of new developments in computer science and information technology research, teaching, and education.

LNCS enjoys close cooperation with the computer science R & D community, the series counts many renowned academics among its volume editors and paper authors, and collaborates with prestigious societies. Its mission is to serve this international community by providing an invaluable service, mainly focused on the publication of conference and workshop proceedings and postproceedings. LNCS commenced publication in 1973.

Qingshan Liu · Hanzi Wang · Zhanyu Ma ·
Weishi Zheng · Hongbin Zha · Xilin Chen ·
Liang Wang · Rongrong Ji
Editors

Pattern Recognition and Computer Vision

6th Chinese Conference, PRCV 2023
Xiamen, China, October 13–15, 2023
Proceedings, Part V

Editors
Qingshan Liu
Nanjing University of Information Science and Technology
Nanjing, China

Zhanyu Ma
Beijing University of Posts and Telecommunications
Beijing, China

Hongbin Zha
Peking University
Beijing, China

Liang Wang
Chinese Academy of Sciences
Beijing, China

Hanzi Wang
Xiamen University
Xiamen, China

Weishi Zheng
Sun Yat-sen University
Guangzhou, China

Xilin Chen
Chinese Academy of Sciences
Beijing, China

Rongrong Ji
Xiamen University
Xiamen, China

ISSN 0302-9743 ISSN 1611-3349 (electronic)
Lecture Notes in Computer Science
ISBN 978-981-99-8468-8 ISBN 978-981-99-8469-5 (eBook)
https://doi.org/10.1007/978-981-99-8469-5

This Springer imprint is published by the registered company Springer Nature Singapore Pte Ltd.
The registered company address is: 152 Beach Road, #21-01/04 Gateway East, Singapore 189721, Singapore

Paper in this product is recyclable.

Preface

Welcome to the proceedings of the Sixth Chinese Conference on Pattern Recognition and Computer Vision (PRCV 2023), held in Xiamen, China.

PRCV is formed from the combination of two distinguished conferences: CCPR (Chinese Conference on Pattern Recognition) and CCCV (Chinese Conference on Computer Vision). Both have consistently been the top-tier conference in the fields of pattern recognition and computer vision within China's academic field. Recognizing the intertwined nature of these disciplines and their overlapping communities, the union into PRCV aims to reinforce the prominence of the Chinese academic sector in these foundational areas of artificial intelligence and enhance academic exchanges. Accordingly, PRCV is jointly sponsored by China's leading academic institutions: the Chinese Association for Artificial Intelligence (CAAI), the China Computer Federation (CCF), the Chinese Association of Automation (CAA), and the China Society of Image and Graphics (CSIG).

PRCV's mission is to serve as a comprehensive platform for dialogues among researchers from both academia and industry. While its primary focus is to encourage academic exchange, it also places emphasis on fostering ties between academia and industry. With the objective of keeping abreast of leading academic innovations and showcasing the most recent research breakthroughs, pioneering thoughts, and advanced techniques in pattern recognition and computer vision, esteemed international and domestic experts have been invited to present keynote speeches, introducing the most recent developments in these fields.

PRCV 2023 was hosted by Xiamen University. From our call for papers, we received 1420 full submissions. Each paper underwent rigorous reviews by at least three experts, either from our dedicated Program Committee or from other qualified researchers in the field. After thorough evaluations, 522 papers were selected for the conference, comprising 32 oral presentations and 490 posters, giving an acceptance rate of 37.46%. The proceedings of PRCV 2023 are proudly published by Springer.

Our heartfelt gratitude goes out to our keynote speakers: Zongben Xu from Xi'an Jiaotong University, Yanning Zhang of Northwestern Polytechnical University, Shutao Li of Hunan University, Shi-Min Hu of Tsinghua University, and Tiejun Huang from Peking University.

We give sincere appreciation to all the authors of submitted papers, the members of the Program Committee, the reviewers, and the Organizing Committee. Their combined efforts have been instrumental in the success of this conference. A special acknowledgment goes to our sponsors and the organizers of various special forums; their support made the conference a success. We also express our thanks to Springer for taking on the publication and to the staff of Springer Asia for their meticulous coordination efforts.

We hope these proceedings will be both enlightening and enjoyable for all readers.

October 2023

Qingshan Liu
Hanzi Wang
Zhanyu Ma
Weishi Zheng
Hongbin Zha
Xilin Chen
Liang Wang
Rongrong Ji

Organization

General Chairs

Hongbin Zha	Peking University, China
Xilin Chen	Institute of Computing Technology, Chinese Academy of Sciences, China
Liang Wang	Institute of Automation, Chinese Academy of Sciences, China
Rongrong Ji	Xiamen University, China

Program Chairs

Qingshan Liu	Nanjing University of Information Science and Technology, China
Hanzi Wang	Xiamen University, China
Zhanyu Ma	Beijing University of Posts and Telecommunications, China
Weishi Zheng	Sun Yat-sen University, China

Organizing Committee Chairs

Mingming Cheng	Nankai University, China
Cheng Wang	Xiamen University, China
Yue Gao	Tsinghua University, China
Mingliang Xu	Zhengzhou University, China
Liujuan Cao	Xiamen University, China

Publicity Chairs

Yanyun Qu	Xiamen University, China
Wei Jia	Hefei University of Technology, China

Local Arrangement Chairs

Xiaoshuai Sun	Xiamen University, China
Yan Yan	Xiamen University, China
Longbiao Chen	Xiamen University, China

International Liaison Chairs

Jingyi Yu	ShanghaiTech University, China
Jiwen Lu	Tsinghua University, China

Tutorial Chairs

Xi Li	Zhejiang University, China
Wangmeng Zuo	Harbin Institute of Technology, China
Jie Chen	Peking University, China

Thematic Forum Chairs

Xiaopeng Hong	Harbin Institute of Technology, China
Zhaoxiang Zhang	Institute of Automation, Chinese Academy of Sciences, China
Xinghao Ding	Xiamen University, China

Doctoral Forum Chairs

Shengping Zhang	Harbin Institute of Technology, China
Zhou Zhao	Zhejiang University, China

Publication Chair

Chenglu Wen	Xiamen University, China

Sponsorship Chair

Yiyi Zhou	Xiamen University, China

Exhibition Chairs

Bineng Zhong	Guangxi Normal University, China
Rushi Lan	Guilin University of Electronic Technology, China
Zhiming Luo	Xiamen University, China

Program Committee

Baiying Lei	Shenzhen University, China
Changxin Gao	Huazhong University of Science and Technology, China
Chen Gong	Nanjing University of Science and Technology, China
Chuanxian Ren	Sun Yat-Sen University, China
Dong Liu	University of Science and Technology of China, China
Dong Wang	Dalian University of Technology, China
Haimiao Hu	Beihang University, China
Hang Su	Tsinghua University, China
Hui Yuan	School of Control Science and Engineering, Shandong University, China
Jie Qin	Nanjing University of Aeronautics and Astronautics, China
Jufeng Yang	Nankai University, China
Lifang Wu	Beijing University of Technology, China
Linlin Shen	Shenzhen University, China
Nannan Wang	Xidian University, China
Qianqian Xu	Key Laboratory of Intelligent Information Processing, Institute of Computing Technology, Chinese Academy of Sciences, China
Quan Zhou	Nanjing University of Posts and Telecommunications, China
Si Liu	Beihang University, China
Xi Li	Zhejiang University, China
Xiaojun Wu	Jiangnan University, China
Zhenyu He	Harbin Institute of Technology (Shenzhen), China
Zhonghong Ou	Beijing University of Posts and Telecommunications, China

Contents – Part V

Face Recognition and Pose Recognition

Biometric Recognition

Spoof-Guided Image Decomposition for Face Anti-spoofing

Bin Zhang[1,2], Xiangyu Zhu[3,4], Xiaoyu Zhang[1,2], Shukai Chen[6], Peng Li[7], and Zhen Lei[3,4,5](✉)

[1] Institute of Information Engineering, Chinese Academy of Sciences, Beijing, China
{zhangbin1998,zhangxiaoyu}@iie.ac.cn
[2] School of Cyber Security, University of Chinese Academy of Sciences, Beijing, China
[3] State Key Laboratory of Multimodal Artificial Intelligence Systems, Institute of Automation, Chinese Academy of Sciences, Beijing, China
{xiangyu.zhu,zlei}@nlpr.ia.ac.cn
[4] School of Artificial Intelligence, University of Chinese Academy of Sciences, Beijing, China
[5] Centre for Artificial Intelligence and Robotics, Hong Kong Institute of Science & Innovation, Chinese Academy of Sciences, Hong Kong, China
[6] ZKTeco Co., Ltd., Dongguan, China
richard.chen@zkteco.com
[7] China University of Petroleum (East China), Dongying, China
lipeng@upc.edu.cn

Abstract. Face spoofing attacks have become an increasingly critical concern when face recognition is widely applied. However, attacking materials have been made visually similar to real human faces, making spoof clues hard to be reliably detected. Previous methods have shown that auxiliary information extracted from the raw RGB data, including depth map, rPPG signal, HSV color space, etc., are promising ways to highlight the hidden spoofing details. In this paper, we consider extracting novel auxiliary information to expose hidden spoofing clues and remove scenarios specific, so as to help the neural network improve the generalization and interpretability of the model's decision. Considering that presenting faces from spoof mediums will introduce 3D geometry and texture differences, we propose a spoof-guided face decomposition network to disentangle a face image into the components of normal, albedo, light, and shading, respectively. Besides, we design a multi-stream fusion network, which effectively extracts features from the inherent imaging components and captures the complementarity and discrepancy between them. We evaluate the proposed method on various databases, i.e. CASIA-MFSD, Replay-Attack, MSU-MFSD, and OULU-NPU. The results show that our proposed method achieves competitive performance in both intra-dataset and inter-dataset evaluation protocols.

Supplementary Information The online version contains supplementary material available at https://doi.org/10.1007/978-981-99-8469-5_1.

Q. Liu et al. (Eds.): PRCV 2023, LNCS 14429, pp. 3–15, 2024.
https://doi.org/10.1007/978-981-99-8469-5_1

Keywords: Face Anti-spoofing · Face Presentation Attack Detection · Imaging Components · Face Decomposition

1 Introduction

With the development of mobile devices and embedded devices, face authentication technology has infiltrated all aspects of our lives. Face authentication systems widely adopt RGB cameras as acquisition devices, but they are easily deceived by identity attacks. Face spoofing [1–3] is one of the most easily implemented identity attacks. Attackers fool face authentication systems by presenting the target faces from spoof mediums, such as printed photos and video replay. To secure face authentication systems, both the industry and academia have been paying great attention to the problem of face anti-spoofing, which aims to discriminate spoofing attacks from bonafide attempts of genuine users.

In this paper, we aim to find the spoofing clues from a single RGB image by analyzing the imaging process of presentation attack instruments. Advances in making spoofing materials have been able to reduce the spoofing signals to a low magnitude, making anti-spoofing an extremely challenging task. For example, the appearance of the high-resolution recorded sensors, high-precision color laser printers, and retina screens have made it difficult for traditional anti-spoofing methods [1,2] to achieve satisfactory results. Despite the success of recent deep learning techniques in face anti-spoofing, training a vanilla CNN with binary supervision to predict the spoofness of an RGB input will easily overfit the training data leading to poor performance on unseen data [4]. To solve the problem, works show that the combination of auxiliary information extracted from raw RGB images effectively improves the generalization of the face anti-spoofing methods, including complementary color space [5,6], rPPG [7], noise pattern [8], reflectance [9,10], depth map [11–13], etc. This auxiliary information proves that although the subtle spoofing clues in the original image are difficult to detect, they can be highlighted in some auxiliary information extracted by specially designed preprocessing methods.

Considering that the spoofing images are obtained through secondary imaging, it will inevitably introduce imaging components differences from the genuine face. Compared with the genuine faces, the printed photos or digital displays adopted by presentation attack have different 3D geometry, which is more like a flat surface. Therefore, we use the surface normal to better represent the intra-structure and depth variation of the scene. In addition, the material textures of the printed photos and display devices are different from the human skin, and this material difference can be reflected in the inherent imaging components of the face, like albedo. We compare the albedo difference between genuine face and spoofing mediums in Appendix Fig. 1. Besides, the meaning and advantages of normal and albedo are detailed in Appendix B.

To capture and magnify this difference, we proposed a learnable decomposition network called Spoof-guided Decomposition Network (SgDN), which can disentangle an RGB face image into the imaging components normal, albedo, light, and shading maps. This is a challenging objective due to the lack of the ground-truth components of real-world data and spoofing data during the model learning. To enable the network to disentangle both genuine and spoof faces, we first train SgDN with a mixture of labeled synthetic and unlabeled real-world images to simulate the physical model of Lambertian image generation. Then, we set the normal channel of spoofing samples to 0, and push the albedo to encode the artifacts that Lambertian imaging model cannot resolve. To further utilize the above auxiliary features effectively, we design a multi-stream network to fuse the information from different components at different scales.

In summary, our contributions are:

1) we propose a Spoof-guided Decomposition Network to disentangle an image into normal, albedo, light, and shading maps, either on real or spoof data.
2) A multi-stream fusion network is developed to capture the complementarity and discrepancy of imaging components for face anti-spoofing.
3) Our method not only outperforms the state-of-the-art methods on the intra-testing of OULU-NPU dataset, but also demonstrates better performance on MICO (with initial letters from the four datasets) and a variant of the MICO benchmark without using domain knowledge.

2 Related Work

2.1 Deep Learning-Based Face Anti-spoofing

As deep learning has proven to be more effective than the traditional methods in many computer vision problems, there are many recent attempts at CNN-based methods in face anti-spoofing. At first, most of the works regard face anti-spoofing as a simple binary classification problem by applying softmax loss. For example, Yang et al. [3] use CNN as a feature extractor, and train an SVM classifier with deep features to discriminate genuine and spoofing faces. Then, some methods [11,14] propose to use pixel-wise labels for the supervision of network training, proving that pixel-wise supervision can improve the performance of spoofing detection. Among them, the depth map is widely used in various face anti-spoofing methods, which is more informative than binary labels since it indicates one of the fundamental differences between genuine and spoofing faces. Yu et al. [13] propose a central difference operator to extract inherent spoofing patterns, and combine it with depth supervision to significantly improve the performance of face anti-spoofing.

2.2 Auxiliary Information-Based Face Anti-spoofing

Since spoofing images are obtained by secondary imaging, the inherent imaging components of faces are changed when they are compared with genuine faces. To explore the distortion of spoofing faces, Boulkenafet et al. [5] propose to extract color distortion from the YCrCb or HSV color spaces. Chen et al. [6] propose a two-stream CNN that works on two complementary spaces: RGB space and

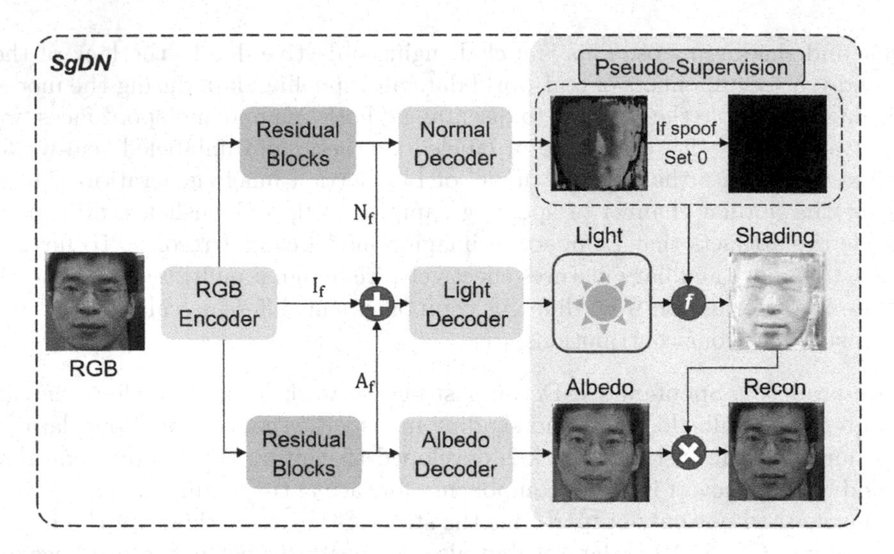

Fig. 1. The overall architecture of Spoof-guided Decomposition Network. RGB image is fed into the decomposition network for disentangling the imaging components into normal, albedo, and shading maps.

an illumination-invariant space called MSR. Bian et al. [15] propose a framework to learn multiple generalizable cues from the boundary of spoof medium, moiré pattern, reflection artifacts, and facial depth. Recent literature [5,10] have shown that exploring the texture and 3D geometry information can improve the detection performance. Many methods draw on the idea of 3D geometry by sensing depth changes [7,11,13]. The reflectance of the face image is another widely used cue for face anti-spoofing [9,10] as the material differences between genuine and spoofing faces. Mishra et al. [16] also identify the contribution of albedo in enhancing presentation attack detection, but their method does not leverage the potential of the normal map. Moreover, their model trained exclusively on genuine data tend to excessively focus on the semantic information of genuine faces, which consequently compromises its robustness in decomposing spoofing samples. See more related work in Appendix A.

3 Approach

3.1 Spoof-Guided Decomposition Network (SgDN)

Estimating normal and albedo for genuine and spoofing faces from single images is the key to our method. An intuitive solution is fitting an imaging model defined in computer graphics, like the 3D morphable model (3DMM) [17] and Phong reflection model [18], to the image. However, the fitted results are unsatisfactory because the highly simplified imaging model cannot cover the complicated appearance variations of human faces, especially in our task where faces are possibly unreal.

To model the structural and textured difference, SgDN is designed to reflect the Lambertian reflectance model [19], where the formation process of face image $\mathbf{I}$ can be represented as:

$$\mathbf{I} = f(\mathbf{N}, \mathbf{L}) \odot \mathbf{A}, \tag{1}$$

where $\mathbf{N} \in \mathbb{R}^{n \times n \times 3}$, $\mathbf{L} \in \mathbb{R}^{3 \times 9}$, $\mathbf{A} \in \mathbb{R}^{n \times n \times 3}$ are normal, lighting and albedo, $\odot$ represents the element-wise product and f is the Lambertian shading function.

To approximate the local behavior of light on the face, we adopt the spherical harmonics as a decomposition basis. Specifically, the lighting $\mathbf{L}$ is defined as nine-dimensional second order spherical harmonics coefficients $\gamma = [\gamma_1, \gamma_2, ..., \gamma_9]$ for each RGB channels, and the normal map $\mathbf{N}$ is utilized to construct spherical harmonic basis [19]:

$$B = [B_{00}, B_{10}, B_{11}^e, B_{11}^o, B_{20}, B_{21}^e, B_{21}^o, B_{22}^e, B_{22}^o]. \tag{2}$$

Then the shading map can be calculated as:

$$\mathbf{S} = B(\mathbf{N}) \odot \gamma. \tag{3}$$

To reconstruct the face, the model is further required to estimate the albedo map. According to the constraint of Lambertian assumption, the albedo is obtained as:

$$\mathbf{A} = \frac{\mathbf{I}}{\mathbf{S}}. \tag{4}$$

Finally, with the estimated Normal and Albedo maps, the reconstructed face $\mathbf{R} \in \mathbb{R}^{n \times n \times 3}$ becomes:

$$\mathbf{R} = \mathbf{S} \odot \mathbf{A}. \tag{5}$$

As the image reconstruction is an end-to-end process in our SgDN, the face image formation process defined in Eq. 1 is a differentiable function. We propose a deep learning-based network to regress texture and shape parameters directly from a single image, which is shown in the diagram of Fig. 1. The detailed construction of different components and training settings of SgDN are deferred to sections C.1 and C.2 in Appendix.

The SgDN can be trained by synthetic data [20] and CelebA [21] with pseudo-supervision to produce a precise reconstruction. However, the SgDN trained in this stage cannot be directly applied to spoofing data as the Lambertian imaging model cannot cover the complicated appearance variations of human faces, especially the unreal faces in our task. We show the decomposition results of SgDN without discriminative supervision in Appendix Fig. 2. As the presentation attack instruments are like a flat surface, we refine the pseudo-supervision of spoofing samples by setting the normal channel to zero, which can enforce the SgDN to not only reconstruct the spoofing images but also capture the spoofing artifacts that the Lambertian model cannot resolve, simultaneously. In Sect. 4, we will prove the effectiveness of the pseudo-supervised training method.

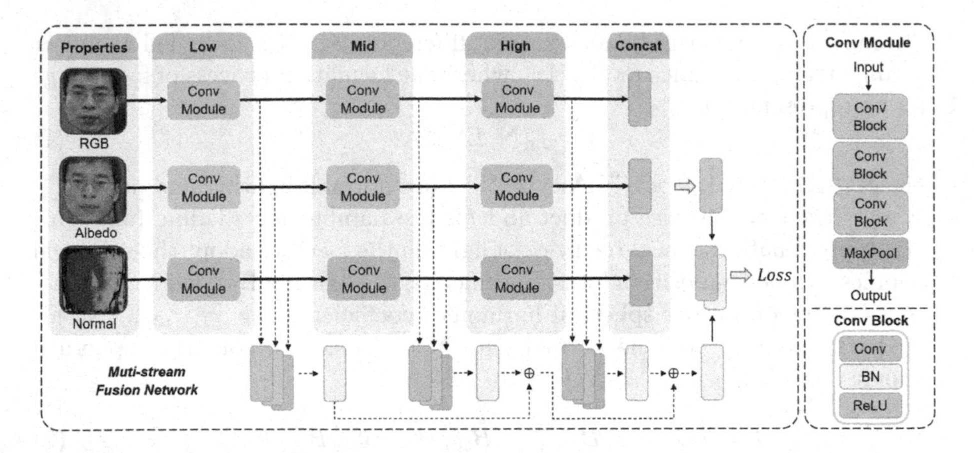

Fig. 2. Imaging components of RGB, normal, and albedo maps are transmitted into the three-stream fusion network to extract features from different components and capture the complementarity and discrepancy between them.

3.2 Fusion Framework

To capture the complementarity and discrepancy of imaging components, we propose a three-stream fusion network as shown in Fig. 2. The three-stream fusion network consists of three identical sub-networks with different inputs and extracts the learned features derived from RGB, albedo, and normal images following different convolution modules of the three subnetworks. Instead of aggregating different levels of information separately, an aggregation block is utilized to aggregate the extracted features and the output from previous layers, which is explained in detail in Appendix D.

4 Experiments

4.1 Settings

Datasets including MSU-MFSD [22], Replay-Attack [23], CASIA-MFSD [24], and OULU-NPU [25] are used to evaluate our proposed method with print and replay attacks. We strictly follow the evaluation benchmarks for data partitioning and the implementation details of SgDN and the three-stream fusion network are given in Appendix C, D and E.

4.2 Ablation Study

1) Pseudo-supervised training: In Sect. 3.1, we propose to achieve a more differentiated decomposition by refining the pseudo-supervision process. Different from traditional Shape from Shading algorithm [26,27], our face decomposition network is a data-driven model and relies on preprocessing to achieve a better

Table 1. The ablation study results (%) of the pseudo-supervised training and other decomposition algorithm on the Protocol-2 of OULU-NPU.

Methods	APCER ↓	BPCER ↓	ACER ↓
SgDN w/o zero	5.6	5.8	5.7
SgDN w/ zero (Our)	**1.9**	**0.5**	**1.2**
ADMM [26]	4.7	4.7	4.7
3DMM [17]	2.8	3.1	2.9

decomposition performance. As shown in Table 1, we can see that our pseudo-supervised training achieves lower HTER than the method without setting the normal channel to zero.

2) Comparison with Traditional Decomposition Method: This section compares our method with other traditional Shape from Shading methods. We employ a PDEs-based method ADMM [26] and a 3DMM-based method to estimate albedo and normal maps as a comparison. All experiments are carried out under the same fusion model, and the 3DMM-based method is realized by fitting a classical face reconstruction model 3DMM [17]. As shown in Table 1, our proposed method achieves significant improvement over the traditional Shape from Shading methods.

3) Efficacy of Each Components: As discussed in Appendix B, we argue that RGB, normal, and albedo contain complementary information for spoofing detection. To further understand the characteristics of these components, we list all possible combinations of the imaging components in Appendix Table 1, including albedo, normal, and shading, proving that RGB fused with albedo and Normal can achieve the best performance.

4) Multi Levels: In our multi-stream fusion framework, features from different levels play an important part in spoofing detection, so we concatenate these features to fully explore the spoofing clues. We present the results of fusion at different levels in Appendix Table 2.

4.3 Intra Testing

The intra-testings are carried out on the OULU-NPU, a large-scale face anti-spoofing dataset with four testing protocols. Table 2 shows the result of our method on these protocols and our method surpasses all state-of-the-art approaches on three or more protocols. When compared with the convMLP [30] model, our method is slightly weaker on protocol 2. Moreover, compared to the IDCL [16] method, we are not only able to perform differential decomposition but also further utilize normal information, and exceed their method's performance, especially under challenging scenarios such as protocols 3 and 4.

Table 2. The results of intra-testing on four protocols of OULU-NPU.

Prot	Method	APCER↓	BPCER↓	ACER↓
1	STASN [28]	1.2	2.5	1.9
	Auxiliary [7]	1.6	1.6	1.6
	FAS-TD [12]	2.5	**0**	1.3
	CDCN [13]	0.4	1.7	1
	PGSTD [29]	**0**	0.8	**0.4**
	Conv-MLP [30]	2.5	3.2	2.8
	IDCL [16]	0.7	0.6	0.6
	Ours	0.6	0.8	0.7
2	Auxiliary [7]	2.7	2.7	2.7
	STASN [28]	4.2	**0.3**	2.2
	FAS-TD [12]	1.7	2.0	1.9
	CDCN [13]	1.5	1.4	1.5
	PGSTD [29]	1.2	1.3	1.3
	Conv-MLP [30]	**0**	1.6	**0.8**
	IDCL [16]	1.3	1.1	1.2
	Ours	1.9	0.5	1.2
3	FAS-TD [12]	5.9 ± 1.9	5.9 ± 3.0	5.9 ± 1.0
	Auxiliary [7]	2.7 ± 1.3	3.1 ± 1.7	2.9 ± 1.5
	STASN [28]	4.7 ± 3.9	$\mathbf{0.9 \pm 1.2}$	2.8 ± 1.6
	CDCN [13]	2.4 ± 1.3	2.2 ± 2.0	2.3 ± 1.4
	PGSTD [29]	1.7 ± 1.4	2.2 ± 3.5	1.9 ± 2.3
	Conv-MLP [30]	2.5 ± 1.0	2.0 ± 0.8	2.2 ± 0.6
	IDCL [16]	1.7 ± 1.4	1.8 ± 1.1	1.7 ± 0.7
	Ours	$\mathbf{1.4 \pm 0.6}$	1.7 ± 1.5	$\mathbf{1.5 \pm 1.0}$
4	Auxiliary [7]	9.3 ± 5.6	10.4 ± 6.0	9.5 ± 6.0
	FAS-TD [12]	14.2 ± 8.7	4.2 ± 3.8	9.2 ± 3.4
	STASN [28]	6.7 ± 10.6	8.3 ± 8.4	7.5 ± 4.7
	CDCN [13]	4.6 ± 4.6	9.2 ± 8.0	6.9 ± 2.9
	PGSTD [29]	$\mathbf{2.3 \pm 3.6}$	4.2 ± 5.4	3.6 ± 4.2
	Conv-MLP [30]	6.4 ± 4.5	3.4 ± 5.1	4.9 ± 4.8
	IDCL [16]	3.4 ± 1.5	5.5 ± 4.4	4.5 ± 2.7
	Ours	2.5 ± 2.7	$\mathbf{3.3 \pm 2.6}$	$\mathbf{2.9 \pm 1.0}$

4.4 Inter Testing

1) Result on MICO: MICO benchmark is a widely used evaluation about domain generalization, which follows the 'Leave-one-out' protocol with four face anti-spoofing datasets. Table 3 compares our method with the state-of-the-art meth-

Table 3. The results of inter-testing on MICO benchmark. The proposed method is compared with the methods without utilizing domain knowledge.

Method	O&C&I to M		O&M&I to C		O&C&M to I		I&C&M to O	
	HTER ↓	AUC ↑	HTER ↓	AUC ↑	HTER ↓	AUC ↑	HTER ↓	AUC ↑
Binary CNN [3]	29.25	82.87	34.88	71.94	34.47	65.88	29.61	77.54
IDA [22]	66.67	27.86	55.17	39.05	28.35	78.25	54.20	44.69
Color Texture [31]	28.09	78.47	30.58	76.89	40.40	62.78	63.59	32.71
Auxiliary [32]	22.72	85.88	33.52	73.15	29.14	71.69	30.17	77.61
NAS-FAS [33]	19.53	88.63	16.54	90.18	14.51	93.84	**13.80**	**93.43**
DTN [34]	19.40	86.87	22.03	87.71	21.43	88.81	18.26	89.40
Ours	**10.46**	**94.43**	**10.95**	**94.73**	**9.95**	**95.57**	17.78	88.93

Table 4. The results of inter-testing with limited source domains. The proposed method is compared with other methods in terms of HTER(%) and AUC(%).

Method	M&I to C		M&I to O	
	HTER↓	AUC↑	HTER↓	AUC↑
IDA [22]	45.16	58.80	54.52	42.17
LBP-TOP [36]	45.27	54.88	47.26	50.21
MADDG [37]	41.02	64.33	39.35	65.10
SSDG-M [35]	31.89	71.29	36.01	66.88
DR-MD-Net [38]	31.67	75.23	34.02	72.65
ANRL [39]	31.06	72.12	30.73	74.10
SSAN-M [40]	30	76.20	29.44	76.62
Ours	**26.67**	**81.34**	**22.50**	**83.64**

ods trained without domain information, and we achieve the best results in three testing tasks. Meanwhile, our method outperforms most of the domain generalization methods as shown in Appendix Table 3.

2) Result on limited source domains: When the number of source data domains is limited, the performance of specially designed domain generalization methods may degrade. To further evaluate the proposed methods, a variant of the MICO benchmark [35] is proposed to conduct domain generalization experiments. We also evaluate our method on this benchmark, and the comparison results are shown in Table 4. It can be seen that our method's HTER and AUC performance is comparable with that of the state-of-the-art methods, which has a significant improvement over other domain generalization methods.

3) Result on CASIA-MFSD and Replay-Attack: In this experiment, there are two cross-dataset testing protocols. The first protocol 'CR' is trained on CASIA-MFSD and tested on Replay-Attack. The second one 'RC' is trained on Replay-

Attack and tested on CASIA-MFSD. It can be seen from Appendix Table 4 that the HTER of our proposed method is 11.7% on C2R and 28.4% on R2C. Our method outperforms the prior state-of-the art method CDCN [13] over 24.5% on C2R protocol.

4.5 Performance Analysis

1) The Effectiveness of SgDN: To further illustrate the effectiveness of the pseudo-supervised training method for the SgDN, we show the decomposition results of different spoofing samples in Appendix Fig. 3. Because of the spoof-guided supervision, SgDN is able to decompose the genuine face into reasonable normal and albedo, and the normal obtained from the spoofing face is inclined to a plane. Under careful comparison, we can observe that the albedo maps of spoofing faces have a blurred visual appearance compared with genuine faces.

2) The interpretability of our Fusion Network: We convert the final-layer features into heatmaps to produce a visual explanation for our three-stream fusion network, which is illustrated and discussed in Appendix Fig. 4 and section G.4. Through the multi-stream fusion network, the subtle artifacts can be harvested from RGB, normal, and albedo map.

5 Conclusion and Future Work

This work proposes a novel method for face anti-spoofing by designing a spoof-guided face decomposition network and harvesting spoofing clues from the imaging components of RGB, albedo, and normal. With the spoofing guidance, our decomposition network can push the imaging components of spoofing samples to encode the artifacts that the Lambertian imaging model cannot resolve. Besides, we devise a multi-stream network to fuse the information from different components and capture the complementarity and discrepancy between them. Moreover, extensive experiments are performed to demonstrate that our method can achieve state-of-the-art performances in the intra-testing protocol of OULU-NPU and three domain generalization benchmarks. We note that the study of spoof-guided image decomposition is still at an early stage. Future directions include: 1) designing a more generalized decomposition method for spoofing samples. 2) exploring other auxiliary information for presentation attack detection.

Acknowledgement. This work was supported in part by Chinese National Natural Science Foundation Projects #62276254, #62176256, #62106264, #62206280, #U2003111), Beijing Natural Science Foundation under no. L221013, the Defense Industrial Technology Development Program (#JCKY2021906A001), Shandong Provincial Natural Science Foundation under Project ZR2021MF066 and the InnoHK program.

References

1. de Freitas Pereira, T., Anjos, A., De Martino, J.M., Marcel, S.: $LBP-TOP$ based countermeasure against face spoofing attacks. In: Park, J.-I., Kim, J. (eds.) ACCV 2012. LNCS, vol. 7728, pp. 121–132. Springer, Heidelberg (2013). https://doi.org/10.1007/978-3-642-37410-4_11
2. Komulainen, J., Hadid, A., Pietikäinen, M.: Context based face anti-spoofing. In: 2013 IEEE Sixth International Conference on Biometrics: Theory, Applications and Systems (BTAS), pp. 1–8. IEEE (2013)
3. Yang, J., Lei, Z., Li, S.Z.: Learn convolutional neural network for face anti-spoofing, arXiv preprint arXiv:1408.5601 (2014)
4. Cai, R., Li, Z., Wan, R., Li, H., Hu, Y., Kot, A.C.: Learning meta pattern for face anti-spoofing. IEEE Trans. Inf. Forensics Secur. **17**, 1201–1213 (2022)
5. Boulkenafet, Z., Komulainen, J., Hadid, A.: Face spoofing detection using colour texture analysis. IEEE Trans. Inf. Forensics Secur. **11**(8), 1818–1830 (2016)
6. Chen, H., Hu, G., Lei, Z., Chen, Y., Robertson, N.M., Li, S.Z.: Attention-based two-stream convolutional networks for face spoofing detection. IEEE Trans. Inf. Forensics Secur. **15**, 578–593 (2019)
7. Liu, Y., Jourabloo, A., Liu, X.: Learning deep models for face anti-spoofing: binary or auxiliary supervision. In: Proceedings of the IEEE Conference on Computer Vision and Pattern Recognition, pp. 389–398 (2018)
8. Jourabloo, A., Liu, Y., Liu, X.: Face de-spoofing: anti-spoofing via noise modeling. In: Proceedings of the European Conference on Computer Vision (ECCV), pp. 290–306 (2018)
9. Pinto, A., Pedrini, H., Schwartz, W.R., Rocha, A.: Face spoofing detection through visual codebooks of spectral temporal cubes. IEEE Trans. Image Process. **24**(12), 4726–4740 (2015)
10. Pinto, A., Goldenstein, S., Ferreira, A., Carvalho, T., Pedrini, H., Rocha, A.: Leveraging shape, reflectance and albedo from shading for face presentation attack detection. IEEE Trans. Inf. Forensics Secur. **15**, 3347–3358 (2020)
11. Atoum, Y., Liu, Y., Jourabloo, A., Liu, X.: Face anti-spoofing using patch and depth-based CNNs. In: 2017 IEEE International Joint Conference on Biometrics (IJCB), pp. 319–328. IEEE (2017)
12. Wang, Z., et al.: Exploiting temporal and depth information for multi-frame face anti-spoofing, arXiv preprint arXiv:1811.05118 (2018)
13. Yu, Z., et al.: Searching central difference convolutional networks for face anti-spoofing. In: Proceedings of the IEEE/CVF Conference on Computer Vision and Pattern Recognition, pp. 5295–5305 (2020)
14. George, A., Marcel, S.: Deep pixel-wise binary supervision for face presentation attack detection. In: 2019 International Conference on Biometrics (ICB), pp. 1–8. IEEE (2019)
15. Bian, Y., Zhang, P., Wang, J., Wang, C., Pu, S.: Learning multiple explainable and generalizable cues for face anti-spoofing. In: ICASSP 2022-2022 IEEE International Conference on Acoustics, Speech and Signal Processing (ICASSP), pp. 2310–2314. IEEE (2022)
16. Mishra, S.K., Sengupta, K., Chu, W.S., Horowitz-Gelb, M., Bouaziz, S., Jacobs, D.: Improved presentation attack detection using image decomposition. In: 2022 IEEE International Joint Conference on Biometrics (IJCB), pp. 1–10. IEEE (2022)
17. Blanz, V., Vetter, T.: Face recognition based on fitting a 3D morphable model. IEEE Trans. Pattern Anal. Mach. Intell. **25**(9), 1063–1074 (2003)

18. Phong, B.T.: Illumination for computer generated pictures. Commun. ACM **18**(6), 311–317 (1975)
19. Basri, R., Jacobs, D.W.: Lambertian reflectance and linear subspaces. IEEE Trans. Pattern Anal. Mach. Intell. **25**(2), 218–233 (2003)
20. Trigeorgis, G., Snape, P., Kokkinos, I., Zafeiriou, S.: Face normals "in-the-wild" using fully convolutional networks. In: Proceedings of the IEEE Conference on Computer Vision and Pattern Recognition, pp. 38–47 (2017)
21. Liu, Z., Luo, P., Wang, X., Tang, X.: Deep learning face attributes in the wild. In: IEEE International Conference on Computer Vision (2016)
22. Wen, D., Han, H., Jain, A.K.: Face spoof detection with image distortion analysis. IEEE Trans. Inf. Forensics Secur. **10**(4), 746–761 (2015)
23. Chingovska, I., Anjos, A., Marcel, S.: On the effectiveness of local binary patterns in face anti-spoofing. In: 2012 BIOSIG-Proceedings of the International Conference of Biometrics Special Interest Group (BIOSIG), pp. 1–7. IEEE (2012)
24. Zhang, Z., Yan, J., Liu, S., Lei, Z., Yi, D., Li, S.Z.: A face antispoofing database with diverse attacks. In: 2012 5th IAPR international conference on Biometrics (ICB), pp. 26–31. IEEE (2012)
25. Boulkenafet, Z., Komulainen, J., Li, L., Feng, X., Hadid, A.: OULU-NPU: a mobile face presentation attack database with real-world variations. In: 2017 12th IEEE International Conference on Automatic Face & Gesture Recognition (FG 2017), pp. 612–618. IEEE (2017)
26. Quéau, Y., Mélou, J., Castan, F., Cremers, D., Durou, J.-D.: A variational approach to shape-from-shading under natural illumination. In: Pelillo, M., Hancock, E. (eds.) EMMCVPR 2017. LNCS, vol. 10746, pp. 342–357. Springer, Cham (2018). https://doi.org/10.1007/978-3-319-78199-0_23
27. Shah, P.S.: Shape from shading using linear approximation. Image Vis. Comput. **12**(8), 487–498 (1994)
28. Yang, X., et al.: Face anti-spoofing: model matters, so does data. In: Proceedings of the IEEE/CVF Conference on Computer Vision and Pattern Recognition, pp. 3507–3516 (2019)
29. Liu, Y., Liu, X.: Physics-guided spoof trace disentanglement for generic face anti-spoofing, arXiv preprint arXiv:2012.05185 (2020)
30. Wang, W., Wen, F., Zheng, H., Ying, R., Liu, P.: Conv-MLP: a convolution and MLP mixed model for multimodal face anti-spoofing. IEEE Trans. Inf. Forensics Secur. **17**, 2284–2297 (2022)
31. Boulkenafet, Z., Komulainen, J., Hadid, A.: Face antispoofing using speeded-up robust features and fisher vector encoding. IEEE Signal Process. Lett. **24**(2), 141–145 (2016)
32. Gan, J., Li, S., Zhai, Y., Liu, C.: 3D convolutional neural network based on face anti-spoofing. In: 2017 2nd International Conference on Multimedia and Image Processing (ICMIP) (2017)
33. Yu, Z., Wan, J., Qin, Y., Li, X., Li, S.Z., Zhao, G.: NAS-FAS: static-dynamic central difference network search for face anti-spoofing. IEEE Trans. Pattern Anal. Mach. Intell. **43**(9), 3005–3023 (2021)
34. Wang, Y., Song, X., Xu, T., Feng, Z., Wu, X.-J.: From RGB to depth: domain transfer network for face anti-spoofing. IEEE Trans. Inf. Forensics Secur. **16**, 4280–4290 (2021)
35. Jia, Y., Zhang, J., Shan, S., Chen, X.: Single-side domain generalization for face anti-spoofing. In: Proceedings of the IEEE/CVF Conference on Computer Vision and Pattern Recognition, pp. 8484–8493 (2020)

36. de Freitas Pereira, T., et al.: Face liveness detection using dynamic texture. EURASIP J. Image Video Process. **2014**(1), 1–15 (2014)
37. Qin, Y., et al.: Learning meta model for zero-and few-shot face anti-spoofing. In: Proceedings of the AAAI Conference on Artificial Intelligence, vol. 34, no. 07, pp. 11916–11923 (2020)
38. Wang, G., Han, H., Shan, S., Chen, X.: Cross-domain face presentation attack detection via multi-domain disentangled representation learning. In: Proceedings of the IEEE/CVF Conference on Computer Vision and Pattern Recognition, pp. 6678–6687 (2020)
39. Liu, S., et al.: Adaptive normalized representation learning for generalizable face anti-spoofing. In: Proceedings of the 29th ACM International Conference on Multimedia, pp. 1469–1477 (2021)
40. Wang, Z., et al.: Domain generalization via shuffled style assembly for face anti-spoofing. In: Proceedings of the IEEE/CVF Conference on Computer Vision and Pattern Recognition, pp. 4123–4133 (2022)

TransFCN: A Novel One-Stage High-Resolution Fingerprint Representation Method

Yanfeng Xiao[1], Feng Liu[1(✉)], and Xu Tan[2]

[1] School of Computer Science and Technology, Shenzhen University, Shenzhen, China
feng.liu@szu.edu.cn

[2] School of Software Engineering, Shenzhen Institute of Information Technology, Shenzhen, China
tanx@sziit.edu.cn

Abstract. Since pores are widely used to represent high-resolution fingerprint images, the detection and representation of pores are essential for high-resolution fingerprint recognition. The latest method uses only one fully convolutional network to represent high-resolution fingerprint images for subsequent recognition by combining pore detection and pore representation into one stage, showing good generalization and pore detection ability. Nevertheless, it still has limitations in feature learning and pore detection due to its network architecture and the loss used. To tackle the limitations, in this paper, we propose a novel network architecture, namely TransFCN, for one-stage high-resolution fingerprint representation. We introduce the transformer and attention module into our network architecture and combine them with the fully convolutional network to effectively learn both global and local information. In addition, we employ the adaptive wing loss and weighted loss map to further improve the pore detection capability. Experimental results on the PolyU HRF dataset demonstrate the effectiveness of our proposed method in pore detection and feature learning. Furthermore, the experimental results on an in-house dataset demonstrate the excellent generalization capability of our proposed method when compared to the state-of-the-art two-stage method.

Keywords: High-Resolution Fingerprint Recognition · Transformer · Fingerprint Representation · Pore Detection

1 Introduction

Fingerprint is the most widely deployed biometrics characteristic because of its well-known distinctiveness and permanence [1]. With the development of fingerprint sensors, high-resolution fingerprint images become available, accompanied

This work was supported in part by the National Natural Science Foundation of China under Grant 62076163, and The Innovation Team Project of Colleges in Guangdong Province (2020KCXTD040).

Q. Liu et al. (Eds.): PRCV 2023, LNCS 14429, pp. 16–28, 2024.
https://doi.org/10.1007/978-981-99-8469-5_2

by the emergency of level-3 features. The addition of level-3 features has driven researchers toward the use of new features for more accurate and more secure fingerprint recognition, namely high-resolution fingerprint recognition. Among various level-3 features, sweat pores have many excellent properties, including high distinctiveness, natural anti-spoof ability, and large quantities, which have attracted the most attention of researchers [2–14]. Current high-resolution fingerprints are almost entirely represented by sweat pores.

High-resolution fingerprint representation based on sweat pores involves two important parts: pore detection and pore representation. These two parts are usually two separate stages. For pore detection, traditional methods based on image processing technique [2,3] and learning-based methods [4–8] have been investigated. Learning-based methods demonstrate improved adaptability and robustness to various image qualities and have become the mainstream method. For pore representation, hand-crafted feature based [9–11], and deep feature based [12,13] have been proposed. In [9], the pore representation is directly built from the pixel values in the local neighborhood to the pore. In [10,11], sparse representation methods are proposed to represent pore. As deep learning has evolved, Zhao et al. [12,13] propose the DeepPoreID method, which involves training a classification network to classify between different pores. Subsequently, the deep features obtained from the classification network are utilized as pore representations. After the two-stage fingerprint representation, the commonly used hierarchical coarse-to-fine DP framework [9] is used for recognition. In the coarse matching step, coarse pore correspondences are established by pore representation. In the fine matching step, RANdom SAmple Consensus (RANSAC) is used for refinement to obtain the final result.

To avoid information loss and improve robustness, Liu et al. [14] propose a novel high-resolution fingerprint representation method that uses only one fully convolutional network (FCN) to provide both pore and deep matching features simultaneously to represent high-resolution fingerprint images. By combining pore detection and pore representation into one stage, the method can achieve the best pore detection capability and have better generalization ability than the state-of-the-art two-stage method [12] when using the DP framework for recognition. However, it still has some limitations: (1) the deep features learned from a single fully convolutional network may struggle to cope with pore changes and thus many true correspondences may be missed. (2) the mean square error (MSE) loss used for pore detection treats foreground and background pixels equally, which makes it easy for training to be dominated by a large number of meaningless background pixels, resulting in poor regression accuracy.

To tackle the above limitations, in this paper, we propose a novel one-stage high-resolution fingerprint representation method, namely TransFCN. Firstly, we utilize the transformer-based encoder to learn the preliminary patch representation of the high-resolution fingerprint image, leveraging the self-attention module to incorporate global information for a more robust pore representation. Subsequently, the fully convolutional network is employed to detect pores and further learn fingerprint representation for one-stage high-resolution fingerprint representation. An attention module is utilized to optimize the learning of fingerprint representation, while the adaptive wing loss and weighted loss map are applied to prioritize the regression of foreground pixels, specifically the pore locations, to improve pore detection accuracy. To sum up, our main contributions can be summarized as follows:

- We propose a novel network architecture for one-stage high-resolution fingerprint representation, which combines the transformer and the fully convolutional network to learn global and local information. To the best of our knowledge, the transformer is the first time to be used for pore detection and pore representation.
- We propose to utilize the attention module and apply adaptive wing loss for facial landmark localization to fingerprint pore detection with the weighted loss map to improve the pore detection and feature learning ability.
- Extensive experimental results on the public PolyU HRF dataset and an in-house dataset have demonstrated the effectiveness of the proposed method.

2 Methodology

2.1 Overall Framework

The overall framework is illustrated in Fig. 1, consisting of three parts: input, proposed network, and DP framework. Among them, the proposed network adopts an encoder-decoder structure. Firstly, the input fingerprint image is fed into the proposed network and outputs the corresponding pore map and reconstructed fingerprint image. Similar to DeepPore [4], the output pore map is also a Gaussian heat map for pore detection. After obtaining the pore map, the sliding window algorithm is used to obtain the pore locations. The sliding window algorithm scans the entire pore map using a window of $k \times k$ and a threshold of P_t. When the center of the window is the maximum value of the entire window and is greater than the threshold P_t, the location is then judged to be a pore. In addition, we fill the boundary of the pore map with a value of 0 to deal with the pores at the boundary. By guiding the network to learn the reconstruction of the fingerprint image and the prediction of pore positions, discriminative pore features can be learned.

Subsequently, the DP framework is employed for fingerprint recognition. Specifically, the detected pore locations, the original image, and the intermediate feature map ($Feat$) of the proposed network are input into the DP framework. Given N and M detected pores in the query image and the template image, we utilize these pore locations to extract pore representations from the original image or $Feat$, resulting in $N \times D_p$ and $M \times D_p$ pore representations, respectively. We start by using each detected pore location to slice the corresponding patch from the original image or $Feat$. For the multi-channel $Feat$ patch, we perform the channel-wise sum to obtain the single-channel patch.

Finally, the resulting patch is flattened into the D_p dimension vector to represent the pore. The size of dimension D_p is 961, consistent with [9,14]. These pore representations are used for subsequent coarse-to-fine pore matching. In the coarse matching step, a $M \times N$ similarity matrix can be obtained by calculating pair-wise pore distances in the query and template images using Eq. 1. In the similarity matrix, a lower value means more similarity. Coarse one-to-one pore correspondences are then established by this similarity matrix. The rule for matching pore A and pore B is that both pore A and pore B are the most matched pores with each other. In the fine matching step, the RANSAC algorithm is used to refine the coarse matched results. The number of the final matched pore pairs is considered as the match score for recognition.

$$S_{i,j} = 1 - \frac{\sum_{k=1}^{961} P_{i,k} P'_{j,k}}{\sqrt{\sum_{k=1}^{961} P_{i,k}^2}\sqrt{\sum_{i=1}^{961} P'^2_{j,k}}}, i \in [1, N], j \in [1, M] \quad (1)$$

where $S_{i,j}$ represents the similarity of the ith pore in the query image to the jth pore in the template image. $P_{i,k}$ and $P'_{j,k}$ represent the kth dimension of the ith pore representation in the query image and the jth pore representation in the template image, respectively.

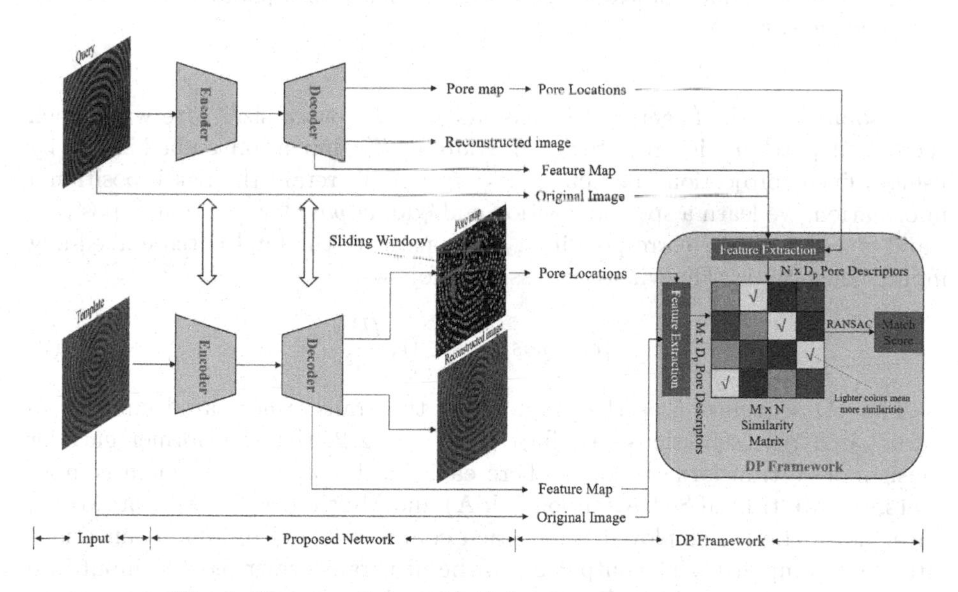

Fig. 1. The overall framework for fingerprint recognition.

2.2 Transformer-Based Encoder

As shown in Fig. 2, given a 2D high-resolution fingerprint image $x \in \mathbb{R}^{3 \times H \times W}$, the fingerprint image is first divided into a grid of $\frac{H}{p} \times \frac{W}{p}$ patches and then these patches are flattened into a $\frac{HW}{p^2} \times (p^2 \cdot 3)$ sequence, where (H, W) is

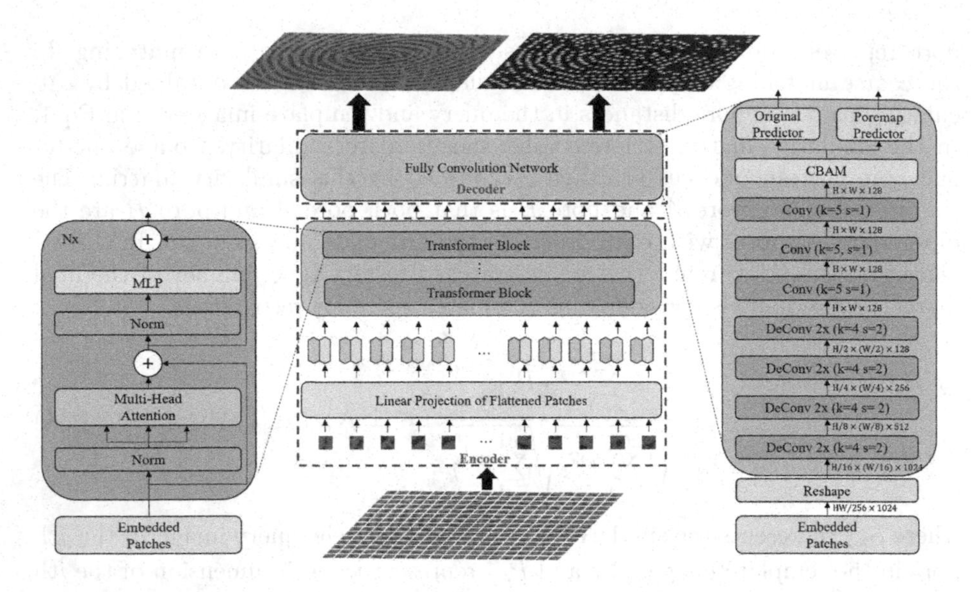

Fig. 2. The detailed architecture of the proposed network. It consists of a Transformer-based encoder and an FCN-based decoder to regress the pore map and reconstruct the original image with different losses. The CBAM module is incorporated to enhance the feature learning ability.

the resolution of the fingerprint image and p is the patch size. Afterward, each vectorized patch v_p is projected into a latent D_e dimension embedding e by using a linear projection function f_e: $e = f_e(v_p)$. To retain the patch positional information, we learn a specific position embedding pos_i for each patch position i and is added to the corresponding e_i to constitute the final patch embedding input sequences E. The whole process is as follows:

$$E = \{e_i + pos_i, \quad i \in [1, \frac{HW}{p^2}]\} \tag{2}$$

The 1D sequence E is then input into the transformer-based encoder to learn patch representations. As illustrated in Fig. 2, the transformer encoder consists of N transformer blocks where each block contains two main components, the Multi-head Self-Attention (MSA) and MultiLayer Perceptron (MLP). Layer norm (LN) is employed before each component and residual connections after each component. The output Z_n of the nth transformer block is input into the next transformer block (Eq. 3, 4, 5). With the Self-Attention (SA) module, each transformer block has a global receptive field, which tackles the limitations of the pure CNN's receptive field and enables global information to be learned.

$$Z_0 = E, \quad E \in \mathbb{R}^{\frac{HW}{p^2} \times D_e} \tag{3}$$

$$Z_n^{'} = MSA(LN(Z_{n-1})) + Z_{n-1}, \quad n \in [1, N] \tag{4}$$

$$Z_n = MLP(LN(Z_n^{'})) + Z_n^{'}, \quad n \in [1, N] \tag{5}$$

2.3 FCN-Based Decoder

To generate the pore map and reconstruct the original image, the transformer encoder output Z_N is first projected to a higher dimension D_d and the corresponding learnable position embedding is added. The resulting sequence is then reshaped into a 2D feature map, as shown in Fig. 2. However, due to the patch operation in the encoder, the resolution of the reshaped feature map is only $\frac{1}{p}$ of the original image. To restore the feature map to its original resolution, we design an upsampling module that utilizes $\frac{p}{2}$ double upsampling operations. Specifically, we employ the deconvolution (DeConv) operation, which is a commonly used technique for upsampling in deep learning, to increase the resolution of the feature map. After restoring the feature map to its original resolution, we further use the convolution (Conv) operation to learn feature representation. In addition, we use the convolutional block attention module (CBAM) [15] at the end of the convolution operation to enhance feature learning. Finally, two predictors utilize convolution operations to generate the pore map and reconstructed image. The feature with both low correlation [16] and high discrimination [17] is useful for differentiating between different pores. To this end, we select the output feature map of the CBAM module to represent the pore.

2.4 Loss Function

The overall loss consists of two parts. The first part $\mathcal{L}_{reconstructed}$ is derived from the fingerprint reconstruction branch. We employ the MSE as the loss function for this branch, which is widely used in most image reconstruction tasks. The formula is as follows:

$$\mathcal{L}_{reconstructed} = \frac{1}{HW} \sum_{i=1}^{H} \sum_{j=1}^{W} (I_{ij} - \hat{I}_{ij})^2 \tag{6}$$

where I_{ij} and $\hat{I}_{ij}$ represent the pixel values at coordinates (i, j) of the original image and the reconstructed image, respectively. The second part $\mathcal{L}_{poremap}$ is derived from the pore detection branch. To further improve the pore detection ability, we propose to use adaptive wing (AW) [18] loss for pore detection instead of MSE loss because of its better adaptability. Furthermore, we use the weighted loss map (WLM) to increase the penalty on the regression of sweat pore locations, enabling the network to focus more on the regression of the pixels corresponding to the pore locations. The WLM is generated by a mask matrix $Mask$ with a value of 0 or 1 for each position. The positions with a value of 1 are those with a value greater than a certain threshold in the ground truth pore map. The formula is as follows:

$$AW(y,\hat{y}) = \begin{cases} \omega \ln(1 + \left|\frac{y-\hat{y}}{\epsilon}\right|^{\alpha-y}), & \text{if } |(y-\hat{y})| < \theta \\ A\,|y-\hat{y}| - C, & \text{otherwise} \end{cases} \quad (7)$$

$$WLM = (w \cdot Mask + 1) \odot AW(P, \hat{P}) \quad (8)$$

$$\mathcal{L}_{poremap} = \frac{1}{HW}\sum_{i=1}^{H}\sum_{j=1}^{W} WLM_{ij} \quad (9)$$

where ω, α, θ and ϵ are positive hyperparameters. $A = \omega(1 + (\frac{\theta}{\epsilon})^{\alpha-y}))^{-1}(\alpha - y)(\frac{1}{\epsilon})(\frac{\theta}{\epsilon}^{\alpha-y-1})$ and $C = (\theta A - \omega \ln(1 + (\frac{\theta}{\epsilon})^{\alpha-y})$ are utilized to ensure the loss function remains continuous and smooth at $|y - \hat{y}| = \theta$. $\odot$ represents element-wise production. P and $\hat{P}$ are ground truth pore map and prediction pore map. w is a scaling factor to control the degree of weighting. Finally, the overall loss is formulated as follows:

$$\mathcal{L}_{all} = \lambda \mathcal{L}_{reconstructed} + \beta \mathcal{L}_{poremap} \quad (10)$$

where λ and β are used to balance two loss.

3 Experiments

3.1 Datasets and Evaluation Protocols

Two high-resolution fingerprint datasets are used to evaluate the performance of our proposed method. The first dataset is the most used PolyU HRF dataset. The second dataset is an in-house HRF dataset to further evaluate the generalization of the proposed method, which is the same as [14].

PolyU HRF dataset contains two sub-datasets, denoted as DBI and DBII. Both the DBI and DBII datasets contain 1480 high-resolution fingerprints (~1200 dpi) from 148 fingers. Each finger is collected in two sessions, with five fingerprint images collected in each session. The difference between DBI and DBII is the size of the fingerprint image, which is 240×320 pixels in DBI and 640×480 pixels in DBII. In addition, the PolyU HRF dataset contains 30 manually annotated pore images from the DBI dataset which are used to evaluate the effectiveness of pore detection.

In-house HRF dataset is collected using the same equipment as the PolyU HRF dataset, which contains 1000 high-resolution fingerprints from 250 fingers. Each finger is collected in two sessions, with two fingerprint images collected in each session. The image size of the in-house dataset is 640×480 pixels.

Evaluation protocols. For pore detection evaluation, the true detection rate (R_T) and the false detection rate (R_F) are employed to evaluate the effectiveness of pore detection. R_T is the ratio of the number of truly detected pores to the number of all ground truth pores. R_F is the ratio of the number of falsely detected pores to the number of all detected pores. The equal error rate (EER) is employed to evaluate the recognition performance. The recognition protocol on DBI and DBII is the same as the previous method [6,9,14,19], which includes 3,700 genuine matches and 21,456 imposter matches. Genuine matches consist of pairwise matching fingerprints of the same finger from different sessions. Imposter matches consist of matching the first images of different fingers between the first session and the second session. In the in-house dataset, we use the same recognition protocol as [14]. The four images of the same finger are matched pairwise, leading to 1500 genuine matches. The first image of the first session of one finger is matched with the first image of the second session of the other fingers, leading to 62,250 imposter matches.

3.2 Implementation Details

Our network is implemented using the PyTorch framework and trained on a single NVIDIA Tesla V100 GPU with a batch size of 32 for 50 epochs. We use the VIT-Base [20] as our encoder, initialing it with pre-trained model parameters. During training, the input image size is 240 × 320. We use the same fingerprint image as [14] and apply data augmentation to construct the training data. During inference, when the input image size is 480 × 640, we obtain the corresponding position embedding through interpolation. The D_d is 1024. The hyperparameters are determined based on experimental results. Specifically, ω, α, θ and ϵ in the AW loss are set to 14, 2.1, 0.5, and 1. The threshold used to generate Mask is set to 0.3 and the scaling factor w is set to 10. The λ and β are set to 0.8 and 1. We apply linear warmup to adjust the learning rate, with a maximum value of 1e-4 and a warmup step size equal to one-tenth of the total number of iterations. The network is optimized using AdamW [21].

3.3 Ablation Study

To analyze the impact of AW loss plus WLM and CBAM used in the proposed network, we conducted pore detection ablation experiments on 30 manually annotated fingerprint images from the PolyU DBI dataset with different slide window parameters, as well as pore matching ablation experiments on both the PolyU DBI and DBII datasets using the original image and $Feat$ to represent pores, as shown in Table 1 and Table 2. Baseline refers to using the traditional MSE loss and without attention module. Observing Table 1, we can see that smaller sliding windows achieve higher R_T but also result in higher R_F under the same conditions. For the same threshold, we believe that larger sliding windows have a larger field of view, rendering more reliable results and thus have lower R_F. However, smaller sliding windows offer a more complete detection of pores in areas with a dense distribution of pores, leading to higher R_T. Moreover, it's clear that the complete network structure achieves the highest R_T. Additionally, the results show that the combination of AW loss and WLM can significantly enhance the R_T. For all subsequent experiments, we adopt a sliding window size of 5 × 5 and a threshold of 0.35 for pore detection. The recognition results, as presented in Table 2, indicate that the ability to detect and represent

Table 1. Effects of two components on pore detection with different sliding window parameters (window size and threshold) on PolyU DBI dataset.

Baseline	CBAM	AW+WLM	5 × 5						7 × 7					
			0.25		0.3		0.35		0.25		0.3		0.35	
			R_T	R_F	R_T	R_F	R_T	R_F	R_T	R_F	R_T	R_F	R_T	R_F
✓	×	×	91.74	6.90	90.54	5.64	88.87	4.77	90.89	6.70	89.81	5.55	88.29	4.69
✓	✓	×	93.13	8.11	91.64	6.65	89.78	5.32	92.30	7.79	90.97	6.50	89.34	5.22
✓	×	✓	97.26	24.13	97.08	21.34	96.68	17.88	96.04	21.34	95.96	19.20	95.70	16.49
✓	✓	✓	**98.15**	23.59	**97.93**	20.06	**97.61**	16.33	**97.13**	20.19	**97.00**	17.94	**96.77**	15.16

Table 2. Effects of two components on pore matching on PolyU DBI and DBII dataset.

Baseline	CBAM	AW+WLM	DBI		DBII	
			Feat	original image	*Feat*	original image
✓	×	×	16.18	6.47	10.63	1.45
✓	✓	×	11.49	6.46	6.22	1.50
✓	×	✓	11.89	5.29	8.22	1.64
✓	✓	✓	**4.85**	**5.22**	**0.89**	**1.35**

pores is key to recognition performance. By incorporating both components, we observe significant improvements in these aspects, leading to the best recognition performance. These findings provide strong evidence for the effectiveness of these components within our proposed network.

3.4 Comparison with Other Pore Detection Methods

Accurate pore detection is vital for high-resolution fingerprint representation and recognition. Therefore, we evaluated the performance of pore detection compared with other existing pore detection methods on the same dataset used in the pore ablation experiment, as shown in Table 3. As seen in the table, a state-of-the-art R_T can be achieved by our proposed method. Although the proposed method also have a higher R_F than some of the other methods, we believe that even a slight increase in R_F can improve recognition performance as long as R_T is high enough. This is because accurately detecting more sweat pores provides greater opportunities to establish correct pore correspondences. Thus, we believe that achieving a high R_T is crucial for subsequent pore matching and the proposed method effectively achieves this goal. We will further demonstrate this in subsequent experiments.

3.5 Comparison with Other Pore Matching Methods

To further evaluate the effectiveness of the proposed method, several comparative recognition experiments have been carried out. Firstly, we compared the recognition performance with some classical methods on the PolyU HRF dataset

Table 3. Pore detection results of R_T (%) and R_F (%) compared with other methods.

	R_T	R_F
Gabor Filter [2]	75.90	23.00
Adapt. DoG [3]	80.80	22.20
DAPM [3]	84.80	17.60
Labati et al. [5]	84.69	15.31
Xu et al. [22]	85.70	11.90
DeepPore [4]	93.09	8.64
Zhao et al. [7]	93.14	**4.39**
Gabriel et al. [6]	92.00	16.90
Shen et al. [8]	93.20	8.10
Liu et al. [14]	94.88	24.25
Ours	**97.61**	16.33

using a simple pore representation, namely the original image. Table 4 shows the recognition results. Observing Table 4, it can be seen that our method achieves the best recognition performance compared with other methods. The results also demonstrate the effectiveness of our method in pore detection. Then, we compared the recognition performance with the latest one-stage method [14] on the same dataset using both the original image and deep features to represent pores, as shown in Table 5. The results reveal a significant improvement in recognition performance, particularly when utilizing deep features. Notably, the recognition performance of our method's deep features outperforms the original image. Furthermore, the fusion results show that the fusion of the original image and deep features can further improve recognition performance and the deep features play an important role. Finally, we compared the recognition performance with the state-of-the-art two-stage high-resolution representation method, DeepPoreID [12], and the latest one-stage method [14] on an in-house dataset to evaluate the generalization ability of the proposed one-stage method, as shown in Table 6. The results show that the one-stage method can achieve better generalization and our proposed one-stage method outperforms the latest one-stage method. Figure 3 shows the recognition results of genuine and imposter matching examples using our proposed method and the latest one-stage method on three datasets, with deep features being used to represent pores. The results show that our proposed method can match more pore pairs for challenging genuine matching pairs, while yielding close or fewer pore pairs for imposter matching pairs, further demonstrating the effectiveness of our proposed method.

Table 4. The EERs (%) of original image on PolyU DBI and DBII dataset.

	DP [9]	DAPM [3]	Gabriel et al. [6]	Shen et al. [8]	Ours
DBI	15.42	9.6	9.2	7.7	**5.22**
DBII	7.05	4.4	4.5	4.4	**1.35**

Table 5. The EERs (%) of the original image, deep feature, and their fusion on PolyU DBI and DBII dataset.

	original image		*Feat*		fusion	
	Liu et al. [14]	Ours	Liu et al. [14]	Ours	Liu et al. [14]	Ours
DBI	5.73	**5.22**	7.93	**4.85**	5.55	**4.24**
DBII	1.64	**1.35**	1.84	**0.89**	1.27	**0.65**

Table 6. The EERs (%) compared with a typical two-stage and the latest one-stage representation method on three datasets.

	DeepPoreID [12]	Liu et al. [14]	Ours
DBI	**1.42**	5.55	<u>4.24</u>
DBII	**0.51**	1.27	<u>0.65</u>
In-house Dataset	1.02	<u>0.50</u>	**0.37**

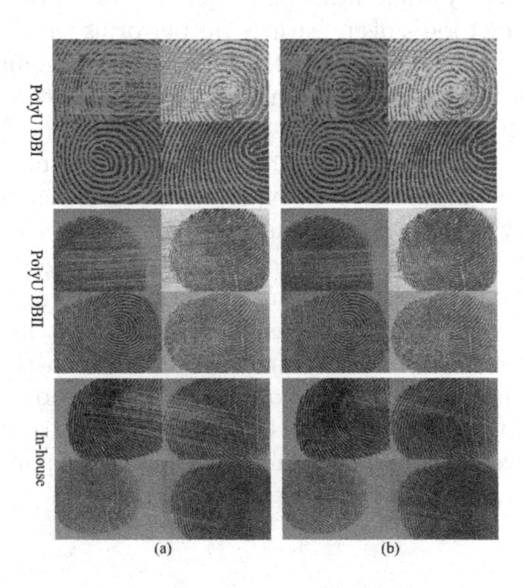

Fig. 3. Visualization of genuine (top) and imposter (bottom) matching pairs using deep features on three datasets: (a) Our method (b) The latest one-stage method.

4 Conclusion

The latest one-stage high-resolution fingerprint representation method outperforms the current two-stage method in robustness and generalization. However, the pore detection and feature learning ability are still limited by the network architecture and the loss used. To further improve the one-stage high-resolution fingerprint representation method, in this paper, we propose to combine the transformer-based encoder and the fully convolutional network to learn both local and global information, providing a more robust pore representation. Moreover, we incorporate the attention module and apply adaptive wing loss with a weighted loss map to further improve pore detection and feature learning capability. The experimental results on two high-resolution fingerprint datasets demonstrate the effectiveness of our proposed method. In our future work, we will continue to focus on better integrating various information related to pores and designing more lightweight models.

References

1. Maltoni, D., Maio, D., Jain, A.K., Prabhakar, S., et al.: Handbook of Fingerprint Recognition, vol. 2. Springer, London (2009). https://doi.org/10.1007/978-1-84882-254-2
2. Jain, A., Chen, Y., Demirkus, M.: Pores and ridges: fingerprint matching using level 3 features. In: 18th International Conference on Pattern Recognition (ICPR 2006), vol. 4, pp. 477–480. IEEE (2006)
3. Zhao, Q., Zhang, D., Zhang, L., Luo, N.: Adaptive fingerprint pore modeling and extraction. Pattern Recogn. **43**(8), 2833–2844 (2010)
4. Jang, H.U., Kim, D., Mun, S.M., Choi, S., Lee, H.K.: Deeppore: fingerprint pore extraction using deep convolutional neural networks. IEEE Signal Process. Lett. **24**(12), 1808–1812 (2017)
5. Labati, R.D., Genovese, A., Munoz, E., Piuri, V., Scotti, F.: A novel pore extraction method for heterogeneous fingerprint images using convolutional neural networks. Pattern Recogn. Lett. **113**, 58–66 (2018)
6. Dahia, G., Segundo, M.P.: Improving fingerprint pore detection with a small FCN. arXiv preprint arXiv:1811.06846 (2018)
7. Zhao, Y., Liu, F., Shen, L.: Fingerprint pore extraction using convolutional neural networks and logical operation. In: Zhou, J., et al. (eds.) CCBR 2018. LNCS, vol. 10996, pp. 38–47. Springer, Cham (2018). https://doi.org/10.1007/978-3-319-97909-0_5
8. Shen, Z., Xu, Y., Lu, G.: CNN-based high-resolution fingerprint image enhancement for pore detection and matching. In: 2019 IEEE Symposium Series on Computational Intelligence (SSCI), pp. 426–432. IEEE (2019)
9. Zhao, Q., Zhang, L., Zhang, D., Luo, N.: Direct pore matching for fingerprint recognition. In: Tistarelli, M., Nixon, M.S. (eds.) ICB 2009. LNCS, vol. 5558, pp. 597–606. Springer, Heidelberg (2009). https://doi.org/10.1007/978-3-642-01793-3_61
10. Liu, F., Zhao, Q., Zhang, L., Zhang, D.: Fingerprint pore matching based on sparse representation. In: 2010 20th International Conference on Pattern Recognition, pp. 1630–1633. IEEE (2010)

11. Liu, F., Zhao, Q., Zhang, D.: A novel hierarchical fingerprint matching approach. Pattern Recogn. **44**(8), 1604–1613 (2011)
12. Zhao, Y., Liu, G., Liu, F., Shen, L., Li, Q.: Deepporeid: an effective pore representation descriptor in direct pore matching. In: 2019 IEEE International Conference on Image Processing (ICIP), pp. 1690–1694. IEEE (2019)
13. Liu, F., Zhao, Y., Liu, G., Shen, L.: Fingerprint pore matching using deep features. Pattern Recogn. **102**, 107208 (2020)
14. Liu, F., Liu, G., Zhang, W., Wang, L., Shen, L.: A novel high-resolution fingerprint representation method. IEEE Trans. Biom. Behav. Identity Sci. **4**(2), 289–300 (2022)
15. Woo, S., Park, J., Lee, J.-Y., Kweon, I.S.: CBAM: convolutional block attention module. In: Ferrari, V., Hebert, M., Sminchisescu, C., Weiss, Y. (eds.) ECCV 2018. LNCS, vol. 11211, pp. 3–19. Springer, Cham (2018). https://doi.org/10.1007/978-3-030-01234-2_1
16. Leng, L., Zhang, J.: Palmhash code vs. palmphasor code. Neurocomputing **108**, 1–12 (2013)
17. Leng, L., Li, M., Kim, C., Bi, X.: Dual-source discrimination power analysis for multi-instance contactless palmprint recognition. Multimedia Tools Appl. **76**, 333–354 (2017)
18. Wang, X., Bo, L., Fuxin, L.: Adaptive wing loss for robust face alignment via heatmap regression. In: Proceedings of the IEEE/CVF International Conference on Computer Vision, pp. 6971–6981 (2019)
19. Shen, Z., Xu, Y., Li, J., Lu, G.: Stable pore detection for high-resolution fingerprint based on a CNN detector. In: 2019 IEEE International Conference on Image Processing (ICIP), pp. 2581–2585. IEEE (2019)
20. Dosovitskiy, A., et al.: An image is worth 16x16 words: transformers for image recognition at scale. arXiv preprint arXiv:2010.11929 (2020)
21. Loshchilov, I., Hutter, F.: Decoupled weight decay regularization. arXiv preprint arXiv:1711.05101 (2017)
22. Xu, Y., Lu, G., Liu, F., Li, Y.: Fingerprint pore extraction based on multi-scale morphology. In: Zhou, J., et al. (eds.) CCBR 2017. LNCS, vol. 10568, pp. 288–295. Springer, Cham (2017). https://doi.org/10.1007/978-3-319-69923-3_31

A Video Face Recognition Leveraging Temporal Information Based on Vision Transformer

Hui Zhang[1], Jiewen Yang[2](✉), Xingbo Dong[3](✉), Xingguo Lv[1], Wei Jia[4], Zhe Jin[3], and Xuejun Li[1]

[1] Anhui Provincial International Joint Research Center for Advanced Technology in Medical Imaging, School of Computer Science and Technology, Anhui University, Hefei 230093, China

[2] Department of Electronic and Computer Engineering, Hong Kong University of Science and Technology, Hong Kong, China
jyangcu@connect.ust.hk

[3] Anhui Provincial Key Laboratory of Secure Artificial Intelligence, School of Artificial Intelligence, Anhui University, Hefei 230093, China
xingbo.dong@ahu.edu.cn

[4] School of Computer Science and Information, Hefei University of Technology, Hefei 230009, China

Abstract. Video face recognition (VFR) has gained significant attention as a promising field combining computer vision and artificial intelligence, revolutionizing identity authentication and verification. Unlike traditional image-based methods, VFR leverages the temporal dimension of video footage to extract comprehensive and accurate facial information. However, VFR heavily relies on robust computing power and advanced noise processing capabilities to ensure optimal recognition performance. This paper introduces a novel length-adaptive VFR framework based on a recurrent-mechanism-driven Vision Transformer, termed TempoViT. TempoViT efficiently captures spatial and temporal information from face videos, enabling accurate and reliable face recognition while mitigating the high GPU memory requirements associated with video processing. By leveraging the reuse of hidden states from previous frames, the framework establishes recurring links between frames, allowing the modeling of long-term dependencies. Experimental results validate the effectiveness of TempoViT, demonstrating its state-of-the-art performance in video face recognition tasks on benchmark datasets including iQIYI-ViD, YTF, IJB-C, and Honda/UCSD.

Keywords: Video face recognition · Vision Transformer · Temporal information

1 Introduction

Video face recognition (VFR) has emerged as a dynamic and promising field at the intersection of computer vision and artificial intelligence, revolutionizing the way individuals are identified and authenticated [7]. Unlike traditional

Q. Liu et al. (Eds.): PRCV 2023, LNCS 14429, pp. 29–43, 2024.
https://doi.org/10.1007/978-981-99-8469-5_3

image-based approaches, video face recognition utilizes the temporal dimension inherent in video footage to extract more comprehensive and accurate facial information. As surveillance systems, social media platforms, and video communication technologies continue to proliferate, the demand for robust and efficient VFR systems has grown significantly. However, VFR receives far less attention than image-based face recognition [7,10].

To date, two prevalent approaches have emerged for VFR. The first approach treats the frames of a video sequence as a collection of individual images, disregarding the temporal order of the frames, e.g., [23]. However, such methods lack computational efficiency as they necessitate comparing similarities across all feature vectors between two face videos. Another approach involves aggregating feature vectors from each frame in a sequential manner, preserving the temporal order. This method allows for the comprehensive capture of facial dynamics over time, leading to a more accurate representation of the evolving face, e.g., [11,17]. In practice, the sequential aggregation approach may be preferred as it can accurately capture temporal dynamics and ultimately boost the recognition performance even under an uncontrolled environment.

However, using videos for face recognition has two sides of the same coin. On the one hand, face videos offer the potential for enhanced performance due to their enriched information content, encompassing valuable temporal dynamics and multi-view perspectives. On the other hand, processing hundreds or thousands of frames typically takes a lot of computing power, and noise frames may also impair performance. Therefore, a good VFR system should make full use of video data, avoid noisy distortion, and be efficient and accurate.

To achieve the mentioned objective, one can employ state-of-the-art sequential data modeling tools like ConvLSTM [29], ConvGRU [2], 3DCNN [16], and Vision Transformer (ViT) [3,6]. Among these options, ViT stands out as a promising candidate for capturing temporal information in videos. ViT has proven its effectiveness in modeling long-distance relationships [26] and has shown success in various video understanding tasks [3] when compared to models based on LSTM and ConvNets.

Yet, it is crucial to acknowledge that face videos in real-world scenarios can have varying lengths. This presents a formidable obstacle for ViT-based approaches, as transformers are inherently limited in their ability to capture dependencies within input sequences due to the fixed input size employed during training [31]. For instance, if the maximum sentence size is set to 256 words, the transformer model will be unable to capture dependencies between words that occur beyond this limit.

Inspired by the above discussion, we propose a new video face recognition framework, termed **TempoViT**, based on the sequential temporal information aggregation concept. Specifically, the proposed framework centers around a carefully crafted recurrent Vision Transformer for temporal information extraction. The TempoViT consists of a stack of units that operate by taking the current input frame, denoted as $x^{(t)}$, and the hidden state $h^{(t-1)}$ from the previous frame as inputs. From this current TempoViT unit, an output $O^{(t)}$ and an updated hidden state $h^{(t)}$ are generated. Then, a joint attention mechanism is proposed for the TempoViT unit, which is used to draw global dependencies between image patches and temporal dependencies between neighboring frames through a hidden state. Length-adaptive spatial-temporal feature extraction can be done by recursively running the TempoViT unit on individual frames.

An important aspect to highlight is that we leverage the reuse of hidden states from previous frames instead of computing them from scratch for each new frame. The reused hidden states act as a memory for the current frame, forming a recurring link between them. As a result, modeling very long-term dependencies becomes possible because information can be propagated through recurrent connections. Additionally, the frame-wise recursive mechanism can help mitigate the high GPU memory requirements associated with video processing.

The contributions of our work can be summarized as follows: 1) We introduce an end-to-end ViT-based length-adaptive VFR framework, termed TempoViT. By directly inputting face videos, our system can output the corresponding face embeddings. The TempoViT framework efficiently captures spatial and temporal information from videos, allowing for accurate and reliable face recognition. Furthermore, we address the high GPU memory requirements typically associated with video processing, ensuring that our framework is suitable for applications requiring both accuracy and efficiency in face video identification; 2) We evaluate the performance of our face recognition framework on benchmark datasets such as iQIYI-ViD, YTF, IJB-C, and Honda/UCSD. Our results demonstrate the effectiveness of TempoViT in achieving state-of-the-art performance in face recognition tasks.

2 Related Works

Video Face Recognition. In comparison to image-based face recognition, videos offer a richer source of information as they inherently capture faces of the same individual in diverse poses and lighting conditions. This abundance of data enhances the robustness and accuracy of face recognition algorithms, enabling them to better handle variations in facial appearance and improve overall performance.

The primary difficulty in video face recognition lies in constructing a suitable representation for the video face, which can successfully combine information from multiple frames while filtering out noisy or irrelevant data.

Convolutional neural networks (CNNs) have emerged as one of the most widely used and successful tools for video face recognition. In 2017, a Neural Aggregation Network (NAN) was proposed in [30] for video face recognition, which consists of two modules that can be trained sequentially or individually. The first is a deep CNN feature embedding module that extracts frame-level features. The other is the aggregation module, which performs the adaptive fusion of feature vectors from all video frames.

For dealing with bad frames, [25] proposed a method for discarding undesirable frames using a Markov decision process and trained an attention model through a deep reinforcement learning framework. In [5], a trunk-branch ensemble CNN model was proposed to solve the illumination and low-resolution problems. It has been shown to have competitive performance compared to conventional CNN networks.

A recurrent regression neural network (RRNN) framework was introduced in [19] for cross-pose face recognition tasks, specifically targeting still images and videos. RRNN explicitly builds the potential dependencies of sequential images and adaptively memorizes and forgets information that benefits final classification by performing progressive transforms on adjacent images sequentially.

In [9], a novel approach called the component-wise feature aggregation network (C-FAN) was introduced. C-FAN is designed to handle a set of face images belonging to a particular individual as input and generates a single feature vector as the representation for face recognition. The key aspect of C-FAN is its ability to automatically learn the significance of different face features by assigning quality scores, which enhances the overall face representation for improved performance in recognition tasks.

Vision Transformer. Recently, transformer-based networks have gained significant attention and have been increasingly applied to computer vision tasks.

Dosovitskiy et al. [6] introduced the Visual Transformer (ViT) model and made a groundbreaking observation that the transformer framework, even without convolutional layers, can achieve impressive performance on image processing tasks [6].

Zhong et al. [33] conducted the first study to investigate the performance of Transformer models in face recognition. They trained and evaluated a ViT model on several mainstream benchmarks. The results showed that Face Transformer models trained on a large-scale database can perform similarly to CNN models. However, it is important to note that the proposed ViT model was only tested on still images [33].

Transformer-based models have also been increasingly utilized for video-related tasks [1,3,8,24]. The ViViT model [1] employs two transformer encoders, one for processing spatial information and the other for temporal information. This approach effectively captures both spatial and temporal features in videos.

The TimeSformer [3] introduces a convolution-free methodology that extends self-attention to incorporate joint spatial-temporal attention. It allows the model to capture both spatial and temporal dependencies in videos. The VTN model [24] combines a 2D spatial feature extraction model with a temporal-attention-based encoder. This combination results in an efficient architecture specifically designed for video understanding tasks.

The MViT model [8] introduces multi-head pooling attention with a focus on specific spatial-temporal resolutions, which leads to promising performance. Unlike traditional RNN and LSTM methods, transformer-based approaches are designed to process batches of frames in parallel for video tasks. However, these approaches often require significant GPU memory due to their parallel processing nature. Additionally, the extraction of temporal features is typically performed within the batch, which can limit the amount of information captured in the temporal domain. These considerations highlight the trade-offs and challenges involved in designing efficient and informative transformer-based models for video analysis tasks.

To address these challenges, recursive methods, such as ConvLSTM [29], have been successfully employed in video tasks. These recursive models have demonstrated effectiveness in capturing temporal dependencies and modeling video sequences. However, recent studies suggest that transformer-based methods, with their self-attention mechanisms designed specifically for video tasks, can establish interactions between the spatial and temporal domains [1,3,8,24].

In summary, current face recognition approaches for videos primarily rely on convolutional neural networks (CNNs), while limited efforts have been made to extend the usage of Vision Transformer (ViT) models to video face recognition. However, incorporating recursive mechanisms and Transformer self-attention mechanisms hold promise in improving performance.

3 Methods

3.1 Overview

The proposed TempoViT is a video face recognition framework that utilizes a carefully designed recurrent Vision Transformer. In the following subsections we introduce the specific details of preprocessing, recurrent unit and joint attention gate respectively, as shown in Fig. 1. The main idea is to capture both spatial and temporal information from face videos while addressing the challenges associated with varying video lengths and high computational requirements.

The TempoViT consists of a stack of units that operate on individual frames of a video sequence in a sequential manner. Specifically, given the current input frame $x^{(t)}$ and the hidden state $h^{(t-1)}$ obtained from the previous frame, the TempoViT unit processes this information and produces an output $O^{(t)}$ along with a new hidden state $h^{(t)}$ for the current frame. To enhance performance, a custom-designed multi-head temporal-spatial joint attention module is employed, which facilitates interaction between the current frame input and the previous hidden state.

3.2 Preprocessing

The input frame $X^{(t)} \in \mathbb{R}^{H \times W \times C}$, with dimensions of height (H), width (W), and channel number (C), will undergo a decomposition process, where it is divided into non-overlapping patches of size $P \times P$. These patches are subsequently flattened into vectors $x_p^{(t)} \in \mathbb{R}^{P^2 \times D}$ and $D = \frac{H}{P} \cdot \frac{W}{P} \cdot C$. In order to generate the input vector for the TempoViT unit, the patched vector $x_p^{(t)}$ undergoes a convolutional embedding layer. This layer applies a convolution operation to extract relevant features from the patch vector. Additionally, a position encoding vector is added to the embedded representation:

$$x^{(t)} = \ell(x_p^{(t)}) + Pos_p \tag{1}$$

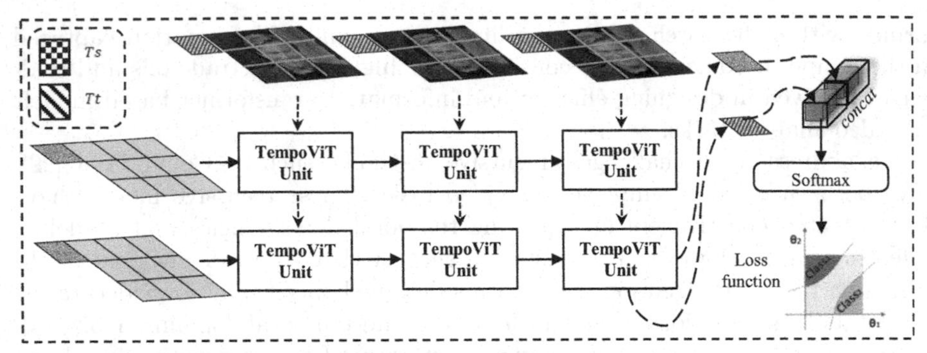

(a) Overview of the proposed ViT-based video face recognition

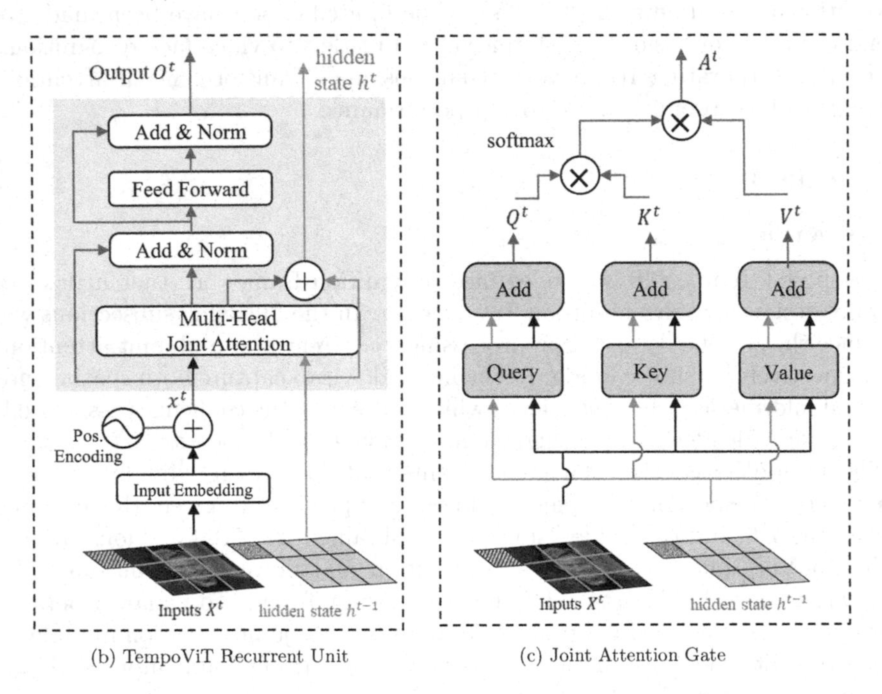

(b) TempoViT Recurrent Unit (c) Joint Attention Gate

Fig. 1. TempoViT. (a) provides an overview of the TempoViT framework, showcasing a two-layer TempoViT architecture as an example. In (b), a recurrent unit is depicted, representing the core component of our framework. (c) illustrates the joint attention gate, which operates on both the current input $x^{(t)}$ and the hidden state $h^{(t-1)}$.

where ℓ denotes the convolution embedding function and $Pos_p \in \mathbb{R}^{P^2 \times D}$ is a learnable positional encoding vector. The positional encoding vector is the same for all frames.

In a manner similar to the conventional ViT approach, Fig. 1a demonstrates the usage of learnable tokens $Ts \in \mathbb{R}^D$ and $Tt \in \mathbb{R}^D$ to represent spatial and temporal directions, respectively. Thereby, Ts tokens are added at the beginning of each input frame $x(t)$ to aggregate spatial features, while Tt tokens are appended to the initial hidden state in each layer to aggregate temporal features. The aggregated Tt tokens and the Ts tokens from the last frame's output are concatenated. This concatenated representation is then fed into a linear classification layer W_{class}, enabling the model to generate a prediction result. Hence a prediction result can be given as:

$$P = W_{class}(Concat(Tt_0^{(t)}, ..., Tt_n^{(t)}, Ts_n^{(t)})). \tag{2}$$

3.3 TempoViT Recurrent Unit

Figure 1b provides an overview diagram of a single TempoViT unit, illustrating its key components and their interactions. The input frame $x(t)$ and hidden state $h(t-1)$ collaborate to compute the current hidden state $h(t)$ and output $O(t)$ in a TempoViT unit. The attention unit in the TempoViT framework takes as input the current frame $x(t)$ and the previous hidden state $h(t-1)$, both of which are first passed through a Layer Normalization layer. Following that, the attention preserves previous attention while appending new information from the current frame. Finally, the hidden state $h^{(t)}$ of the current frame, belonging to $R^{P^2 \times D}$, is generated in a residual manner.

$$h^{(t)} = h^{(t-1)} + A^{(t)}, \tag{3}$$

The output $O^{(t)} \in \mathbb{R}^{P^2 \times D}$ of the current unit is generated by a Feed-Forward Network (FFN) with a residual connection. This can be defined as follows:

$$O^{(t)} = f(o^{(t)}) + o^{(t)}, \tag{4}$$

where $o^{(t)}$ is the intermediate output defined as $o^{(t)} = x^{(t)} + A^{(t)}$ and $f(\cdot)$ represents the FFN.

Unlike existing approaches that often process a batch of frames, our method takes a different approach by utilizing a recurrent unit to process face videos frame by frame. This sequential processing allows us to effectively reduce redundant computation, especially when dealing with long video sequences. In contrast to methods such as 3D-ResNet and TimeSformer which require the entire video sequence for inference and training, our approach focuses on individual frames, resulting in more efficient and streamlined processing.

3.4 Temporal-Spatial Joint Attention

As shown in Fig. 1c, a joint attention gate based on [31] is specifically designed to establish an interaction between the current frame $x^{(t)}$ and the hidden state $h^{(t-1)} \in \mathbb{R}^{P^2 \times D}$ from the previous frame. It takes these inputs and generates an attended vector $a^{(t)} \in \mathbb{R}^{P^2 \times D}$, which represents a fused representation that captures the relevant information from both the current frame and the previous hidden state. By utilizing the joint attention gate, the TempoViT framework enhances the integration of temporal and spatial information, enabling more effective and accurate analysis of the video frames.

Table 1. Datasets of face videos

Datasets	#Identity	# Video	Remarks
iQIYI-VID [21]	10034	200k	length of the video clip ranges from 1 s to 30 s
YTF [28]	1,595	3,425	length of the video clip ranges from 48 to 6,070 frames
IJB-C [22]	3,531	11779	video with a length <6 frame will be excluded
Honda/UCSD [18]	37	1480	captured indoors with natural light, fixed distance

$$a^{(t)} = (\sigma(Q^{(t)}K^{(t)}))V^{(t)} \tag{5}$$

The activation function $\sigma(\cdot)$, in this case, represents the ELU (Exponential Linear Unit) function $elu(\cdot)$. $Q^{(t)}$, $K^{(t)}$, $V^{(t)}$ are matrices denoting the Query, Key, and Value, respectively. These matrices are defined as follows:

$$\begin{aligned} Q^{(t)} &= x^{(t)}W_x^Q + h^{(t-1)}W_h^Q \\ K^{(t)} &= x^{(t)}W_x^K + h^{(t-1)}W_h^K \\ V^{(t)} &= x^{(t)}W_x^V + h^{(t-1)}W_h^V \end{aligned} \tag{6}$$

Note that the attention vector $a(t)$ captures a joint attention mechanism between the current frame input $x(t)$ and the previous hidden state $h(t-1)$. This joint attention is achieved because the matrices $Q(t)$, $K(t)$, and $V(t)$ are computed from both the current frame input and the previous hidden state. Furthermore, to achieve multi-head attention, q attention heads are concatenated together, resulting in a combined representation that incorporates the contributions from multiple attention heads.

$$A^{(t)} = \text{Concat}(a_1^{(t)}, \cdots, a_q^{(t)})W_{proj} \tag{7}$$

where $a_q^{(t)} = (\sigma(Q_q^{(t)}K_q^{(t)}))V_q^{(t)}$ and $a_q^{(t)} \in \mathbb{R}^{P^2 \times \frac{D}{q}}$. A linear layer $W_{proj} \in \mathbb{R}^{D \times D}$ is applied to project the attended vector to a desired output dimension.

4 Experiments and Results

4.1 Experiment Setting

Datasets. As given in Table 1, four public video face benchmark datasets are used in our experiments to verify the effectiveness of the proposed TempoViT, including iQIYI-VID-FACE [21], YouTube Face (YTF) [28], IJBC [22] and Honda/UCSD [18].

The iQIYI-VID dataset is the largest open-source video celebrity recognition dataset composed of more than 200k video clips of 10, 034 celebrities. The training set is 90% samples of each user in the iQIYI-VID dataset and the remaining samples are composed of a testing set. YFT is a comprehensive database of labeled videos of faces in challenging, uncontrolled conditions, which contains

3425 videos of 1595 subjects. IJBC is composed of 3531 subjects with 117. 5k images from 11779 videos. Still images and frames with insufficient length (<6 frames) are not used in our experiments. YTF and IJB-C are used in the training of the multilayer perceptron head (MLP), and follow the training protocol from [4]. The Honda/UCSD video database contains 1480 videos of 37 subjects and all are recorded indoors with natural light. The training set consists of one video of each of 20 subjects, while the 39 videos of 17 subjects are used in the testing set. Some examples are shown in Fig. 2.

Implementation Details. We first verify the performance of the proposed TempoViT on the iQIYI-VID dataset. Then we use the pretained model on the iQIYI-VID dataset to evaluate the performance of YTF, IJB-C, and Honda/UCSD datasets which is in line with practical application scenarios.

We apply a random initialization to train our proposed TempoViT on the iQIYI-VID dataset with Adam optimizer and Cross-entropy loss. The initial learning rate and weight decay are set to $3e^{-4}$ and $5e^{-4}$, respectively. Both the training and testing phases randomly sample 8 frames from each video with a sampling rate of 1/1 to 1/8. The frames from the sample are first resized to 112×112. Additionally, a random horizontal flipping is applied to each frame.

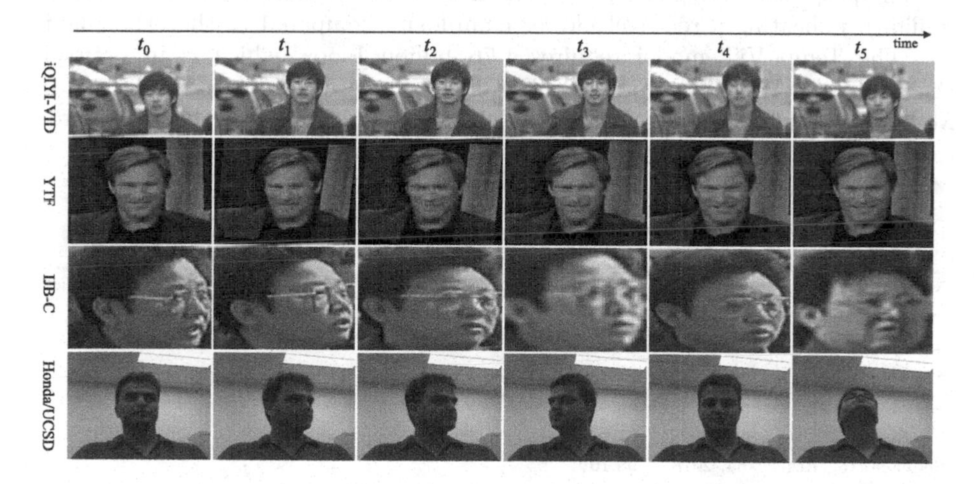

Fig. 2. Dataset examples.

Table 2. Performance on iQIYI-ViD dataset

Method	mAP@1(%)	mAP@100(%)	Params(M)	Flops(G)
ArcFace,Res100 [4]	N/A	79.80	59.27	2.04
MobileNetV3+LSTM [14]	78.19	80.42	15.82	0.50
ResNet3D-50 [12]	82.65	86.13	48.71	5.29
Ours(TempoViT)	**85.07**	**87.11**	**1.49**	1.76

After these transformations, we set P to 8. Each frame is divided into 8×8 non-overlapping patches and flattened for further processing.

4.2 Quantitative Evaluation

In this section, we adopt mAP@1 and mAP@100 as the retrieval performance indicator to evaluate the proposed TempoViT on iQIYI-VID. We compare the recognition performance of different methods, including ArcFace [4], MobileNetV3+LSTM [14], ResNet3D-50 [12] and ours, the result is shown in Table 2.

As can be seen from Table 2, ArcFace achieves an mAP@100 accuracy of 79.80%. MobileNetV3+LSTM achieves an mAP@1 accuracy of 78.19% and an mAP@100 accuracy of 80.42%. ResNet3D-50 achieves an mAP@1 accuracy of 82.65% and an mAP@100 accuracy of 86.13%. Our proposed approach achieves an mAP@1 accuracy of 85.07% and an mAP@100 accuracy of 87.11%.

The number of parameters (Params) in a model refers to the number of learnable weights that the model contains. Floating-point operations (Flops) represent the number of mathematical computations during its forward pass. Flops are indicative of the computational complexity or workload required by the model.

The proposed TempoViT model has the lowest number of parameters (1.49 million), indicating a relatively lower complexity compared to the other models. The TempoViT model requires 1.76 billion Flops, which is in between the MobileNetV3+LSTM and ResNet3D-50 models. Compared to Arcface, MobileNetV3+LSTM and ResNet3D-50, our TempoViT can be considered extremely lightweight and achieve the best performance.

Table 3. Performance(%) on YTF, IJB-C and Honda/UCSD datasets

Method	YTF accuracy	IJB-C TPR@FPR=1e-4	Honda Top-1 accuracy
DSR-Full len. [11], 2017	92.55	-	100.0
TBE-CNN [5], 2017	94.96	-	-
NAN [30], 2017	95.72	-	-
ADRL [25], 2017	96.52	-	-
CosFace [27], 2018	97.60	-	-
SeqFace [15], ResNet-64, 2018	98.10	-	-
C-FAN [9], 2019	96.50	-	-
Hörmann et al. [13], 2021	96.62	-	-
R100, ArcFace [4], 2019	98.02	95.60	-
Lin et al. [20] (50 frames), 2020	-	-	97.44
DDL [32], 2020	98.18	96.41	-
Ours(TempoViT)	86.19	84.47	100.0
Ours(TempoViT)+MLP	**99.40**	**99.58**	**100.0**

We then evaluate the cross-dataset recognition performance on YTF, IJB-C, and Honda /UCSD, which is consistent with data deployment in real systems.

Specifically, the TempoViT model is pre-trained on the large-scale iQIYI-ViD dataset and then used to infer the YTF and IJB-C datasets. As shown in Table 3, we calculate the accuracy of the normalized feature embeddings. For the Honda/UCSD dataset, a multi-layer perceptron (MLP) head is added to the backbone architecture and fine-tuned using the training set of the Honda/UCSD dataset to improve the model's performance. As observed from the table, the pre-trained TempoViT achieves an acceptable accuracy of 86.19% on YTF and 84.47% TPR@FPR1e^{-4} on IJB-C, while achieving 100% accuracy on the Honda/UCSD dataset.

Clearly, in the cross-dataset scenario, the accuracy on the YTF and IJB-C datasets is lower than that of most existing works. Therefore, inspired by [4], we attach an MLP layer to the pre-trained TempoViT model and fine-tune it on each testing dataset. As a result, the accuracy of TempoViT improves to 99.27% on YTF and 98.75% on IJB-C. The performance of TempoViT is greatly improved, and a series of results demonstrate its strong feature extraction ability.

4.3 Ablation Study

Since the video length of the TempoViT input is not fixed, we conduct an ablation study to assess accuracy and time cost for different video lengths in this stage.

First of all, the Honda/UCSD dataset is chosen for this experiment due to its videos being primarily captured under similar situations, resulting in fewer variations. This selection ensures a more controlled and consistent environment for conducting the study. During training, TempoViT is trained using input sequences of 8, 50, and 70 frames for each corresponding setting. However, for input sequences with 100 frames and full frames, we directly use TempoViT trained with 70 frames for inference due to the limitation of memory resources.

Table 4 shows the classification accuracy of different methods for video lengths ranging from 8 to full frames. We can clearly see that as the video clip length increases, the performance of face recognition increases. Moreover, compared with existing methods, our proposed TempoViT achieves better performance on the Honda/UCSD dataset. Specifically, proposed TempoViT without pre-training achieves 100% accuracy at 50 frame lengths, while TempoViT with pre-training achieves an accuracy of 100% even at 8 frame lengths. This shows that our TempoViT has the ability to extract the spatial and temporal information of the faces in the video, which is conducive to the construction of a high accuracy video-based face recognition system.

To demonstrate the efficiency of the TempoViT, Fig. 3 records the inference time cost of TempoViT for processing face videos of different lengths. The inference time takes only 48 ms when the video length is extended to all frames. The observations confirm that the proposed method can be used for real-time video face recognition.

In addition, we use attention visualization of the transformer to ascertain the relationship of the attention maps among frames at the inference stage. As we can observe in Fig. 4, face areas such as eyes, hair, and nose are activated in

Table 4. Classification accuracy(%) vs. video length on the Honda/UCSD

Method	Video clip Length (Frame)				
	8	50	70	100	Full
LBP+AdaBoost [11]	-	82. 75	88. 52	92. 63	96. 10
Pose [20]	-	97. 44	-	100.0	-
DSR [11]	-	98. 74	100.0	100.0	100.0
TempoViT w/o Pretrained	98. 69	100.0	100.0	100.0	100.0
TempoViT Pretrained	100.0	100.0	100.0	100.0	100.0

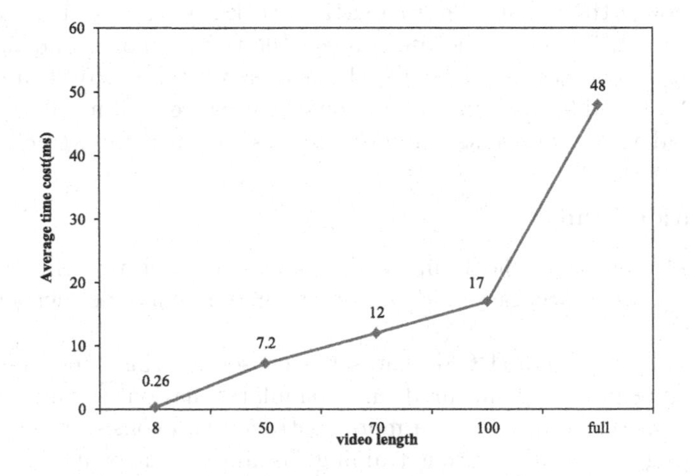

Fig. 3. Inference time cost for different video lengths.

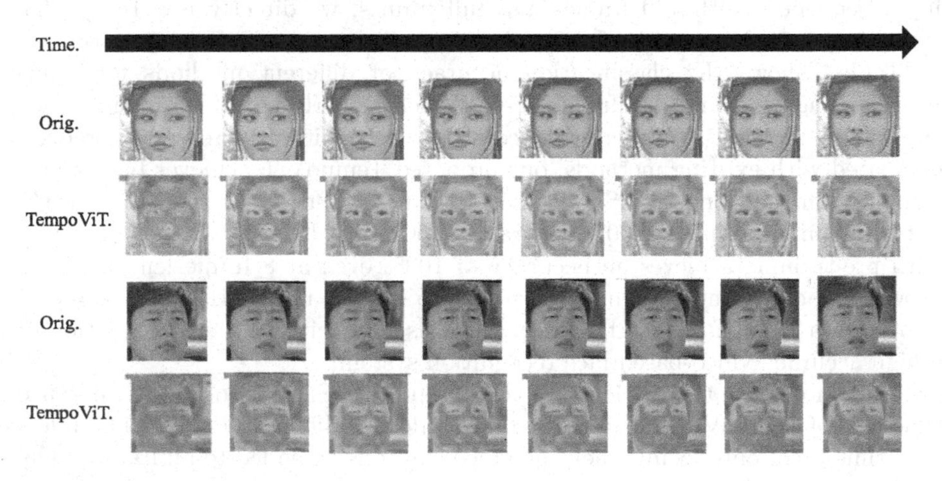

Fig. 4. Attention visualisation for TempoViT.

proposed TempoViT model, which suggests that the model can focuses on face areas for identity recognition under typical situations.

5 Conclusions

While image-based face recognition has received more attention, the demand for robust and efficient VFR systems has grown significantly with the proliferation of surveillance systems, social media platforms, and video communication technologies. Two prevalent approaches have emerged for VFR: treating video frames as individual images and aggregating feature vectors sequentially. The latter approach, which preserves temporal dynamics, has been shown to be more accurate in capturing the evolving face. However, processing video frames can be computationally demanding, and noisy frames can impair performance.

To address these challenges, we proposed a video face recognition framework based on the sequential aggregation approach. The framework leverages the power of Vision Transformer (TempoViT) and carefully crafted recurrent connections to capture both spatial and temporal information from face videos. By reusing hidden states and recursively running the TempoViT unit on individual frames, our framework achieves efficient processing and mitigates high GPU memory requirements.

Our evaluation on benchmark datasets demonstrates the effectiveness of TempoViT, showcasing its state-of-the-art performance in face recognition tasks. Overall, TempoViT offers a robust, accurate, and efficient solution for video face recognition, meeting the increasing demands of real-world applications in various domains.

Acknowledgments. This work was supported by the National Natural Science Foundation of China (Grant Nos. 62376003, 62306003, 62372004, 62302005).

References

1. Arnab, A., Dehghani, M., Heigold, G., Sun, C., Lučić, M., Schmid, C.: Vivit: a video vision transformer. In: Proceedings of the IEEE/CVF International Conference on Computer Vision (2021)
2. Ballas, N., Yao, L., Pal, C., Courville, A.: Delving deeper into convolutional networks for learning video representations. In: 4th International Conference on Learning Representations, ICLR 2016 (2015)
3. Bertasius, G., Wang, H., Torresani, L.: Is space-time attention all you need for video understanding? In: Proceedings of the International Conference on Machine Learning (ICML), vol. 2, p. 4 (2021)
4. Deng, J., Guo, J., Xue, N., Zafeiriou, S.: Arcface: additive angular margin loss for deep face recognition. In: Proceedings of the IEEE/CVF Conference on Computer Vision and Pattern Recognition, pp. 4690–4699 (2019)
5. Ding, C., Tao, D.: Trunk-branch ensemble convolutional neural networks for video-based face recognition. IEEE Trans. Pattern Anal. Mach. Intell. **40**(4), 1002–1014 (2017)

6. Dosovitskiy, A., et al.: An image is worth 16x16 words: transformers for image recognition at scale. In: ICLR (2021)
7. Du, H., Shi, H., Zeng, D., Zhang, X.P., Mei, T.: The elements of end-to-end deep face recognition: a survey of recent advances. ACM Comput. Surv. (CSUR) **54**(10s), 1–42 (2022)
8. Fan, H., et al.: Multiscale vision transformers. In: Proceedings of the IEEE International Conference on Computer Vision (2021)
9. Gong, S., Shi, Y., Kalka, N.D., Jain, A.K.: Video face recognition: component-wise feature aggregation network (C-FAN). In: 2019 International Conference on Biometrics (ICB), pp. 1–8. IEEE (2019)
10. Guo, G., Zhang, N.: A survey on deep learning based face recognition. Comput. Vis. Image Underst. **189**, 102805 (2019)
11. Hajati, F., Tavakolian, M., Gheisari, S., Gao, Y., Mian, A.S.: Dynamic texture comparison using derivative sparse representation: application to video-based face recognition. IEEE Trans. Hum.-Mach. Syst. **47**(6), 970–982 (2017)
12. Hara, K., Kataoka, H., Satoh, Y.: Can spatiotemporal 3D CNNs retrace the history of 2D CNNs and imagenet? In: Proceedings of the IEEE Conference on Computer Vision and Pattern Recognition, pp. 6546–6555 (2018)
13. Hörmann, S., Cao, Z., Knoche, M., Herzog, F., Rigoll, G.: Face aggregation network for video face recognition. In: 2021 IEEE International Conference on Image Processing (ICIP), pp. 2973–2977. IEEE (2021)
14. Howard, A.G., et al.: Mobilenets: efficient convolutional neural networks for mobile vision applications. CoRR abs/1704.04861 (2017)
15. Hu, W., Huang, Y., Zhang, F., Li, R., Li, W., Yuan, G.: Seqface: make full use of sequence information for face recognition. arXiv preprint arXiv:1803.06524 (2018)
16. Kim, S.T., Kim, D.H., Ro, Y.M.: Spatio-temporal representation for face authentication by using multi-task learning with human attributes. In: 2016 IEEE International Conference on Image Processing (ICIP), pp. 2996–3000. IEEE (2016)
17. Kim, S.T., Ro, Y.M.: Facial dynamics interpreter network: what are the important relations between local dynamics for facial trait estimation? In: Proceedings of the European Conference on Computer Vision (ECCV), pp. 464–480 (2018)
18. Lee, K., Ho, J., Yang, M., Kriegman, D.: Visual tracking and recognition using probabilistic appearance manifolds. Comput. Vis. Image Underst. **99**(3), 303–331 (2005)
19. Li, Y., Zheng, W., Cui, Z., Zhang, T.: Face recognition based on recurrent regression neural network. Neurocomputing **297**, 50–58 (2018)
20. Lin, J., Xiao, L., Wu, T., Bian, W.: Image set-based face recognition using pose estimation with facial landmarks. Multimedia Tools Appl. **79**(27), 19493–19507 (2020)
21. Liu, Y., et al.: iQIYI-VID: a large dataset for multi-modal person identification. arXiv preprint arXiv:1811.07548 (2018)
22. Maze, B., et al.: IARPA Janus benchmark-C: face dataset and protocol. In: 2018 International Conference on Biometrics (ICB), pp. 158–165. IEEE (2018)
23. Mokhayeri, F., Granger, E.: A paired sparse representation model for robust face recognition from a single sample. Pattern Recogn. **100**, 107129 (2020)
24. Neimark, D., Bar, O., Zohar, M., Asselmann, D.: Video transformer network. arXiv preprint arXiv:2102.00719 (2021)
25. Rao, Y., Lu, J., Zhou, J.: Attention-aware deep reinforcement learning for video face recognition. In: Proceedings of the IEEE International Conference on Computer Vision, pp. 3931–3940 (2017)

26. Vaswani, A., et al.: Attention is all you need. In: Advances in Neural Information Processing Systems, vol. 30 (2017)
27. Wang, H., et al.: CosFace: large margin cosine loss for deep face recognition. In: Proceedings of the IEEE Conference on Computer Vision and Pattern Recognition, pp. 5265–5274 (2018)
28. Wolf, L., Hassner, T., Maoz, I.: Face recognition in unconstrained videos with matched background similarity. In: CVPR 2011, pp. 529–534. IEEE (2011)
29. Xingjian, S., Chen, Z., Wang, H., Yeung, D.Y., Wong, W.K., Woo, W.C.: Convolutional LSTM network: a machine learning approach for precipitation nowcasting. In: Advances in Neural Information Processing Systems, pp. 802–810 (2015)
30. Yang, J., et al.: Neural aggregation network for video face recognition. In: Proceedings of the IEEE Conference on Computer Vision and Pattern Recognition, pp. 4362–4371 (2017)
31. Yang, J., Dong, X., Liu, L., Zhang, C., Shen, J., Yu, D.: Recurring the transformer for video action recognition. In: Proceedings of the IEEE/CVF Conference on Computer Vision and Pattern Recognition, pp. 14063–14073 (2022)
32. Zhang, M., Song, G., Zhou, H., Liu, Yu.: Discriminability distillation in group representation learning. In: Vedaldi, A., Bischof, H., Brox, T., Frahm, J.-M. (eds.) ECCV 2020. LNCS, vol. 12355, pp. 1–19. Springer, Cham (2020). https://doi.org/10.1007/978-3-030-58607-2_1
33. Zhong, Y., Deng, W.: Face transformer for recognition. arXiv preprint arXiv:2103.14803 (2021)

Where to Focus: Central Attention-Based Face Forgery Detection

Jinghui Sun[1,2], Yuhe Ding[1], Jie Cao[3,4], Junxian Duan[3,4], and Aihua Zheng[2,5,6](✉)

[1] School of Computer Science and Technology, Anhui University, Hefei 230601, China
[2] Anhui Provincial Key Laboratory of Multimodal Cognitive Computation, Anhui University, Hefei, China
ahzheng214@foxmail.com
[3] Center for Research on Intelligent Perception and Computing (CRIPAC), Beijing, China
jie.cao@cripac.ia.ac.cn
[4] Institute of Automation, Chinese Academy of Sciences, Beijing, China
junxian.duan@ia.ac.cn
[5] Information Materials and Intelligent Sensing Laboratory of Anhui Province, Hefei, China
[6] School of Artificial Intelligence, Anhui University, Hefei 230601, China

Abstract. Face forgery detection in compressed images is an active area of research. However, previous frequency-based methods are subject to two limitations. One aspect to consider is that they apply the same weight to different frequency bands. Moreover, they exhibit an equal treatment of regions that contain distinct semantic information. To address these limitations above, we propose the Central Attention Network (CAN), a multi-modal architecture comprising two bright components: Adaptive Frequency Embedding (AFE) and Central Attention (CA) block. The AFE module adaptively embeds practical frequency information to enhance forged traces and minimize the impact of redundant interference. Moreover, the CA block can achieve fine-grained trace observation by concentrating on facial regions where indications of forgery frequently manifest. CAN is efficient in extracting forgery traces and robust to noise. It effectively reduces the unnecessary focus of our model on irrelevant factors. Extensive experiments on multiple datasets validate the advantages of CAN over existing state-of-the-art methods.

Keywords: Face Forgery Detection · Multi-level Frequency Fusion · Attention Mechanism

1 Introduction

Deep learning advancements and the widespread availability of online resources make tools like deepfakes [1] and face2face [2] easily accessible, allowing individuals without professional training to easily manipulate facial expressions,

Q. Liu et al. (Eds.): PRCV 2023, LNCS 14429, pp. 44–56, 2024.
https://doi.org/10.1007/978-981-99-8469-5_4

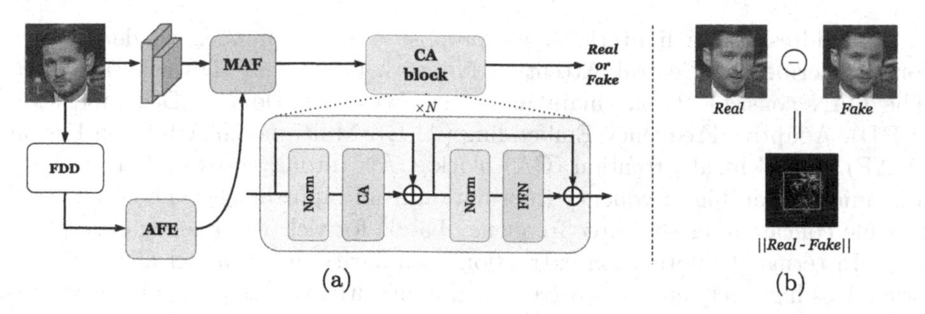

Fig. 1. (a) Overview of our proposed CAN. Combining *FDD* with *AFE* allows for extracting fine-grained frequency information and highlighting the components most useful for forgery detection. The *CA* block enables the network to focus more on key central areas. (b) Illustration of the differences between *Real* and *Fake*. The forgery traces are clustered in the central region (in the red box), indicating that the center is more important than the other areas. (Color figure online)

attributes, and identities within images. However, criminals misuse these technologies, resulting in a proliferation of high-quality fake photographs on social media, making it difficult to distinguish between genuine and modified faces.

The above issues prompt the development of face forgery detection based on deep neural networks [3–11]. However, they perform poorly in compressed images. Recent works [12–15] highlight the effectiveness of capturing forgery traces in the frequency domain under high compression. While decent detection results are achieved by combining RGB and frequency information, their method of information processing is coarse-grained, which causes two limitations.

For one thing, previous studies usually obtain frequency domain information through Discrete Cosine Transform and then use hand-crafted filters to extract it into high, middle, and low frequency bands. According to [15], the low and middle frequency preserve rich semantic information, such as human faces and backgrounds, which is highly consistent with RGB input. Meanwhile, the high frequency reveals small-scale details, often related to forging sensitive edges and textures. These show that the role and importance of these three frequency bands are completely different. Previous works show excellent performance by combining frequency information. They apply the same weight for different frequency bands, which may not be optimal for using frequency information and may lead to magnifying irrelevant noise and ignoring the more valuable components.

For another thing, the equal treatment of regions with different semantic information prevails in existing methods. However, as shown in Fig. 1(b), most of the differences between real image and fake image are obviously clustered in the central region (in the red box). This means that the central region can provide rich traces of forgery compared to other regions (outside the red box). Treating the regions equally not only results in superfluous noise but also neglects significant evidence.

To address these limitations, we propose a new approach to detect face forgery, termed as **C**entral **A**ttention **N**etwork (CAN), as shown in Fig. 1(a). The CAN consists of four main modules: Frequency Domain Decomposition (FDD), Adaptive Frequency Embedding (AFE), Multi-modal Attention Fusion (MAF), and Central Attention (CA) block. CAN initially uses FDD to extract low, middle, and high frequency information from input images. Then our AFE module concatenates the three frequency bands for richer frequency perception cues. In terms of information extraction granularity and channel allocation, it prioritizes high frequency information. Subsequently, the frequency is fused into the RGB branch by the MAF module. Finally, we add the CA block, which is similar to the Transformer block [16], to prevent the network from focusing on irrelevant areas. The module uses different scale attention mechanisms for the central and global regions, enabling the network to prioritize the central region more efficiently.

Extensive experiments have demonstrated that our proposed Central Attention Network effectively captures forgery traces and significantly improves upon the shortcomings of existing detection methods. Our work makes the following primary contributions:

- We propose the AFE module aiming at mining the more valuable fine-grained frequency components to uncover subtle nuances and hidden artifacts.
- We propose the Central Attention mechanism that provides a refined perspective of forged regions and reduces the attention to irrelevant areas.
- Numerous experiments demonstrate that our proposed Central Attention block is highly versatile and can be seamlessly integrated into various existing networks, resulting in a significant enhancement of their detection capabilities.

2 Related Work

Face Forgery Detection. With the rise of deep learning, the adverse effects of image forgery techniques on political credibility, social stability, and personal reputation have increasingly received attention from society.

Therefore, various image forgery detection technologies have developed rapidly in recent years. Previous works [7–11] use deep CNN models to predict whether a face region is real or fake. Unfortunately, they are only partially effective in high compression scenarios.

Inspired by [13], recent studies try to improve detection performance in high compression scenes by incorporating frequency domain information into existing detection techniques. Qian et al. [15] proposes a dual-stream network named F^3-Net, where one branch utilizes three filters to perform frequency decomposition on RGB information. Chen et al. [17] uses the Spatial Rich Model to extract residual noise to guide the RGB features. Li et al. [18] and Gu et al. [14] further decompose fine-grained frequency domain information from the perspective of image compression. While previous methods demonstrate significant effects, they either underutilize frequency information or treat all levels of frequency equally.

In contrast, our method involves decomposing frequency domain information and adaptive embedding to leverage the available frequency fully.

Vision Transformers. Transformers are known for their powerful remote contextual information modeling capabilities and high performance in natural language processing tasks. While various backbones are proposed to handle computer vision tasks, conventional transformers treat each patch at a single scale. Recent works [19–21] introduce multiple scales to focus on objects of different sizes, [22] proposes a multi-modal framework that integrates multi-scale transformer. Nevertheless, these approaches are generic and not tailored to the specific characteristics of forgery image detection. In this paper, we propose a Central Attention block that addresses the fact that fake regions tend to be concentrated in the central area of an image while other areas contain interference information.

3 Proposed Method

3.1 FDD: Frequency Domain Decomposition

For the input $rgb \in \mathbb{R}^{3\times H\times W}$, where H and W are the height and width of the image. First, we apply $\mathcal{DCT}$ as **D**iscrete **C**osine **T**ransform to transform the RGB domain to the frequency domain. Based on [15], we devise $N = 3$ filters that are capable of effectively decomposing the frequency into three distinct frequency bands: high, middle, and low:

$$dct^n = \mathcal{DCT}(rgb) \odot f^n, \qquad n = 1, ..., N. \tag{1}$$

We utilize $\mathcal{ID}$ as **I**nverse **D**iscrete Cosine Transform to transform the frequency domain into RGB domain to obtain the $\tilde{freq} \in \mathbb{R}^{3N\times H\times W}$ which is concatenated by $freq^n$ along the channel dimension. This manipulation helps to preserve the shift invariance and local consistency of natural images.

$$freq^n = \mathcal{ID}(dct^n), \qquad n = 1, ..., N. \tag{2}$$

To achieve a more refined analysis of the frequency information, we apply $\mathcal{M}$ as the median filter to extract noise information from the input features $\tilde{freq}$:

$$\tilde{freq}_{noise} = \tilde{freq} - \mathcal{M}(\tilde{freq}). \tag{3}$$

To magnify subtle forgery clues, we utilize the following formula:

$$freq = \tilde{freq} + Conv_{1\times 1}(Sigmoid(\tilde{freq}_{noise})). \tag{4}$$

Specifically, a 1×1 convolution layer followed by a *Sigmoid* activation function is used to generate a noise mask, which is then added back to the original feature maps to enhance the frequency input.

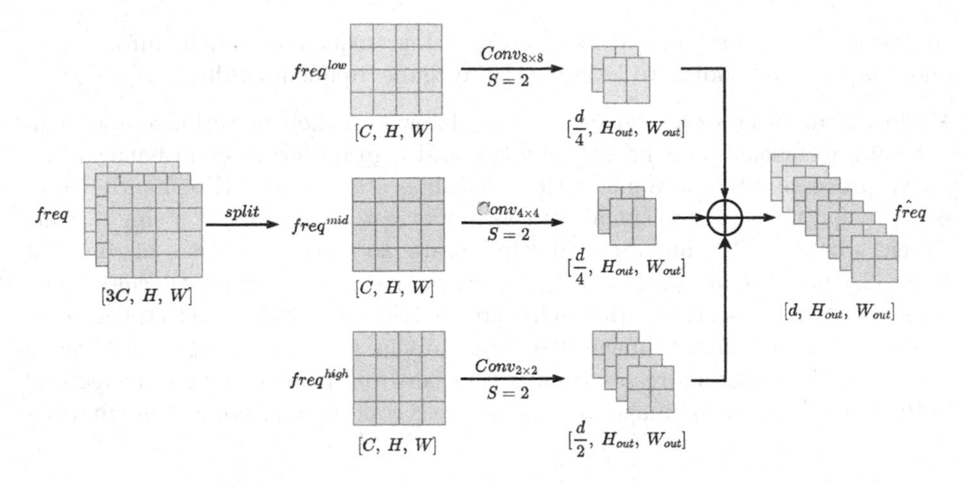

Fig. 2. The illustration of the proposed AFE allocates weight based on the value of frequency levels.

3.2 AFE: Adaptive Frequency Embedding

Previous works show excellent performance by combining frequency information. Applying the same weight to different frequency bands might be the general method in their works. It may not be optimal for using frequency domain information because it may magnify irrelevant noise or misuse the valuable components. To address this point, we propose the AFE module that fully exploits the role of different frequency components, as shown in Fig. 2. The AFE module extracts information from different frequency bands via different convolution kernels. Tampering artifacts reside mainly in the high-frequency spectrum. Therefore, we use a 2×2 convolution kernel to extract fine-grained texture information from it. For middle and low frequency that still contain basic information, which provides a solid foundation for fusing Frequency and RGB, we adopt 4×4 and 8×8 convolution kernels to extract semantic features, respectively. The channel outputs generated by these convolutions are also treated differently based on their importance in different frequency bands. Specifically, $\frac{d}{2}$ channels are allocated for high frequency channels while middle and low frequency each occupy $\frac{d}{4}$ channels. The d represents the number of output feature channels. Ultimately, the three branches are concatenated along the channel to obtain the $\hat{freq}$.

3.3 MAF: Multi-modal Attention Fusion

The complementary relationship between RGB and Freq is acknowledged. The MAF module integrates them by means of an attention mechanism. The RGB feature map is denoted as $\hat{rgb} \in \mathbb{R}^{d \times h \times w}$, while the frequency feature map is represented as $\hat{freq} \in \mathbb{R}^{d \times h \times w}$. We obtain the query vector Q from $\hat{rgb}$ using a 1×1 convolution layer. Similarly, we obtain the key vector K and value vector V from $\hat{freq}$ using 1×1 convolution layers. Then, we flatten them along the spatial dimension to get 2D embeddings Q_e, K_e, V_e. Using the self-attention mechanism, we generate an attention map that represents relevance between the input features $\hat{rgb}$ and $\hat{freq}$:

$$\hat{W} = softmax(\frac{Q_e K_e}{\sqrt{D}})V_e, \tag{5}$$

where $\boldsymbol{D}$ is the dimensionality of the key vectors. After obtaining attention weights, we compute weighted values via a 3×3 convolution. Additionally, we adopt residual connections to add them to the original input, alleviating the potential gradient vanishing issue during the training process.

$$f = \hat{rgb} + Conv_{3\times3}(\hat{W}). \tag{6}$$

3.4 CA Block: Central Attention Block

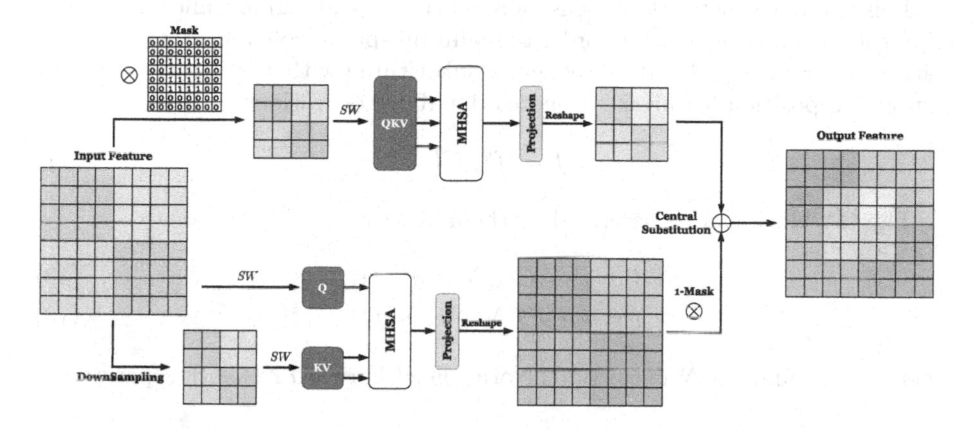

Fig. 3. The proposed Central Attention mechanism when α is 0.5.

The conventional transformer models treat all patches of an image equally without taking into account the relative significance of distinct areas. Recent studies [20, 22] show that incorporating multi-scale information can improve detection accuracy. Yet these models are not optimized for detecting forged face images. Our observation is that forged regions tend to cluster around the centre of input images. Based on this insight, we propose Central Attention, which aids the network in concentrating on key regions.

For the input global feature $f^g \in \mathbb{R}^{c\times h\times w}$, we commence by initializing a $Mask$ of size $h \times w$. Subsequently, we selectively filled the central region, characterized by dimensions of $\alpha h \times \alpha w$, with the value 1. The surrounding area is then filled with the value 0 to complete the mask initialization process. α is the proportion that determines the size of the central region. We then apply this $Mask$ to the input f^g, resulting in a central feature map $f^c = f^g \odot mask$.

Figure 3 illustrates the framework of the Central Attention mechanism, with a value of 0.5 for parameter α.

For the global feature f^g, we downsample it into $\frac{h}{2} \times \frac{w}{2}$ by convolution to obtain f^d. We obtain the embedding Q_g from f^g, the embeddings K_g and V_g from f^d. Inspired by [21], we define the operation of dividing the input into $G \times G$ patches through sliding windows and grouping as $SW^G(\cdot)$.

$$Q_g = SW^g(Q_g), \quad K_g, V_g = SW^{\frac{g}{2}}(K_g, V_g), \tag{7}$$

$$f^g = MHSA(Q_g, K_g, V_g). \tag{8}$$

Similarly, for the central feature f^c, we embed f^c into Q_c, K_c, V_c.

$$Q_c, K_c, V_c = SW^c(Q_c, K_c, V_c), \tag{9}$$

$$f^c = MHSA(Q_c, K_c, V_c), \tag{10}$$

where $MHSA$ represents Multi-Head Self-Attention.

This allows the network to focus more on the central region while still considering the surrounding areas. In order to maintain spatial coherence, the grouping features are rearranged and subsequently substituted with f^c to replace the corresponding position features. $[\cdot]$ denotes the above operations.

$$f = [f^g, f^c]. \tag{11}$$

The CA block can be described mathematically:

$$f = f^g + CA(Norm(f^g)), \tag{12}$$

$$f = f + FFN(Norm(f)), \tag{13}$$

where $Norm$ and FFN mean BatchNorm, Feed Forward Network separately.

3.5 Overall Loss

After passing through several CA blocks, the feature is sent into the remaining backbone network to extract richer features f. Then a fully connected layer and a *sigmoid* function are used to obtain the final prediction probability y. So the Binary cross-entropy loss is defined as:

$$\mathcal{L}_{Bce}(y) = y \log \hat{y} + (1-y)\log(1-\hat{y}), \tag{14}$$

where y is set to 1 if the face image has been manipulated, otherwise it is set to 0. To ensure feature consistency, we use the Consistency loss function $\mathcal{L}_{Cos}$ in [23] to constrain the feature distribution. f_1 and f_2 are the final features obtained from the same input image after through distinct data augmentation and being passed through the network. Mathematically:

$$\mathcal{L}_{Cos}(f_1, f_2) = \left(1 - \tilde{f_1} \cdot \tilde{f_2}\right)^2, \tag{15}$$

where $\tilde{f} = \frac{f}{\|f\|_2}$ denotes the normalized vector of the representation vector f.

So we combine the Binary cross-entropy loss and the Consistency loss function linearly with $\beta = 2$.

$$\mathcal{L}_{all} = \mathcal{L}_{Bce}(y_1) + \mathcal{L}_{Bce}(y_2) + \beta\mathcal{L}_{Cos}(f_1, f_2). \quad (16)$$

Table 1. Quantitative results on Celeb-DF dataset and FF++ dataset.

Methods	FF++(HQ)		FF++(LQ)		Celeb-DF	
	Acc	AUC	Acc	AUC	Acc	AUC
MesoNet [6]	83.10	-	70.47	-	-	-
Xception [24]	95.73	96.30	86.86	89.30	97.90	99.73
Face X-ray [7]	-	87.40	-	61.60	-	-
Two-branch [25]	96.43	98.70	86.34	86.59	-	-
RFM [11]	95.69	98.79	87.06	89.83	97.96	99.94
Add-Net [9]	96.78	97.74	87.5	91.01	96.93	99.55
F3-Net [15]	97.52	98.10	90.43	93.30	95.95	98.93
FDFL [18]	96.69	99.30	89.00	92.40	-	-
Multi-Att [8]	97.60	99.29	88.69	90.40	97.92	99.94
SIA [26]	97.64	99.35	90.23	93.45	-	-
PEL [14]	97.63	99.32	**90.52**	94.28	-	-
Ours	**97.65**	**99.44**	90.40	**95.09**	**99.36**	**99.98**

4 Experiments

4.1 Experimental Setup

Datasets. We adopt two widely-used public datasets in our experiments, *i.e.*, FaceForensics++ [27], Celeb-DF [28].

1) **FaceForensics++** (FF++) [27] is a large forensics dataset containing 1000 original video sequences and 4000 manipulated video sequences produced by four automated face manipulation methods: *i.e.*, Deepfakes [1], Face2Face [2], FaceSwap [29], NeuralTextures [30]. Raw videos are compressed, resulting in two versions: high quality (HQ) and low quality (LQ). Following the official splits, we utilized 720 videos for training, 140 for validation, and 140 for testing.

2) **Celeb-DF** [28] dataset comprises 590 authentic videos sourced from YouTube, featuring individuals of varying ages, ethnicities, and genders. Additionally, the dataset includes 5639 corresponding DeepFake videos.

Implementation Detail. The EfficientNet-B4 [31] pre-trained on ImageNet is adopted as the backbone of our network. We insert several CA blocks respectively after the second and third convolutional blocks with $\alpha = 0.5$. The input images are resized to 320×320. The whole network is trained with Adam optimizer with the learning rate of 2×10^{-4}, $\beta_1 = 0.9$, $\beta_2 = 0.999$. The batch size is 48 split on $4 \times$ RTX 3090 GPUs.

Evaluation Metrics. Following the convention [10,14,15,22,27], we apply Accuracy score (Acc), Area Under the Receiver Operating Characteristic Curve (AUC) as our evaluation metrics.

Comparing Methods. We compare our methods with several advanced methods: MesoNet [6], Xception [24], Face X-ray [7], Two-branch [25], RFM [11], Add-Net [9], F^3-Net [15], FDFL [18], Multi-Att [8], SIA [26], PEL [14].

Table 2. The effect of each component. The CAB represents CA blocks.

RGB	Freq	AFE	CAB	Acc	AUC
✓				88.70	92.87
	✓			88.49	92.63
✓	✓			88.89	92.89
✓	✓	✓		89.94	93.57
✓	✓		✓	90.36	94.15
✓	✓	✓	✓	**90.40**	**95.09**

Table 3. Ablation study of other backbones with our CA blocks.

Model		Acc	AUC
PF	+None	66.79	69.28
	+CAB	78.79	80.31
CNX	+None	76.45	77.92
	+CAB	80.43	80.64
PF*	+None	86.93	90.09
	+CAB	87.22	90.34
CNX*	+None	87.57	90.77
	+CAB	87.93	91.07

4.2 Comparison to the State-of-the-Arts

Following [15,27], we compare our method with various advanced techniques on the FF++ dataset with different quality settings (*i.e.*, HQ and LQ), and further evaluate the performance of our approach on the Celeb-DF dataset. In Table 1 the best, second, third results are shown in Red, Blue, Green. The performance of our proposed method, especially under high compression, is comparable or superior to existing methods, as evidenced by the Acc and AUC metrics. It is worth noting that the method PEL [14] is a two-stream network with twice as many parameters as ours. We achieve competitive results using only half the parameters. These gains mainly come from the CAN's ability to utilize frequency information and fully reduce interference from irrelevant information.

4.3 Ablation Study and Architecture Analysis

Components. As shown in Table 2, we develop several variants and conduct a series of experiments on the FF++ (LQ) dataset to explore the impact of different components in our proposed method. Using only RGB or frequency as input in the single-stream setting leads to similar results. Combining both original streams can slightly improve performance, which demonstrates that frequency and RGB are unique and complementary. Adding an AEF module or CA blocks can significantly improve performance, achieving optimal results using the overall CAN framework. It shows that each module is effective: the AFE module fully mines frequency domain information and filters noise, and the CA blocks strengthen the network to focus on forged regions.

Validity of the CA Block. We insert the CA block into Transformer and CNN to further examine its validity and universality. PoolFormer-S (PF) [32] and ConvNeXt-S (CNX) [33] are chosen as the backbone. The results on FF++ (LQ) are displayed in Table 3, where * means loading pre-trained weight. Embedding CA blocks significantly improves the performance of both baseline networks due to their critical attention to central regions.

Convolution Kernel Size. In the AFE module, we conduct experiments with several convolution kernel combinations under the same settings. The specific results are shown in Table 4. The combination of [2, 4, 8] performs best.

Table 4. Quantitative results of different convolution kernel sizes in AFE.

Kernel	Acc	AUC
[2, 4, 8]	**90.40**	**95.09**
[2, 8, 16]	90.09	94.04
[4, 8, 16]	89.79	94.10

Table 5. The results on FF++ (LQ) with different α.

α	Acc	AUC
0.5	**90.40**	**95.09**
0.6	90.11	94.59
0.7	90.13	94.26

Hyperparameter α. The hyperparameter α has a significant impact on the CA block's performance by restricting the size of the central area. In Table 5, we conduct experiments with different value of α and find that the optimal performance is achieved when the α is 0.5. It means that the inclusion of too much irrelevant information would weaken the performance, and the center area can supply adequate forgery traces.

4.4 Visualizations

To further understand how our method makes decisions, we use Grad-CAM [34] to show the attention maps of input samples for both the baseline and CAN. Figure 4 demonstrates that all four forgery methods have their faked areas centered in the center. The baseline network is significantly disturbed due to increased noise information after compression. However, with the AFE module filtering out noise information and Central Attention emphasis focused on central areas, the CAN can more reliably capture forgery traces.

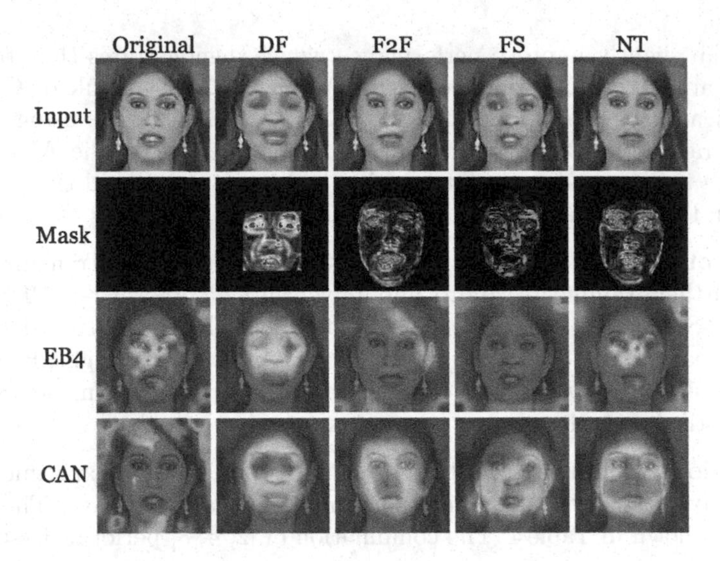

Fig. 4. The attention maps for different kinds of faces

4.5 Limitations

When applying improper masks, the performance drops significantly, suggesting that a more meticulous attention mechanism is required. Focusing on specific facial components may lead to better results, which we will explore in the future.

5 Conclusion

The paper proposes a Central Attention Network (CAN) framework for detecting forged images. We conduct a comprehensive analysis of the frequency amplification forgery traces, which has laid a strong foundation for the network's optimal performance. The Central Attention block effectively filters out irrelevant background noise, ensuring the network concentrates primarily on capturing forgery traces. Visualizing class activation mapping explains the internal mechanism and demonstrates the effectiveness of our methodology.

Acknowledgment. This research is supported by National Natural Science Foundation of China (Grant No. 62206277) and the University Synergy Innovation Program of Anhui Province (No. GXXT-2022-036). The authors would like to thank Ran He (Professor at CASIA) and Jiaxiang Wang (Ph.D. at AHU) for their valubale suggestions.

References

1. Tora: Deepfakes (2018). https://github.com/deepfakes/faceswap/tree/v2.0.0
2. Thies, J., Zollhofer, M., Stamminger, M., Theobalt, C., Nießner, M.: Face2face: real-time face capture and reenactment of RGB videos. In: Proceedings of CVPR (2016)
3. Yang, X., Li, Y., Lyu, S.: Exposing deep fakes using inconsistent head poses. In: IEEE International Conference on Acoustics, Speech and Signal Processing (2019)
4. Matern, F., Riess, C., Stamminger, M.: Exploiting visual artifacts to expose deepfakes and face manipulations. In: IEEE Winter Applications of Computer Vision Workshops (2019)
5. Haliassos, A., Vougioukas, K., Petridis, S., Pantic, M.: Lips don't lie: a generalisable and robust approach to face forgery detection. In: Proceedings of CVPR (2021)
6. Afchar, D., Nozick, V., Yamagishi, J., Echizen, I.: Mesonet: a compact facial video forgery detection network. In: IEEE International Workshop on Information Forensics and Security (2018)
7. Li, L., et al.: Face X-ray for more general face forgery detection. In: Proceedings of CVPR (2020)
8. Zhao, H., Zhou, W., Chen, D., Wei, T., Zhang, W., Yu, N.: Multi-attentional deepfake detection. In: Proceedings of CVPR (2021)
9. Zi, B., Chang, M., Chen, J., Ma, X., Jiang, Y.G.: Wilddeepfake: a challenging real-world dataset for deepfake detection. In: Proceedings of ACM-MM (2020)
10. Dang, H., Liu, F., Stehouwer, J., Liu, X., Jain, A.K.: On the detection of digital face manipulation. In: Proceedings of CVPR (2020)
11. Wang, C., Deng, W.: Representative forgery mining for fake face detection. In: Proceedings of CVPR (2021)
12. Chen, S., Yao, T., Chen, Y., Ding, S., Li, J., Ji, R.: Local relation learning for face forgery detection. In: Proceedings of AAAI (2021)
13. Frank, J., Eisenhofer, T., Schönherr, L., Fischer, A., Kolossa, D., Holz, T.: Leveraging frequency analysis for deep fake image recognition. In: Proceedings of ICML (2020)
14. Gu, Q., Chen, S., Yao, T., Chen, Y., Ding, S., Yi, R.: Exploiting fine-grained face forgery clues via progressive enhancement learning. In: Proceedings of AAAI (2022)
15. Qian, Y., Yin, G., Sheng, L., Chen, Z., Shao, J.: Thinking in frequency: face forgery detection by mining frequency-aware clues. In: Vedaldi, A., Bischof, H., Brox, T., Frahm, J.-M. (eds.) ECCV 2020. LNCS, vol. 12357, pp. 86–103. Springer, Cham (2020). https://doi.org/10.1007/978-3-030-58610-2_6
16. Vaswani, A., et al.: Attention is all you need. In: Proceedings of NeurIPS (2017)
17. Luo, Y., Zhang, Y., Yan, J., Liu, W.: Generalizing face forgery detection with high-frequency features. In: Proceedings of CVPR (2021)
18. Li, J., Xie, H., Li, J., Wang, Z., Zhang, Y.: Frequency-aware discriminative feature learning supervised by single-center loss for face forgery detection. In: Proceedings of CVPR (2021)
19. Chen, C.F.R., Fan, Q., Panda, R.: Crossvit: cross-attention multi-scale vision transformer for image classification. In: Proceedings of ICCV (2021)
20. Ren, S., Zhou, D., He, S., Feng, J., Wang, X.: Shunted self-attention via multi-scale token aggregation. In: Proceedings of CVPR (2022)
21. Wang, W., et al.: Crossformer: a versatile vision transformer hinging on cross-scale attention. In: Proceedings of ICLR (2022)

22. Wang, J., et al.: M2TR: multi-modal multi-scale transformers for deepfake detection. In: Proceedings of ICMR (2022)
23. Ni, Y., Meng, D., Yu, C., Quan, C., Ren, D., Zhao, Y.: Core: consistent representation learning for face forgery detection. In: Proceedings of CVPR Workshops (2022)
24. Chollet, F.: Xception: deep learning with depthwise separable convolutions. In: Proceedings of CVPR (2017)
25. Masi, I., Killekar, A., Mascarenhas, R.M., Gurudatt, S.P., AbdAlmageed, W.: Two-branch recurrent network for isolating deepfakes in videos. In: Vedaldi, A., Bischof, H., Brox, T., Frahm, J.-M. (eds.) ECCV 2020. LNCS, vol. 12352, pp. 667–684. Springer, Cham (2020). https://doi.org/10.1007/978-3-030-58571-6_39
26. Sun, K., et al.: An information theoretic approach for attention-driven face forgery detection. In: Avidan, S., Brostow, G., Cissé, M., Farinella, G.M., Hassner, T. (eds.) ECCV 2022. LNCS, vol. 13674, pp. 111–127. Springer, Cham (2022). https://doi.org/10.1007/978-3-031-19781-9_7
27. Rossler, A., Cozzolino, D., Verdoliva, L., Riess, C., Thies, J., Nießner, M.: Faceforensics++: learning to detect manipulated facial images. In: Proceedings of ICCV (2019)
28. Li, Y., Yang, X., Sun, P., Qi, H., Lyu, S.: Celeb-DF: a large-scale challenging dataset for deepfake forensics. In: Proceedings of CVPR (2020)
29. Kowalski, M.: Faceswap (2018). https://github.com/marekkowalski/faceswap
30. Thies, J., Zollhöfer, M., Nießner, M.: Deferred neural rendering: image synthesis using neural textures. ACM Trans. Graph. **38**(4), 1–12 (2019)
31. Tan, M., Le, Q.: Efficientnet: rethinking model scaling for convolutional neural networks. In: Proceedings of ICML (2019)
32. Yu, W., et al.: Metaformer is actually what you need for vision. In: Proceedings of CVPR (2022)
33. Liu, Z., Mao, H., Wu, C.Y., Feichtenhofer, C., Darrell, T., Xie, S.: A convnet for the 2020s. In: Proceedings of CVPR (2022)
34. Li, H., Huang, J.: Localization of deep inpainting using high-pass fully convolutional network. In: Proceedings of ICCV (2019)

Minimum Assumption Reconstruction Attacks: Rise of Security and Privacy Threats Against Face Recognition

Dezhi Li[1], Hojin Park[2], Xingbo Dong[1](✉), YenLung Lai[1], Hui Zhang[1], Andrew Beng Jin Teoh[3], and Zhe Jin[1]

[1] Anhui Provincial Key Laboratory of Secure Artificial Intelligence, School of Artificial Intelligence, Anhui University, Hefei 230093, China
xingbo.dong@ahu.edu.cn

[2] Hanwha Vision, Seongnam-si 10285, Republic of Korea

[3] School of Electrical and Electronic Engineering, College of Engineering, Yonsei University, Seoul 120749, Republic of Korea

Abstract. Facial Recognition (FR), despite its remarkable precision and advancements achieved through deep learning, exhibits vulnerability to security threats, specifically originating from deep generative models proficient in synthesizing deceptive face images. Generative Adversarial Networks (GANs) present substantial risks by showcasing the capacity to exploit potential vulnerabilities within FR systems. While the existing research primarily focuses on the scenario of a compromised database facilitating facial reconstruction attacks, it often overlooks more realistic threats where adversaries attack with a limited number of queries without breaching the database. This work introduces Minimum Assumption Reconstruction Attacks (MARA), offering a realistic attack framework against FR systems. MARA treats an attacker as a regular user interacting with the FR system's user interface and observing the matching scores. We formulate the MARA attack as an optimization problem, aiming to find a latent vector in the $\mathcal{W}^+$ latent space of StyleGAN for generating adversarial face images that can bypass the targeted FR system. A latent space mining strategy is also proposed to enhance attack performance by obtaining 'good' initial guesses in the latent space. Our experiments show that MARA achieves performance comparable to false accept attacks while adhering to query limits and mimicking user-like interaction behavior. This study highlights the importance of considering attack models requiring minimal effort from the adversary, an essential perspective for adversarial research that seeks to guard against powerful and less resource-intensive attacks.

Keywords: Face Recognition Attack · Reconstruction Attack · Black-box Attack · White-box Attack

This work was supported by the National Natural Science Foundation of China (Grant Nos. 62376003, 62306003).

Q. Liu et al. (Eds.): PRCV 2023, LNCS 14429, pp. 57–73, 2024.
https://doi.org/10.1007/978-981-99-8469-5_5

1 Introduction

Human Facial recognition (FR) has gained significant popularity as a biometric trait, comparable to fingerprint and iris recognition, owing to its non-intrusive nature, convenience, user-friendliness, contactless operation, and high accuracy. Following the recent advancement of deep learning, face recognition systems using deep neural network encoders are being rapidly deployed in practice. During registration, these systems convert a facial image into a feature vector (a template). The template is subsequently stored in a database as a reference for future matching purposes.

On the Vulnerable Points of FR: However, several points in the FR system can be vulnerable to attacks. Follows Fig. 1: 1) attackers may exploit the image acquisition stage in facial recognition systems by using fake or manipulated facial features, such as printed photos, masks, or 3D models; 2) attackers may target the face recognition system's database, aiming to gain unauthorized access and manipulate identity information, compromising the system's integrity; 3) attackers may exploit the system's interface by analyzing output elements like matching decisions or scores to reverse-engineer the user's facial image.

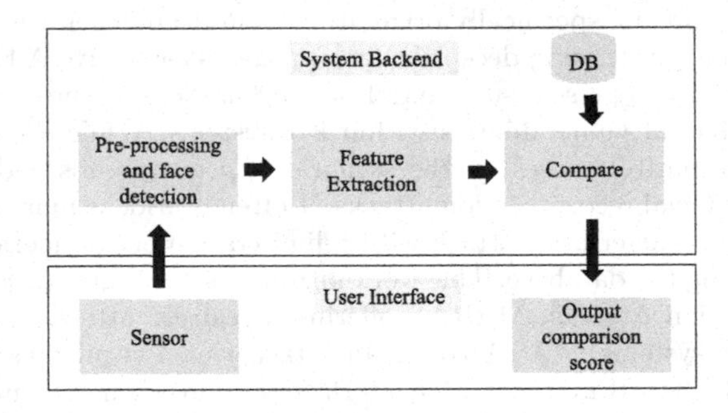

Fig. 1. Components of a typical face recognition system.

Table 1. Comparison of attack assumptions and aims of different image-level attacks.

Attack	Access to sth.		Know sth.		Goal	Related reference
	User interface	DB in backend	Victim's photo	Face distribution		
FAA	✓			✓	Deceive/impersonate	Palmprint FAA [3]
PA	✓		✓	✓	Deceive/impersonate	See survey [2,9,10]
TI		✓		✓	Template inversion	NBNet [1], Mapping [7]
MARA	✓			✓	Reconstruct face	-

On Various Attacks over FR System: Threat actors are developing sophisticated attack strategies over attack above points, including template inversion attacks (TI) [1], presentation attacks (PA) [2] and false acceptance attack (FAA) [3], to exploit potential vulnerabilities (see Table 1).

Presentation attacks involve the creation of fake biometric images on spoofing carriers like paper, monitors, masks, or gloves to deceive biometric systems (see survey [2]). The false acceptance attack leverages image generation techniques such as GANs [4] and VAEs [5] to create a vast collection of fraudulent biometric images or utilize existing real datasets. In FAA, attackers search for an image within this set that closely matches the target feature template, ensuring similarity in the feature domain without requiring genuine user images.

Template inversion attacks aim to create fake biometric images that resemble genuine ones from the stored biometric template. In recent studies, multiple efforts have focused on reconstructing facial images using deep features extracted from the face. These endeavors have predominantly employed convolutional networks [6] or de-convolutional neural networks [1]. More recent work [7] trained a fully connected neural network that projects a feature vector into the latent space of a pre-trained StyleGAN [8] such that a closely matching face is generated. Notably, the work presented in [1] is dedicated explicitly to reconstructing facial images based on their deep features. The resultant face images generated by the method proposed in [1] have demonstrated a remarkable attack success rate, setting a new benchmark in this domain. In this work, we focus on especially attacks of reconstruction face images.

Adopting a More Realistic Attack Environment: While the attacks above have demonstrated promising results, they often predicate strong assumptions about the attacker's capabilities. These include the attacker having excessive access to the target victims' facial photos or the ability to breach the database unauthorizedly. For example, approaches like [1,7] require millions of queries and access to the template in the database, making them impractical for real-world scenarios.

On the other hand, although these assumptions represent worst-case scenarios and underscore the need for secure data storage methodologies, they lack comprehensive insight into evaluating face recognition systems' security and privacy preservation. Indeed, the objective of evaluating face recognition systems should be to contemplate a broader spectrum of realistic scenarios, where attackers may not necessarily possess the victims' facial photos or even require breaching the system storage.

By adopting a less assumption-driven approach, researchers can gain more realistic insights into the robustness and effectiveness of face recognition systems against a broader array of potential threats. This evaluation methodology could guide the development of more reliable and secure systems, focusing on privacy preservation and risk mitigation associated with face recognition technology.

In this work, we propose Minimum Assumption Reconstruction attacks (MARA), which adopt a more realistic approach. MARA assumes that the attacker, with limited resources, can only interact with the user interface and observe the similarity scores provided by the FR system (see Table 1). We simulate a challenging scenario where the attacker has a limited number of query attempts. This scenario is feasible since an adversary could mimic typical user behavior, interact with the system through queries, and glean potentially exploitable information from the returned matching scores.

The implications of the above exploration are significant. Not only does it shed light on the minimum requirements for an attacker to exploit the system, but it also has immediate relevance in combating database attacks, which can be viewed as a 'stronger version' of our case (since, in our case, we do not necessarily require the attacker to compromise a template or database). We can develop prompt and effective solutions to mitigate attacks on face recognition systems by uncovering the minimum query threshold necessary for an adversary to gather exploitable information.

The highlighted contributions in this paper are as follows:

1. We propose the Minimum-assumption-based Reconstruction Attack, which offers significant less number of query attempts (from 4k to 2.6k) while gaining comparative attack performance, which prevents the attacker from interacting with the FR system unlimited.
2. We formulate the attack problem as an optimization task to determine the optimal latent in $\mathcal{W}^+$ latent space of StyleGAN. To prioritize query efficiency, we adopt a gradient estimation technique [11,12] to estimate gradients from the black-box face feature encoder. More importantly, we introduce a latent space mining strategy to obtain optimal initial guesses in $\mathcal{W}^+$ latent space, thus gaining the attack performance.
3. Through systematic experiments, we demonstrate that our proposed reconstruction attack achieves comparable performance to the false accept attack, using the same number of queries. This implies that the attacker's behavior resembles a regular user, who adheres to query limits and interacts with the FR interface accordingly.

2 Related Works

Over the past five years, advancements in computing hardware, big data, and innovative algorithms have fueled the growth of startups leveraging deep learning-based facial recognition (FR) techniques. These startups have developed practical applications that have gained significant attention due to the widespread deployment of FR systems using deep learning models such as FaceNet [13], ArcFace [14], and CosFace [15]. However, this increased usage has also raised concerns about privacy and security, particularly regarding the potential for reconstructing face images from features extracted from deep learning models (deep features) [16,17].

In a study by Zhmoginov and Sandler [18], a method was introduced to invert the face embedding generated by FaceNet [13] back into face images. The inversion process was formulated as a minimization problem, aiming to reduce the template difference between the original and reconstructed images. To accomplish this, a regularization function was employed, utilizing the intermediate layer of the feature extractor. However, in real-world scenarios, obtaining the detailed parameters of the feature extractor may not be feasible.

Another approach proposed by Cole et al. [6] focused on generating face images using FaceNet features. Although the motivation behind this method differed from reconstructing faces from features, it relied on differentiable image warping by combining landmark and texture information. However, this approach required both landmark and texture information, and it also utilized the last convolutional layer of the feature extractor, which could be impractical for face image reconstruction in reality.

In the paper by Mai et al. [1], a de-convolutional neural network (DCNN) called the Neighborly de-convolutional Neural Network (NBNet) is utilized for face image reconstruction. The choice of DCNN is motivated by its effective up-sampling capability. Notably, the authors assume that the feature extractor used in their approach is a black box, meaning that the adversary may not have access to its specific parameters.

The NBNet consists of multiple stacked de-convolution blocks and a convolution block. This network architecture is specifically designed to generate output face images. To train the NBNet, face image datasets are employed along with a Generative Adversarial Network (GAN) to synthesize face images. The performance of the generated face images is evaluated using two benchmark face datasets. The experimental results show that 95.20% of the generated face images can successfully bypass a face recognition system that has enrolled the same face image, with a False Acceptance Rate (FAR) of 0.1% (type-I attack in [1]).

In a subsequent work by Keller et al. [19], the NBNet is further utilized for face image reconstruction from a binary template produced by a given binarization method. This approach leverages the capabilities of the NBNet to reconstruct the original face image from a simplified binary representation. In [20], a Bijective Generative Adversarial Network in a Distillation framework (DiBiGAN) is proposed to tackle the challenging task of generating face images from high-level representations obtained from a black box Face Recognition encoder. The method incorporates a bijective metric learning process for image reconstruction, utilizes a distillation process to maximize information from the black box encoder, and introduces a Feature-Conditional Generator Structure with Exponential Weighting Strategy for robust face synthesis while preserving the person's identity.

The paper by Razzhigaev et al. [21] introduces a method for reconstructing face images from features using a zero-order iterative optimization technique in the linear space of 2D Gaussian functions. The proposed approach involves iteratively updating the current state image. Each iteration generates a batch of random Gaussian blobs and adds to the current state image. The feature extractor, treated as a black box, is then used to calculate the embeddings of this augmented image batch. The loss function is computed based on the embeddings, and the image from the batch with the lowest loss value is selected as the updated current image. This process continues until the loss function converges, indicating the reconstruction of the face image from the given features.

The paper by Dong et al. [7] presents a framework to reconstruct high-quality face images from deep features. The approach involves establishing a neural network that learns a mapping between the latent vector space of StyleGAN2 [24] and the feature vector space of a face feature extraction model. Given a feature vector as input, the model predicts the corresponding latent vector, which can then be used to generate face images. However, using a vanilla fully connected neural network in the proposed method has limitations for this task. The stochastic gradient descent algorithm can get trapped in local minima, hindering reconstruction. Consequently, the method achieves a relatively low successful attack rate of only 10% on the Labeled Faces in the Wild (LFW) dataset under a type-I attack at a 1% False Acceptance Rate (FAR). This implies that most reconstructed face images exhibit poor visual similarity compared to genuine face images.

Later, Dong et al. propose using the genetic algorithm [22] to search for the latent vector that generates a synthetic face close to the target feature in the feature space of the target encoder. This modification yields state-of-the-art performance in face image reconstruction, improving upon the limitations of the initial vanilla fully connected neural network approach.

Significant advancements have been made in the field of face reconstruction attacks. However, several issues persist, including strong assumptions made by the attacker and the high number of queries required. Approaches like [1,19,20] utilize de-convolutional neural networks that operate directly on the feature template. However, this assumes that the attacker has access to the database

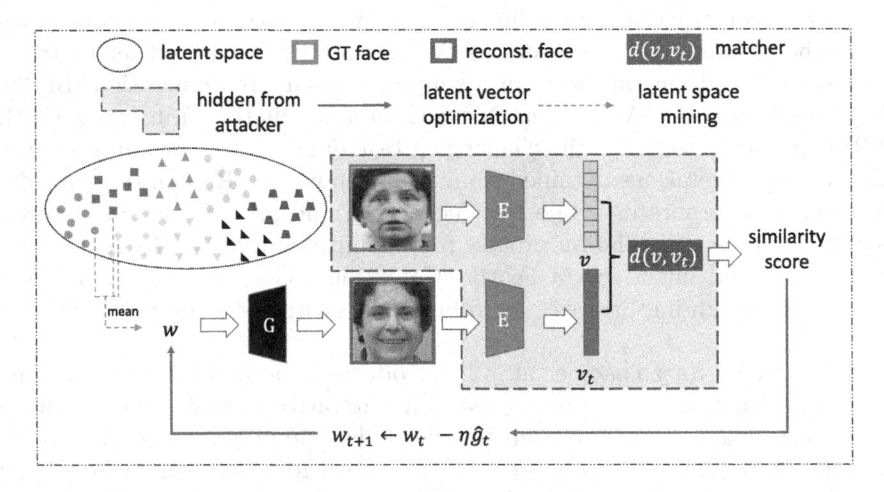

Fig. 2. Overview of the proposed minimum assumption reconstruction attack. We optimize the latent vector w of a pre-trained StyleGAN G to reconstruct the target face x, such that the distance between its feature v and the reconstructed feature v^* is minimized. An latent space mining is utilized to find the best initial guess for the optimization algorithm. We treat the attacker as a regular user interacting with the system's user interface and observing the matching scores.

in the system's backend, which inadvertently neglects the potential for 'weaker' attackers who may compromise the system without the necessity of breaching the database.

Furthermore, most of these methods require a substantial number of queries in the system to generate the training dataset. This training process, whether it involves de-convolutional neural network training or mapping network training, as shown in [7], can be challenging for attackers due to the high query requirement. While [23] offers an alternative that does not involve training, their optimization process is inefficient in terms of query usage as it updates multiple latent vectors simultaneously.

Motivations: Given the preceding discussion, it becomes evident that conventional scenarios may not accurately represent the most plausible threat models. In reality, an adversary may not have the capability or intent to launch a large number of queries or to compromise a database. They may instead favor more covert and efficient tactics that exploit systemic vulnerabilities. Therefore, while we recognize the efficacy of current techniques, it is vital to continually explore sophisticated attack strategies that encapsulate a broader range of feasible real-world adversarial scenarios. This understanding underscores the need for an encompassing threat model that acknowledges not only 'strong' adversaries capable of database compromise but also 'weaker' adversaries who could exploit the system via less conspicuous avenues.

3 Methods

3.1 Overview of MARA

We consider a scenario where the attacker's objective is to reconstruct a target face image, represented as x, using the corresponding similarity score observable from the user interface. In the system's back-end, a feature vector $v = E(x)$ is derived by encoding the face image x via a target encoder E, denoted as $E : \mathcal{X} \rightarrow \mathcal{V}$. It's important to note that the attacker **does not have access to** v stored in the back-end database. Instead, the attacker's actions are constrained to querying the user interface for the similarity score between an input face image and the stored template, a computationally expensive process.

Based on the above scenario, the attacker must exploit the restricted access to reconstruct a face image x that closely resembles the original face ones x^*. The reconstructed face image x will then be utilized to launch an attack against another face recognition system that employs a different encoder E'. It is worth mentioning that the encoder E' may not necessarily be the same as the target encoder E. For the attack to be deemed successful, the distance between x and x^* in the feature space of E' must be smaller than a predefined threshold denoted as τ. Please refer to Fig. 2 for a visual representation of the attack pipeline.

3.2 Latent Vector Optimization by Gradient Estimation

Based on the approach described in [23] by Dong et al., we adopt a similar method that utilizes a pre-trained StyleGAN as the generator for face image reconstruction. This approach avoids the need to train a generator from scratch, which typically requires many queries, as shown in prior works such as [1,20]. By leveraging a pre-trained generator, an attacker can iteratively search for the optimal latent vector that produces a face image close to the original face in the feature space of the target encoder. We can rewrite the optimization problem as follows:

$$z^* = \arg\min_{z \in \mathcal{Z}} d(E(G(z)), v) \tag{1}$$

where $E(\cdot)$ represents the target encoder that maps a face image to its corresponding features, $G(z)$ generates a face image based on the latent vector z using the pre-trained StyleGAN, $dist(\cdot)$ is a distance metric that quantifies the dissimilarity between two features.

To tackle this optimization problem, [23] employed the genetic algorithm (GA) [22], which is an evolutionary algorithm. GA starts by initializing a set of random samples and then updates them through selection and mutation based on their fitness values (distance). However, using GA for optimization cannot make precise adjustments due to its reliance on the random evolution of latent vectors. This randomness can hinder the attack's success, as face recognition encoders rely on capturing fine details to differentiate between different faces. Additionally, this approach requires numerous queries to update multiple latent vectors simultaneously in each iteration.

Instead, we propose to optimize the latent vector using gradient descent directly:

$$z_{t+1} \leftarrow z_t - \eta \frac{\partial \mathcal{L}}{\partial z_t}, \quad \mathcal{L}(z) := d(E(G(z)), v), \tag{2}$$

where t refers to the current iteration and α is a fixed learning rate. To address the gradient computation challenge in the black-box setting, where the parameters of E are unknown, a detour is taken by employing zeroth-order optimization based on our previous preliminary study [12]. This approach allows us to estimate the gradient $\frac{\partial \mathcal{L}}{\partial z}$, denoted as g for simplicity. The zeroth-order optimization technique (ZOGE) [11,12] provides a formalized way to perform this estimation:

$$\hat{g} = \frac{n}{m} \sum_{i=1}^{m} \frac{\mathcal{L}(z + \epsilon u_i) - \mathcal{L}(z)}{\epsilon} u_i, \quad u_i \sim \mathcal{U}(\mathcal{S}^{n-1}), \tag{3}$$

here the dimension of the latent vector is denoted as n, and u_i represents a random direction sampled from a n-dimensional unit sphere, denoted as $\mathcal{S}^{n-1}$. Additionally, ϵ is a small positive constant known as the smoothing parameter. When estimating the gradient $\hat{g}$ using Eq. (3), there is inherent variance due to the randomness in the sampling process. This variance can be reduced by increasing the sampling parameter m, corresponding to the number of random directions sampled. However, increasing m also leads to extra querying in each iteration of the optimization process. To compute the gradient estimate $\hat{g}$ using Eq. (3), $m + 1$ queries are required. To this end, our optimization method works by combining (2) and (3) can be applied in a black-box setting, allowing us to optimize the feature $\hat{v}$.

3.3 Optimize in $\mathcal{W}$ Latent Space

StyleGAN differs from conventional GANs in mapping the latent input vector $z \in \mathcal{Z}$ to the output space. Instead of a direct mapping based on the training distribution, StyleGAN introduces a non-linear mapping function $f : \mathcal{Z} \rightarrow \mathcal{W}$ that transforms z into an intermediate latent vector $w \in \mathcal{W}$.

Recent studies [8,24] have demonstrated that the $\mathcal{W}$ space in StyleGAN learns a more disentangled representation compared to the $\mathcal{Z}$ space. In other words, the $\mathcal{W}$ space is less influenced by biases from the training distribution. This characteristic is advantageous for reconstruction tasks, mainly when the target face lies far from the training distribution. Therefore, we propose a latent vector optimization approach that operates in the $\mathcal{W}$ latent space of StyleGAN:

$$w^* = \underset{w \in \mathcal{W}}{\arg\min}\ dist(E(G_{\mathcal{W}}(w)), v),\ \ G_{\mathcal{W}}(f(\cdot)) = G(\cdot). \tag{4}$$

3.4 Latent Space Mining and Cascade Filtering

The convergence, efficiency, and solution quality of an optimization algorithm are heavily dependent on the initial guess [25]. A good initial guess leads to faster convergence and better solutions, while a poor guess can cause the algorithm to get stuck or take longer. Overall, a well-informed initial guess enhances performance and increases the likelihood of finding optimal solutions.

However, it is usually challenging to find a good initial guess. In [26], the "master faces" notion in the context of facial recognition systems provides good inspiration for optimizing black-box attacks on initial guess generation. The term "master faces" is used in analogy to "master keys", which refers to specific face images with the unique characteristic of successfully authenticating a significant number of identities in a given dataset. Inspired by the concept of "master faces," we propose the idea of discovering "master latents" that can generate face images resembling multiple input face images. These master latents can serve as an initial guess for optimization, thereby minimizing the number of queries required in subsequent tasks.

Specifically, our methodology consists of three main steps. Firstly, we generate l initial latent vectors using a random generation process. Then, we apply the k-means clustering algorithm [27,28] to partition the l latent vectors into k clusters in a mining manner. The detailed algorithm is as below:

1. Initialization: Choose the number of clusters, k. Randomly initialize k centroids, denoted by $\mu_1, \mu_2, \ldots, \mu_k$. Assign each data point z_i to the nearest centroid based on the Euclidean distance, which can be denoted as $c_i = \arg\min_j ||z_i - \mu_j||^2$
2. Assignment Step: For each data point z_i, find the nearest centroid and update its assignment: $c_i = \arg\min_j ||z_i - \mu_j||^2$
3. Update Step: Update the centroids by computing the mean of all the data points assigned to each cluster: $\mu_j = \frac{1}{n_j}\sum_{i=1}^{n}[z_i]$ where n_j is the number of data points assigned to cluster j.

4. Repeat Steps 2 and 3 until convergence. Return the final assignments of data points to clusters and the centroids.

Finally, a cascade filtering is applied to the obtained k clusters. Specifically, k centroid latent vectors are utilized to generate corresponding face images and face features in a sequential manner. The top r centroids with the lowest distance, as obtained by the user interface, are selected. Next, all latents belonging to those r clusters are filtered out and used to find the top-5 latent vectors with the lowest distance. These vectors are then averaged to create a single latent vector. This averaged latent vector is utilized for subsequent latent optimization through gradient estimation. In this latent mining and cascade filtering process, the query count is determined by $k + r * \frac{l}{k}$.

4 Experiments and Results

4.1 Datasets, Encoders, and Configurations

We evaluated our method using two widely recognized face recognition benchmarking datasets: Labeled Faces in the Wild (LFW) [29] and Celebrities in Frontal-Profile (CFP-FP) [30]. We reconstruct the initial image from each positive pair listed officially to evaluate our method. This process yields 2,551 reconstructed images for LFW and 2,772 for CFP-FP.

Four different encoders were used, with VGGNet19 trained using CosFace loss (VGG19-Cos), ResNet50 trained using AdaFace (Res50-Ada), SwinTransformer trained using AM-Softmax loss (Swin-Softmax) and InceptionResnet trained using FaceNet triplet loss (InRes-Facenet). Using different encoders to simulate encoders in a realistic database, an attacker can use these four encoders to reconstruct face images corresponding to different feature vector spaces. We compared our approach with three baselines:NBNet [1], GA [23], and ZOGE [11]. For NBNet, we utilized the pre-training weights from its FaceNet counterpart and employed the face generator DCGAN [31]. For GA and ZOGE, we also utilized 4k queries for face reconstruction. In contrast, our proposed method achieved superior results, with face reconstruction completed in approximately 2.6k queries. Identical to GA, we employed StyleGAN2, trained on the FFHQ dataset, as the face generator for our approach.

In this paper, $k = 1024, l = 8000, r = 80$ are used as the default parameter in the latent mining process empirically; we utilized ZOGE with parameters $\epsilon = 0.1$ and $m = 9$ to perform face reconstruction. We employed the Adam optimizer [32] with a fixed learning rate of 0.1. As Fig. 3 illustrates, an optimal ZOGE performance is achieved when the iteration number reaches 100, thus, 100 iterations are adopted in our subsequent experiments. Therefore, the query count for ZOGE optimization process is determined as $(m + 1) * 100 = 1000$, and the total query count can be computed as:

$$Q = k + r * \frac{l}{k} + 1000. \quad (5)$$

Figure 4 shows the final feature cosine distance between the reconstructed face images and target images obtained from 30, 100, 230, and 630 iterations (equivalent to 2k, 2.6k, 4k, and 8k queries, respectively). Our method demonstrates a lower distance to ZOGE, thereby validating that our approach outperforms ZOGE and GA with an equal number of queries.

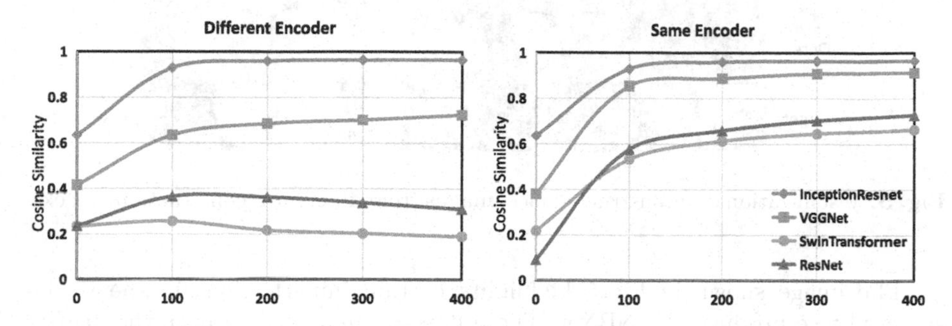

Fig. 3. Optimal performance is achieved with an iteration number of 100. Cosine similarity computation for the test set LFW-200 using different encoders on the left and the same encoder on the right.

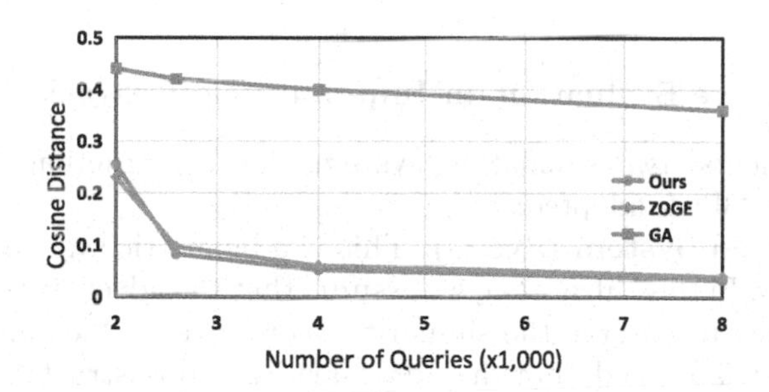

Fig. 4. Our method can lead to lower cosine distance.

4.2 Qualitative Evaluation

In this subsection, we showcase the robustness of our method in handling high-resolution images without introducing artifacts. To demonstrate this, we provide a visual comparison with two existing works: NBNet [1], GA [23] and ZOGE [11], specifically on the LFW subset. However, it is essential to note that other existing works were evaluated on different datasets or settings, making a fair comparison difficult. As a result, we have excluded a direct visual comparison with those works in this section.

Fig. 5. Visualization of reconstructed face images from different reconstruction attacks.

The image shown in Fig. 5 highlights certain imperfections in the reconstructed face produced by NBNet. These flaws primarily stem from the utilization of DCNN, which decreases image quality from features to a generated face image during the mapping process. On the other hand, GA and ZOGE demonstrate good visual quality by leveraging StyleGAN. However, artifacts may still be present due to sub-optimal latent codes generated by GA. Although ZOGE achieves comparable visual quality, it requires a higher query count than our method.

4.3 Quantitative Evaluation on Impersonation Attack

In our experiments, two simulation systems for construction impersonation attack from [23,33] are adopted:

Compromised System (Sys C): This is a biometric system compromised by the adversary. In this subsystem, we assume that the adversary can only query the user interface to observe the similarity score given a face image in Sys C. The query count is limited, and the adversary can reconstruct a face image in Sys C.

Targeted System (Sys T): This is a biometric system vulnerable to impersonation attacks. An adversary exploits the reconstructed image from Sys C to compromise Sys T in this system. The feature extractor $E(\cdot)$ utilized in Sys T may differ from that of Sys C, and the user's enrolled image might also be distinct from that used in Sys C. We consider both scenarios in our subsequent experiments.

In our experiment, we evaluate impersonation attacks using the Successful Attack Rate (SAR). SAR is calculated as the proportion of mated-attack attempts that are falsely declared to match the template of the same user at the given similarity threshold under different FAR, i.e., the ratio of mated-attack scores above the similarity threshold. The threshold is determined based on normal genuine and imposter scores. A higher SAR indicates better reconstruction attack performance, while a lower SAR suggests the opposite. We categorize attacks into type-I (same compromised face image in Sys C) and type-II (different face images of the same person), following prior research [1,23].

We quantitatively evaluate the performance of the proposed reconstruction scheme based on type-I and type-II attacks with the same/different feature extractors. We compute the true accept rate (TAR) at FAR under the normal situation in Sys T. The SARs of type I and II attacks are evaluated subsequently.

NBNet [1]**.** When both Sys T and Sys C use the same face feature extractor, such as InRes-Facenet as presented in Table 2, the NBNet method achieves a performance on the LFW dataset with an SAR of 80.83% and 46.9% for type-I and type-II attacks, respectively. Similarly, for the CFP-FP dataset, when InRes-Facenet is used as the feature extractor for both Sys T and Sys C, the NBNet method achieves an SAR of 89.97% and 28.8% for type-I and type-II attacks, respectively. When Sys C and Sys T use different face feature extractors, such as in Table 1 where Sys C uses InRes-Facenet and Sys T uses Res50-Ada, the NBNet method achieves a type-I SAR of 73.2% and a type-II SAR of 49.03%. Similarly, on the CFP-FP dataset, targeting the Res50-Ada method, NBNet achieves a type-I SAR of 83.89% and a type-II SAR of 32.77%. It is important to note that these success rates are obtained by utilizing **25600 attack queries**.

GA [23] **and ZOGE** [12]**.** Both GA and ZOGE have shown higher efficiency than NBNet. They both achieve remarkable performance with 4k queries, as demonstrated in Table 2. In the case where both Sys T and Sys C use the same face feature extractor, InRes-Facenet, the GA attack achieves a success rate (SAR) of 90.47% and 49.57% for type-I and type-II attacks on LFW, respectively. On the other hand, the ZOGE method achieves a SAR of 100% and 92.1% for type-I and type-II attacks under the same setting.

When Sys C and Sys T use different face feature extractors, such as InRes-Facenet for Sys C and Res50-Ada for Sys T, the GA attack achieves a SAR of 54.53% and 33.13% for type-I and type-II attacks on LFW. In comparison, the ZOGE method achieves a SAR of 95.1% and 77.8% for type-I and type-II attacks under the same setting. Based on the findings presented in Table 2, it can be concluded that ZOGE outperforms GA in terms of performance when the same number of queries is used.

Ours Latent Mining Method. By using different values for the parameters k and r in the latent mining process, we can achieve varying results. In our default setting, we use $k = 1024$ and $r = 80$, which results in a query of $Q = 2649$ (approximately 2.6k). However, to ensure a fair comparison, we also adopt the values $k = 1024$ and $r = 256$, which leads to a query of $Q = 4024$ (approximately 4k). The performance of our proposed method using these parameters is presented in Table 2. Based on Table 2, we can observe that:

1. When comparing GA, ZOGE, and our method using 4k queries, our method consistently outperforms the others in most settings. Table 2 demonstrates that our method excels when System T and System C use different face feature extractors. For instance, when InRes-Facenet is used for System C and VGG19-Cos is used for System T on the LFW dataset, our method achieves 1.2% and 41.77% higher SAR (Success Attack Rate) compared to

Table 2. Successful attack rate (%) at 0.1% FAR of each reconstruction attack on LFW and CFP-FP dataset. The columns colored in blue show the SAR when attacking the target encoder. Other columns show the SAR when attacking different encoders. The true accept rate (TAR) is measured instead of SAR for the genuine face. The number of queries is displayed in units of 1,000 (K).

Dataset	Sys C	Method	Attack Queries	Sys T									
				InRes-Facenet		VGG19-Cos		Swin-Softmax		Res50-Ada		Average	
				type-I	type-II	type-I	type-II	type-I	type-II	type-I	type-II	type-I	type-II
LFW	Genuine Face			96.7		96.93		99.73		99.73		98.27	
	InRes-Facenet	NBNet	25600k	80.83	46.9	66.3	36	3.83	1.43	73.2	49.03	56.04	33.34
		GA	4k	90.47	49.57	54.53	33.13	22	14.2	24.87	15.5	47.97	28.1
		ZOGE	4k	**100**	**92.1**	95.1	77.8	69.87	51.13	82.4	**60.8**	86.84	70.46
		Ours	4k	**100**	91.57	**96.3**	**78.77**	70.8	**52.7**	**82.77**	60	**87.47**	**70.76**
		Ours	2.6k	**100**	92	95.57	77.67	**71.33**	51.5	81.27	59.4	87.04	70.14
	VGG19-Cos	GA	4k	34.63	19.87	94.93	54.27	20.97	15	23.13	14.8	43.42	25.99
		ZOGE	4k	90.77	69.13	**100**	**91.63**	78.2	59.17	84.73	64.2	88.43	71.03
		Ours	4k	**91.77**	**70.93**	99.93	91	**80.2**	**60.6**	**85.87**	**66.17**	**89.44**	**72.18**
		Ours	2.6k	89.27	68.6	99.9	88.87	76.5	58.4	83.6	62.7	87.32	69.64
	Swin-Softmax	GA	4k	12.1	9.03	18.93	12.37	92.73	52.07	38.77	20.37	40.63	23.46
		ZOGE	4k	58.8	39.57	72.87	54.47	**100**	**96.43**	94.2	80.63	81.47	67.78
		Ours	4k	**62.23**	**43.13**	**75.37**	**58**	**100**	96.23	**95.83**	**82.1**	**83.36**	**69.87**
		Ours	2.6k	60.47	40.23	72.97	53.5	99.9	94.17	93.73	79.63	81.77	66.88
	Res50-Ada	GA	4k	7.63	4.4	11.47	7	15.07	7.83	49.37	20.6	20.89	9.96
		ZOGE	4k	38.03	23.93	46.83	30.6	**57.53**	40.1	**91.37**	**74**	58.44	42.16
		Ours	4k	**41.5**	**24.23**	**48.5**	**31.77**	57.3	**41.37**	89.93	72.13	**59.31**	**42.38**
		Ours	2.6k	37.5	22.63	43.7	29.3	54.03	37.33	89.9	68.5	56.28	39.44
CFP-FP	Genuine Face			74.46		78.46		85.34		88.91		81.79	
	InRes-Facenet	NBNet	25600k	89.97	28.8	85.34	27.6	6.71	2.09	83.89	32.77	66.48	22.82
		GA	4k	95.66	29.83	51.97	17.03	22.31	8.86	23.46	9.2	48.35	16.23
		ZOGE	4k	**100**	**66.8**	**95.69**	**50.29**	70.11	**28.54**	**82.4**	33.69	**87.05**	**44.83**
		Ours	4k	**100**	64.37	95.29	49.06	**70.86**	27.34	80.34	**34.09**	86.62	43.72
		Ours	2.6k	**100**	63.94	94.46	48.54	69.8	27.51	78.94	33.2	85.8	43.3
	VGG19-Cos	GA	4k	42.29	14.54	95.51	31.26	20.69	8.37	23.74	10.06	45.56	16.06
		ZOGE	4k	**88**	42.6	99.97	**65.23**	73.74	32.03	80.66	**38.34**	85.59	**44.55**
		Ours	4k	87.86	**42.66**	99.97	64.2	**73.83**	**32.77**	**81.09**	37.4	**85.69**	44.26
		Ours	2.6k	84.43	40.23	**100**	62.14	70.74	31.06	77.74	37.03	83.23	42.62
	Swin-Softmax	GA	4k	14.09	5.69	18.94	6.89	93.26	24.8	41	10.71	41.82	12.02
		ZOGE	4k	62.77	22.49	72.54	31.03	**100**	**69.37**	92.97	49.03	82.07	42.98
		Ours	4k	**64.57**	**23.69**	**74.17**	**31.57**	99.91	67.69	**93.4**	**50.57**	**83.01**	**43.38**
		Ours	2.6k	62.06	21.09	70.94	30.2	99.94	65.6	91.91	48.06	81.21	41.24
	Res50-Ada	GA	4k	18.8	4.66	6.23	1.83	21.77	6.51	26.31	7.63	18.28	5.16
		ZOGE	4k	18.2	**6.46**	21.54	**8.94**	24.83	**9.66**	34.94	**17.69**	24.88	**10.69**
		Ours	4k	**19.4**	6.06	**22.31**	8.37	**25**	9.14	**35.06**	16.94	**25.44**	10.13
		Ours	2.6k	17.03	5.14	20.34	7.29	23.09	8.37	34.71	15.66	23.79	9.12

ZOGE and GA, respectively. However, we have observed that our method only achieves comparable performance to ZOGE when System T and System C use the same face feature extractor. This is due to the fact that ZOGE performs 400 iterations, which can lead to overfitting the final reconstructed face images and consequently result in poor attack performance in settings involving different face feature extractors.

2. When using 2.6k queries, our method achieved comparable performance with only a slight degradation. For example, when both System T and System C use the same face feature extractor, InRes-Facenet, our method achieves a Success Attack Rate (SAR) of 100% and 92% for type-I and type-II attacks,

respectively, on the LFW dataset. On the CFP-FP dataset, our method achieves a SAR of 100% and 63.94% for type-I and type-II attacks, respectively. These results indicate that our method successfully attacked a system with a reconstructed face image from another system using identical encoders. When System C and System T use different face feature extractors, such as InRes-Facenet for System C and VGG19-Cos for System T, our method achieves a SAR of 95.57% for type-I attacks and 77.67% for type-II attacks on the LFW dataset. Furthermore, when we consider Swin-S for System C and Res50-Ada for System T, a SAR of 93.73% for type-I attacks and 79.63% for type-II attacks on LFW can be achieved. Similar competitive SARs are also observed across all encoders on the CFP-FP datasets. These outcomes demonstrate the efficacy of our method in attacking a system by employing a reconstructed face image from another system that uses different encoders with only 2.6k queries.

5 Conclusion

We proposed Minimum Assumption Reconstruction Attacks (MARA) that assume that the attacker, with limited resources, can only interact with the user interface and observe similarity scores provided by the face recognition system. We aimed to mimic typical user behavior and evaluate the system's robustness against potential threats by simulating a challenging scenario with limited query attempts.

Our proposed attack methodology uncovered the minimum query threshold necessary for an attacker to gather exploitable information. This information can be used to develop effective solutions to mitigate attacks on face recognition systems, even without compromising the system's template or database. Our experiments demonstrated that MARA achieves comparable performance to false accept attacks, highlighting the resemblance between attacker behavior and a regular user.

References

1. Mai, G., Cao, K., Yuen, P.C., Jain, A.K.: On the reconstruction of face images from deep face templates. IEEE Trans. Pattern Anal. Mach. Intell. **41**(5), 1188–1202 (2018)
2. Ramachandra, R., Busch, C.: Presentation attack detection methods for face recognition systems: a comprehensive survey. ACM Comput. Surv. (CSUR) **50**(1), 1–37 (2017)
3. Wang, F., Leng, L., Teoh, A.B.J., Chu, J.: Palmprint false acceptance attack with a generative adversarial network (GAN). Appl. Sci. **10**(23), 8547 (2020)
4. Wang, Z., She, Q., Ward, T.E.: Generative adversarial networks in computer vision: a survey and taxonomy. ACM Comput. Surv. (CSUR) **54**(2), 1–38 (2021)
5. Zhai, J., Zhang, S., Chen, J., He, Q.: Autoencoder and its various variants. In: IEEE International Conference on Systems, Man, and Cybernetics (SMC), pp. 415–419. IEEE (2018)

6. Cole, F., Belanger, D., Krishnan, D., Sarna, A., Mosseri, I., Freeman, W.T.: Synthesizing normalized faces from facial identity features. In: Proceedings of the IEEE Conference on Computer Vision and Pattern Recognition, pp. 3703–3712 (2017)
7. Dong, X., Jin, Z., Guo, Z., Teoh, A.B.J.: Towards generating high definition face images from deep templates. In: International Conference of the Biometrics Special Interest Group (BIOSIG), pp. 1–11. IEEE (2021)
8. Karras, T., Laine, S., Aila, T.: A style-based generator architecture for generative adversarial networks. In: Proceedings of the IEEE/CVF Conference on Computer Vision and Pattern Recognition, pp. 4401–4410 (2019)
9. Jia, S., Guo, G., Xu, Z.: A survey on 3D mask presentation attack detection and countermeasures. Pattern Recogn. **98**, 107032 (2020)
10. Sousedik, C., Busch, C.: Presentation attack detection methods for fingerprint recognition systems: a survey. IET Biometrics **3**(4), 219–233 (2014)
11. Liu, S., Chen, P.-Y., Kailkhura, B., Zhang, G., Hero, A.O., III., Varshney, P.K.: A primer on zeroth-order optimization in signal processing and machine learning: principals, recent advances, and applications. IEEE Signal Process. Mag. **37**(5), 43–54 (2020)
12. Park, H., Park, J., Dong, X., Teoh Beng-Jin, A.: Query efficient and generalizable black-box face reconstruction attack. In: 2023 IEEE International Conference on Image Processing (in press)
13. Schroff, F., Kalenichenko, D., Philbin, J.: FaceNet: a unified embedding for face recognition and clustering. In: Proceedings of the IEEE Conference on Computer Vision and Pattern Recognition, pp. 815–823 (2015)
14. Deng, J., Guo, J., Xue, N., Zafeiriou, S.: ArcFace: additive angular margin loss for deep face recognition. In: Proceedings of the IEEE/CVF Conference on Computer Vision and Pattern Recognition, pp. 4690–4699 (2019)
15. Wang, H., et al.: CosFace: large margin cosine loss for deep face recognition. In: Proceedings of the IEEE Conference on Computer Vision and Pattern Recognition, pp. 5265–5274 (2018)
16. Ranjan, R., et al.: Deep learning for understanding faces: machines may be just as good, or better, than humans. IEEE Signal Process. Mag. **35**(1), 66–83 (2018)
17. Wang, M., Deng, W.: Deep face recognition: a survey. Neurocomputing **429**, 215–244 (2021)
18. Zhmoginov, A., Sandler, M.: Inverting face embeddings with convolutional neural networks. arXiv preprint arXiv:1606.04189 (2016)
19. Keller, D., Osadchy, M., Dunkelman, O.: Inverting binarizations of facial templates produced by deep learning (and its implications). IEEE Trans. Inf. Forensics Secur. **16**, 4184–4196 (2021)
20. Duong, C.N., et al.: Vec2Face: unveil human faces from their blackbox features in face recognition. In: Proceedings of the IEEE/CVF Conference on Computer Vision and Pattern Recognition, pp. 6132–6141 (2020)
21. Razzhigaev, A., Kireev, K., Kaziakhmedov, E., Tursynbek, N., Petiushko, A.: Black-box face recovery from identity features. In: Bartoli, A., Fusiello, A. (eds.) ECCV 2020. LNCS, vol. 12539, pp. 462–475. Springer, Cham (2020). https://doi.org/10.1007/978-3-030-68238-5_34
22. Srinivas, M., Patnaik, L.M.: Genetic algorithms: a survey. Computer **27**(6), 17–26 (1994)
23. Dong, X., et al.: Reconstruct face from features based on genetic algorithm using GAN generator as a distribution constraint. Comput. Secur. **125**, 103026 (2023)

24. Karras, T., Laine, S., Aittala, M., Hellsten, J., Lehtinen, J., Aila, T.: Analyzing and improving the image quality of StyleGAN. In: Proceedings of the IEEE/CVF Conference on Computer Vision and Pattern Recognition, pp. 8110–8119 (2020)
25. Boyd, S.P., Vandenberghe, L.: Convex Optimization. Cambridge University Press (2004)
26. Shmelkin, R., Friedlander, T., Wolf, L.: Generating master faces for dictionary attacks with a network-assisted latent space evolution. In: 2021 16th IEEE International Conference on Automatic Face and Gesture Recognition (FG 2021), pp. 01–08 (2021)
27. Bock, H.-H.: Clustering methods: a history of k-means algorithms. In: Brito, P., Cucumel, G., Bertrand, P., de Carvalho, F. (eds.) Selected Contributions in Data Analysis and Classification, pp. 161–172. Springer, Heidelberg (2007). https://doi.org/10.1007/978-3-540-73560-1_15
28. Ahmed, M., Seraj, R., Islam, S.M.S.: The k-means algorithm: a comprehensive survey and performance evaluation. Electronics **9**(8), 1295 (2020)
29. Huang, G.B., Mattar, M., Berg, T., Learned-Miller, E.: Labeled faces in the wild: a database for studying face recognition in unconstrained environments. In: Workshop on Faces in 'Real-Life' Images: Detection, Alignment, and Recognition (2008)
30. Sengupta, S., Chen, J.-C., Castillo, C., Patel, V.M., Chellappa, R., Jacobs, D.W.: Frontal to profile face verification in the wild. In: IEEE Winter Conference on Applications of Computer Vision (WACV), pp. 1–9. IEEE (2016)
31. Radford, A., Metz, L., Chintala, S.: Unsupervised representation learning with deep convolutional generative adversarial networks. arXiv preprint arXiv:1511.06434 (2015)
32. Kingma, D.P., Ba, J.: Adam: a method for stochastic optimization. arXiv preprint arXiv:1412.6980 (2014)
33. Dong, X., Jin, Z., Teoh, A.B.J., Tistarelli, M., Wong, K.: On the security risk of cancelable biometrics. arXiv preprint arXiv:1910.07770 (2019)

Emotion Recognition via 3D Skeleton Based Gait Analysis Using Multi-thread Attention Graph Convolutional Networks

Jiachen Lu[1], Zhihao Wang[1], Zhongguang Zhang[1], Yawen Du[2], Yulin Zhou[2,3], and Zhao Wang[2,3](✉)

[1] School of Software Technology, Zhejiang University, Hangzhou, China
[2] Ningbo Innovation Center, Zhejiang University, Ningbo, China
zhao_wang@zju.edu.cn
[3] Ningbo Tech University, Ningbo, China

Abstract. Human gait is a manner of individual's walking that observers are able to learn useful information through daily walking activities. Recently, gait skeletons-based emotion recognition has attracted much attention, while many methods have been proposed gradually. Skeleton-based representations offer several advantages for recognition tasks. In particular, such representation is extremely lightweight and could be directly extracted from video data using off-the-shelf algorithms. Moreover, skeleton data is not tied to any specific cultural or ethnic context, it hence become increasingly popular in recent years for cross-cultural studies and other related applications. To effectively process this type of data, many researchers have turned to Graph Convolutional Networks (GCNs) to leverage the topological structure of the data, which improves performance by modeling the relationships between different joints and body parts as a graph. This allows GCNs to capture complex spatial and temporal patterns. In this work, we have constructed an efficient multi-stream GCN framework for emotion recognition task. We have identified the complementary effect among streams using a multi-thread attention method (MTA), which is able to improve the emotion recognition performance. In addition, the proposed MTA graph convolution layer is able to extract effective features from the topology of the graph to further improve recognition performance. The proposed method outperforms state-of-art methods on challenging benchmark dataset.

Keywords: Emotion Recognition · 3D Skeleton · Action Recognition · Gait Analysis · GCN

1 Introduction

Human emotion recognition aims to understand the physiological and psychological reactions produced by people's cognition and perception of the outside world. Emotion recognition has attracted lots of attention, especially in mental

Q. Liu et al. (Eds.): PRCV 2023, LNCS 14429, pp. 74–85, 2024.
https://doi.org/10.1007/978-981-99-8469-5_6

state monitoring and human computer interaction (HCI) [24], which is heavily involved in psycho-physiological computing and Internet of Things (IoT).

Existing emotion recognition approaches usually utilize visual signals from facial cues or aural signals from speech to capture emotion features [15,32,34]. However, perceiving emotions from facial expression and speech could be unreliable since these signals are difficult to detect from remote locations. In addition, further difficulties could occur in complex situations [1], such as imitation expressions and concealed expressions. On other hand, emotion expression is embedded in the body languages, including gait and postural features. Compare with facial expression data which usually need to be recorded in short distance, the subjects for monitoring gait patterns could be relatively far from the cameras, e.g. several meters in practice, where most other bio-metrics are no longer observable or can only provide low resolution. Therefore, gait of human walking has received much attentions since it's able to utilize non-verbal cues and reduce the interference of subjective factors.

Gait data is a complex sequence with both spatial and temporal features that are critical for accurate recognition and classification of different gait patterns. These features are important for capturing the dynamic nature of gait and can provide valuable information about an individual's walking pattern. Several existing approaches for gait-based analysis leverage these spatial and temporal features to improve performance [7,8,12]. The graph convolutional network (GCN) that appeared in recent years can make full use of the connection relationship between nodes to model data, which have been used in gait-based emotion recognition. The physical structure of the human skeleton is used as a spatial graph named as natural link in STEP proposed by [1]. After that, many researchers have tried to construct skeleton graph according to physical structure of the human body, and uses GCNs o extract the features between emotion and gait [2,4,28]. Recently, GCN-based methods become an essential role and improve the capabilities of emotion recognition using skeleton-based gait.

Although existing methods for skelton-gait based emotion recognition achieve promising results, there are still some major drawbacks. The aggregation of joints and effective features might be rigid, resulting in insufficient exploitation of complementary information. This can limit the ability of these models to capture complex emotional states and may lead to sub-optimal performance on certain tasks.

In order to tackle above issues, a multi-stream GCN approach is proposed in this work, where a multi-thread attention graph layer is designed to alleviate the receptive field imbalance problem. Meanwhile, the self-attention mechanism used by transformers allows them to selectively attend to different parts of the input sequence, enabling them to capture complex relationships between distant elements [19,20,23]. The probabilities of human emotions are predicted at the end of the module. Generally, the main contribution are as follows:

- We have proposed an efficient method for gait-based human emotion recognition, learning the deep features with a fused multi-stream architecture. The

experimental results have demonstrated that fused streams offer more distinctive and complementary information for recognition.
- We have designed an adaptive multi-thread attention graph convolution layer. This approach alleviates the receptive field imbalance problem while also capturing the non-local dependencies among different joints.
- We have conducted comprehensive evaluations on Emotion-Gait benchmark dataset and outperforms SOTA approaches. Notably, the proposed mechanism is generic and robust to seamlessly work with existing GCN framework, where reasonable performance could be achieved in a simple way.

2 Related Work

2.1 Gait Based Human Emotion Recognition

Great effort has been made by many researchers to enable computers to observe, understand, and even express various emotions like humans. Hand-crafted skeleton-based gait features are commonly utilized in early emotion recognition research. For example, Cren et al. have employed features such as speed, arm swinging and angle among body joints for body expression recognition [7]. Li et al. have used frequency-domain features and principal component analysis (PCA) to analyze gait sequences for distinguishing different emotions [12,14]. Daoudi et al. have leveraged a covariance descriptor to obtain the dynamics of the body and analyzed geometric means and geodesic distances for emotion recognition [8]. However, these hand-crafted features often require careful design based on the data characteristics, which may need to be redesigned when the data source changes.

Deep learning has recently achieved significant success in computer vision, while many methods now use deep learning techniques for gait emotion perception [18,21,22]. For instance, Randhavane et al. adopted a time sequence-based approach that utilized an Long Short-term Memory (LSTM) to extract temporal features [27]. They then combined these features with hand-extracted emotional features and used a random forest classifier for classification. Gated recursive units (GRUs) have been employed to extract features from joint coordinates at a single time step then temporal analysis has been taken to identify emotions [3]. The paradigm of these sequence-based methods is to construct a sequence deep model based on skeleton sequences to predict the emotion. An image-based method named ProxEmo that encoded skeleton sequences by converting 3D joint point data into images has been proposed, where convolutional neural networks (CNN) were then used to extract features related to emotions and identify them [25]. Another image-based method proposed a two-stream network with transformer-based complementarity (TNTC) to extract effective features from joint skeletons and classify emotions [11].

The relationship of joints were represented through a spatio-temporal graph convolutional network (ST-GCN) to extract features from the spatio-temporal graph and identified emotions [33]. A graph-based method that exploited the characteristic that skeletons are naturally graphs in non-Euclidean space has been proposed in [2]. An attention-enhanced temporal convolutional network (AT-GCN) has been adopted to capture discriminate features in spatial dependencies and temporal dynamics for sentiment classification [28]. Chai et al. designed a multi-head pseudo nodes technique that using a series of extra pseudo nodes with GCN models to effectively obtain the global information for emotion gait recognition [4]. Lu et al. have utilized more synthesised data and reconstruction strategy to improve the performance of emotion recognition [17]. These approaches demonstrate the effectiveness of graph-based methods in gait emotion perception tasks.

2.2 Graph-Based Methods for 3D Skeleton-Based Action Recognition

Graph-based methods have achieved great success in skeleton-based action recognition studies recently. In ST-GCN, the skeleton is treated as a graph, with joints as nodes and bones as edges [33]. Following ST-GCN, several approaches have explored the relationship between distant joints [29,35]. In addition, multi-scale structural feature representation methods have been developed using higher-order polynomials of the skeleton adjacency matrix [13]. Inspired by [16], Chen et al. introduced a sub-graph convolution cascaded by residual connection with enriched temporal receptive field [6]. Song et al. designed a combination approach named Efficient GCN and proposed a compound scaling strategy to expand the model's width and depth synchronously [31]. Chen et al. designed a multi-granular GCN-based method on the temporal domain to capture both short-term and long-term temporal dependencies [5]. Qin et al. fused angle information extracted from manually specified joint groups to a GCN model, improving its ability to capture fine-grained motion information [26].

In addition, many existing studies employ multi-stream pipelines to enhance the model's ability to learn more expressive features for skeleton-based action recognition. For example, Shi et al. utilized joints and bones input in their two-channel framework called 2s-AGCN [30]. Song et al. considered joints, bone, and velocity to increase the capacity of their model named EfficientGCN [31]. Hou et al. studied the effectiveness of using multiple streams [10]. A prospective shifting approach has been designed which transforms an action into many views and is based on the angle representation in skeletons data [9]. These studies demonstrate that incorporating multiple streams of information can improve the ability of graph-based models to capture both spatial and temporal features, leading to better performance on skeleton-based action recognition tasks.

3 Methodology

We aim to exploit complementary information of the multi streams. Meanwhile, existing works have shown that bone information generated from joint information promote the model to extract discriminative features in space and time dimensions. However, the impact between the local frame stream and the global

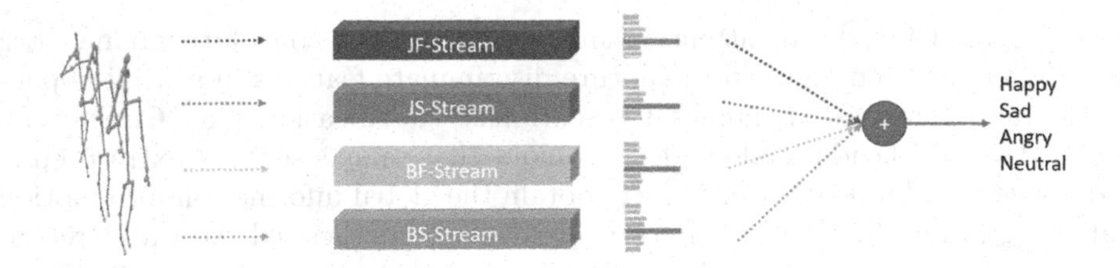

Fig. 1. Multi-Stream Input for the Proposed Graph Model

sequence stream has not been comprehensively investigated in Emotion Recognition. Thus, in the following paragraph, we will illustrate the pre-processing methods of the global streams and local stream and the generation mechanism of bone information. Then, we will introduce the designed multi-thread attention graph layer and model architecture to combine the multi-stream input.

3.1 Multi-stream Input

For the data pre-processing, we apply the local frame stream and global sequence stream of the joint information and bone information.

Local Frame Stream of Joint. For the local frame stream, the coordinates of joints in each frame refer to the fixed joint within its own frame. The fixed joint is often the centre of the spine. The orientation of the body in the different frames are same by setting the spine and the shoulder as Y-axis and X-axis respectively. Such input could help the model reduce the direction bias and focus more on changes in local patterns of postures between different frames. We denote the local frame channel as J^F, which can be obtained as Eq. (1).

$$J^F_{i,t} = (J_{i,t} - J_{root,t}) \times R_t \tag{1}$$

In this equation, $J_{root,t}$ is the center joint in the t-th frame, R_t is the rotation matrix that defined as

$$R_t = R^x_{t,\alpha} R^z_{t,\gamma}$$

$$R^x_{t,\alpha} = \begin{bmatrix} 1 & 0 & 0 \\ 0 & \cos(\alpha_t) & \sin(\alpha_t) \\ 0 & -\sin(\alpha_t) & \cos(\alpha_t) \end{bmatrix} \tag{2}$$

$$R^z_{t,\gamma} = \begin{bmatrix} \cos(\gamma_t) & 0 & -\sin(\gamma_t) \\ 0 & 1 & 0 \\ \sin(\gamma_t) & 0 & \cos(\gamma_t) \end{bmatrix}$$

where $R^x_{t,\alpha}, R^z_{t,\gamma,}$, denote rotating the camera coordinate system around the X-axis by α_t radians, and the Z-axis by γ_t radians anticlockwise. α_t is the angle

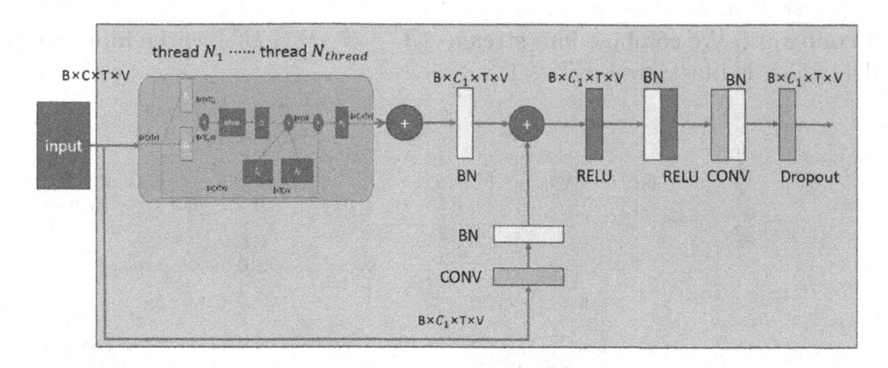

Fig. 2. Multi-thread Attention Graph Convolution layer. B, C, T, V denotes the number of samples, the dimension of each node, the number of time steps, the number of nodes respectively.

between the X-axis of the camera coordinate and shoulder, γ_t is for the Z-axis and spine.

Global Sequence Stream of Joint. During data preprocessing of the global sequence stream, all joints in the action sequence are referenced to the same global coordinate system. This approach enables the model to capture global information, such as movement trajectory between different frames. We denote the joint action sequence as J. $J_{i,t}$ represents the i-th joint in the t-th frame of the joint action sequence. J^C represents the sequence under the camera coordinate system, the global sequence stream of Joint J^S can be obtained as Eq. (3) and B^S can also be obtained like this.

$$J^S_{i,t} = J_{i,t} - J_{root,t=0} \tag{3}$$

The positions where different actions start could be identical after such operation.

Local and Global Stream of Bone. The bone stream provides information on bone length and direction, focusing on the original topology structure of the body to help the model extract kinetic features. This stream is generated from joint information. The direction of the bone is defined as from the center to the outside, with the black node representing the center of the body. Since there is one less edge than nodes in the body graph, we set the center joint of the body as zero. The bone stream is denoted as B, with $B_{i,t}$ similar to $J_{i,t}$. The function for calculating the bone channel is defined by Eq. (4).

$$B_{i,t} = J_{i,t} - J_{near,t} \tag{4}$$

In this equation, $B_{i,t}$ is the i-th bone in the t-th frame, $J_{i,t}$ is the i-th joint in the t-th frame, and $J_{near,t}$ is the joint adjacent with the i-th joint in the t-th frame. Since the bone stream could be extracted with either frame-level joints or sequence-level joints, we further denote these two as local stream of bone B^F and global stream of bone B^S.

In our work, We combine four streams (J^F, J^S, B^F, B^S) as the input to the model, which is illustrated in Fig. 1.

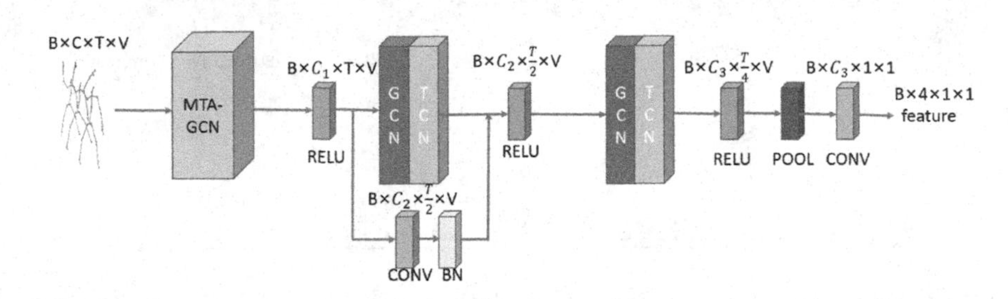

Fig. 3. An illustration of graph network for a single stream. B, C, T, V denotes the number of samples, the dimension of each node, the number of time steps, the number of nodes respectively.

3.2 Multi-thread Attention Graph Layer

In order to improve the spatio-temporal graph convolution for the skeleton data, we have proposed a multi-thread attention graph convolution layer (MTA-GCN). The attention block is defined as following Eq. 5.

$$\begin{aligned} f_{atten} &= \sum_{i}^{N_{thread}} \sum_{k}^{N_{kernel}=3} w_{i,k} f_{in}(A_k + B_{i,k} + C_{i,k}) \\ C_{i,k} &= G(\theta_{i,k}(f_{in})\phi_{i,k}(f_{in})) \end{aligned} \tag{5}$$

where N_{thread} denotes the number of the thread in the MTA-GCN. N_{kernel} denotes the kernel size of the spatial dimension. With the partition strategy designed in the early work, N_{kernel} is set to 3. A_k is the original normalized $V \times V$ adjacency matrix. It represents the physical structure of the human body and it keep same in each head. $B_{i,k}$ is an $V \times V$ adjacency matrix in the i th head. In contrast to A_k, the elements of $B_{i,k}$ are parameterized and optimized together with the other parameters in the training process. There are no constraints on the value of $B_{i,k}$, which means that the graph is completely learned according to the training data. $C_{i,k}$ is a data-dependent graph which learn a unique graph for each sample. To determine whether there is a connection between two vertexes and how strong the connection is, we apply the normalized embedded Gaussian function to calculate the similarity of the two vertexes. Since Gaussian function is equivalent to a softmax operation in practice. Therefore, $C_{i,k}$ could be rewritten as

$$C_{i,k} = \mathrm{softmax}(f_{in}^T W_{\theta_{i,k}}^T W_{\phi_{i,k}} f_{in}) \tag{6}$$

In detail, given the input feature map f_{in} whose size is $C \times T \times V$, we first embed it into $C_e \times T \times V$ with two embedding functions, i.e., θ and ϕ. Here, through extensive experiments, we choose one 1×1 convolutional layer as the embedding function. The two embedded feature maps are rearranged and reshaped to an $V \times C_e T$ matrix and a $C_e T \times V$ matrix. They are then multiplied to obtain an $V \times V$ similarity matrix $C_{i,k}$. $w_{i,k}$ is the is the $C_{out} \times C_{in} \times 1 \times 1$ weight vector of the 1×1 convolution operation. Finally, f_{atten} gets from the summation of the convolution results of all kernels of all heads.

After that, a batch norm operation is applied to f_{atten} with a residual branch, which is defined as

$$f_g = \mathrm{BN}(f_{atten}) + \mathrm{BN}(W_{in} f_{in}) \tag{7}$$

Then a temporal averaging process is taken via a series of *Relu-BN-Relu-Conv-Dropout* operation. The whole process of proposed multi-thread attention graph convolution is illustrated in Fig. 2.

3.3 Network Architecture

The overview of proposed method is illustrated in Fig. 3. It consists of MTA-GCN, a residual ST-GCN block, a normal ST-GCN block, Average Pool, Conv2D and fully connected(FC) layers respectively. The input is the human gaits processed from walking videos. Then the channel of the feature keep same after the first layer. Through the last two levels, the number of channels is halved. Then the predicted label is taken via a series of *Relu-Pool-Conv-fc* and a softmax operation.

Table 1. Comparison of our method with the state-of-the-art on Emotion-Gait. The best result of accuracy is highlighted in **bold**.

Methods		Publisher	Top-1 Acc
STEP [2]	Graph-based	AAAI 2020	0.7824
ST-TR*	Graph-based	CVIU 2021	0.7882
DGNN* [29]	Graph-based	CVPR 2019	0.7919
MS-G3D* [16]	Graph-based	CVPR 2020	0.8130
TEW [3]	GRU-based	ECCV 2020	0.8189
G-GCSN [36]	Graph-based	ACCV 2020	0.8150
ProxEmo [25]	Image-based	IROS 2020	0.8240
2s-AGCN*	Graph-based	CVPR 2019	0.8140
MS-AAGCN*	Graph-based	TIP 2020	0.8190
PN($CTR-GCN$) [4]	Graph-based	Neurocomputing 2022	0.8319
TNTC [11]	Image-based	ICASSP 2022	0.8597
Our method	Graph-based	-	**0.8685**

* The values are reproduced results on Emotion-Gait dataset reported in [4].

4 Experimental Results and Discussions

4.1 Experimental Settings

Dataset. The proposed method is evaluated on Emotion-Gait dataset, which consists of 2177 real gait sequences separately annotated into one of four emotion categories including happy, sad, angry, or neutral. The gait is defined as the 16-joint-skeleton. Each gait sample in the consolidated dataset was labeled as one of the four emotional categories by domain experts. The steps of gait sequences are maintained via duplication to 240 which is the maximum length of gait sequence in the dataset.

Evaluation Metrics. Top-1 accuracy has been adopted to measure the quality of proposed classification model, which is defined as

$$\text{Accuracy} = \frac{T}{S} \tag{8}$$

where T denotes the number of successfully classified gait sequences, and S denotes the number of test samples. Following [11], we employ 5-fold cross-validation. The average accuracy of 5-fold cross validation is recorded along its standard deviation.

Implementation Details. The experiments are conducted on an NVIDIA RTX 3090 GPU. We apply Adam optimizer with momentum of 0.9 and a weight decay of 1×10^{-4}. We train our model for 200 epochs with the initial learning rate of 0.01 then divided by 10 after 100, 150 epochs and a batch size of 8.

4.2 Comparison with SOTA Methods

We compare our method with several state-of-the-art methods lately reported on Emotion-Gait. Table 1 presents the experiment results on Emotion-gait dataset and the comparison with SOTA skeleton-based methods. According to the results, we observe 0.88% improvement in accuracy (85.97% to 86.85%) and our method outperform competitors.

4.3 Ablation Study

An ablation study has been conducted on EmotionGait dataset with a 3 ST-GCN block network as the backbone to illustrate the effectiveness of multi-stream combinations. The result is shown in Table 2. For settings of 1 stream and 4 streams, each input type is listed individually. For other settings, only combinations with top performance are listed. From Table 2, local stream of joint J^S achieve top accuracy for single stream input. The performance could be benefited with additional kinematic information with local stream Bone B^F. The combination of the local and global channels is able to improve performance and provide improvement in multi-stream input settings. Generally, combining different streams can further improve recognition accuracy, while the complementary effect between local frame information and global sequence information is obvious. Additionally, the proposed MTA-GCN outperform normal GCN block significantly.

Table 2. Ablation study on the effect of multi-stream input mechanism and proposed MTA-GCN

Methods	Top-1 Acc
1s (J^F)	0.8264
2s ($J^F + B^F$)	0.8329
3s ($J^F + B^F + B^S$)	0.8352
4s ($J^F + J^S + B^F + B^S$)	0.8361
Our method (4s MTA-GCN)	0.8685

5 Conclusion

To sum up, we address the skeleton-gait based emotion recognition task with a multi-stream input strategy using multi-thread attention graph convolution. We have presented a simple but impressive method to construct effective baselines for such task. Extensive experimental results reveal that using the global sequence channel and local frame channel could boost performance. We have conducted comprehensive evaluations on Emotion-Gait benchmark dataset and outperforms SOTA approaches. Notably, the proposed mechanism is generic and robust to seamlessly work with existing GCN framework.

Acknowledgements. This research has been supported by National Key Research and Development Project of China (Grant No. 2021ZD0110505), Natural Key Research and Development Project of Zhejiang Province (Grant No. 2023C01043) and Ningbo Natural Science Foundation (Grant 2022Z072, 2023Z236).

References

1. Barrett, L.F.: How Emotions are Made: The Secret Life of the Brain. Pan Macmillan (2017)
2. Bhattacharya, U., Mittal, T., Chandra, R., Randhavane, T., Bera, A., Manocha, D.: STEP: spatial temporal graph convolutional networks for emotion perception from gaits. In: Proceedings of the AAAI Conference on Artificial Intelligence, vol. 34, pp. 1342–1350 (2020)
3. Bhattacharya, U., et al.: Take an emotion walk: perceiving emotions from gaits using hierarchical attention pooling and affective mapping. In: Vedaldi, A., Bischof, H., Brox, T., Frahm, J.-M. (eds.) ECCV 2020. LNCS, vol. 12355, pp. 145–163. Springer, Cham (2020). https://doi.org/10.1007/978-3-030-58607-2_9
4. Chai, S., et al.: A multi-head pseudo nodes based spatial-temporal graph convolutional network for emotion perception from gait. Neurocomputing **511**, 437–447 (2022)
5. Chen, T., et al.: Learning multi-granular spatio-temporal graph network for skeleton-based action recognition. In: Proceedings of the 29th ACM International Conference on Multimedia, pp. 4334–4342 (2021)

6. Chen, Z., Li, S., Yang, B., Li, Q., Liu, H.: Multi-scale spatial temporal graph convolutional network for skeleton-based action recognition. In: Proceedings of the AAAI Conference on Artificial Intelligence, vol. 35, pp. 1113–1122 (2021)
7. Crenn, A., Khan, R.A., Meyer, A., Bouakaz, S.: Body expression recognition from animated 3D skeleton. In: 2016 International Conference on 3D Imaging (IC3D), pp. 1–7. IEEE (2016)
8. Daoudi, M., Berretti, S., Pala, P., Delevoye, Y., Del Bimbo, A.: Emotion recognition by body movement representation on the manifold of symmetric positive definite matrices. In: Battiato, S., Gallo, G., Schettini, R., Stanco, F. (eds.) ICIAP 2017. LNCS, vol. 10484, pp. 550–560. Springer, Cham (2017). https://doi.org/10.1007/978-3-319-68560-1_49
9. Hou, R., Li, Y., Zhang, N., Zhou, Y., Yang, X., Wang, Z.: Shifting perspective to see difference: a novel multi-view method for skeleton based action recognition. In: Proceedings of the 30th ACM International Conference on Multimedia, pp. 4987–4995 (2022)
10. Hou, R., Wang, Z., Ren, R., Cao, Y., Wang, Z.: Multi-channel network: constructing efficient GCN baselines for skeleton-based action recognition. Compu. Graph. **110**, 111–117 (2023)
11. Hu, C., Sheng, W., Dong, B., Li, X.: TNTC: two-stream network with transformer-based complementarity for gait-based emotion recognition. In: ICASSP 2022-2022 IEEE International Conference on Acoustics, Speech and Signal Processing (ICASSP), pp. 3229–3233. IEEE (2022)
12. Li, B., Zhu, C., Li, S., Zhu, T.: Identifying emotions from non-contact gaits information based on microsoft kinects. IEEE Trans. Affect. Comput. **9**(4), 585–591 (2016)
13. Li, B., Li, X., Zhang, Z., Wu, F.: Spatio-temporal graph routing for skeleton-based action recognition. In: Proceedings of the AAAI Conference on Artificial Intelligence, vol. 33, pp. 8561–8568 (2019)
14. Li, S., Cui, L., Zhu, C., Li, B., Zhao, N., Zhu, T.: Emotion recognition using kinect motion capture data of human gaits. PeerJ **4**, e2364 (2016)
15. Liu, W., Zheng, W.-L., Lu, B.-L.: Emotion recognition using multimodal deep learning. In: Hirose, A., Ozawa, S., Doya, K., Ikeda, K., Lee, M., Liu, D. (eds.) ICONIP 2016. LNCS, vol. 9948, pp. 521–529. Springer, Cham (2016). https://doi.org/10.1007/978-3-319-46672-9_58
16. Liu, Z., Zhang, H., Chen, Z., Wang, Z., Ouyang, W.: Disentangling and unifying graph convolutions for skeleton-based action recognition. In: Proceedings of the IEEE/CVF Conference on Computer Vision and Pattern Recognition, pp. 143–152 (2020)
17. Lu, H., Xu, S., Zhao, S., Hu, X., Ma, R., Hu, B.: EPIC: emotion perception by spatio-temporal interaction context of gait. IEEE J. Biomed. Health Inf. (2023)
18. Ma, R., Hu, H., Xing, S., Li, Z.: Efficient and fast real-world noisy image denoising by combining pyramid neural network and two-pathway unscented Kalman filter. IEEE Trans. Image Process. **29**, 3927–3940 (2020)
19. Ma, R., Li, S., Zhang, B., Fang, L., Li, Z.: Flexible and generalized real photograph denoising exploiting dual meta attention. IEEE Trans. Cybern. (2022)
20. Ma, R., Li, S., Zhang, B., Hu, H.: Meta PID attention network for flexible and efficient real-world noisy image denoising. IEEE Trans. Image Process. **31**, 2053–2066 (2022)
21. Ma, R., Li, S., Zhang, B., Li, Z.: Towards fast and robust real image denoising with attentive neural network and PID controller. IEEE Trans. Multimedia **24**, 2366–2377 (2021)

22. Ma, R., Li, S., Zhang, B., Li, Z.: Generative adaptive convolutions for real-world noisy image denoising. In: Proceedings of the AAAI Conference on Artificial Intelligence, vol. 36, pp. 1935–1943 (2022)
23. Ma, R., Zhang, B., Zhou, Y., Li, Z., Lei, F.: PID controller-guided attention neural network learning for fast and effective real photographs denoising. IEEE Trans. Neural Netw. Learn. Syst. **33**(7), 3010–3023 (2021)
24. Muhammad, G., Hossain, M.S.: Emotion recognition for cognitive edge computing using deep learning. IEEE Internet Things J. **8**(23), 16894–16901 (2021)
25. Narayanan, V., Manoghar, B.M., Dorbala, V.S., Manocha, D., Bera, A.: ProxEmo: gait-based emotion learning and multi-view proxemic fusion for socially-aware robot navigation. In: 2020 IEEE/RSJ International Conference on Intelligent Robots and Systems (IROS), pp. 8200–8207. IEEE (2020)
26. Qin, Z., et al.: Fusing higher-order features in graph neural networks for skeleton-based action recognition (2021)
27. Randhavane, T., Bhattacharya, U., Kapsaskis, K., Gray, K., Bera, A., Manocha, D.: Identifying emotions from walking using affective and deep features. arXiv preprint arXiv:1906.11884 (2019)
28. Sheng, W., Li, X.: Multi-task learning for gait-based identity recognition and emotion recognition using attention enhanced temporal graph convolutional network. Pattern Recogn. **114**, 107868 (2021)
29. Shi, L., Zhang, Y., Cheng, J., Lu, H.: Skeleton-based action recognition with directed graph neural networks. In: Proceedings of the IEEE/CVF Conference on Computer Vision and Pattern Recognition, pp. 7912–7921 (2019)
30. Shi, L., Zhang, Y., Cheng, J., Lu, H.: Two-stream adaptive graph convolutional networks for skeleton-based action recognition. In: Proceedings of the IEEE/CVF Conference on Computer Vision and Pattern Recognition, pp. 12026–12035 (2019)
31. Song, Y.F., Zhang, Z., Shan, C., Wang, L.: Constructing stronger and faster baselines for skeleton-based action recognition. arXiv preprint arXiv:2106.15125 (2021)
32. Vu, M.T., Beurton-Aimar, M., Marchand, S.: Multitask multi-database emotion recognition. In: Proceedings of the IEEE/CVF International Conference on Computer Vision, pp. 3637–3644 (2021)
33. Yan, S., Xiong, Y., Lin, D.: Spatial temporal graph convolutional networks for skeleton-based action recognition. In: Proceedings of the AAAI Conference on Artificial Intelligence, vol. 32 (2018)
34. Zhang, J., Yin, Z., Chen, P., Nichele, S.: Emotion recognition using multi-modal data and machine learning techniques: a tutorial and review. Inf. Fusion **59**, 103–126 (2020)
35. Zhang, X., Xu, C., Tao, D.: Context aware graph convolution for skeleton-based action recognition. In: Proceedings of the IEEE/CVF Conference on Computer Vision and Pattern Recognition, pp. 14333–14342 (2020)
36. Zhuang, Y., Lin, L., Tong, R., Liu, J., Iwamot, Y., Chen, Y.W.: G-GCSN: global graph convolution shrinkage network for emotion perception from gait. In: Proceedings of the Asian Conference on Computer Vision (2020)

Cross-Area Finger Vein Recognition via Hierarchical Sparse Representation

Xiufeng Shi, Lu Yang(✉), Jie Guo, and Yuling Ma

School of Computer Science and Technology, Shandong Jianzhu University, Jinan 250101, Shandong, China
yangluhi@163.com

Abstract. The prior studies in finger vein recognition have mainly focused on personal identification based on images with same area. However, the upgrade of finger vein acquisition devices is inevitable, and therefore the scale variation of acquisition windows among various devices may cause the cross-area finger vein recognition problem. To address this problem, a hierarchical sparse representation-based cross-area finger vein recognition method is proposed in this paper. In the proposed method, the first layer locates the potential corresponding regions in each full training image for the small-area testing image based on coding coefficients on the image-specific dictionary, and the small-area testing image is classified in the second layer by its reconstruction error on the compact dictionary. In addition, the method is performed on the original and down-sampled images, and the weighted sum of the reconstruction errors on two kinds of images are used in recognition. The experiments are performed on two widely used finger vein databases, and the experimental results show that the proposed method achieves 91.35% and 78.94% recognition rates on cross-area finger vein recognition.

Keywords: Biometrics · Finger vein recognition · Cross-area recognition · Hierarchical sparse representation

1 Introduction

Finger vein recognition performs personal identification based on the internal vessels of human finger [1]. It has attracted lots of research interests from researchers owing to its living body identification and high security. Many kinds of methods, *e.g.*, compact multi-representation feature descriptor [2], locality constrained consistent dictionary learning method [3], and explicit and implicit feature fusion network [4], have been developed for finger vein recognition, and promising recognition performance has been reported.

In recent years, finger vein recognition has been used in various applications like intelligent door locks and attendance machines. In practical applications, the upgrade of finger vein imaging devices is inevitable, and the scale of acquisition windows among various devices may vary largely. For one finger, images with

Q. Liu et al. (Eds.): PRCV 2023, LNCS 14429, pp. 86–96, 2024.
https://doi.org/10.1007/978-981-99-8469-5_7

varied areas are captured by different devices. For example, one full finger vein image was enrolled by an imaging device, and one partial finger vein image may be captured by another updated imaging device. The area of enrolled finger vein image may be different with the area of testing finger vein image, causing the cross-area finger vein recognition problem.

The prior studies mainly focused on finger vein recognition based on images with same area, and the recognition methods may not be well performed on cross-area finger vein images. For example, a convolutional neural network, trained on images who have 50×50 pixels, may not deal well with images with 60×60 pixels. Another example is about the overlapped point computation between the enrolled and testing vein patterns, in which an incorrect matching score between unaligned full and partial vein patterns may be generated. So, a prompt solution to deal with cross-area finger vein recognition is needed.

This paper proposes a cross-area finger vein recognition method, in which a hierarchical sparse representation method is used for matching of the full training finger vein image and partial testing finger vein image. In first-layer sparse representation, image-specific dictionaries are built by partitioning each training image into multiple blocks, and a testing image is sparsely presented by each image-specific dictionary in sequence. The atoms in each image-specific dictionary with top T coding coefficients constitute a compact dictionary. This compact dictionary is used for sparsely representing the testing image in the second-layer sparse representation. The hierarchical sparse representation method is performed on the original images and their down-sampled images, the weighted sum of the reconstruction errors on two kinds of images are used in recognition. The experiments are performed on finger vein databases from the Hong Kong Polytechnic University and the Shandong University. The experimental results show that the proposed method achieves 91.35% and 78.94% recognition rates, which exhibits a good performance on cross-area finger vein recognition.

The rest of this paper is organized as follows. Firstly, we review the state-of-the-art finger vein recognition literatures in Sect. 2. Secondly, we present the details of the hierarchical sparse representation based cross-area finger vein recognition method in Sect. 3. Thirdly, Sect. 4 reports the experimental results. Finally, we conclude our work in Sect. 5.

2 Related Work

In this section, sparse representation based finger vein recognition methods are firstly reviewed, and then non-sparse representation based recognition methods are presented.

Sparse Representation Based Recognition Method. Shazeeda *et al.* introduced a nearest centroid neighbor based sparse representation method for finger vein recognition [5]. In this method, the k nearest training images were selected for a testing image by the nearest centroid neighbor classification, and the testing image was classified by sparse representation on the selected k nearest training images. The experimental results showed that the nearest centroid neighbor based sparse representation method achieved better finger vein recognition performance over the traditional sparse representation method. Shazeeda *et al.* also introduced a mutual sparse representation method for finger vein recognition [6]. This method classified a testing image based on the coding coefficients of this testing image and its nearest training images. An adaptive sparse representation method with distance-based dictionary learning was proposed for finger vein image denoising [7]. The dictionary atoms were classified into the high-information group and the low-information group, and different weights were assigned to atoms. Mei *et al.* presented a weighted sparse representation method for finger vein recognition [8]. The Euclidean distance between a training image and a testing image was used as the weight for the coding coefficient on this training image in classification. Recently, Zhao *et al.* [9] proposed a progressive sparse representation method for single-sample finger vein recognition, in which a progressive strategy was used for representation refinement of sparse representation classification.

Non-sparse Representation Based Recognition Method. There are some typical local descriptor based methods and vein pattern based methods in finger vein recognition. Typical local descriptor based methods include local binary pattern [10], Weber local descriptors [11], cross-section asymmetrical coding [12], and physiological characteristic based local descriptor [13]. For further improving feature discrimination, some methods were proposed to learn binary code from local descriptor, for example, anchor-based manifold binary pattern [14], joint discriminative feature learning [15]. Typical vein pattern based methods are the weighted spatial curve filter [16] and the anatomy structure analysis-based vein extraction [17]. Deep neural networks, like convolutional neural network [18] and generative adversarial network [19], were also used in vein pattern segmentation. Moreover, in finger vein recognition, some researchers employed deep neural networks for deep feature extraction and classification [20–23].

Above finger vein recognition methods achieved promising recognition performance on full finger vein images. However, there is still a lack of effective method for cross-area finger vein recognition.

3 Hierarchical Sparse Representation-Based Cross-Area Finger Vein Recognition

Here, we present our hierarchical sparse representation-based cross-area finger vein recognition method. There are two-layer sparse representations, *i.e.*, image-specific dictionary based sparse representation and compact dictionary based sparse representation. In first layer, each full training image is partitioned into multiple blocks to establish image-specific dictionary and represent the small-area testing image, and the blocks with top T largest coding coefficients are seen as the candidate matching regions for the testing image. The candidate matching regions from all training images are then used as the compact dictionary for sparse representation in the second layer. The reconstruction error of the testing

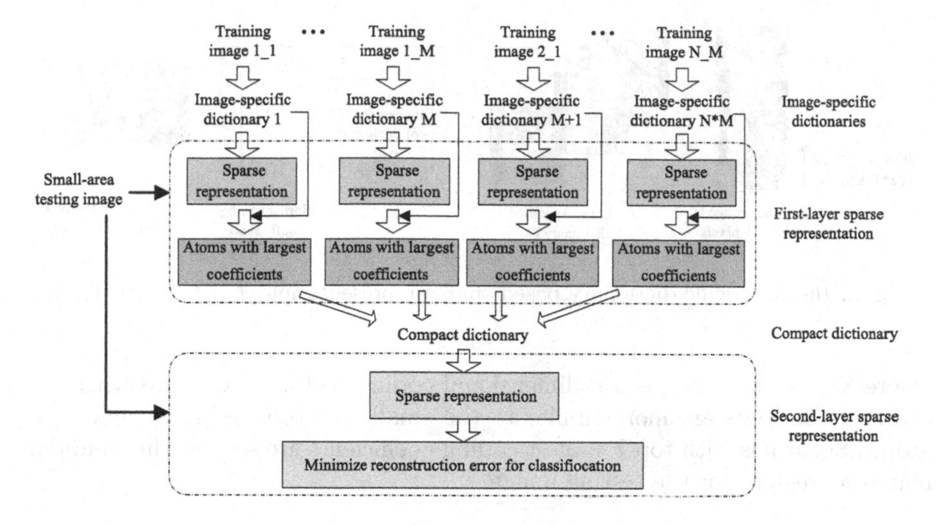

Fig. 1. The framework of the proposed method.

image on the compact dictionary will be used for classification. The framework of the proposed method is illustrated in Fig. 1.

In the following, we will give three keyparts of our method, *i.e.*, image-specific dictionary based sparse representation, compact dictionary based sparse representation and two-scale hierarchical sparse representation.

3.1 Image-Specific Dictionary Based Sparse Representation

For correctly measuring the similarity between the full training image and the small-area testing image, locating the candidate matching regions in the training image is crucial for the small-area testing image. Hence, we propose an image-specific dictionary based sparse representation to deal with this problem.

In detail, each training image is partitioned into multiple blocks with a fixed step, and each block has a same size with the small-area testing image. The step size is 3 pixles in our experiments. The blocks from a training image are used to establish an image-specific dictionary and represent the small-area testing image. The blocks with top T largest coding coefficients are seen as the candidate matching regions for the testing image. This process is given in Fig. 2.

Assuming one training image and one testing image are denoted by G and P, and the image-specific dictionary of this training image is denoted by $D_G = [g_1, g_2, ..., g_n]$, in which $g_i, i = 1, 2, ..., n$ is the blocks from the image G. The testing image P can be linearly represented by D_G:

$$P = s_1 g_1 + s_2 g_2 + ... + s_n g_n \tag{1}$$

The sparse coding coefficients of the above representation can be calculated by solving the following $L1 - norm$ minimization problem as:

$$\min_{S} ||P - D_G S||_F^2 + \alpha ||S||_1 \tag{2}$$

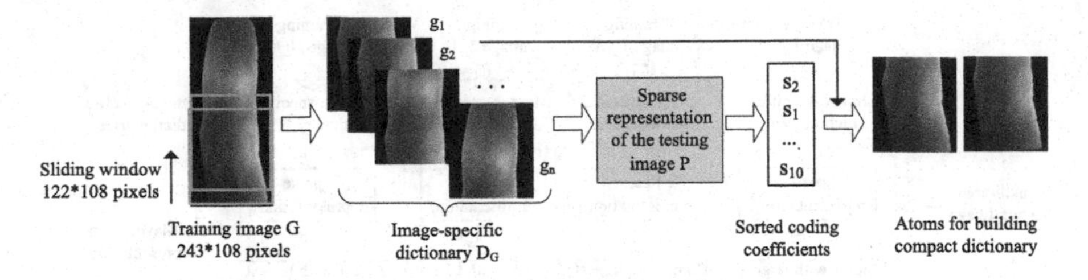

Fig. 2. Image-specific dictionary based sparse representation. $T = 2$ in this figure.

where $S = [s_1, s_2, ..., s_n]$ is a n-dimensional coding vector. The atoms with large coding coefficients are more similar to the small-area testing image than other atoms. So, atoms with top T largest coding coefficients are seen as the candidate matching regions for the testing image.

3.2 Compact Dictionary Based Sparse Representation

By the image-specific dictionary based sparse representation, we can locate the candidate matching regions from each full training images for the small-area testing image, and the candidate matching regions from all training images are then used as the compact dictionary for sparse representation in the second layer.

In detail, we use the atoms with top T coding coefficients from each image-specific dictionary to build the compact dictionary. $T = 7$ is used in our experiments. Assuming that there are N fingers, and each finger has M finger vein images. We can denote the compact dictionary by $D = [D_{1,1}, D_{1,2}, ..., D_{i,j}, ..., D_{N,M}]$, in which $D_{i,j} = [d_{i,j,1}, d_{i,j,2}, ..., d_{i,j,t}, ..., d_{i,j,T}]$. The coding vector of the testing image P can be obtained by:

$$\min_{W} ||P - DW||_F^2 + \lambda ||W||_1 \tag{3}$$

And then, we can compute the reconstruction error of the testing image on each training image:

$$r_{i,j} = ||P - \sum_{t=1}^{T} d_{i,j,t} W_{i,j,t}||_2^2 \tag{4}$$

The class label of the testing image is predicted by:

$$i = \arg\max(r_{i,j}) \tag{5}$$

3.3 Two-Scale Hierarchical Sparse Representation

To explore more discriminative information from small-area images, we further propose two-scale hierarchical sparse representation. In detail, a Gaussian filter, whose size is 5×5, is defined, and the convolution between the filter and each finger vein image is performed for denoising. The convolution result are further

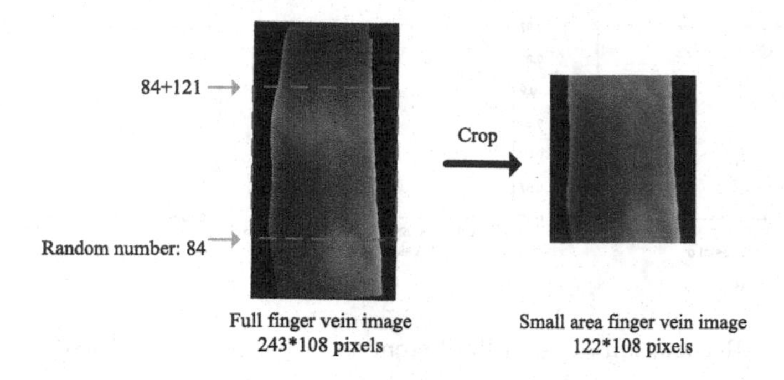

Fig. 3. Cropping for obtaining small-area finger vein image.

down-sampled by removing the even rows and even columns to obtain the small scale image. The full and partial finger vein images are down-sampled to 122×54 pixels and 61×54 pixels respectively.

The hierarchical sparse representation is used firstly on the original images and then on their down-sampled images, and two reconstruction error vectors of a testing image can be computed. We denote the reconstruction error vectors on the original images and the down-sampled images by r_{ij} and r_{ij}^{d}. Two reconstruction error vectors are fused by:

$$r_{ij}^{f} = c_1 r_{ij} + c_2 r_{ij}^{d} \tag{6}$$

in which c_1 and c_2 are weights, and $c_1 + c_2 = 1$. The class label of the testing image is predicted by:

$$i = \arg\max(r_{i,j}^{f}) \tag{7}$$

4 Experiment

4.1 Experimental Setting

Two finger vein databases from the Hong Kong Polytechnic University (HKPU) [24] and the Shandong University (SDU) [25] are used in experiments. The 1,872 finger vein images from 312 fingers captured in the first session on HKPU database and 3,816 finger vein images from 636 fingers on SDU database are used. The images from the first and second databases are normalized into 243×108 pixels and 160×120 pixels respectively. On each database, first three images from each finger are used as full training finger vein images, and each of last three images is cropped based on a random reference row to be used as small-area testing finger vein image. An example is given in Fig. 3. All images are compacted by principal component analysis for dimensionality reduction.

All involved methods are tested in MATLAB R2016a on a PC with an Intel(R) Core(TM) i5-10500 CPU at 3.10 GHz and 16G RAM. The recognition rates and cumulative match curves (CMCs) of the recognition methods are reported.

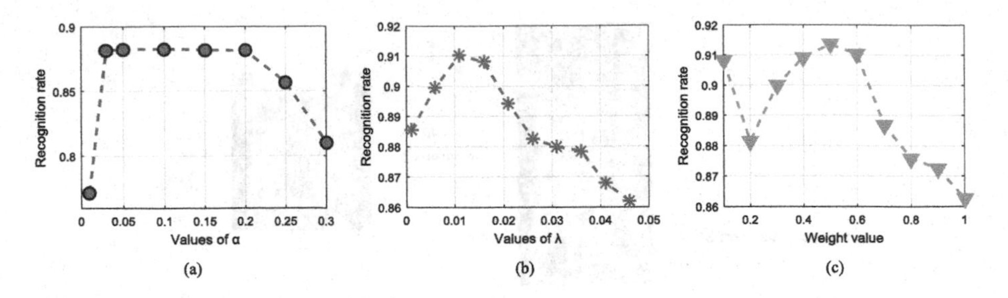

Fig. 4. Recognition rates with different values of (a) α, (b) λ and (c) c_1.

4.2 Parameter Analysis

This experiment is performed on HKPU database for best parameter values. There are two parameters in our hierarchical sparse representation, *i.e.*, α and λ. We firstly fix $\lambda = 0.1$ and vary α. The recognition rates are given in Fig. 4(a). We can see that the best recognition rate is achieved when $\alpha = 0.05$, and there is no big variation on recognition rate when α varies from 0.05 to 0.2. And then, we fix $\alpha = 0.05$ and vary λ. The curve of the recognition rates is illustrated in Fig. 4(b). The figure shows that λ has a larger impact on the recognition rate than α. When $\lambda = 0.01$, our hierarchical sparse representation achieves the best recognition rate. There is also an important parameter in our two-scale hierarchical sparse representation, *i.e.*, weight in fusion of reconstruction errors. We adjust the weights of different scales when α and λ are fixed. We give the recognition rate curve with different values of c_1 in Fig. 4(c). The figure shows that $c_1 = 0.5$ is best for the recognition rate.

4.3 Effectiveness of Two-Scale Hierarchical Sparse Representation

This experiment aims to evaluate our two-scale hierarchical sparse representation method on HKPU database and SDU database. To achieve this, a comparison of hierarchical sparse representation on original images, hierarchical sparse representation on down-sampled images and two-scale hierarchical sparse representation is tested. The recognition rates are given in Table 1. The results in the table show that two-scale hierarchical sparse representation performs better than hierarchical sparse representation on single scale images. It proves that two-scale hierarchical sparse representation can enhance the recognition performance.

4.4 Recognition Performance Testing

The recognition performance of our proposed method is tested on HKPU database and SDU database. Firstly, we illustrate the candidate matching regions located by the image-specific dictionary based sparse representation. One full training finger vein image, one small-area testing finger vein image and the located candidate matching regions are listed in Fig. 5. From the figure we can

Table 1. Comparison of recognition rates (%) from the original images and the down-sampled images.

Method	HKPU database	SDU database
Hierarchical sparse representation on original images	85.36	76.52
Hierarchical sparse representation on down-sampled images	90.81	77.88
Two-scale hierarchical sparse representation	91.35	78.94

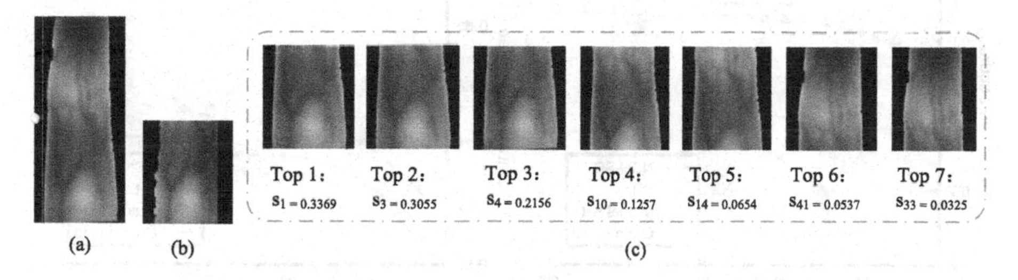

Fig. 5. Candidate matching regions located by the image-specific dictionary based sparse representation. (a) Training image, (b) Testing image, and (c) Candidate matching regions with top T largest coding coefficients.

see that, the image regions with larger coding coefficients are more similar to the testing image than the other regions. For example, the image region with top 1 coding coefficient is more similar to the testing image than the image regions with top 4 to 7 coding coefficients. The figure indicates that our image-specific dictionary based sparse representation can search candidate matching regions from full training finger vein image for the partial testing image.

Secondly, we report the recognition rates of our method and three sparse representation based finger vein recognition methods. The state-of-the-art literatures paid attention to finger vein recognition based on images with same area, and most methods are inappropriate for cross-area finger vein recognition. For the compared sparse representation methods, the training image is also partitioned into multiple blocks to represent the small-area testing image. Table 2 lists the recognition rates of all involved methods on two databases, and the corresponding CMC curves are presented in Fig. 6.

The experimental results from two databases consistently show that, our two-scale hierarchical sparse representation achieves a superior recognition performance over the compared methods. The superior performance of our method may be attributed to the alignment of the small area testing image to the full training image by the image-specific dictionary based sparse representation and the discriminative information explored by two-scale hierarchical sparse representation method.

Table 2. Comparison of recognition rates (%) from sparse representation methods.

Method	HKPU database	SDU database
Sparse representation (SR)	86.20	66.70
Nearest centroid neighbor based sparse representation (NCN-SR) [5]	81.20	69.60
Mutual sparse representation (Mutual SR) [6]	61.02	65.20
Our method	91.35	78.94

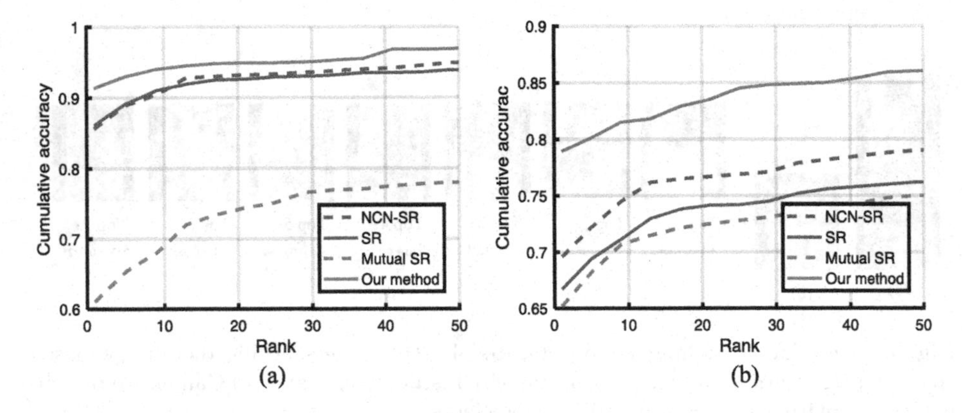

Fig. 6. CMC curves of sparse representation methods on (a) HKPU database and (b) SDU database.

5 Conclusion

This paper pays attention to cross-area finger vein recognition problem caused by the upgrade of finger vein acquisition devices, and proposes a hierarchical sparse representation method for matching of full training finger vein image and small-area testing finger vein image. By the image-specific dictionary based first layer sparse representation, the candidate matching regions can be located from full training images for small-area testing image. The located regions are then used to build the compact dictionary for the second layer sparse representation classification of small-area testing image. In addition, the cross-area finger vein recognition performance is further enhanced by two-scale hierarchical sparse representation, which explores the discriminative information from the original images and their down-sampled images. In experiments, last three images per finger on two public finger vein databases are cropped as small-area testing images. The experimental results on two databases show that the superior performance of the proposed method over the compared methods on cross-area finger vein recognition.

Acknowledgement. This work was supported in part by the National Natural Science Foundation of China under Grants 62076151 and 62177031, in part by the Taishan Scholar Project of Shandong Province under Grant tsqn202211182 and Youth Innovation Team of Shandong Province Higher Education Institutions under Grant 2022KJ205, and in part by the Shandong Provincial Natural Science Foundation under Grants ZR2021QF119 and ZR2021MF044.

References

1. Qin, H., Hu, R., El-Yacoubi, M.A., Li, Y., Gao, X.: Local attention transformer-based full-view finger-vein identification. IEEE Trans. Circuits Syst. Video Technol. **33**(6), 2767–2782 (2022)
2. Li, S., Ma, R., Fei, L., Zhang, B.: Learning compact multirepresentation feature descriptor for finger-vein recognition. IEEE Trans. Inf. Forensics Secur. **17**, 1946–1958 (2022)
3. Yang, L., Liu, X., Yang, G., Wang, J., Yin, Y.: Small-area finger vein recognition. IEEE Trans. Inf. Forensics Secur. **18**, 1914–1925 (2023)
4. Song, Y., Zhao, P., Yang, W., Liao, Q., Zhou, J.: EIFNet: an explicit and implicit feature fusion network for finger vein verification. IEEE Trans. Circuits Syst. Video Technol. **33**(5), 2520–2532 (2023)
5. Shazeeda, S., Rosdi, B.A.: Nearest centroid neighbor based sparse representation classification for finger vein recognition. IEEE Access **7**, 5874–5885 (2019)
6. Shazeeda, S., Rosdi, B.A.: Finger vein recognition using mutual sparse representation classification. IET Biometrics **8**(1), 49–58 (2019)
7. Lei, L., Xi, F., Chen, S., Liu, Z.: A sparse representation denoising algorithm for finger-vein image based on dictionary learning. Multimedia Tools Appl. **80**, 15135–15159 (2021)
8. Mei, X., Ma, H.: Finger vein recognition algorithm based on improved weighted sparse representation. In: International Conference on Information Technology and Computer Application, pp. 6–8. IEEE (2019)
9. Zhao, P., et al.: Single-sample finger vein recognition via competitive and progressive sparse representation. IEEE Trans. Biometrics Behav. Identity Sci. **5**(2), 209–220 (2022)
10. Lee, E.C., Jung, H., Kim, D.: New finger biometric method using near infrared imaging. Sensors **11**(3), 2319–2333 (2011)
11. Wang, H., Du, M., Zhou, J., Tao, L.: Weber local descriptors with variable curvature Gabor filter for finger vein recognition. IEEE Access **7**, 108261–108277 (2019)
12. Yang, W., Chen, Z., Qin, C., Liao, Q.: α-trimmed Weber representation and cross section asymmetrical coding for human identification using finger images. IEEE Trans. Inf. Forensics Secur. **14**(1), 90–101 (2019)
13. Zhang, L., Li, W., Ning, X., Sun, L., Dong, X.: A local descriptor with physiological characteristic for finger vein recognition. In: International Conference on Pattern Recognition, pp. 4873–4878. IEEE (2021)
14. Liu, H., Yang, G., Yang, L., Su, K., Yin, Y.: Anchor-based manifold binary pattern for finger vein recognition. Sci. China Inf. Sci. **62**, 1–16 (2019)
15. Li, S., Zhang, B., Fei, L., Zhao, S.: Joint discriminative feature learning for multimodal finger recognition. Pattern Recogn. **111**, 1–11 (2021)
16. Yang, J., Shi, Y., Jia, G.: Finger-vein image matching based on adaptive curve transformation. Pattern Recogn. **66**, 34–43 (2017)
17. Yang, L., Yang, G., Yin, Y., Xi, X.: Finger vein recognition with anatomy structure analysis. IEEE Trans. Circuits Syst. Video Technol. **28**(8), 1892–1905 (2018)

18. Qin, H., El-Yacoubi, M.A.: Deep representation-based feature extraction and recovering for finger-vein verification. IEEE Trans. Inf. Forensics Secur. **12**(8), 1816–1829 (2017)
19. Yang, W., Hui, C., Chen, Z., Xue, J.-H., Liao, Q.: FV-GAN: finger vein representation using generative adversarial networks. IEEE Trans. Inf. Forensics Secur. **14**(9), 2512–2524 (2019)
20. Das, R., Piciucco, E., Maiorana, E., Campisi, P.: Convolutional neural network for finger-vein-based biometric identification. IEEE Trans. Inf. Forensics Secur. **14**(2), 360–373 (2018)
21. Xie, C., Kumar, A.: Finger vein identification using convolutional neural network and supervised discrete hashing. Pattern Recogn. Lett. **119**, 148–156 (2019)
22. Shaheed, K., et al.: DS-CNN: a pre-trained Xception model based on depth-wise separable convolutional neural network for finger vein recognition. Expert Syst. Appl. **191**, 1–18 (2022)
23. Huang, J., Zheng, A., Shakeel, M.S., Yang, W., Kang, W.: FVFSNet: frequency-spatial coupling network for finger vein authentication. IEEE Trans. Inf. Forensics Secur. **18**, 1322–1334 (2023)
24. Kumar, A., Zhou, Y.: Human identification using finger images. IEEE Trans. Image Process. **21**(4), 2228–2244 (2012)
25. Yin, Y., Liu, L., Sun, X.: SDUMLA-HMT: a multimodal biometric database. In: Sun, Z., Lai, J., Chen, X., Tan, T. (eds.) CCBR 2011. LNCS, vol. 7098, pp. 260–268. Springer, Heidelberg (2011). https://doi.org/10.1007/978-3-642-25449-9_33

Non-local Temporal Modeling for Practical Skeleton-Based Gait Recognition

Pengyu Peng, Zongyong Deng, Feiyu Zhu, and Qijun Zhao(✉)

School of Computer Science and Technology, Sichuan University, Chengdu, China
qjzhao@scu.edu.cn

Abstract. Gait, a unique biometric identifier for recognizing individual identity at a distance, plays an important role in practical applications. Existing gait recognition methods utilize either a gait set or a sequence. However, these methods ignore the periodic characteristic of gait, where actions at one moment are related to actions at another moment. As a result, their recognition accuracy in real scenes can significantly decrease due to noise and frame loss. To deal with this issue, we design a NLGait network to explore the temporal relation among gait frames, which adaptively leverages both local and non-local relations to achieve practical gait recognition. Specifically, we design multi-scale temporal information extractor (MTIE) to capture these relations. Furthermore, we design an attention based adaptive frame fuser (AFF) to aggregate the features of frames in a gait sequence. Extensive experiments have verified the competitive accuracy and robustness of our method. The accuracy of the counterpart methods is degraded by 8.9% and 19.3%, respectively, due to noise and temporal loss, while ours is degraded by only 3.6% and 2.7%.

Keywords: Gait Recogniton · Non-local Temporal Relations · Key Frames

1 Introduction

Gait is a sort of dynamic biometric traits which represents the walking patterns of people. Unlike other static biometric traits such as face, iris and fingerprint, gait is a unique biometric feature that can be captured in long-distance conditions without the cooperation of subjects. Moreover, it's difficult to imitate a person's walking style and body shape. Therefore, gait recognition plays an important role in various applications, *e.g.*, crime investigation, social security and intelligent transportation. However, in practical applications, the performance of gait recognition is vulnerable to the loss of temporal information and noise caused by uncontrollable factors such as occlusion, carrying, view variation and speed changing. So, it's of great significance for gait recognition methods to obtain robust and distinctive spatial-temporal representations.

Q. Liu et al. (Eds.): PRCV 2023, LNCS 14429, pp. 97–109, 2024.
https://doi.org/10.1007/978-981-99-8469-5_8

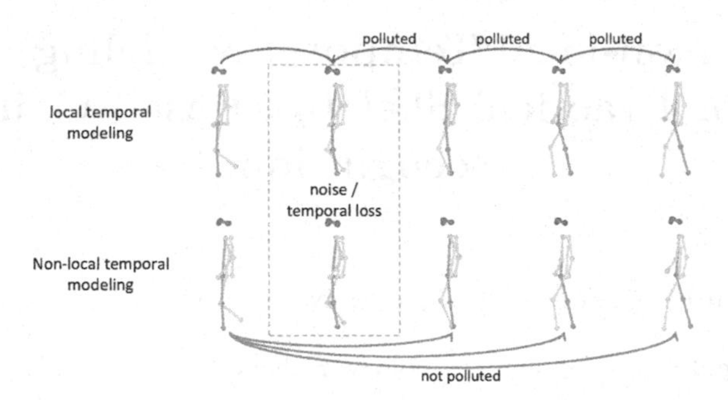

Fig. 1. Compared with temporal modeling based on CNN and LSTM, non-local temporal modeling is more robust for practical applications since it can skip local noise and temporal loss while extracting non-local temporal information.

With the proposal of robust pose estimation methods, a lot of skeleton-based methods [10,11,20,21,26] have been developed to alleviate those issues. PTSN [10] uses a combination of LSTM and CNN models to extract temporal and spatial features, respectively. PoseGait [11] employs handcrafted features, including angles, limbs, and motion, in combination with pose to extract features using a CNN model. Teepe *et al.* [20,21] and AGCN [26] utilize graph convolution blocks to obtain spatial temporal features.

As a periodic motion, there are certain connections between frames in a gait cycle. Therefore, non-local and local temporal relation should be both captured. Existing gait recognition methods obtained the temporal information via either convolutional [6,9] or recurrent [13,14] operations both of which are defined in a local neighborhood; thus, the long-range dependencies in gait can be captured by repeatedly applying these operations. These methods are easily affected by local noise and temporal loss since errors will accumulate during forward propagation. Fortunately, non-local [25] operations could be another way to capture long-range dependencies directly. The comparison is illustrated in Fig. 1.

Motivated by above observations, we propose a novel network called Non-Local Gait (NLGait) to explore the temporal relation among gait frames. The input of NLGait is a sequence of gait skeletons where nodes represent human parts. Specifically, we build a new component called multi-scale temporal information extractor (MTIE) to attain local and non-local relations simultaneously. MTIE is composed of nodes interaction module (NIM) and temporal self-attention [22] module. In MTIE, temporal relations are extracted within each node. NIM is designed to achieve the interaction among all nodes. Temporal pooling in gait recognition often applies max pooling or average pooling which ignore the importance of different frames for the final gait representation. To address this issue, we design an attention based adaptive frame fuser (AFF) to aggregate the features of frames in gait sequences. The major contributions of this paper can be summarized as follows:

- We propose a new component which explores local and non-local temporal relations simultaneously, in comparison to set, convolutional, and recurrent networks.
- We design an attention-based temporal pooling that emphasizes the importance of key frames, as opposed to traditional temporal pooling.
- We apply our method to the widely used CASIA-B dataset [28] and the effectiveness of our method is verified. Visualizations and extensive studies are conducted to further prove the validity of our idea and robustness of proposed method.

2 Related Work

Most current methods have taken spatial feature extraction and temporal modeling as the focus [1–4,12,23,27]. For the spatial feature extraction, there are two categories: appearance-based and skeleton-based approaches. Appearance-based approaches take binary silhouettes as input extracted from the original RGB gait video [24]. While most early works [4,17,27] implement the spatial information extraction on entire feature map, recent methods focus on specific body parts. GaitPart [5] developed frame-level part feature extractor (FPFE) to obtain several fine-grained part-level spatial features. Lin *et al.* [12] applied 3D-CNN to achieve the extraction of global and local features. Huang *et al.* [7] proposed a novel 3D local operations to obtain accurate body parts with adaptive spatial and temporal scales, locations and lengths. Skeleton-based approaches apply human skeleton extracted from the raw input images via pose estimation network. Liao *et al.* [10] extracted spatial feature from joints data via CNN. PoseGait [11] took pose, angle, limb and motion as a combination input to obtain spatial feature. Teepe *et al.* [20,21] applied ResGCN [18] to simulate the spatial structure among joints of human body.

For temporal modeling, the approaches of modeling temporal changes of gait can be generally divided into: template-based [15], CNN-based [5,7,12], LSTM-based [10,29] and set-based [4]. Template based methods are unable to obtain accurate temporal information since they compress the silhouettes of a sequence into one image. Set-based methods believe that the appearance of a silhouette frame contains its position information. Accordingly, they are easily affected by the noise in gait silhouette. Also, they are likely to ignore the fine-grained short-term temporal information. CNN-based and LSTM-based methods extract temporal information starting from the interaction between adjacent frames. Therefore, long-term temporal information is captured when these operations are applied repeatedly. Moreover, the LSTM-based methods have to retain unnecessary sequential constraints for periodic gait.

3 Method

As shown in Fig. 2, our method consists of three components, *i.e.*, spatial feature extractor, multi-scale temporal information extractor, and adaptive frame fuser.

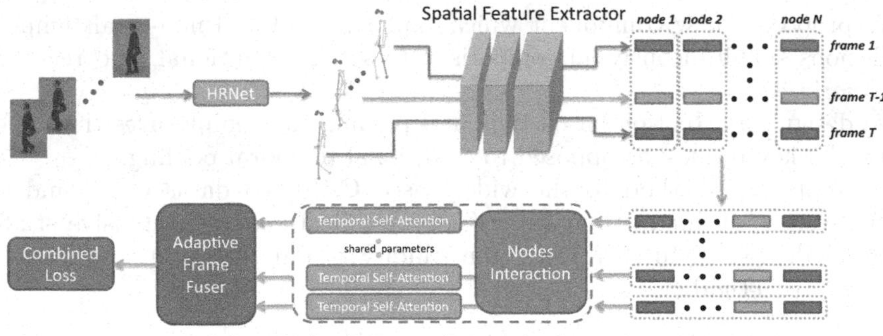

Fig. 2. The pipeline of NLGait. The gait pose sequence is extracted by HRNet [19]. Then, the spatial and temporal features are sequentially extracted by spatial feature extractor and multi-scale temporal information extractor. After that, the final feature aggregated by adaptive frame fuser is obtained.

We use HRNet [19] as a human pose estimator to extract 2D poses from the raw gait video, which is further sent to spatial feature extractor to inference spatial features. After that, the multi-scale temporal information extractor takes spatial features as input to exploit the temporal feature. Finally, adaptive frame fuser (AFF) is used to aggregate the frame-wise features within each node for final representation.

3.1 Preliminaries

Notation. The human skeleton can be illustrated as a graph $G = (V, \xi)$. The set $V = \{v_i | i = 0, 1, ..., N-1\}$ contains N nodes representing human joints in a single image. ξ is the set of edges representing bones constructed by an adjacency matrix $A \in \{0,1\}^{N \times N}$ which denotes the connection between nodes. If there is a connection between nodes v_i and v_j, then $A_{i,j} = 1$, and vice versa. Every node consists of three channels $v_i = (x_i, y_i, c_i)$ where x_i, y_i is the estimated joint coordinate and c_i is the keypoint confidence.

Gait is a sequence of these graphs which is defined as $\mathbf{X} = \{v_{t,n} | t = 0, 1, ..., T-1,\ i = 0, 1, ..., N-1\}$ with temporal dimension T. The input of our network is composed of A structurally and $\mathbf{X}$ feature-wise with $\mathbf{X_t} \in \mathbb{R}^{N \times C}$ being a pose at time t. C is a tuple of 2D coordinate and its confidence and N is the number of joints, which is set at 17 of our work.

Graph Convolutions. Based on the feature $\mathbf{X_t}$ and adjacent matrix A, the frame-wise graph convolution operation can be described as:

$$\mathbf{X}^{l+1} = \sigma\left(\check{\mathrm{D}}^{-\frac{1}{2}} \tilde{\mathrm{A}} \check{\mathrm{D}}^{-\frac{1}{2}} \mathbf{X}^{l} \theta^{l}\right) \tag{1}$$

where $\tilde{A} = A + I$, I is the identity matrix. $\check{D}$ is the diagonal degree matrix of $\tilde{A}$ and θ^l is a learnable weight matrix at layer l. Equation 1 indicates the process of extracting spatial feature by information exchange between adjacent nodes.

3.2 Spatial Feature Extractor

We extract spatial features of gait by performing graph convolution to the nodes and their connection relationships included in the graphs. Specifically, we applied the spatial block of ResGCN [18] to obtain frame-wise spatial feature. The spatial block consists of graph convolution with an optional bottleneck structure. Moreover, we utilize a bottleneck convolution block achieved by two 1×1 convolutional layers to reduce the redundant information. At first, we use basic spatial block to extract gait feature from input 2D pose. With the expansion of channel dimension, we utilize spatial block with bottleneck structure to simplify the model parameters (see Table 1 for detailed configuration).

Table 1. Overview of spatial feature extractor architecture for a pose with 17 nodes and sequence length of 60.

Block	Module	Output Dimensions
Block0	BatchNorm	$60 \times 117 \times 3$
Block1	Spatial-Basic	$60 \times 117 \times 64$
	Spatial-Bottleneck	$60 \times 117 \times 64$
	Spatial-Bottleneck	$60 \times 117 \times 32$
Block2	Spatial-Bottleneck	$60 \times 117 \times 128$
	Spatial-Bottleneck	$60 \times 117 \times 128$
	Spatial-Bottleneck	$60 \times 117 \times 256$
	Spatial-Bottleneck	$60 \times 117 \times 256$

3.3 Multi-scale Temporal Information Extractor

We propose MTIE to capture local and non-local temporal information concurrently. To be more specific, MTIE is composed of multiple temporal self-attention modules and nodes interaction module (NIM) alternately.

Temporal Self-attention Module. Transformer [22] can perform attention operations between two tokens at any distance in a sequence. By treating gait frames as a sequence of tokens, we utilize the self-attention module in transformer to capture temporal relations within each node.

Given an input gait feature $x_t^i \in \mathbb{R}^c$, which represents the i-th node of t-th frame, we first compute a query vector $q_{h,t}^i$, a key vector $k_{h,t}^i$ and a value vector $v_{h,t}^i$ for each head h of the total H heads:

$$\left(q_{h,t}^i, k_{h,t}^i, v_{h,t}^i\right) = x_t^i \left(W_h^q, W_h^k, W_h^v\right) \tag{2}$$

where $W_h^q \in \mathbb{R}^{c \times d_q}$, $W_h^k \in \mathbb{R}^{c \times d_k}$, $W_h^v \in \mathbb{R}^{c \times d_v}$. Then we calculate attention score of each frame by applying scaled dot-product attention function to each query and key. By this way, the correlation between each frame in a gait sequence is captured. To gather local and non-local temporal information, attention score standardized by the softmax function is utilized to weight value vectors. This process can be formulated as:

$$head_h^i = \text{Attention}\left(q_h^i, k_h^i, v_h^i\right) = \sum_s \text{softmax}_s\left(\frac{q_{h,t}k_{h,s}^T}{\sqrt{d_k}}\right) v_{h,s} \tag{3}$$

Eventually, node $x^i \in \mathbb{R}^{d_o}$ is updated by concatenating the value vectors from H heads and multiplying them with a learnable matrix:

$$x^i = Concat\left(\ head_1^i, head_2^i, \ldots, head_H^i\right) W^o \tag{4}$$

where $W^o \in \mathbb{R}^{H * d_v \times d_o}$.

However, since each frame is not aware of its position in the gait sequence, the order of adjacent frames is ambiguous. Thus, local temporal relation is not precisely captured. To tackle this issue, we use positional encoding [22] generated with sinusoidal functions of different frequencies.

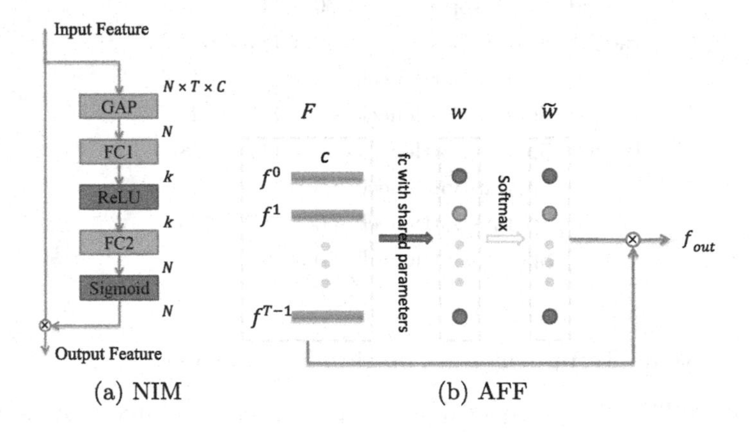

(a) NIM (b) AFF

Fig. 3. The detailed structure of NIM and AFF

Nodes Interaction Module. NIM is proposed to capture the certain correlation between different nodes during walking.

It should be noted that the interactions between nodes can not break the unique motion pattern of each node. Considering of this, interactions are conducted in a node-specific manner but not on frame-level or channel-level. Thus, NIM is designed to learn the correlation between the representations of each node obtained by statistic function.

The detailed structure of NIM is illustrated in Fig. 3a. Average pooling operation is used to compress the feature on temporal and channel dimension. Thus, we get a vector $h \in \mathbb{R}^N$ where every single element represents the motion information of each node respectively. Then, NIM achieved the interaction between each node by two fully connected layers:

$$h^c = \sigma\left(W_2 \delta\left(W_1 h\right)\right) \tag{5}$$

where $W_1 \in \mathbb{R}^{k \times N}$ and $W_2 \in \mathbb{R}^{N \times k}$ are the learnable matrices of the two fully connected layers, k is set to reduce redundant information, $\sigma(\cdot)$ and $\delta(\cdot)$ are Sigmoid activation function and ReLU activation function. By now, NIM has obtained the importance of the role played by each node during walking. Finally, NIM weights the input feature node by node via multiplying h^c with the input feature.

3.4 Adaptive Frame Fuser

Temporal pooling is applied to gather the features of a whole sequence in gait recognition. Existing methods utilize either max pooling or average pooling for the final gait representation. However, this will lead to the situation that the contribution of key frame will be weakened and some poor informative frame will be highlighted. To tackle this issue, AFF is designed to learn the importance of each frame during walking and properly weight them.

As shown in Fig. 3b, given the input feature $F \in \mathbb{R}^{T \times C}$, AFF first applies a fully connected layer as attention pooling on each f^j to get a frame-wise weight w^j. Then the final representation f_{out} is calculated by $\sum_{j=0}^{T-1} \tilde{w}^j f^j$, where $\tilde{w}^j$ is the softmax-normalized frame-wise weight.

4 Experiments

4.1 Dataset and Training Details

Dataset. The CASIA-B [28] dataset encompasses 124 subjects, each demonstrated under 11 distinct viewing angles (ranging from 0° to 180°) and three different walking conditions. These conditions include normal walking (NM, represented by 6 sequences), walking while carrying a bag (BG, represented by 2 sequences), and walking while wearing a coat (CL, represented by 2 sequences). Consequently, each subject is represented by 110 sequences, calculated as 11 multiplied by the total number of sequences from each walking condition (6 for NM, 2 for BG, and 2 for CL).

This paper follows the popular protocol [27] and uses the widely called large-sample training (LT) partition. In this scheme, the first 74 subjects (labelled 001-074) constitute the training set, while the remaining 50 (labelled 075-124) subjects constitute the test set. The test set is further divided into gallery set and probe set. The gallery set comprises the first four sequences of the Normal walking (NM) condition. The remaining six sequences are subdivided into three probe subsets: the NM subset (consisting of sequences NM#5-6), the Bag-carrying (BG) subset (comprising sequences BG#1-2), and the Coat-wearing (CL) subset (including sequences CL#1-2).

Trainging Details. For NLGait, we set the input sequence length $T = 60$ frames. We set the reduction rate of the bottleneck in spatial feature extractor to 8. The number of MTIE is 3. In NIM, we set $k = 5$. Adam optimizer is used with a *1-cycle* learning rate schedule [16] with a maximum learning rate of 1e-4 and a weight decay penalty of 1e-5. We use a combination of two loss functions: cross-entropy loss and supervised contrastive loss [8]. The batchsize is 64.

4.2 Results

Performance Comparison. Table 2 presents a comparative analysis of proposed NLGait with several existing methods, including PoseGait [11], AGCN [26], GaitGraph [21] and GaitGraph2 [20]. Except for NLGait, other results were directly taken from their original papers. All the results were averaged on the 11 gallery views and the identical views were excluded. For instance, the accuracy of probe view 54° was averaged on 10 gallery views, excluding gallery view 54°.

Table 2. Average Rank-1 accuracy on CASIA-B on three test subsets, excluding identical-view cases.

Gallery NM#1-4		0°–180°											mean
Probe		0°	18°	36°	54°	72°	90°	108°	126°	144°	162°	180°	
NM#5-6	PoseGait [11]	55.3	69.6	73.9	75.0	68.0	68.2	71.1	72.9	76.1	70.4	55.4	68.7
	AGCN [26]	72.4	81.2	85.6	80.4	79.4	85.0	81.0	77.6	82.5	79.1	80.2	80.4
	GaitGraph [21]	85.3	88.5	91.0	92.5	87.2	86.5	88.4	89.2	87.9	85.9	81.9	87.7
	GaitGraph2 [20]	78.5	82.9	85.8	85.6	83.1	81.5	84.3	83.2	84.2	81.6	71.8	82.0
	Ours	83.8	85.4	88.4	90.3	89.6	91.6	90.1	91.8	88.5	86.2	80.7	**87.9**
BG#1-2	PoseGait [11]	35.3	47.2	52.4	46.9	45.5	43.9	46.1	48.1	49.4	43.6	31.1	44.5
	AGCN [26]	62.5	68.7	69.4	64.8	62.8	67.2	68.3	65.7	60.7	64.1	60.3	65.0
	GaitGraph [21]	75.8	76.7	75.9	76.1	71.4	73.9	78.0	74.7	75.4	75.4	69.2	74.8
	GaitGraph2 [20]	69.9	75.9	78.1	79.3	71.4	71.7	74.3	76.2	73.2	73.4	61.7	73.2
	Ours	78.7	79.8	81.3	81.2	81.3	76.6	75.3	73.9	76.9	80.4	72.1	**78.0**
CL#1-2	PoseGait [11]	24.3	29.7	41.3	38.8	38.2	38.5	41.6	44.9	42.2	33.4	22.5	36.0
	AGCN [26]	57.8	63.2	68.3	64.1	66.0	64.8	67.7	60.2	66.0	68.3	60.3	64.2
	GaitGraph [21]	69.6	66.1	68.8	67.2	64.5	62.0	69.5	65.6	65.7	66.1	64.3	66.3
	GaitGraph2 [20]	57.1	61.1	68.9	66.0	67.8	65.4	68.1	67.2	63.7	63.6	50.4	63.6
	Ours	70.7	70.0	70.2	68.0	69.9	69.0	73.0	68.7	70.2	63.2	64.6	**68.9**

As shown in Table 2, NLGait brings about significant improvements throughout all walking conditions.

In NM condition, people walk normally without any carrying or occlusion. The moving patterns of the human body and legs are explicitly illustrated. Our network achieves satisfactory performance on the NM subset with an accuracy of 87.9. The improvements on BG and CL suggest that our network can obtain more discriminative and robust representations, and prove the importance of non-local temporal information in gait recognition.

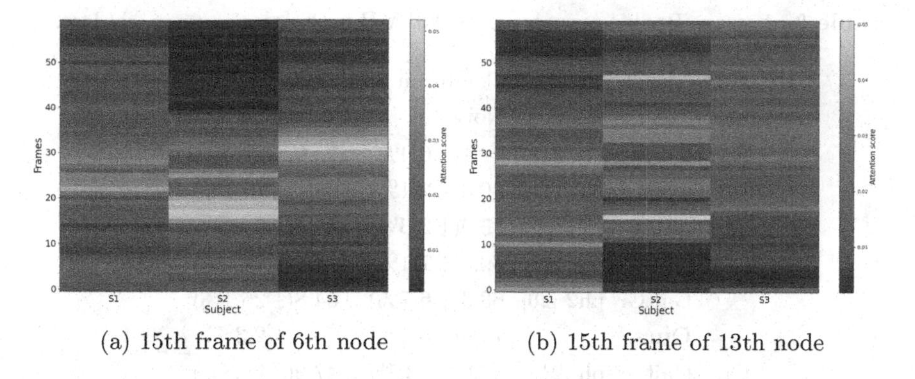

(a) 15th frame of 6th node (b) 15th frame of 13th node

Fig. 4. The visulization of attention scores

Visualization. To verify that NLGait can capture local and non-local temporal relations, we visualized the attention scores of three subjects standardized by softmax function in the first temporal self-attention module. As shown in Fig. 4a, the 15th frame of S2 appears to be strongly associated with local frames, whereas the 15th frame of S3 is highly related to non-local frames. Figures 4a and 4b indicates that different nodes of the same subject behave variously during walking. The diversity in attention scores indicates that, thanks to the multi-scale temporal information extractor, our proposed method is capable of effectively learning the unique gait patterns of each individual.

Robustness Study. In practical applications, due to the influence of some uncontrollable factors, the obtained gait sequences usually suffer from noise or temporal loss. A series of experiments were conducted to testify the robustness of NLGait against these situations. GaitGraph [21] and GaitGraph2 [20] represent a small subset of recent advancements in skeleton-based gait recognition, notable for their open-sourced code. Importantly, they utilize the same pose estimation methodology that we have implemented in our approach. As such, they serve as valuable points of comparison for our robustness experiments. The outcomes of these experiments are presented in Table 3, where the figures enclosed in parentheses denote the rates of accuracy degradation.

Table 3. Average Rank-1 accuracy on CASIA-B with noise or temporal loss.

Probe		Condition	
		Noise	Temporal Loss
NM	GaitGraph [21]	82.1(↓ 6.4%)	74.6(↓ 14.9%)
	GaitGraph2 [20]	80.3(↓ 2.1%)	68.1(↓ 17.0%)
	Ours	85.9(↓ 2.3%)	86.1(↓ 2.0%)
BG	GaitGraph [21]	69.5(↓ 7.1%)	64.2(↓ 14.2%)
	GaitGraph2 [20]	68.3(↓ 6.7%)	54.8(↓ 25.1%)
	Ours	75.6(↓ 3.1%)	76.2(↓ 2.3%)
CL	GaitGraph [21]	57.6(↓ 13.1%)	47.2(↓ 28.8%)
	GaitGraph2 [20]	55.9(↓ 12.1%)	45.8(↓ 28.0%)
	Ours	65.1(↓ 5.5%)	66.2(↓ 3.9%)

Noise. To simulate the disturbances in the expression of some human motion information, we add three kinds of uniform noises to the estimated 2D coordinates of keypoints along with their confidence in test set. Compared with GaitGraph [21] and GaitGraph2 [20], NLGait is more robust because of the captured non-local relation and the reduction of redundant information by bottleneck convolution block and NIM.

Temporal Loss. We also randomly drop the frames of the input sequence in test set to simulate the loss of temporal information. Specifically, given a input sequence with 80 frames, we randomly drop 20 of them. Since MTIE directly obtains non-local temporal information, NLGait demonstrates a considerable level of robustness to the loss of temporal information which is of great significance for the practical application of gait recognition.

4.3 Ablation Study

To validate the contribution of proposed components in NLGait, corresponding ablation experiments were conducted on CASIA-B. Results of the ablation experiments are presented in Table 4.

Table 4. Ablation study results on the CASIA-B dataset.

SFE	MTIE	Concurrency	NIM	Temporal Pooling			Accuracy		
				Mean	Max	AFF	NM	BG	CL
√				√			64.5	52.4	40.0
√						√	71.7	58.1	44.7
√	√		√			√	83.9	74.0	59.6
√	√	√				√	85.6	75.6	64.7
√	√	√	√	√			87.4	76.5	66.0
√	√	√	√		√		82.8	71.0	58.3
√	√	√	√			√	87.9	78.0	68.9

The Impact of MTIE. The results from the second to last rows clearly demonstrate that the frame-wise feature becomes considerably challenging in the absence of both local and non-local temporal information for identifying a person's identity. With MTIE, The performance increased by 16.2% under NM, 19.9% under BG and 24.2% under CL.

The Impact of Concurrency. In NLGait, we simultaneously capture local and non-local temporal information. To highlight the importance of the concurrency, we refer to the manner that convolutional operation expands the receptive field to extract the non-local temporal information. Specifically, we initially perform self attention operation on each local part of the sequence, and as we repeat this operation, the length of the local part gradually extends until the entire sequence is covered. Thus, local and non-local temporal information are not captured concurrently by this way.

The third row and the last row show that the accuracy is decreased under three walking conditions. The result indicates that the non-local temporal information which is not captured directly is vulnerable and redundant.

The Impact of NIM. The fourth row and the last row show that NIM improves all test subsets. This result is consistent with our experience mentioned in Sect. 3.3, *i.e.*, that the certain correlation among different nodes of human body during walking can not be ignored.

The Impact of AFF. To verify the effectiveness of the AFF, we compare the performance of using this module with other two most applied temporal pooling. As shown in the last three rows, though global average pooling behaves better than global max pooling, it is still defective. By weighting each frame in a sequence, AFF improves the performance under all walking conditions.

5 Conclusion

In this paper, we introduce an innovative perspective on gait sequences, highlighting the significance of exploring non-local temporal relation. As a solution, we propose NLGait, which combines the adaptive frame fuser and the multi-scale temporal information extractor, comprising the temporal self-attention module and nodes interaction module. These two parts work together to obtain discriminative and robust gait representations enriched with non-local temporal information. The effectiveness and robustness of NLGait are validated through extensive experimental studies.

Acknowledgements. This research is supported by the National Natural Science Foundation of China (No. 62176170, 61971005) and the Science and Technology Department of Tibet (Grant No. XZ202102YD0018C).

References

1. Ariyanto, G., Nixon, M.S.: Model-based 3d gait biometrics. In: 2011 international joint conference on biometrics (IJCB), pp. 1–7. IEEE (2011)
2. Ariyanto, G., Nixon, M.S.: Marionette mass-spring model for 3d gait biometrics. In: 2012 5th IAPR International Conference on Biometrics (ICB), pp. 354–359. IEEE (2012)
3. Chai, T., Li, A., Zhang, S., Li, Z., Wang, Y.: Lagrange motion analysis and view embeddings for improved gait recognition. In: Proceedings of the IEEE/CVF Conference on Computer Vision and Pattern Recognition, pp. 20249–20258 (2022)
4. Chao, H., Wang, K., He, Y., Zhang, J., Feng, J.: GaitSet: cross-view gait recognition through utilizing gait as a deep set. IEEE Trans. Pattern Anal. Mach. Intell. **44**(7), 3467–3478 (2021)
5. Fan, C., et al.: Gaitpart: temporal part-based model for gait recognition. In: Proceedings of the IEEE/CVF Conference on Computer Vision and Pattern Recognition, pp. 14225–14233 (2020)
6. Fukushima, K., Miyake, S.: Neocognitron: a self-organizing neural network model for a mechanism of visual pattern recognition. In: Competition and cooperation in neural nets, pp. 267–285 (1982)
7. Huang, Z., et al.: 3d local convolutional neural networks for gait recognition. In: Proceedings of the IEEE/CVF International Conference on Computer Vision, pp. 14920–14929 (2021)
8. Khosla, P., et al.: Supervised contrastive learning. Adv. Neural. Inf. Process. Syst. **33**, 18661–18673 (2020)
9. LeCun, Y., et al.: Backpropagation applied to handwritten zip code recognition. Neural Comput. **1**(4), 541–551 (1989)
10. Liao, R., Cao, C., Garcia, E.B., Yu, S., Huang, Y.: Pose-based temporal-spatial network (PTSN) for gait recognition with carrying and clothing variations. In: Zhou, J., et al. (eds.) CCBR 2017. LNCS, vol. 10568, pp. 474–483. Springer, Cham (2017). https://doi.org/10.1007/978-3-319-69923-3_51
11. Liao, R., Yu, S., An, W., Huang, Y.: A model-based gait recognition method with body pose and human prior knowledge. Pattern Recogn. **98**, 107069 (2020)

12. Lin, B., Zhang, S., Yu, X.: Gait recognition via effective global-local feature representation and local temporal aggregation. In: Proceedings of the IEEE/CVF International Conference on Computer Vision, pp. 14648–14656 (2021)
13. Rumelhart, D.E., Hinton, G.E., Williams, R.J.: Learning representations by back-propagating errors. Nature **323**(6088), 533–536 (1986)
14. Schmidhuber, J., Hochreiter, S., et al.: Long short-term memory. Neural Comput. **9**(8), 1735–1780 (1997)
15. Shiraga, K., Makihara, Y., Muramatsu, D., Echigo, T., Yagi, Y.: Geinet: view-invariant gait recognition using a convolutional neural network. In: 2016 International Conference on Biometrics (ICB), pp. 1–8. IEEE (2016)
16. Smith, L.N., Topin, N.: Super-convergence: very fast training of neural networks using large learning rates. In: Artificial Intelligence and Machine Learning for Multi-domain Operations Applications, vol. 11006, pp. 369–386. SPIE (2019)
17. Song, C., Huang, Y., Huang, Y., Jia, N., Wang, L.: Gaitnet: an end-to-end network for gait based human identification. Pattern Recogn. **96**, 106988 (2019)
18. Song, Y.F., Zhang, Z., Shan, C., Wang, L.: Stronger, faster and more explainable: a graph convolutional baseline for skeleton-based action recognition. In: proceedings of the 28th ACM International Conference on Multimedia, pp. 1625–1633 (2020)
19. Sun, K., Xiao, B., Liu, D., Wang, J.: Deep high-resolution representation learning for human pose estimation. In: Proceedings of the IEEE/CVF Conference on Computer Vision and Pattern Recognition, pp. 5693–5703 (2019)
20. Teepe, T., Gilg, J., Herzog, F., Hörmann, S., Rigoll, G.: Towards a deeper understanding of skeleton-based gait recognition. In: Proceedings of the IEEE/CVF Conference on Computer Vision and Pattern Recognition, pp. 1569–1577 (2022)
21. Teepe, T., Khan, A., Gilg, J., Herzog, F., Hörmann, S., Rigoll, G.: Gaitgraph: graph convolutional network for skeleton-based gait recognition. In: 2021 IEEE International Conference on Image Processing (ICIP), pp. 2314–2318. IEEE (2021)
22. Vaswani, A., et al.: Attention is all you need. In: Advances in Neural Information Processing Systems, vol. 30 (2017)
23. Wang, C., Zhang, J., Wang, L., Pu, J., Yuan, X.: Human identification using temporal information preserving gait template. IEEE Trans. Pattern Anal. Mach. Intell. **34**(11), 2164–2176 (2011)
24. Wang, L., Tan, T., Ning, H., Hu, W.: Silhouette analysis-based gait recognition for human identification. IEEE Trans. Pattern Anal. Mach. Intell. **25**(12), 1505–1518 (2003)
25. Wang, X., Girshick, R., Gupta, A., He, K.: Non-local neural networks. In: Proceedings of the IEEE Conference on Computer Vision and Pattern Recognition, pp. 7794–7803 (2018)
26. Wang, Z., Tang, C., Su, H., Li, X.: Model-based gait recognition using graph network with pose sequences. In: Ma, H., et al. (eds.) PRCV 2021. LNCS, vol. 13021, pp. 491–501. Springer, Cham (2021). https://doi.org/10.1007/978-3-030-88010-1_41
27. Wu, Z., Huang, Y., Wang, L., Wang, X., Tan, T.: A comprehensive study on cross-view gait based human identification with deep CNNs. IEEE Trans. Pattern Anal. Mach. Intell. **39**(2), 209–226 (2016)
28. Yu, S., Tan, D., Tan, T.: A framework for evaluating the effect of view angle, clothing and carrying condition on gait recognition. In: 18th International Conference on Pattern Recognition (ICPR 2006), vol. 4, pp. 441–444. IEEE (2006)
29. Zhang, Z., et al.: Gait recognition via disentangled representation learning. In: Proceedings of the IEEE/CVF Conference on Computer Vision and Pattern Recognition, pp. 4710–4719 (2019)

PalmKeyNet: Palm Template Protection Based on Multi-modal Shared Key

Xinxin Liu, Huabin Wang(✉), Mingzhao Wang, and Liang Tao

Anhui Provincial Key Laboratory of Multimodal Cognitive Computation, School of Computer Science and Technology, Anhui University, Hefei, China
wanghuabin@ahu.edu.cn

Abstract. The distinct ridge features of palmvein and palmprint images, among other palm-related images, make them vulnerable to reversible attacks that can reconstruct the original structure, leading to permanent leakage of biometric features. Additionally, existing multi-modal template protection schemes treat the feature data of each modality as independent, failing to fully capture the inter-modality correlation. Therefore, this paper proposes a multi-modal shared biometric key generation network called PalmKeyNet. By designing keys unrelated to the original palm images as biometric templates, the irreversibility of features is achieved. Additionally, by constructing a multi-modal biometric key generation network, we transform the palm images of different modalities into a unified feature-key space, enhancing the inter-modal correlation. Furthermore, LDPC coding is introduced for multi-modal key error correction to reduce noise interference and improve key discriminability. The proposed approach simultaneously enhances the discriminability, correlation, and security of multi-modal features. The trained PalmKeyNet can be deployed in four modes: single-modal matching (palmprint vs. palmprint and palmvein vs. palmvein), multi-modal matching, and cross-matching. Experimental results on four publicly available palm databases consistently demonstrate the superiority of the proposed method over state-of-the-art approaches.

Keywords: Key generation · Deep learning · Palm print and palm vein

1 Introduction

Biometric recognition has been widely applied due to its convenience. It's pretty remarkable that biometric features are permanent and unchangeable, and once leaked, they can cause irreversible damage to the user. Therefore, researches on protecting biometric template have emerged, known as biometric template

This work was supported by the Natural Science Foundation for the Higher Education Institutions of Anhui Province (Grant No. 2022AH050091).

Q. Liu et al. (Eds.): PRCV 2023, LNCS 14429, pp. 110–121, 2024.
https://doi.org/10.1007/978-981-99-8469-5_9

protection (BTP), which includes cancellable biometrics (CB) and biometric cryptosystem (BCS).

Biometric cryptosystem [1,2] is initially developed to protect encryption keys using biometric features or directly generate encryption keys from biometric features. However, they can also be used as template protection mechanisms. In [3], a framework was designed that uses a fuzzy extractor to generate keys and employs hash functions for protection, ensuring that no information can be retrieved from the hash key output by attackers. In [4], a new approach to a fuzzy vault biometric cryptosystem utilizing palmprint features was proposed, creating a fuzzy vault by combining reference points with randomly generated impostor points or chaff points. However, current traditional key-binding schemes [5,10] require carefully designed auxiliary data based on specific biometric features and the nature of related user internal variations. Additionally, key generation schemes are difficult to achieve both high stability and high entropy.

Recently, deep learning has gained significant attention, and the field of biometric cryptosystems has also been influenced by deep learning-based approaches. Ma et al. [6] proposed a deep learning-based BCS method that utilizes different layers of a convolutional neural network (CNN) to extract features and obtain an optimal feature set as input for BCS. Kumar et al. [7] introduced a facial template protection method using deep CNN, which learns a robust mapping from a user's facial image to a unique binary code assigned during the enrollment phase through single and multiple registrations. Roh et al. [8] proposed an approach combining CNN and necurrent neural network (RNN), where CNN extracts feature vectors from facial images and RNN generates keys based on these feature vectors. Roy et al. [9] propose to design and implement a retinal biometric key generation framework with deep neural network in order to to replace the semi-automated or automated retinal vascular feature identification methods.

Currently, most deep learning-based BCS methods are not end-to-end and require feature extraction followed by key generation based on the feature set. To address this issue, an end-to-end key generation network is proposed that can directly generate keys from biometric images. The key mentioned in this paper is the encrypted biometric feature, which is used for identification. Additionally, considering the improved recognition performance of multi-modal biometric features, we propose a template protection scheme that combines palmprint and palmvein modalities. Furthermore, the existing methods do not fully utilize the key space and overlook the impact of noise during the feature transformation process, therefore the proposed method introduces the utilization of pre-allocated keys to fully exploit the key space, along with LDPC(Low-Density Parity-Check) encoding for error correction to enhance recognition performance.

The PalmKeyNet in this paper is a palm-based multi-modal shared key generation network, which consists of two sub-networks: one for palmprint and another for palmvein. The trained PalmKeyNet can generate keys based on input palmprint and/or palmvein images. Given a query palmprint and/or palmvein image, PalmKeyNet generates LDPC-encoded keys. The effectiveness of PalmKeyNet using backbone structures such as ResNet18 [11], MobileNetV2 [12], and EfficientNet-B5 [13] is demonstrated. The contributions of this work are as follows:

1. A feature learning-based key generation methods is proposed. By constructing a multi-modal shared key generation network, PalmKeyNet transform multi-modal palm features into a unified feature space. This approach enables end-to-end protection of biometric templates and enhances the correlation between different modalities.
2. This paper have studied the encoding and error correction methods for multi-modal shared key. The proposed method introduces encryption algorithms to protect the keys and further introduces LDPC coding to correct errors and reduces noise interference, thereby enhancing the distinctiveness of the keys.
3. The proposed key generation scheme, PalmKeyNet, can be applied to cross-modality, multi-modality, and single-modality scenarios, providing flexible deployment options that are suitable for various application scenarios.

2 Methodology

Existing handcrafted key generation methods do not simultaneously consider the inter-modality correlations within the same class and the intra-modality distinctiveness between different classes. However, multi-modal palm images such as palmveins and palmprints exhibit evident correlated features, and the recognition performance is limited by the quality of the original biometric features. Therefore, we propose an innovative approach to design a multi-modal shared biometric key and constructs a deep learning network for key generation. This network encrypts and transforms the multi-modal biometric templates into the same key space, simultaneously enhancing the inter-modality feature correlation and template security. As an encrypted biometric feature, the keys generated by PalmKeyNet are ultimately used for identity recognition.

The proposed approach in this paper mainly consists of key encoding, key binding, and key generation, as illustrated in Fig. 1. Firstly, a set of randomly generated keys that are orthogonal to each other and have sufficient distance are pre-generated. These keys are encrypted using SHA-256 and stored as biometric templates in the database. Secondly, the random keys are LDPC encoded and assigned to users, and the encoded shared keys are simultaneously bound to the users' palmprints and palm veins. Finally, a network is trained to take palmprint and/or palm vein images as input and generate LDPC encoded keys. During the verification phase, before matching with the templates in the database, the network's output undergoes decoding for error correction, resulting in improved recognition performance.

2.1 Key Encoding

During the process of feature data transformation, noise interference is inevitably introduced, which can affect recognition performance. Inspired by traditional key

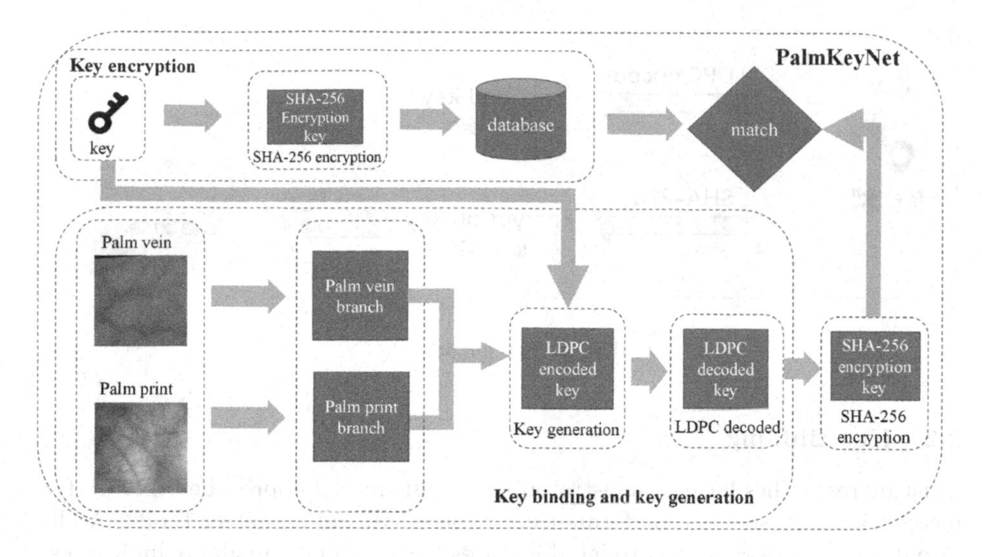

Fig. 1. Overall structure of the scheme in this paper

generation schemes, error correction code is incorporated into the key to mitigate the impact of noise and LDPC was chosen after consideration. LDPC are a type of error correction coding technique widely used in communication and storage systems, known for their excellent error correction performance and relatively low decoding complexity.

First, pre-generate m (where m is the number of registered users) orthogonal keys $k_i \in R^n$, $i = 1, 2, \cdots, m$ with sufficient mutual distances. Then, LDPC is used to encode the key k_i for error correction, resulting in encoded key $k_i^{''}$. The error correction information is only used during the verification stage to effectively mitigate the impact of noise on recognition performance and is immediately discarded after that, without occupying additional storage space.

An important indicator for evaluating the security of biometric recognition is the irreversibility of the features, meaning that an attacker cannot reverse-engineer the original features from the biometric template. The pre-generated keys are independent of the features, making it infeasible to derive the original biometric features from the keys. To further enhance the security of the keys, SHA-256 is introduced to provide additional security assurance. SHA-256 is a secure and reliable hash algorithm known for its uniqueness and irreversibility. The pre-generated keys are encrypted using SHA-256, and the encrypted keys $k_i^{'}$ are stored in the database. This process is also irreversible, thereby enhancing the system's resistance against attacks.

In summary, the key encoding process consists of two branches: LDPC coding and SHA-256 encryption, as shown in Fig. 2. LDPC coding is utilized to reduce the impact of noise during the data transformation process, while SHA-256 encryption provides additional security measures to ensure both the recognition performance and security of the system.

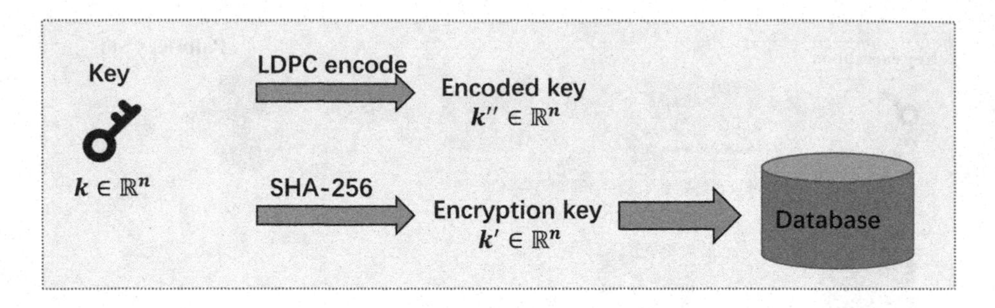

Fig. 2. key encoding

2.2 Key Binding

Existing researches have demonstrated that multi-modal approaches offer better recognition and security performance compared to uni-modal methods. Additionally, there are evident correlated features between multi-modal palm images, such as palm veins and palm prints. Therefore, this paper aims to bind palm prints and palm veins into the same key space.

Specifically, an LDPC-encoded key $k_i^{''}$ is assigned to each registered identity, which is independent of the biometric modality but correlated with the identity. In other words, each identity has only one key associated with multiple biometric modalities simultaneously. During training, the encoded key $k_i^{''}$ is bound with the palmprint image $x_p \in R^{C\times H\times W}$ and palmvein image $x_v \in R^{C\times H\times W}$, meaning that the palm print and palm vein images of the same identity are bound to the same key. We use $x_p^{i,j} \in R^{C\times H\times W}$ to represent the j-th palm print image of the i-th identity, where $i = 1, 2, \cdots, m$ and $j = 1, 2, \cdots, d$ (d is the number of feature images for a particular modality of the same identity), and $x_v^{i,j} \in R^{C\times H\times W}$ represents the j-th palm vein image of the i-th identity. Then, $k_i^{''}$ is bound with x_p^i and x_v^i. From another perspective, $k^{''}$ can be regarded as the label shared by the palm print and palm vein images during training.

Feature fusion is implemented by obligatorily binding palmveins and palmprints into the same key space, which not only increases the intra-class correlation between the two modalities but also enables the introduction of feature-independent keys to enhance inter-class discrimination.

2.3 Key Generating

The goal of PalmKeyNet is to map the input space to the key space such that $S(x^i, x^j) = S(k^i, k^j)$, where S is a similarity function. To learn a robust mapping from user's palm images to specified binary codes, the deep CNN is introduced, which maximizes inter-user variations while minimizing intra-user variations and thereby providing high matching performance.

To handle palmprint and palmvein images, PalmKeyNet consists of two subnetworks of deep CNNs that map palmprint and palmvein images to the same key space. After training, these networks generate keys with LDPC encoding.

The design and structure of the deep CNN networks depend on the application requirements and computational resource constraints. ResNet18 is suitable for scenarios with larger computational resources and higher accuracy requirements. EfficientNet is suitable for scenarios that balance performance and computational resources. MobileNet, on the other hand, focuses on lightweight and computationally constrained devices.

The two sub-networks of PalmKeyNet independently process palmprint image x_p and palmvein image x_v, generating their respective keys κ. These keys are then binarized as $b_p = sign(\kappa^{print}) \in \{0,1\}^n$ and $b_v = sign(\kappa^{vein}) \in \{0,1\}^n$, where $\kappa^{print} = f_p(x_p)$ and $\kappa^{vein} = f_v(x_v)$, and f_p and f_v represent the feature extraction backbones. During training, the keys κ generated from PalmKeyNet should be consistent with the keys $k_i^{''}$ bound to the same identity.

For query, PalmKeyNet takes input palmvein and/or palmprint images and generates LDPC-encoded key. The key is then subjected to LDPC decoding for error correcting. The decoded key is encrypted using SHA-256 and used for matching against the key in the database for verification.

2.4 Deployment

The trained PalmKeyNet can be flexibly deployed in the following operational modes:

Multi-modality mode: This is the most secure mode. During enrollment, the key is assigned to each identity, which is encoded with SHA-256 and stored in the database along with the parameters of PalmKeyNet. For query, both palm print and palm vein images need to be provided, which can be obtained from a single image. They are transformed into their respective LDPC-encoded keys and binarized as $b_{print} = sign(f_p(X_p))$ and $b_{vein} = sign(f_v(X_v))$. The keys b_{print} and b_{vein} are decoded by LDPC to obtain $b_{print}^{'}$ and $b_{vein}^{'}$. Then, $b_{print}^{'}$ and $b_{vein}^{'}$ are transformed into $b_{print}^{''}$ and $b_{vein}^{''}$ using SHA-256 encoding. The decision is reached if $D_H(b_{print}^{''}, k_i^{'})$ and $D_H(b_{vein}^{''}, k_i^{'})$ are both equal to 0, where D_H represents the Hamming distance.

Single-modality mode: This is the simplest mode. Similar to the multi-modality mode, a key is assigned to each identity during enrollment, which is encoded with SHA-256 and stored in the database. Whether a palm print or palm vein image is provided, for query, the transformed key $b^{''}$ is directly matched with the stored key using $D_H(b^{''}, k_i^{'})$ for decision-making. A result of 0 indicates a successful match.

Cross-modality mode: This is the most flexible mode. This mode simulates cross-modal matching between palm print and palm vein. By calculating the distance between the transformed keys, i.e., $D_H(b_{print}^{''}, b_{vein}^{''})$, cross-modal matching between palm print and palm vein can be performed. If the Hamming distance is below a predefined threshold, a decision to grant the corresponding permission can be made.

Table 1. Datasets summarization

Dataset	Acquisition	Palm vein	Palm print	Samples	Classes
PolyU-M	Contact	NIR	RGB	12	500
IITD	Contactless	-	Gray	5	460
CasiaM	Contactless	RGB	NIR	6	200
Tongji-P	Contactless	Gray	Gray	20	600
Tongji-PV	Contactless	Gray	Gray	20	600

Table 2. Performance of PalmKeyNet on four public benchmark datasets

Dataset	Backbone	Cross far	Print IR	Print far	Vein IR	Vein far	Fusion IR	Fusion far
PolyU-M	EffNetb5	**0.100**	**99.967**	**0.000**	**99.967**	**0.000**	**99.933**	**0.000**
	ResNet18	0.433	99.433	0.167	99.867	0.067	99.433	0.006
	MobileNetV2	0.367	99.600	0.267	99.867	0.067	99.567	0.001
Tongji	EffNetb5	0.067	99.950	0.050	99.883	0.067	99.917	0.001
	ResNet18	0.333	99.733	0.117	99.700	0.283	99.700	0.030
	MobileNetV2	0.333	99.800	0.067	99.667	0.300	99.800	0.045
CasiaM	EffNetb5	2.667	96.000	1.333	98.000	0.500	95.000	0.312
	ResNet18	12.833	71.333	11.167	88.667	3.167	69.667	6.612
	MobileNetV2	5.500	84.500	5.333	95.333	1.167	84.000	1.802
IIT-D	EffNetb5	-	99.020	0.652	-	-	-	-
	ResNet18	-	96.630	2.500	-	-	-	-
	MobileNetV2	-	96.196	3.261	-	-	-	-

3 Experiments and Conclusions

3.1 Database

To evaluate the performance of the proposed method, four publicly available datasets were used in the experiments, as shown in Table 1: 1) PolyU-multi-spectral (PolyU-M) [14]; 2) IITD [15]; 3) CASIA-multi-spectral-palmprintV1 (CASIA-M) [16]; and 4) Tongji palmprint/palmvein(Tongji) [17], [18]. Please note that the left and right hands of the same individual do not belong to the same identity.

If the original dataset does not provide palmprint regions of interest (ROIs), the palmseg tool [22], [23] is used to extract the ROIs. For training, the image inputs are normalized. Additionally, for data augmentation, we apply random brightness, rotation, and affine transformations to the images.

3.2 Experiment Settings

The basic architecture of ResNet18 is ResNet, which has a depth of 18 layers and powerful representation capability. EfficientNet, compared to other networks, has fewer parameters and higher accuracy, making a qualitative breakthrough.

MobileNet network has smaller size, fewer computations, and higher accuracy, making it highly advantageous in lightweight neural networks. Based on these advantages, we select ResNet18, EfficientNet, and MobileNet as backbone networks.

For the allocation of training and testing data, we choose a 1:1 ratio. For example, in CASIA-M, three randomly selected samples per identity are used for training, while the remaining three samples are used for testing. In IITD, three samples are randomly selected for training, and the remaining two samples are used for testing.

Table 3. Performance (%) comparison of coding with and without(Y/N) LDPC

Dataset	Backbone	LDPC	Print IR	Vein IR	Fusion IR
PolyU-M	EffNetb5	Y	99.967	99.967	99.933
		N	99.800	99.900	99.800
	ResNet18	Y	99.433	99.867	99.433
		N	81.133	69.333	81.133
	MobileNetV2	Y	99.600	99.867	99.567
		N	85.967	88.867	85.967
Tongji	EffNetb5	Y	99.950	99.883	99.917
		N	99.850	99.767	99.850
	ResNet18	Y	99.733	99.700	99.700
		N	87.567	88.000	87.567
	MobileNetV2	Y	99.800	99.667	99.800
		N	90.450	92.650	90.450
CasiaM	EffNetb5	Y	96.000	98.000	95.000
		N	89.333	94.500	89.333
	ResNet18	Y	71.333	88.667	69.667
		N	47.333	71.167	47.333
	MobileNetV2	Y	84.500	95.333	84.000
		N	57.667	75.500	57.667

3.3 Quantitative Evaluation

To evaluate the recognition and authentication performance of the proposed method in four application modes, experiments is conducted on four datasets. We use the identification rate (IR) to evaluate the recognition performance and the false acceptance rate (FAR) to evaluate the authentication performance. The proposed method includes four modes: two single-modality matching, multi-modality matching, and cross-modality matching. For cross-modality one vs. one matching, we only report the verification identification rate.

Table 2 summarizes the performance of PalmKeyNet on four databases using three backbone networks, with a key length of 128 bits. As shown in the table, the proposed PalmKeyNet performs well on different models and datasets. It is worth noting that PalmKeyNet based on EfficientNet achieves excellent results in single-modal and multi-modal matching on the PolyU-M, with both single-modal recognition rates reaching 99.967% and a false acceptance rate of 0%. This demonstrates that the design of pre-allocated keys artificially increases the inter-class distance and improves key distinctiveness. Additionally, the PolyU-M also achieves good recognition performance with PalmKeyNet based on ResNet18 and MobileNetV2. Furthermore, the Tongji dataset also exhibits good performance with PalmKeyNet.

In single-modal matching, PalmKeyNet based on EfficientNet demonstrates good performance on all three databases. The recognition rates for both PolyU-M and Tongji are almost above 99.5%, with false acceptance rates below 0.3%. Even if multi-modal recognition cannot be implemented due to limited device capabilities, deploying single-modal matching can still achieve the desired functionality.

Compared to single-modal matching, multi-modal matching does not show a significant improvement in recognition rate but exhibits a significant reduction in false acceptance rate. This indicates the effectiveness of the shared key scheme, where multi-modal matching not only achieves high recognition performance but also significantly enhances security compared to single-modal matching.

For cross-modal matching, only the false acceptance rate is reported. Both the PolyU-M and Tongji achieve excellent recognition and authentication performance regardless of the backbone network used, demonstrating the feasibility of cross-modal matching.

Table 3 summarizes the performance of PalmKeyNet based on different backbone networks with and without LDPC encoding on different databases. It is evident that the recognition rate of PalmKeyNet significantly improves when LDPC encoding is applied, indicating that LDPC error correction can reduce the impact of noise and enhance recognition performance.

3.4 Comparison with State-of-the-Art

We have investigated advanced deep learning-based methods, including [19], [20], [21], and presented the experimental results in Table 4, comparing them with the proposed PalmKeyNet in terms of equal error rate (EER) on the same databases. A lower EER indicates a higher recognition rate for the biometric system. We compared the EER for palmprint, palm vein, multimodal, and cross-modal scenarios on the same databases. Among them, Palmnet [19] applies Gabor filters to CNN, DHN [20] uses score-level fusion and the experiments on the IIT-D database are only for palmprints, and LDBC [21] is based on near-infrared fusion.

It can be observed that PalmKeyNet based on EffNetb5 achieves a significant reduction in EER compared to other approaches, especially on the

Table 4. Comparison with state-of-the-art

		Feature length	prints	veins	multi modality	cross modality	Para.#
Palmnet [19]	Tongji	12M	0.19	-	-	-	12K
	IITD		0.71	-	-	-	
	CASIA[35]		0.72	-	-	-	
DHN [20]	PolyU	64	-	-	0.0253	-	113M
	G+R+N+B	128	-	-	0.0013	-	
	Tongji	64	-	-	0.0013	-	
		128	0.3991	0.7265	0.1484	-	
	IIT-D	128	3.1183	-	-	-	
LDBC [21]	PolyU-M	69,62	-	-	0.470	0.96	-
	CasiaM	69,62	-	-	4.280	-	
PalmKeyNet-EffNetb5	PolyU-M	64	**0.000**	**0.000**	**0.000**	**0.000**	57M
		128	0.033	0.017	**0.000**	0.017	
	Tongji	64	0.033	0.100	5.565	0.083	
	IIT-D	64	0.007	-	-	-	
	CasiaM	128	1.250	0.917	0.939	2.750	
PalmKeyNet-ResNet18	PolyU-M	64	0.200	0.033	**0.000**	0.150	22M
		128	0.333	0.067	**0.000**	0.067	
	Tongji	64	0.042	0.233	0.003	0.175	
	IIT-D	64	0.034	-	-	-	
	CasiaM	128	10.833	5.000	5.710	14.833	
PalmKeyNet-MobileNetV2	PolyU-M	64	0.217	0.033	0.000	0.183	4.7M
		128	0.300	0.083	5.567	0.117	
	Tongji	64	0.083	0.217	0.035	0.267	
	IIT-D	64	2.935	-	-	-	
	CasiaM	128	5.333	1.750	1.713	6.750	

PolyU-M. Compared to PalmKeyNet based on the other two backbone networks, PalmKeyNet based on EffNetb5 achieves better accuracy. This is because PalmKeyNet introduces LDPC, which effectively corrects some inevitable noise interference and improves the stability of the generated keys. Additionally, PalmKeyNet based on MobileNetV2 is a lightweight network, offering higher efficiency.

3.5 Security Analysis

Irreversibility is an important criterion for evaluating the protection of biometric templates, referring to the inability to restore the original biometric from the extracted and encoded feature templates. Irreversibility helps ensure the security and privacy of biometric templates.

The keys in this paper are feature-independent, meaning that even if an attacker obtains the keys, they cannot reverse-engineer the original features based on the keys. The system administrator can regenerate new keys and assign them to users, thereby rendering the leaked keys obsolete. Additionally, the keys in the database are encrypted using SHA-256, making them irreversible. For different input data, SHA-256 generates nearly unique hash values, such that even a slight change in the input data will result in different hash values. As a result, it is not possible to reconstruct the original data from the database, ensuring the confidentiality of the data.

4 Conclusion and Future Work

The proposed PalmKeyNet binds biometric images with pre-allocated keys to generate highly discriminative keys. The application of LDPC encoding further reduces noise interference during data transmission. The use of lightweight networks allows PalmKeyNet to be conveniently deployed in IoT, and the stability of the keys ensures high reliability in data transmission. Our experimental results demonstrate that the proposed network achieves excellent performance in single-modality, multi-modality, and cross-modality scenarios.

For future work, in addition to using keys for matching as done in this paper, we hope that PalmKeyNet can also serve as an intermediate step for biometric recognition and verification, combined with other biometric template protection methods, such as cancelable biometric techniques. Furthermore, PalmKeyNet can be further applied to facial biometrics, where facial features can be matched with iris features in a multi-modal manner, which presents significant challenges in feature extraction.

References

1. Uludag, U., Pankanti, S., Prabhakar, S., et al.: Biometric cryptosystems: issues and challenges. Proc. IEEE **92**(6), 948–960 (2004)
2. Cavoukian, A., Stoianov, A.: Biometric encryption: a positive-sum technology that achieves strong authentication, security and privacy. Technical report, Office of the Information and Privacy Commissioner of Ontario, Toronto, Ontario, Canada, March 2007
3. Kaur, T., Kaur, M.: Cryptographic key generation from multimodal template using fuzzy extractor. In: 2017 Tenth International Conference on Contemporary Computing (IC3), pp. 1–6. IEEE (2017)
4. Sujitha, V., Chitra, D.: Highly secure palmprint based biometric template using fuzzy vault. Concurr. Comput. Pract. Exp. **31**(12), e4513 (2019)
5. Asthana, R., Walia, G.S., Gupta, A.: A novel biometric crypto system based on cryptographic key binding with user biometrics. Multimed. Syst. **2021**, 1–15 (2021)
6. Ma, Y., Wu, L., Gu, X., et al.: A secure face-verification scheme based on homomorphic encryption and deep neural networks. IEEE Access **5**, 16532–16538 (2017)
7. Kumar Jindal, A., Chalamala, S., Kumar Jami, S.: Face template protection using deep convolutional neural network. In: Proceedings of the IEEE Conference on Computer Vision and Pattern Recognition Workshops, pp. 462–470 (2018)
8. Roh, J., Cho, S., Jin, S.H.: Learning based biometric key generation method using CNN and RNN. In: 2018 10th International Conference on Information Technology and Electrical Engineering (ICITEE), pp. 136–139. IEEE (2018)
9. Roy, N.D., Biswas, A.: Fast and robust retinal biometric key generation using deep neural nets. Multimed. Tools Appl. **79**(9–10), 6823–6843 (2020)
10. Yang, B., Busch, C.: Privacy-enhanced biometrics-secret binding scheme: U.S. Patent 10,594,688[P]. 2020-3-17 (2020)

11. He, K., Zhang, X., Ren, S., et al.: Deep residual learning for image recognition. In: Proceedings of the IEEE Conference on Computer Vision and Pattern Recognition, pp. 770–778 (2016)
12. Sandler, M., Howard, A., Zhu, M., et al.: Mobilenetv 2: inverted residuals and linear bottlenecks. In: Proceedings of the IEEE Conference on Computer Vision and Pattern Recognition, pp. 4510–4520 (2018)
13. Tan, M., Le, Q.: Efficientnet: rethinking model scaling for convolutional neural networks. In: International Conference on Machine Learning, pp. 6105–6114. PMLR (2019)
14. Polyu palmprint database (version 2.0). https://www.comp.polyu.edu.hk/~biometrics. Accessed 03 Feb 2022
15. Iitd touchless palmprint database (version 1.0). http://www4.comp.polyu.edu.hk/csajaykr/IITD/DatabasePalm.htm. Accessed 03 Feb 2022
16. Center for biometrics and security research. Cassia multispectral palmprint database. http://biometrics.idealtest.org/. Accessed 03 Feb 2022
17. Zhang, L., Cheng, Z., Shen, Y., et al.: Palmprint and palm vein recognition based on DCNN and a new large-scale contactless palm vein dataset. Symmetry **10**(4), 78 (2018)
18. Tongji palmprint image database. https://cslinzhang.github.io/ContactlessPalm/. Accessed 03 Feb 2022
19. Genovese, A., Piuri, V., Plataniotis, K.N., et al.: PalmNet: gabor-PCA convolutional networks for touchless palmprint recognition. IEEE Trans. Inf. Forensics Secur. **14**(12), 3160–3174 (2019)
20. Wu, T., Leng, L., Khan, M.K., et al.: Palmprint-palmvein fusion recognition based on deep hashing network. IEEE Access **9**, 135816–135827 (2021)
21. Cho, S., Oh, B.S., Toh, K.A., et al.: Extraction and cross-matching of palm-vein and palmprint from the RGB and the NIR spectrums for identity verification. IEEE Access **8**, 4005–4021 (2019)
22. Genovese, A., Piuri, V., Plataniotis, K.N., Scotti, F.: PalmNet: gabor-PCA convolutional networks for touchless palmprint recognition. IEEE Trans. Inf. Forensics Secur. 14, 1556 6013 (2019)
23. Genovese, A., Piuri, V., Scotti, F., Vishwakarma, S.: Touchless palm-print and finger texture recognition: a deep learning fusion approach. In: Proceedings of the 2019 IEEE International Conference on Computational Intelligence and Virtual Environments for Measurement Systems and Applications (CIVEMSA 2019), Tianjin, China, June 2019

Full Quaternion Matrix-Based Multiscale Principal Component Analysis Network for Facial Expression Recognition

Hangyu Li[1], Zuowei Zhang[1], Zhuhong Shao[1(✉)], Bin Chen[2], and Yuanyuan Shang[1]

[1] College of Information Engineering, Capital Normal University, Beijing 100048, China
{1201002013,2221002068,zhshao,5528}@cnu.edu.cn

[2] College of Information Science and Engineering, Jiaxing University, Jiaxing 314001, China
chenbin@zjxu.edu.cn

Abstract. To acquire a more discriminative feature of facial expression, we propose a multi-scale principal component analysis network based on full quaternion matrix representation. Firstly, the structure feature and color components of facial image constitute a full quaternion matrix. Subsequently, two-staged quaternion principal component analysis is employed to learn convolutional filters. Among them, the feature maps of both stages are activated via nonlinear function. With binarization and coding, the local histograms are stacked together and fed to the classifier for expression matching. Experiments conducted on the RafD, MMI, NVIE, and KDEF datasets have demonstrated that the proposed method achieves higher recognition accuracy than several existing algorithms.

Keywords: Facial expression recognition · Full quaternion matrix · Quaternion principal component analysis · Multiscale

1 Introduction

Facial expression is an indispensable part of non-verbal behavior, which can assist people's verbal communication in the real-world. Over the past few years, facial expression recognition (FER) has become one of the most concerned topics in the field of computer vision and pattern recognition [1–3]. Furthermore, it has been successfully applied to safe driving, intelligent monitoring, and medical rehabilitation [4].

FER typically involves three steps: (1) image preprocessing, (2) feature extraction and selection, and (3) classification. Feature extraction methods based on convolutional neural networks (CNN) [5] have been widely used in recent years, which can effectively extract high-level features to represent the concise semantics of data. More and deeper neural networks [6–8] have been proposed since Krizhevsky et al. [9] proposed AlexNet. Yu et al. proposed a semi-supervised learning framework for facial expression recognition task, where a dynamic threshold module was designed for generating more accurate pseudo-labels [10]. To address the ambiguity of emotion and noisy labels, Wang et al.

Q. Liu et al. (Eds.): PRCV 2023, LNCS 14429, pp. 122–133, 2024.
https://doi.org/10.1007/978-981-99-8469-5_10

developed emotion ambiguity-sensitive cooperative networks [11]. However, the parameter training time of these network models is too long and special tuning techniques are required for optimal performance.

In contrast, lightweight convolutional neural networks present an appealing alternative. The main objective of this paper is to develop a multiscale principal component analysis network based on full quaternion matrix for facial expression recognition. The contributions of this paper are summarized as: (1) We propose a full quaternion principal component analysis network that outperforms several existing methods for facial expression recognition. (2) We use an activation function to enhance the nonlinearity of feature maps and improve the generalization ability of the model. (3) The outputs of different stages are complementary, which are favourable to recognition.

The rest of this paper is arranged as follows. Section 2 briefly reviews some related work. Section 3 describes the proposed method in detail. Section 4 presents the experimental results and Sect. 5 concludes the paper.

2 Related Work

To simplify the training process, Chan et al. [12] proposed principal component analysis network (PCANet), and experimental results have shown that PCANet can acquire satisfactory performance for most image classification tasks. Qaraei et al. [13] proposed a nonlinear PCA network dubbed RNPCANet, which used explicit kernel PCA to learn convolutional filters. Zhou and Feng [14] proposed multi-scale spatial pyramid second-order pooling principal component analysis network (M3SPCANet) for face recognition.

Quaternions, as a generalization of complex numbers, can encode different color components into a whole when representing one color image. It has been gradually applied to the analysis and recognition of color images [15]. To name a few, Zeng et al. [16] proposed a quaternion principal component analysis network (QPCANet), which extended PCANet from the real domain to the quaternion domain. Zou et al. [17] proposed a data classifier based on quaternion block sparse representation for color face recognition. Shi et al. [18] proposed a quaternion-based Grassmann average network to learn effective features containing color information for histopathological images. Liu et al. [19] applied quaternion scalar and vector norm decomposition to QPCA for color face recognition, which can address the phenomenon of under-fitting of vector part norm approximation. However, when using quaternion matrices to represent color images, the real part is not fully considered.

3 The Proposed Scheme

In this section, we detail the proposed recognition scheme as shown in Fig. 1, which mainly consists of full quaternion matrix representation, features extraction using multiscale QPCANet (MQPCANet).

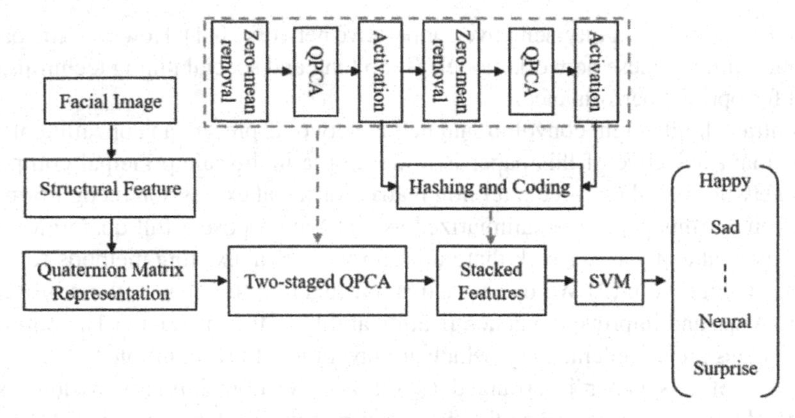

Fig. 1. The flowchart of proposed scheme

3.1 Full Quaternion Matrix Representation

When representing a color facial expression image, we use the way described in [20] to increase the consistency with human perception. In this method, three color components are considered as the imaginary parts of a quaternion matrix, while the structural information of the image is taken as the real part.

Let $\boldsymbol{I}(x, y)$ be a color image represented as a pure quaternion matrix, the local variance is computed in a $n_1 \times m_1$ neighborhood as,

$$\boldsymbol{I}_v(x, y) = \frac{1}{n_1 \times m_1} \sum_{p=1}^{n_1 \times m_1} \left|\eta_p - \overline{\eta}\right|^2 \tag{1}$$

where $|\cdot|$ denotes the magnitude of a quaternion, η_p and $\overline{\eta}$ respectively denote the internal pixels and their average values in the form of pure quaternions.

Afterwards, the red component $\boldsymbol{I}_R(x, y)$, the green component $\boldsymbol{I}_G(x, y)$, the blue component $\boldsymbol{I}_B(x, y)$ and the extracted local variance $\boldsymbol{I}_v(x, y)$ are integrated into a full quaternion matrix as,

$$\mathbf{Q}(x, y) = \boldsymbol{I}_v(x, y) + \boldsymbol{I}_R(x, y)\boldsymbol{i} + \boldsymbol{I}_G(x, y)\boldsymbol{j} + \boldsymbol{I}_B(x, y)\boldsymbol{k} \tag{2}$$

3.2 Feature Extraction

Building upon the QPCANet, the MQPCANet we proposed introduces an activation function following each QPCA layer to amplify the nonlinearity of feature maps. Additionally, it combines the output from each stage to produce the ultimate feature representation.

(1) Suppose that we have N full quaternion images $\left\{\boldsymbol{Q}_i \in \mathbb{Q}^{m\times n}\right\}_{i=1}^{N}$ for training. Let $k_1 \times k_2$ be the patch size. Around each pixel, we gather all overlapping quaternion

patches from the ith full quaternion matrix $\boldsymbol{Q}_i$, and reshape them to column vectors. Then we subtract the mean of each quaternion patch to get zero-mean quaternion patches $\boldsymbol{X}_i=[\boldsymbol{x}_{i,1}, \boldsymbol{x}_{i,2}, \ldots, \boldsymbol{x}_{i,(m-k_1+1)(n-k_2+1)}] \in \mathbb{Q}^{k_1 k_2 \times (m-k_1+1)(n-k_2+1)}$. By repeating the above process for all N training images and combining all quaternion patches together, we obtain

$$\boldsymbol{X}^1=[\boldsymbol{X}_1, \boldsymbol{X}_2, \ldots, \boldsymbol{X}_N] \in \mathbb{Q}^{k_1 k_2 \times (m-k_1+1)(n-k_2+1)N} \quad (3)$$

The covariance matrix of $\boldsymbol{X}^1$ is calculated as,

$$\boldsymbol{C} = \frac{\boldsymbol{X}^1 \boldsymbol{X}^{1^H}}{N(m-k_1+1)(n-k_2+1)} \quad (4)$$

where the superscript H is the conjugate transposition operator.

Next, the QPCA of the training images is given by quaternion eigenvalue decomposition,

$$\boldsymbol{C} = \boldsymbol{W}^1 \boldsymbol{\Omega} \boldsymbol{W}^{1^H} \quad (5)$$

where $\boldsymbol{W}^1 \in \mathbb{Q}^{k_1 k_2 \times k_1 k_2}$ is a unitary matrix that contains eigenvectors of the covariance matrix $\boldsymbol{C}$, and $\boldsymbol{\Omega} \in \mathbb{R}^{k_1 k_2 \times k_1 k_2}$ is a real diagonal matrix with eigenvalues.

Assuming that the number of QPCA filters in the ith layer is L_i. We choose the L_1 principal eigenvectors of $\boldsymbol{W}^1 \in \mathbb{Q}^{k_1 k_2 \times k_1 k_2}$ corresponding to the L_1 largest eigenvalues of $\boldsymbol{\Omega}$. Therefore, the QPCA filter bank is expressed as,

$$\boldsymbol{W}_{l_1}^1 \in \mathbb{Q}^{k_1 \times k_2}, l_1 = 1, 2, \ldots, L_1 \quad (6)$$

Then, we perform a convolution operation using the QPCA filter bank to obtain quaternion feature maps as follows,

$$\boldsymbol{Y}_{i,l_1}^1 = \boldsymbol{Q}_i * \boldsymbol{W}_{l_1}^1, i = 1, 2, \ldots, N, \; l_1 = 1, 2, \ldots, L_1 \quad (7)$$

where $\boldsymbol{Y}_{i,l_1}^1 \in \mathbb{Q}^{m \times n}$ is the l_1 th feature map of $\boldsymbol{Q}_i$ in the first stage. It should be mentioned that zero-padding operation is applied to make sure that the quaternion feature maps have the same size as the input quaternion matrix.

To enhance the nonlinearity of quaternion feature maps, they are activated by employing nonlinear function Tanh [21]. Thus, we have the nonlinear quaternion feature maps as,

$$\boldsymbol{G}_{i,l_1}^1 = \mathrm{Tanh}\left(\boldsymbol{Y}_{i,l_1}^1\right), i = 1, 2, \ldots, N, \; l_1 = 1, 2, \ldots, L_1 \quad (8)$$

Then the outputs are treated as the input of the second stage.

(2) The way of obtaining the QPCA filter bank and other operations are the same as those in the first stage. Firstly, With the input $\left\{\boldsymbol{G}_{i,l_1}^1\right\}_{i,l_1=1,1}^{N,L_1}$, we collect all quaternion patches of the same size as the first stage and get zero-mean quaternion patches $\boldsymbol{X}^2 \in \mathbb{Q}^{k_1 k_2 \times (m-k_1+1)(n-k_2+1)NL_1}$. Then, the QPCA is employed to obtain the QPCA filter bank $\boldsymbol{W}_{l_2}^2 \in \mathbb{Q}^{k_1 \times k_2}, l_2 = 1, 2, \ldots, L_2$. Next, the quaternion feature maps

$Y^2_{i,l_1,l_2} = G^1_{i,l_1} * W^2_{l_2}$, $i = 1, 2, \ldots, N$, $l_1 = 1, 2, \ldots, L_1$, $l_2 = 1, 2, \ldots, L_2$ are computed. Finally, by using the nonlinear function, we get the output of the second stage,

$$G^2_{i,l_1,l_2} = \mathrm{Tanh}\left(Y^2_{i,l_1,l_2}\right),\ i = 1, 2, \ldots, N,\ l_1 = 1, 2, \ldots, L_1, l_2 = 1, 2, \ldots, L_2 \tag{9}$$

(3) To reduce the complexity of quaternion feature maps, we perform a binarization and weighted summation for all quaternion features of the first and second nonlinear layers. Firstly, each nonlinear quaternion feature map is binarized by using the Heaviside step function $H(\cdot)$ to its four parts, which is defined as follows,

$$H(x) = \begin{cases} 0,\ x < 0 \\ 1,\ x \geq 0 \end{cases} \tag{10}$$

For the outputs of the first stage $\left\{G^1_{i,l_1}\right\}^{N,L_1}_{i,l_1=1,1}$, the weighted sum of the L_1 binarized quaternion feature maps in each group is given as,

$$T^1_i = \sum_{l_1=1}^{L_1} 2^{l_1-1}\mathrm{H}\left(G^1_{i,l_1}\right) = \mathrm{S}\left(T^1_i\right) + \mathrm{I}\left(T^1_i\right)\boldsymbol{i} + \mathrm{J}\left(T^1_i\right)\boldsymbol{j} + \mathrm{K}\left(T^1_i\right)\boldsymbol{k} \tag{11}$$

where the pixel values of $\mathrm{S}(T^1_i), \mathrm{I}(T^1_i), \mathrm{J}(T^1_i), \mathrm{K}(T^1_i)$ are integers in the range $\left[0, 2^{L_1-1}\right]$. Similarly, for the outputs of the second stage $\left\{G^2_{i,l_1,l_2}\right\}^{N,L_1,L_2}_{i,l_1,l_2=1,1,1}$, the weighted sum of the L_2 binarized quaternion feature maps is given as,

$$T^2_{i,l_1} = \sum_{l_2=1}^{L_2} 2^{l_2-1}\mathrm{H}\left(G^2_{i,l_1,l_2}\right) = \mathrm{S}\left(T^2_{i,l_1}\right) + \mathrm{I}\left(T^2_{i,l_1}\right)\boldsymbol{i} + \mathrm{J}\left(T^2_{i,l_1}\right)\boldsymbol{j} + \mathrm{K}\left(T^2_{i,l_1}\right)\boldsymbol{k} \tag{12}$$

Afterwards, we need to code each part of T^1_i and T^2_{i,l_1} separately. Taking $\mathrm{S}(T^1_i)$ as an example, $\mathrm{S}(T^1_i)$ is partitioned into B blocks and the histogram (with 2^{L_1} bins) of the decimal values is computed in each block. Then we concatenate all B histograms into one vector and denote this vector as $\mathrm{Bhist}\left(\mathrm{S}(T^1_i)\right)$. Similarly, by repeating the same method on remaining seven parts $\mathrm{I}(T^1_i), \mathrm{J}(T^1_i), \mathrm{K}(T^1_i), \mathrm{S}\left(T^2_{i,l_1}\right), \mathrm{I}\left(T^2_{i,l_1}\right), \mathrm{J}\left(T^2_{i,l_1}\right), \mathrm{K}\left(T^2_{i,l_1}\right)$, we obtain eight vectors. Therefore, for the input full quaternion image Q_i, the final feature vector is expressed as,

$$f_i = \left[f_{i,1}; f_{i,2}\right] \in \mathbb{R}^{4(2^{L_1} + L_1 2^{L_2})B} \tag{13}$$

where
$f_{i,1} = \left[\mathrm{Bhist}(\mathrm{S}(T^1_i)), \mathrm{Bhist}(\mathrm{I}(T^1_i)), \mathrm{Bhist}(\mathrm{J}(T^1_i)), \mathrm{Bhist}(\mathrm{K}(T^1_i))\right]^T \in \mathbb{R}^{4(2^{L_1})B}$,
$f_{i,2} = \left[\mathrm{Bhist}\left(\mathrm{S}\left(T^2_{i,l_1}\right)\right), \mathrm{Bhist}\left(\mathrm{I}\left(T^2_{i,l_1}\right)\right), \mathrm{Bhist}\left(\mathrm{J}\left(T^2_{i,l_1}\right)\right), \mathrm{Bhist}\left(\mathrm{K}\left(T^2_{i,l_1}\right)\right)\right]^T \in \mathbb{R}^{4L_1(2^{L_2})B}$.

Finally, these extracted features are used to train the SVM [22] for expression recognition.

4 Experimental Results

In this section, a series of experiments are performed on several facial expression datasets to demonstrate the performance of the proposed method.

4.1 Datasets

In experiments, the Radbound Faces Database (RafD) [23] is composed of 1407 images from 67 individuals with seven expressions (anger, disgust, fear, happiness, neutral, sadness, surprise) in three different gaze directions. Also containing abovementioned seven expressions, the Karolinska Directed Emotional Faces (KDEF) dataset has 490 images with frontal shooting angle of 70 individuals [24]. The MMI database includes six expressions (anger, disgust, fear, happiness, sadness, surprise) of 17 individuals, with a total of 306 images [25]. The Natural Visible and Infrared Expression Database (NVIE) contains 1374 images of six expressions [26].

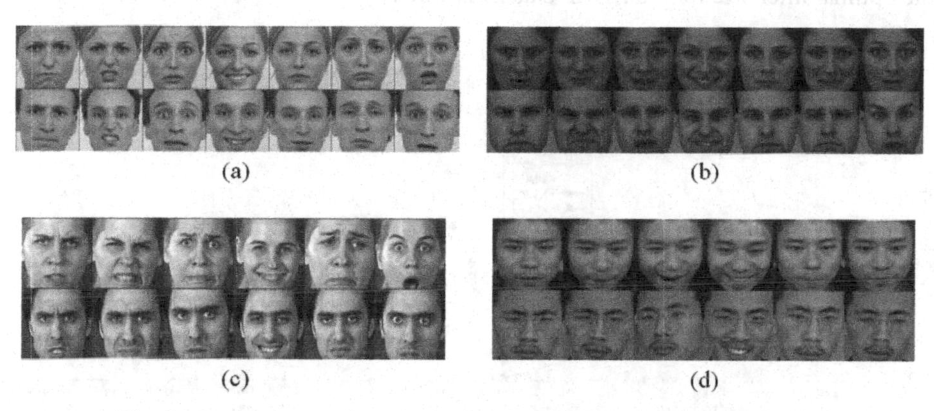

Fig. 2. Sample images from (a) RafD, (b) KDEF, (c) MMI and (d) NVIE

All images are resized to 64 × 64 and Fig. 2 shows some facial images of these datasets. For the RafD, MMI and NVIE, each dataset is divided into three groups as each individual has three images for each facial expression. One group is used as the tested set, and the remaining two groups are used as the training set. Three experiments are conducted separately, and the average value is regarded as the experimental result for each dataset. For KDEF, we randomly select two-thirds of the images in each expression as the training set, and the remaining images are served as the tested set.

4.2 Parameters Selection

For the proposed algorithm, the recognition accuracy is mainly affected by the neighborhood size of local variance, the size and number of filters and block size of histograms. In this subsection, the impact of a single variable on accuracy is quantitatively analyzed by varying one parameter while fixing the others to determine the optimal value, where the overlapping ratio is set at 0.5.

Firstly, the effect of different neighborhood sizes on recognition accuracy is tested, where the filter size is set at 7 × 7, the number of filters is set at L_1=L_2= 8 and the block size of histogram is equal to 8 × 8. The average recognition accuracy of different methods on four datasets is shown in Fig. 3(a), and it can be observed that, besides the NVIE dataset, the recognition accuracy is related to the neighborhood size. For the RafD dataset, the accuracy achieves the highest when the neighborhood size is set at 5 × 5. The MMI dataset has the highest accuracy when the neighborhood size is set at 7 × 7. And the optimal neighborhood size of the KDEF dataset is 3 × 3. This variation in optimal neighborhood size is primarily attributed to the differences in facial samples across datasets, including changes in pose and illumination.

Secondly, experiments are performed to test the impact of filter size on recognition accuracy, where the neighborhood sizes of local variance are fixed to the best fitting for each dataset, the number of filters is set at L_1=L_2= 8 and the block size of histogram is equal to 8 × 8. Figure 3(b) shows the results, the highest recognition accuracy can be obtained on the RafD, MMI, and NVIE datasets when the size of filters is equal to 9 × 9. The optimal filter size for the KDEF dataset is 7 × 7.

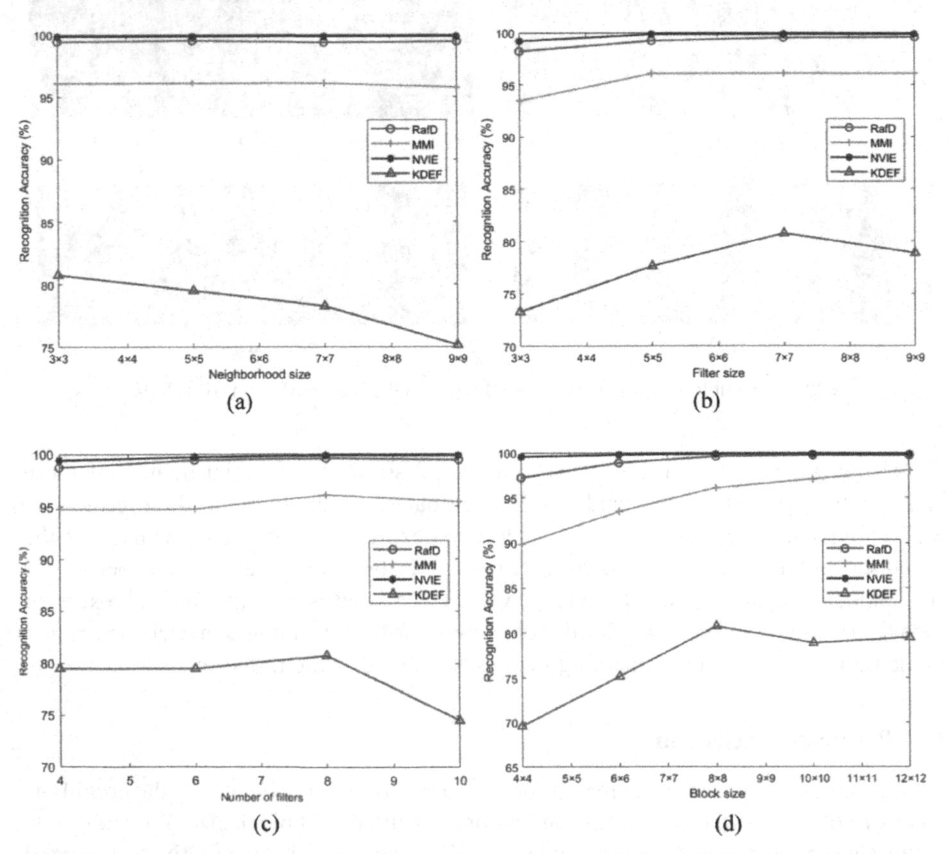

Fig. 3. Recognition accuracy under different (a) neighborhood sizes, (b) filter sizes, (c) number of filters and (d) block sizes

Then, experiments are conducted to determine the number of filters that obtain the optimal recognition rate. The neighborhood sizes of local variance and the filter size are fixed to the best fitting. The block size of histogram is equal to 8 × 8. Figure 3(c) shows the experimental results. We can see that when L_1=L_2= 8, the best accuracy can be achieved on the four datasets. Besides, Fig. 4 shows examples of QPCA filters in two stages. The first four rows are the real part and three imaginary parts of the first stage QPCA filters and the remaining correspond to the four parts of the second stage QPCA filters.

Furthermore, the impact of the block size on recognition accuracy is tested. The neighborhood sizes of local variance and the filter size are fixed to the best fitting. The number of filters is set at L_1=L_2= 8. Figure 3(d) demonstrates the recognition accuracy, it can be observed that the highest recognition accuracy can be obtained on the RafD, MMI, and NVIE datasets when the block size is equal to 12 × 12 while the optimal block size for the KDEF dataset is 8 × 8.

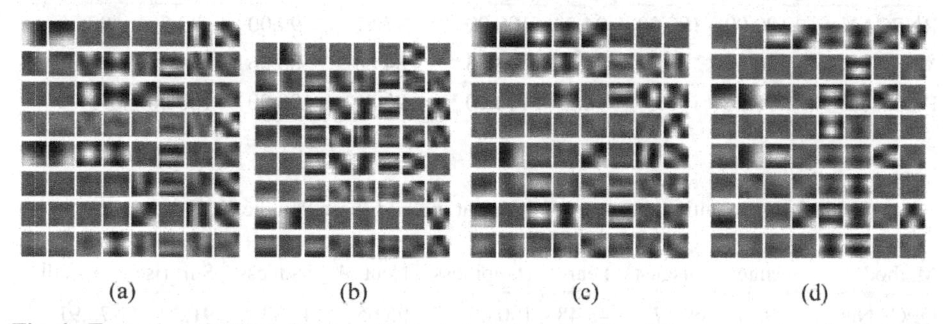

Fig. 4. Examples of QPCA filters in the first stage (rows 1–4) and the second stage (rows 5–8): (a) RafD, (b) KDEF, (c) MMI and (d) NVIE

4.3 Comparison with Other Methods

To further validate the recognition performance of the proposed method, the recognition accuracy of the proposed method is compared with those by using the QPCANet [16] method, the RNPCANet [13] method and the M3SPCANet [14] method. The parameter values for our proposed method are set according to the highest recognition accuracy obtained on each dataset in subsection 4.2.

Tables 1, 2, 3, and 4 show the recognition results for each expression using different methods on the four datasets. From these tables, we see that the average recognition rates of our MQPCANet are higher than the other methods in most cases. Our proposed method achieves overall recognition rates of 99.72% for the RafD dataset, 80.75% for the KDEF dataset, 98.04% for the MMI dataset and 99.85% for the NVIE dataset. Thus, the performance of our MQPCANet is better than the other methods on the four datasets for facial expression recognition.

Figure 5(a)–(d) show the confusion matrices of our MQPCANet for facial expression recognition on four datasets. As shown in Fig. 5(a), our method can precisely recognize most of the facial expressions in the RafD dataset. The recognition rates of fear and neutral are the lowest, at 98.51% and 99.50% respectively, while the others reach 100.00%. As shown in Fig. 5(b), compared to other expressions in the KDEF dataset, happiness and surprise are better recognized, with recognition accuracies of 100.00% and 95.65% respectively. As shown in Fig. 5(c), the recognition rates of anger and surprise reach 100.00%. However, the recognition accuracy for fear is the lowest at 94.12%, which is lower than MMI's overall accuracy of 98.04%. As shown in Fig. 5(d), anger and sadness in the NVIE dataset have the lowest recognition accuracy. Our method achieves recognition accuracies of 99.54% and 99.60%, respectively, for these two categories.

Table 1. Recognition rates (%) of different methods based on the RafD dataset

Methods	Anger	Disgust	Fear	Happiness	Neutral	Sadness	Surprise	Overall
QPCANet	100.00	100.00	94.53	100.00	99.01	99.50	100.00	99.01
RNPCANet	100.00	100.00	94.53	100.00	97.51	99.00	99.50	98.65
M3SPCANet	90.05	93.03	78.11	93.53	84.08	86.56	78.11	86.21
Proposed	100.00	100.00	98.51	100.00	99.50	100.00	100.00	**99.72**

Table 2. Recognition rates (%) of different methods based on the KDEF dataset

Methods	Anger	Disgust	Fear	Happiness	Neutral	Sadness	Surprise	Overall
QPCANet	69.57	69.57	43.48	100.00	95.65	47.83	91.30	73.91
RNPCANet	73.91	78.26	34.78	100.00	82.61	60.87	95.65	75.00
M3SPCANet	4.35	4.35	4.35	34.78	0.00	0.00	26.09	10.56
Proposed	78.26	86.96	52.17	100.00	86.96	65.22	95.65	**80.75**

Table 3. Recognition rates (%) of different methods based on the MMI dataset

Methods	Anger	Disgust	Fear	Happiness	Sadness	Surprise	Overall
QPCANet	92.16	92.16	90.20	94.12	94.12	96.08	93.14
RNPCANet	94.12	96.08	84.31	94.12	94.12	96.08	93.14
M3SPCANet	94.12	94.12	88.24	92.16	94.12	92.16	92.48
Proposed	100.00	98.04	94.12	98.04	98.04	100.00	**98.04**

Table 4. Recognition rates (%) of different methods based on the NVIE dataset

Methods	Anger	Disgust	Fear	Happiness	Sadness	Surprise	Overall
QPCANet	99.54	100.00	98.73	100.00	99.21	100.00	99.56
RNPCANet	99.54	98.57	99.16	100.00	99.21	99.49	99.34
M3SPCANet	99.09	99.52	99.58	100.00	99.21	100.00	99.56
Proposed	99.54	100.00	100.00	100.00	99.60	100.00	**99.85**

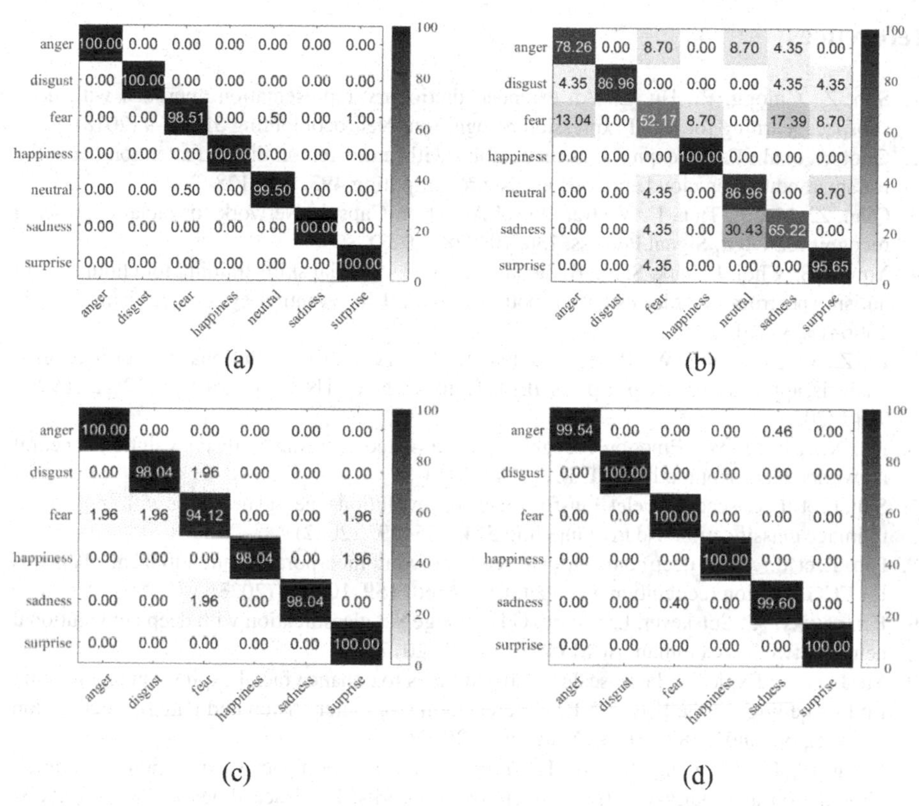

Fig. 5. Confusion matrix of our MQPCANet on (a) RafD, (b) KDEF, (c) MMI and (d) NVIE

5 Conclusion

In this paper, we investigated a new quaternion principal component analysis network for facial expression recognition, which is named as MQPCANet. It takes advantage of full quaternion representation. The activation function can further enhance the nonlinearity of feature maps and generalization ability of the algorithm. Moreover, the stacked features from multiple stages effectively combine global with local semantic information from the original data, resulting in improvement of recognition accuracy. The experimental results performed on several datasets show that the proposed MQPCANet outperforms the QPCANet, the RNPCANet, and the M3SPCANet in facial expression recognition task. Future work will concentrate on new lightweight quaternion convolutional neutral network for large-scale emotion recognition.

Acknowledgement. This work was supported by the National Natural Science Foundation of China (61876112, 61601311) and Science and Technology Planning Project of Jiaxing (2022AY10021).

References

1. Sun, Z., Chiong, R., Hu, Z.: An extended dictionary representation approach with deep subspace learning for facial expression recognition. Neurocomputing **316**, 1–9 (2018)
2. Guo, Y., et al.: Facial expressions recognition with multi-region divided attention networks for smart education cloud applications. Neurocomputing **493**, 119–128 (2022)
3. Qian, Z., Mu, J., Tian, F.: Ventral-Dorsal Attention Capsule Network for facial expression recognition. Digit. Signal Process. **136**, 103978 (2023)
4. Yun, S.-S., Choi, J., Park, S.-K., Bong, G.-Y., Yoo, H.: Social skills training for children with autism spectrum disorder using a robotic behavioral intervention system. Autism Res. **10**, 1306–1323 (2017)
5. Li, Z., Liu, F., Yang, W., Peng, S., Zhou, J.: A survey of convolutional neural networks: analysis, applications, and prospects. IEEE Transact. Neural Netw. Learn. Syst. **33**(12), 6999–7019 (2022)
6. He, X., Zhang, W.: Emotion recognition by assisted learning with convolutional neural networks. Neurocomputing **291**, 187–194 (2018)
7. Sun, J., et al.: Cascade wavelet transform based convolutional neural networks with application to image classification. Neurocomputing **514**, 285–295 (2022)
8. Fan, T., et al.: A new deep convolutional neural network incorporating attentional mechanisms for ECG emotion recognition. Comput. Biol. Med. **159**, 106938 (2023)
9. Krizhevsky, A., Sutskever, I., Hinton, G.E.: ImageNet classification with deep convolutional neural networks. Commun. ACM **60**(6), 84–90 (2017)
10. Yu, J., et al.: Exploring large-scale unlabeled faces to enhance facial expression recognition. In: Proceedings of the IEEE/CVF Conference on Computer Vision and Pattern Recognition (CVPR), pp. 5802–5809. IEEE, Vancouver (2023)
11. Wang, L., Jia, G., Jiang, N., Wu, H., Yang, J.: Ease: robust facial expression recognition via emotion ambiguity-sensitive cooperative networks. In: Proceedings of the 30th ACM International Conference on Multimedia, pp. 218–227. ACM, Lisboa (2022)
12. Chan, T., Jia, K., Gao, S., Lu, J., Zeng, Z., Ma, Y.: PCANet: a simple deep learning baseline for image classification? IEEE Trans. Image Process. **24**(12), 5017–5032 (2015)
13. Qaraei, M., Abbaasi, S., Ghiasi-Shirazi, K.: Randomized non-linear PCA networks. Inf. Sci. **545**, 241–253 (2021)
14. Zhou, D., Feng, S.: M3SPCANet: a simple and effective ConvNets with unsupervised predefined filters for face recognition. Eng. Appl. Artif. Intell. **113**, 104936 (2022)
15. Shao, Z., Liu, X., Yao, Q., Qi, N., Shang, Y., Zhang, J.: Multiple-image encryption based on chaotic phase mask and equal modulus decomposition in quaternion gyrator domain. Signal Process-Image Commun. **80**, 115662 (2020)
16. Zeng, R., et al.: Color image classification via quaternion principal component analysis network. Neurocomputing **216**, 416–428 (2016)

17. Zou, C., Kou, K.I., Wang, Y., Tang, Y.Y.: Quaternion block sparse representation for signal recovery and classification. Signal Process. **179**, 107849 (2021)
18. Shi, J., Zheng, X., Wu, J., Gong, B., Zhang, Q., Ying, S.: Quaternion Grassmann average network for learning representation of histopathological image. Pattern Recogn. **89**, 67–76 (2019)
19. Liu, W., Kou, K.I., Miao, J., Cai, Z.: Quaternion scalar and vector norm decomposition: quaternion PCA for color face recognition. IEEE Trans. Image Process. **32**, 446–457 (2022)
20. Xu, Z., Shao, Z., Shang, Y., Li, B., Ding, H., Liu, T.: Fusing structure and color features for cancelable face recognition. Multimed. Tools Appl. **80**, 14477–14494 (2021)
21. Dubey, S.R., Singh, S.K., Chaudhuri, B.B.: Activation functions in deep learning: a comprehensive survey and benchmark. Neurocomputing **503**, 92–108 (2022)
22. Fan, R.E., Chang, K.W., Hsieh, C.J., Wang, X.R., Lin, C.J.: LIBLINEAR: a library for large linear classification. J. Mach. Learn. Res. **9**, 1871–1874 (2008)
23. Langner, O., Dotsch, R., Bijlstra, G., Wigboldus, D.H.J., Hawk, S.T., van Knippenberg, A.: Presentation and validation of the Radboud Faces Database. Cognit. Emot. **24**(8), 1377–1388 (2010)
24. Goeleven, E., De Raedt, R., Leyman, L., Verschuere, B.: The Karolinska directed emotional faces: a validation study. Cognit. Emot. **22**(6), 1094–1118 (2008)
25. Pantic, M., Valstar, M., Rademaker, R., Maat, L.: Web-based database for facial expression analysis. In: IEEE International Conference on Multimedia and Expo (ICME), pp. 317–321. IEEE, Amsterdam (2005)
26. Wang, S., et al.: A natural visible and infrared facial expression database for expression recognition and emotion inference. IEEE Trans. Multim. **12**(7), 682–691 (2010)

Joint Relation Modeling and Feature Learning for Class-Incremental Facial Expression Recognition

Yuanling Lv[1,2], Yan Yan[1,2](✉), and Hanzi Wang[1]

[1] Fujian Key Laboratory of Sensing and Computing for Smart City, School of Informatics, Xiamen University, Xiamen, China
lvyuanling@stu.xmu.edu.cn, {yanyan,hanzi.wang}@xmu.edu.cn
[2] State Key Laboratory of Integrated Services Networks (Xidian University), Xi'an, China

Abstract. Due to the diversity of human emotions, it is often difficult to collect all the expression categories at once in many practical applications. In this paper, we investigate facial expression recognition (FER) under the class-incremental learning (CIL) paradigm, where we define easily-accessible basic expressions as an initial task and learn new compound expressions continuously. To this end, we propose a novel joint relation modeling and feature learning (JRF) method, which mainly consists of a local nets module (LNets), a dynamic relation modeling module (DRM), and an adaptive feature learning module (AFL) by taking advantage of the relationship between old and new expressions, effectively alleviating the stability-plasticity dilemma. Specifically, we develop LNets to capture subtle distinctions across expressions, where a novel diversity loss is designed to locate informative facial regions in each local net. Then, we introduce DRM to enhance feature representations based on two types of graph convolutional networks (GCNs) (including an image-shared GCN and two image-specific GCNs) from the perspectives of global-local graphs and old-new classes. Finally, we design AFL to explicitly fuse old and new class features via a weight selection mechanism. Extensive experiments on both in-the-lab and in-the-wild facial expression databases demonstrate the superiority of our method in comparison with several state-of-the-art methods for class-incremental FER.

Keywords: Facial expression recognition · Class-incremental learning · Relation modeling · Feature learning

1 Introduction

With the recent development of deep learning, a great number of facial expressions recognition (FER) methods [8,21] have been developed and made remarkable progress. These methods mainly focus on the classification of basic expressions according to Ekman and Friesen's study [6]. However, the limited categories

Q. Liu et al. (Eds.): PRCV 2023, LNCS 14429, pp. 134–146, 2024.
https://doi.org/10.1007/978-981-99-8469-5_11

of basic expressions fail to describe the complexity of human emotions in real scenarios. To comprehensively describe human emotions, Du et al. [5] define compound expressions, which are meaningful combinations of basic expressions. Compared with basic expressions, compound expressions are more fine-grained and involve more subtle distinctions for classification.

Existing FER methods usually train the models based on all the available expression data. On the one hand, in many practical applications, it is difficult to collect all the expression categories simultaneously due to the diversity of human emotions. Notably, we often cannot access all the old data because of the privacy issue of facial data. Therefore, these methods cannot directly apply to these applications. On the other hand, it is expensive to retrain a new model when new data arrive. Therefore, learning incrementally is essential in real-world. Recently, class-incremental learning (CIL) [11,17,20,28], which effectively avoids the heavy burden of retraining models as new classes emerge and the costly storage of old data, has become a hot learning paradigm. Due to storage restrictions, the model easily suffers from catastrophic forgetting (i.e., the model tends to remember new classes while forgetting old classes). Such a problem can be ascribed to the stability-plasticity dilemma (i.e., the conflict between retaining old knowledge and adapting to new concepts) [10].

In this paper, we study FER under the CIL paradigm, where we define easily-accessible basic expressions as initial classes and learn new compound expressions incrementally. For convenience, we call such a task class-incremental FER. Different from natural objects, facial expressions show close connections, which can be well exploited during incremental learning. As a result, we can leverage the intrinsic relationship between old and new expressions to perform class-incremental FER. To this end, we propose a novel joint relation modeling and feature learning (JRF) method for class-incremental FER. JRF mainly consists of a local nets module (LNets), a dynamic relation modeling module (DRM), and an adaptive feature learning module (AFL).

To be specific, LNets aggregate the spatial information across channels and locate informative regions to capture discriminative features. In each local net, an effective diversity loss is introduced to enable the model to discover potential discriminative regions for identifying subtle distinctions across expressions. To exploit the intrinsic relationship between expressions, DRM performs relation modeling to enhance feature representations from the perspectives of global-local graphs and old-new classes. Finally, AFL explicitly inserts the features from the old model into the new one to achieve the adaption of old and new class knowledge via a weight selection mechanism.

The main contributions of this paper are as follows:

- We propose a novel JRF method for class-incremental FER, where relation modeling and feature learning are jointly performed across old and new expressions, effectively alleviating the stability-plasticity dilemma.
- We present an image-shared GCN and two image-specific GCNs to describe the dependency of features in DRM. Such a way benefits maintaining the previously-learned knowledge and adapting to new classes. Moreover, we

incorporate the features from the old model into the learning of the new model in AFL, greatly preventing catastrophic forgetting.
- We conduct extensive experiments on both in-the-lab and in-the-wild facial expression databases to show the effectiveness of our method against several state-of-the-art methods.

2 Related Work

Facial Expression Recognition (FER). A large number of FER methods [8,21], which focus on the classification of basic expressions and train on all the predefined classes, have been proposed in unconstrained environments.

Recent studies have revealed that human beings can express more complex feelings that fall outside of these basic expressions [16]. Notably, Du et al. [5] introduce and define compound expressions to comprehensively describe human emotions and fit into practical applications in real scenarios. Due to the subtle differences across compound expressions, extracting fine-grained expression features is vitally important for identifying compound expressions. Zhang et al. [25] propose a two-stage recognition method to enhance the classification ability of compound expressions, while Li et al. [15] propose a multi-task meta-learning method with a novel alignment loss to learn refined expression representations.

The above methods train the model on all the available predefined classes. Unfortunately, collecting all expression categories at once is often difficult, where samples of new classes often arrive sequentially in many applications. In this paper, we introduce the CIL paradigm for compound FER, achieving a more practical FER learning paradigm.

Class-Incremental Learning (CIL). Existing CIL methods can be broadly classified into three groups: data-centric methods, model-centric methods, and algorithm-centric methods [27]. Data-centric methods [1] preserve samples of old classes for data replay based on different sampling modes or control optimization directions with old data. Model-centric methods [20,28] expand an additional network structure to dynamically adapt to task-specific features from different incremental tasks or estimate the importance of parameters. For instance, FOSTER [20] dynamically expands and compresses the model from the perspective of gradient boosting. MEMO [28] quantitatively measures the influence of different layers on the model and expands the specialized block incrementally. Algorithm-centric methods [11,17] design effective strategies (i.e., knowledge distillation or correction bias) to resist catastrophic forgetting. iCaRL [17] extends a distillation loss with an exemplar set while AFC [11] restricts the update of important features via the distillation loss. Different from natural objects in general CIL tasks, facial expressions show strong connections (i.e., the intrinsic relationship between basic and compound expressions). Thus, we can leverage such a cross-expression relationship to improve the FER performance with the CIL paradigm.

Recently, Zhu et al. [30] first introduce FER in the CIL paradigm. However, this method extracts features via a novel center-expression-distilled loss, and it only works on basic expressions. Unlike this method, we investigate compound FER under the CIL paradigm, which is a more challenging and practical task.

3 Proposed Method

3.1 Problem Formulation

We introduce the CIL paradigm for compound FER, where we define the classification of easily-accessible basic expressions as an initial task and that of compound expressions continuously as incremental tasks.

Assume that there is a sequence of $N+1$ incremental tasks $\{\mathcal{D}^0, D^1, \cdots, \mathcal{D}^N\}$, where $\mathcal{D}^0$ and $\mathcal{D}^n$ denote the initial task and the n-th incremental task, respectively. Following the commonly used rehearsal-based methods [17], we store a tiny number of exemplars from old classes as memory and fix them in the incremental process. At the n-th incremental task, a set of expression samples $\mathcal{B}_n = \{(\mathbf{x}_i^n, y_i^n)\}_{i=1}^B$ are constructed with the exemplars from old classes ever seen and all the samples from new classes at the n-th incremental task. Here, B is the number of samples; $\mathbf{x}_i^n \in \{\mathcal{E}^n \cup \mathcal{D}^n\}$ and $y_i^n \in \mathcal{Y}^n$ denote the i-th input image and its ground-truth label, respectively (note that the classes from different tasks are disjoint, that is, $\mathcal{Y}^t \cap \mathcal{Y}^{t'} = \varnothing$ for $t \neq t'$); $\mathcal{E}^n$ denotes the exemplars of old classes at the n-th incremental task. We evaluate the performance of the classes ever seen so far.

3.2 Overview

An overview of the proposed JRF is shown in Fig. 1. JRF is composed of a backbone, LNets, DRM, and AFL. Specifically, given an input image, the feature map is first extracted from the backbone. Then, LNets aggregate spatial information across channels and capture information from local regions. In each LNet, a novel diversity loss is designed to effectively encourage the model to focus on different discriminative regions, exploiting the subtle distinctions across expressions. Next, DRM performs relation modeling to enhance feature representations from the perspectives of global-local graphs and old-new classes. In particular, we leverage an image-shared GCN (a static GCN) to capture the dependency of features from a global view (the whole training data), and two image-specific GCNs (dynamic GCNs) to exploit the information from a local view (a specific image). Such a way can greatly alleviate the overfitting caused by data imbalance in the CIL paradigm. Finally, AFL selectively fuses information from old and new features to achieve the adaption of old and new class knowledge via a weight selection mechanism.

3.3 Local Nets Module (LNets)

Inspired by LANet [22], we design LNets (consisting of M LNets) to extract discriminative local features. Technically, each LNet is composed of two 1×1 convolutional layers to locate important local facial regions automatically. The first convolutional layer outputs C/r channels with a ReLU operation while the other outputs one channel followed by a Sigmoid operation and generates an

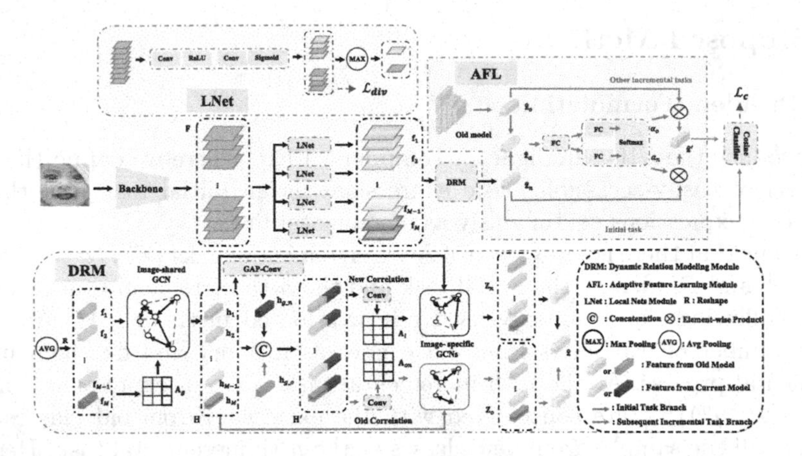

Fig. 1. Overview of our proposed JRF method. It consists of a backbone, a local nets module (LNets), a dynamic relation modeling module (DRM), and an adaptive feature learning module (AFL). We use ResNet-18 as the backbone.

attention map $\mathbf{F} \in \mathbb{R}^{C\times W\times H}$ for the i-th image, where r denotes the reduction ratio. C, W, and H are the channel number, width, and height, respectively.

From each LNet, a number of T-grouped attention maps are used to compute the maximum responses via the cross-channel max pooling (CCMP) operation [9], which can perverse the peaks of feature channels for fine-grained classification. To be specific, we can obtain the i-th group of attention maps $\mathbf{P}_i \in \mathbb{R}^{K\times H\times W}$ $(K = C/T)$, and then we reshape $\mathbf{P}_i$ to obtain grouped feature channels $\mathbf{G}_i \in \mathbb{R}^{K\times HW}$. To encourage the model to locate different parts, inspired by [2], an effective diversity loss for $\mathbf{G}_i$ of each LNet is defined as

$$\mathcal{L}_{div\_j} = \frac{1}{T}\sum_{i=1}^{T} h(\mathbf{G}_i), j = 1, \cdots, M \tag{1}$$

where $h(\mathbf{G}_i)$ can be computed as

$$h(\mathbf{G}_i) = \sum_{k=1}^{WH} \max_{j=1,2,\cdots,K} \left[\frac{e^{\mathbf{G}_{i,j,k}}}{\sum_{k'=1}^{WH} e^{\mathbf{G}_{i,j,k'}}}\right]. \tag{2}$$

The upper-bound of $\mathcal{L}_{div\_j}$ is equal to K when K feature maps focus on different local parts, while the lower bound is 1 when they all locate at the same part. Therefore, we expect to maximize the value of $\mathcal{L}_{div\_j}$ to its upper bound by utilizing a minus sign to $\mathcal{L}_{div\_j}$ through subtraction with the upper bound.

3.4 Dynamic Relation Modeling Module (DRM)

Due to the subtle distinctions between expressions, it is vital to capture and model the intrinsic relationship among expressions. Based on the attention features from LNet, we propose DRM for relation modeling from the perspectives of global-local graphs and old-new classes, enhancing feature representations. Our DRM is based on graph convolutional network (GCN) [12], which can effectively describe the dependency of nodes [3,19,24].

Definition. GCN is an efficient type of convolutional neural network (CNN) on graphs and learns feature representations of nodes over L layers. In LNet, we apply average pooling to $\mathbf{P}_i$ and reshape it into $\mathbf{f}_i \in \mathbb{R}^{1\times K}$. Then, a set of attentive features $\mathbf{F}' = [\mathbf{f}_1; \mathbf{f}_2; \cdots; \mathbf{f}_M] \in \mathbb{R}^{M\times K}$ (K denotes the dimensions of each feature) are obtained from LNets for the i-th facial image. Thus, GCN leverages their correlation to compute the adjacency matrix $\mathbf{A} \in \mathbb{R}^{M\times M}$. Then, the values of vertices are updated via the adjacency matrix $\mathbf{A}$ and the learnable weight matrix $\mathbf{W}$. The updated nodes $\mathbf{F}'^{(l+1)}$ of a single-layer are formulated as

$$\mathbf{F}'^{(l+1)} = \sigma_1(\mathbf{A}\mathbf{F}'^{(l)}\mathbf{W}^{(l)}), \tag{3}$$

where $\sigma_1(\cdot)$ is the Leaky ReLU function; $\mathbf{F}'^{(l)}$ and $\mathbf{W}^{(l)}$ denote the hidden representations and the learnable weight matrix, respectively, at the l-th layer;

We utilize a single-layer GCN to construct the global view and the output of this GCN is defined as $\mathbf{H} = \sigma_1(\mathbf{A}\mathbf{F}'\mathbf{W})$, where $\mathbf{H} = [\mathbf{h}_1, \mathbf{h}_2, \cdots, \mathbf{h}_M] \in \mathbb{R}^{M\times K}$.

Global-Local Dynamic Graph. We introduce two types of GCN (including an image-shared GCN and two image-specific GCNs) to describe the relationship between feature nodes from a global-local perspective. On one side, when the relationship among all the feature nodes is shared by all images of a dataset, the adjacency matrix (denoted as $\mathbf{A}_g$) in the image-shared GCN offers a global perspective to describe the dependency. On the other side, for each image, different local parts may show different contributions to expression recognition. Such a dynamic contribution change exhibits different dependency among these features. Therefore, we introduce an image-specific GCN to adaptively estimate the relationship for each image, whose adjacency matrix (denoted as $\mathbf{A}_l$) takes the image-specific information into consideration and offers a local perspective as a complementary part to the global one.

Old-New Dynamic Graph. Based on the above image-specific GCN, we also adopt another image-specific GCN (i.e., an old-new dynamic graph) to reflect how the model reacts when an image is given and insert this reaction to guide the learning of the new images. Such a way keeps the learned knowledge and navigates the learning of new knowledge based on the old one. Specifically, we compute the adjacency matrix $\mathbf{A}_{on}$ for old-new classes. Then, we leverage $\mathbf{A}_{on}$ and $\mathbf{A}_l$ on the global-local dynamic graph to update the features and fuse them to obtain the relationship-based features.

Dynamic Graph Combination. Inspired by [24], DRM is designed to generate attentive features by considering the relationship between local regions and the global one at the initial task. Based on this, DRM also models the relationship between old features from the old model and the new ones from the current model to resist forgetting in the subsequent incremental tasks. The output $\mathbf{Z} \in \mathbb{R}^{M\times K}$ of the dynamic graph is defined as

$$\mathbf{Z} = \sigma_2(\mathbf{A}'\mathbf{H}\mathbf{W}'), \tag{4}$$

where $\mathbf{A}' = \sigma_2(\mathbf{W}_a\mathbf{H}')$, which σ_2 is the Sigmoid activation function and $\mathbf{H}' = [(\mathbf{h}_1 : \mathbf{h}_g), (\mathbf{h}_2 : \mathbf{h}_g), \cdots, (\mathbf{h}_M : \mathbf{h}_g)] \in \mathbb{R}^{2K\times M}$, which is concatenated by $\mathbf{H}$ and the image-specific global representation $\mathbf{h}_g \in \mathbb{R}^K$. $\mathbf{h}_g$ is obtained via a global average pooling and followed by a convolution layer of $\mathbf{H}$. When it comes from the old model, the adjacency matrix $\mathbf{A}_{on}$ is constructed while the adjacency matrix $\mathbf{A}_l$ is computed when $\mathbf{h}_g$ is from the current model; $\mathbf{W}' \in \mathbb{R}^{K\times K}$ and $\mathbf{W}_a \in \mathbb{R}^{M\times 2K}$ denote the weights of dynamic GCN and the weights of a convolution layer to construct the dynamic correlation matrix $\mathbf{A}$, respectively; The symbol $\mathbf{A}'$ is $\mathbf{A}_l$ ($\mathbf{A}_{on}$) in global-local perspective (old-new perspective) and $\mathbf{h}_g$ that is from the old model (or the current model) is denoted as $\mathbf{h}_{g\_o}$ (or $\mathbf{h}_{g\_n}$) in $\mathbf{H}'$.

After obtaining the features updated by two types of relationships, we aggregate their information via a coefficient to generate enhanced features.

3.5 Adaptive Feature Learning Module (AFL)

To further improve the performance of incremental learning, AFL explicitly inserts the old features into the incremental learning process. Note that the old feature used in DRM is $\mathbf{h}_g$, which represents the global representation of the attentive features for relation modeling rather than the features for classification. Therefore, AFL integrates the final expression features via a weight selection mechanism to adaptively weight the information from multiple branches.

Inspired by [14], we obtain the fused feature $\hat{\mathbf{z}} \in \mathbb{R}^Q$ (i.e.,$Q = MK$) by the concatenation of the attentive features $\mathbf{Z}$ obtained by the DRM (i.e., $\mathbf{Z}_o$ from the old dynamic graph and $\mathbf{Z}_n$ from the new dynamic graph in the subsequent incremental task and add them to obtain $\hat{\mathbf{z}}$). Given the new feature $\hat{\mathbf{z}}_n$ and the old feature $\hat{\mathbf{z}}_o$ for the i-th image learned from the current model and the old model, we first combine the features from two branches via an element-wise summation, where $\hat{\mathbf{z}}_a = \hat{\mathbf{z}}_o + \hat{\mathbf{z}}_n$, and then we leverage a fully-connected (FC) layer to reduce the dimension for the compact representation $\mathbf{s} = FC(\hat{\mathbf{z}}_a) = \sigma_3(\mathbf{W}'_f(\hat{\mathbf{z}}_a))$, where σ_3, $\mathbf{W}'_f \in \mathbb{R}^{Q/r'\times Q}$ and r' are the ReLU operation, learnable weight and reduction ratio, respectively.

Next, we leverage soft attention across channels to selectively fuse information to obtain the final feature $\hat{\mathbf{z}}' = [\hat{\mathbf{z}}'_1, \cdots, \hat{\mathbf{z}}'_Q]$ for classification, where $\hat{\mathbf{z}}'_q = \alpha_{o_q}\hat{\mathbf{z}}_{o_q} + \alpha_{n_q}\hat{\mathbf{z}}_{n_q}$. Note that we use a softmax operator on channel-wise digits $\alpha_{o_q} = \frac{e^{A_{o_q}\mathbf{s}}}{e^{A_{o_q}\mathbf{s}}+e^{A_{n_q}\mathbf{s}}}$, $\alpha_{n_q} = \frac{e^{A_{n_q}\mathbf{s}}}{e^{A_{o_q}\mathbf{s}}+e^{A_{n_q}\mathbf{s}}}$, where α_o and α_n respectively denote the weights of old and new features while A_{o_q} and α_{o_q} are the q-th row of $A_o \in \mathbb{R}^{Q\times Q/r'}$ and the q-th element of α_o, respectively, likewise A_n and α_n.

3.6 Joint Loss and Inference

Classification Loss. We adopt a cosine classifier for classification in the training process, and the classification loss can be formulated as

$$\mathcal{L}_c = -\sum_{j\in\mathcal{Y}_n} \mathbb{1}_{[j=y_c]}\log(\theta(\hat{\mathbf{z}}')), \tag{5}$$

where θ is the cosine classifier. When $j = y_c$, the function $\mathbb{1}_{[j=y_c]}$ is equal to 1; otherwise, its value is 0.

Based on the above formulations, the joint loss is given as

$$\mathcal{L} = \mathcal{L}_d + \lambda_1 \mathcal{L}_c + \lambda_2 \sum_{j=1}^{M} (K - \mathcal{L}_{div\_j}), \tag{6}$$

where $\mathcal{L}_d$ is a simple distillation loss [17]; λ_1 and λ_2 are the balancing parameters.

Inference. We adopt a nearest-mean-of-exemplars classification strategy as [17], which computes a prototype vector for each class. This strategy can be formulated as $\mu_{y_c} = \frac{1}{|y_c|} \sum_{j \in y_c} \hat{\mathbf{z}}'_j$ and then classifies the label with the most similar prototype as $y^* = argmin_{j \in \mathcal{Y}_n} ||\hat{\mathbf{z}}' - \mu_j||$.

4 Experiments

4.1 Facial Expression Databases

In this paper, we evaluate our method on an in-the-lab database (i.e., CFEE [5]) and two in-the-wild databases (i.e., RAF-DB [13] and EmotioNet [7]). CFEE is collected from 230 human subjects, which contains 7 basic expressions (with 1,610 images) and 15 compound expressions (with 3,450 images). RAF-DB contains 7 basic expressions, including 15,339 images (with 12,271 training images and 3,068 test images), and 11 compound expressions, including 3,954 images (with 3,162 training images and 792 test images). For EmotioNet, we use the second track of the EmotioNet Challenge, which contains 2,478 images with 6 basic expressions and 10 compound expressions.

4.2 Implementation Details

Each facial image is first aligned and then resized to the size of 224×224. All the results are reported based on PyCIL [26] (a Python toolbox for CIL). For each incremental task, we store 20 old exemplars as [17] and train the model for 40 epochs with a batch size of 32 via stochastic gradient descent [18] (with the initial learning rate 0.01 and 0.001 at the incremental tasks). We first train our method on basic expressions as the initial task and learn new compound expressions as incremental tasks. The number of incremental classes is $C = 3$ or $C = 5$ in incremental tasks. The parameters λ_1 and λ_2 are empirically set to 1.000 and 0.0001, respectively. We evaluate our method on the test data of the classes ever seen so far and report the results (average accuracy±standard deviation), as done in [23]. We empirically set $T = 4$ and $M = 4$ in LNets.

4.3 Ablation Studies

The ablation study results on RAF-DB are shown in Table 1, where iCaRL is used as our baseline.

Influence of LNets. We evaluate the effectiveness of LNet (denoted as Baseline+LNets (w.o. *div*), which applies LNets into Baseline without the diversity loss. We can see that Baseline+LNets (w.o. *div*) performs better than Baseline since it can encourage the model to focus on the local subtle distinctions and achieves 4.04% improvements in $C=3$. Meanwhile, Baseline+LNets outperforms Baseline+LNets (w.o. *div*), which indicates the effectiveness of the diversity loss.

Table 1. Ablation studies for several variants of our method with the different numbers of incremental classes $C=3$ and $C=5$ on RAF-DB. 'Avg±std' denotes the average accuracy (%) and the standard deviation over the incremental tasks. The best results are marked in **bold**.

Methods	Avg ± std	
	$C=3$	$C=5$
Baseline	$63.33_{\pm 0.79}$	$63.96_{\pm 0.22}$
Baseline+LNets (w.o. *div*)	$67.37_{\pm 0.56}$	$67.74_{\pm 0.12}$
Baseline+LNets	$67.64_{\pm 0.52}$	$67.86_{\pm 0.13}$
Baseline+LNets+DRM (l)	$67.99_{\pm 0.68}$	$69.12_{\pm 0.44}$
Baseline+LNets+DRM (l-g)	$69.88_{\pm 0.82}$	$70.64_{\pm 0.18}$
Baseline+LNets+DRM	$69.97_{\pm 0.85}$	$71.18_{\pm 0.07}$
Baseline+LNets+AFL	$68.15_{\pm 0.87}$	$68.91_{\pm 0.41}$
Baseline+LNets+DRM+AFL	$\mathbf{71.06}_{\pm 0.97}$	$\mathbf{71.23}_{\pm 0.35}$

Influence of DRM. Baseline+LNets+DRM (l), Baseline+LNets+DRM (l-g), and Baseline+LNets+DRM represent the methods that DRM only models the relationship in a global view via an image-specific GCN, DRM only leverages the global-local dynamic graph, and DRM only utilizes the combination of the global-local dynamic graph and the old-new dynamic graph, respectively. Baseline+LNets+DRM (l) performs worse than Baseline+LNets+DRM (l-g). This indicates that not all the images can adapt to the same patterns of dependency shared across images. Moreover, the recognition accuracy is improved when the relationships of the global-local graph and old-new classes are integrated into the model, which validates the importance and effectiveness of relation modeling via GCN. Compared with the method using only one type of relationship, the combination of these two types of relationship performs the best, achieving 2.33% and 3.32% improvements in $C=3$ and $C=5$ between Baseline+LNets+DRM and Baseline+LNets. Since the global-local relationship mainly focuses on subtle identification and the old-new relationship focuses on old and new knowledge adaption, both of them can supplement each other to boost the performance.

Influence of AFL. The performance of JRF is further boosted when AFL is used by exploiting the learned feature information from old classes. In such a case, different information from multiple branches can selectively adapt to the old and new class knowledge via a weight selection mechanism, with 0.51% and 1.05% improvements with the comparison between Baseline+LNets and Baseline+LNets+AFL. Meanwhile, a similar performance can be seen in the comparison of Baseline+LNets+DRM and Baseline+LNets+DRM+AFL. The above results show the effectiveness of AFL via a weight selection mechanism.

4.4 Comparison with State-of-the-Art Methods

Table 2 shows the results obtained by our proposed method and several outstanding CIL methods, and the performance curves are also shown in Fig. 2.

Table 2. Performance comparisons (the average accuracy (%) and the standard deviation over the incremental tasks) between our proposed method and several state-of-the-art methods with the different numbers of incremental classes $C=3$ and $C=5$ on CFEE, RAF-DB, and EmotioNet. The best results are marked in **bold**.

Methods	CFEE		RAF-DB		EmotioNet	
	$C=3$	$C=5$	$C=3$	$C=5$	$C=3$	$C=5$
Finetune	$57.31_{\pm1.84}$	$59.68_{\pm1.75}$	$44.12_{\pm1.15}$	$45.13_{\pm0.69}$	$52.75_{\pm1.39}$	$55.91_{\pm3.03}$
PODNet [4]	$63.82_{\pm1.85}$	$66.31_{\pm1.55}$	$58.36_{\pm1.20}$	$61.02_{\pm0.92}$	$56.11_{\pm0.57}$	$59.73_{\pm1.32}$
COIL [29]	$56.35_{\pm1.26}$	$58.25_{\pm0.47}$	$47.73_{\pm2.65}$	$48.34_{\pm1.13}$	$52.85_{\pm2.21}$	$56.38_{\pm1.62}$
AFC [11]	$65.54_{\pm1.75}$	$66.81_{\pm1.49}$	$68.59_{\pm1.11}$	$66.96_{\pm0.47}$	$59.79_{\pm1.50}$	$61.75_{\pm0.91}$
MEMO [28]	$66.01_{\pm2.28}$	$67.95_{\pm1.97}$	$63.22_{\pm1.47}$	$62.49_{\pm0.72}$	$57.87_{\pm1.85}$	$58.73_{\pm0.93}$
SCN [21]	$46.62_{\pm0.23}$	$57.67_{\pm1.08}$	$51.31_{\pm1.37}$	$40.34_{\pm1.45}$	$50.21_{\pm1.84}$	$55.40_{\pm1.43}$
DACL [8]	$55.90_{\pm1.30}$	$53.73_{\pm1.29}$	$43.34_{\pm2.53}$	$53.14_{\pm1.99}$	$53.99_{\pm1.48}$	$55.67_{\pm1.50}$
Baseline	$67.39_{\pm1.25}$	$68.27_{\pm1.64}$	$63.33_{\pm0.79}$	$63.96_{\pm0.22}$	$59.48_{\pm0.44}$	$61.40_{\pm0.77}$
JRF (Ours)	$\mathbf{67.63}_{\pm1.86}$	$\mathbf{69.31}_{\pm1.37}$	$\mathbf{71.06}_{\pm0.97}$	$\mathbf{71.23}_{\pm0.35}$	$\mathbf{61.39}_{\pm1.56}$	$\mathbf{62.80}_{\pm0.35}$

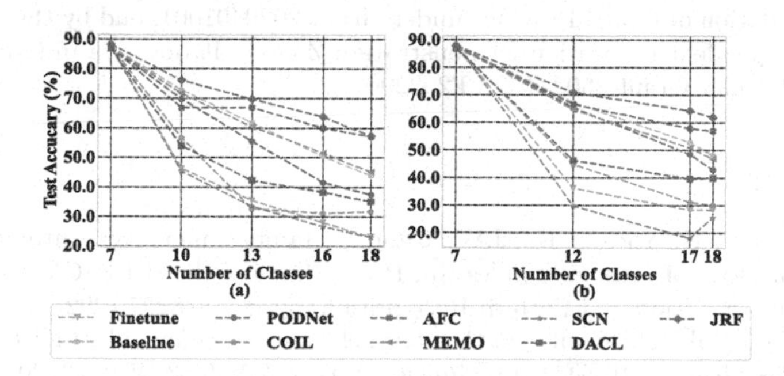

Fig. 2. Test accuracy vs. the number of classes obtained by nine different methods for (a) $C=3$ and (b) $C=5$ on RAF-DB.

The Finetune method, which learns from the new classes without the restriction of learned knowledge, is prone to fit new classes and forgets former knowledge severely. PODNet and AFC explore different ways for distillation to resist forgetting while MEMO investigates the dynamic network structures to adjust old and new knowledge. Different from these methods, our JRF method investigates the relationship between expressions, not only capturing subtle distinctions across expressions but also adapting to old and new class knowledge. JRF achieves the best average accuracy among all computing methods on all databases (67.63% (69.31%), 71.06% (71.23%), and 61.39% (62.80%) on CFEE, RAF-DB, and EmotioNet, respectively, in $C = 3$ ($C = 5$), with comparable standard deviations). Meanwhile, we also represent two FER methods (SCN and DACL). SCN may relabel the old class samples to new classes while DACL performs better but still forgets the old classes. On the contrary, our JRF method considers the global-local relationship to identify the subtle distinctions and explores the old-new relationship to alleviate the stability-plasticity dilemma.

5 Conclusion

In this paper, we develop a novel JRF method for class-incremental FER. In JRF, we design LNets for subtle distinctions discovery and leverage DRM for performing relation modeling to enhance feature representations from the perspectives of global-local graphs and old-new classes. Moreover, we also integrate learned knowledge via AFL to adapt to the old and new class knowledge during incremental learning. Based on the above designs, we effectively alleviate the stability-plasticity dilemma in class-incremental FER. Extensive experiments show the effectiveness of our proposed method.

Acknowledgements. This work was supported by the National Natural Science Foundation of China under Grants 62372388, 62071404, U21A20514, by the Natural Science Foundation of Fujian Province under Grant 2020J01001, and by the Fuxiaquan National Independent Innovation Demonstration Zone Collaborative Innovation Platform Project under Grant 3502ZCQXT2022008.

References

1. Bang, J., Kim, H., Yoo, Y., Ha, J.W., Choi, J.: Rainbow memory: continual learning with a memory of diverse samples. In: Proceedings of the IEEE/CVF Conference on Computer Vision and Pattern Recognition, pp. 8218–8227 (2021)
2. Chang, D., et al.: The devil is in the channels: mutual-channel loss for fine-grained image classification. IEEE Trans. Image Process. **29**, 4683–4695 (2020)
3. Chen, A., Zhou, Y.: An attention enhanced graph convolutional network for semantic segmentation. In: Peng, Y., et al. (eds.) PRCV 2020. LNCS, vol. 12305, pp. 734–745. Springer, Cham (2020). https://doi.org/10.1007/978-3-030-60633-6_61
4. Douillard, A., Cord, M., Ollion, C., Robert, T., Valle, E.: PODNet: pooled outputs distillation for small-tasks incremental learning. In: Vedaldi, A., Bischof, H., Brox, T., Frahm, J.-M. (eds.) ECCV 2020. LNCS, vol. 12365, pp. 86–102. Springer, Cham (2020). https://doi.org/10.1007/978-3-030-58565-5_6

5. Du, S., Tao, Y., Martinez, A.M.: Compound facial expressions of emotion. Proc. Natl. Acad. Sci. **111**(15), E1454–E1462 (2014)
6. Ekman, P., Friesen, W.V.: Constants across cultures in the face and emotion. J. Pers. Soc. Psychol. **17**(2), 124–129 (1971)
7. Fabian Benitez-Quiroz, C., Srinivasan, R., Martinez, A.M.: EmotioNet: an accurate, real-time algorithm for the automatic annotation of a million facial expressions in the wild. In: Proceedings of the IEEE Conference on Computer Vision and Pattern Recognition, pp. 5562–5570 (2016)
8. Farzaneh, A.H., Qi, X.: Facial expression recognition in the wild via deep attentive center loss. In: Proceedings of the IEEE/CVF Winter Conference on Applications of Computer Vision, pp. 2402–2411 (2021)
9. Goodfellow, I., Warde-Farley, D., Mirza, M., Courville, A., Bengio, Y.: Maxout networks. In: International Conference on Machine Learning, pp. 1319–1327 (2013)
10. Grossberg, S.: Adaptive resonance theory: how a brain learns to consciously attend, learn, and recognize a changing world. Neural Netw. **37**, 1–47 (2013)
11. Kang, M., Park, J., Han, B.: Class-incremental learning by knowledge distillation with adaptive feature consolidation. In: Proceedings of the IEEE/CVF Conference on Computer Vision and Pattern Recognition, pp. 16071–16080 (2022)
12. Kipf, T.N., Welling, M.: Semi-supervised classification with graph convolutional networks. arXiv preprint arXiv:1609.02907 (2016)
13. Li, S., Deng, W., Du, J.: Reliable crowdsourcing and deep locality-preserving learning for expression recognition in the wild. In: Proceedings of the IEEE Conference on Computer Vision and Pattern Recognition, pp. 2852–2861 (2017)
14. Li, X., Wang, W., Hu, X., Yang, J.: Selective kernel networks. In: Proceedings of the IEEE/CVF Conference on Computer Vision and Pattern Recognition, pp. 510–519 (2019)
15. Li, X., Deng, W., Li, S., Li, Y.: Compound expression recognition in-the-wild with au-assisted meta multi-task learning. In: Proceedings of the IEEE/CVF Conference on Computer Vision and Pattern Recognition, pp. 5734–5743 (2023)
16. Liu, Y., et al.: MAFW: a large-scale, multi-modal, compound affective database for dynamic facial expression recognition in the wild. arXiv preprint arXiv:2208.00847 (2022)
17. Rebuffi, S.A., Kolesnikov, A., Sperl, G., Lampert, C.H.: iCaRL: incremental classifier and representation learning. In: Proceedings of the IEEE Conference on Computer Vision and Pattern Recognition, pp. 2001–2010 (2017)
18. Robbins, H., Monro, S.: A stochastic approximation method. Ann. Math. Stat. 400–407 (1951)
19. Song, S., Huang, H., Wang, J., Zheng, A., He, R.: Prior-guided multi-scale fusion transformer for face attribute recognition. In: Yu, S., et al. (eds.) PRCV 2022. LNCS, vol. 13534, pp. 645–659. Springer, Cham (2022). https://doi.org/10.1007/978-3-031-18907-4_50
20. Wang, F.Y., Zhou, D.W., Ye, H.J., Zhan, D.C.: FOSTER: feature boosting and compression for class-incremental learning. arXiv preprint arXiv:2204.04662 (2022)
21. Wang, K., Peng, X., Yang, J., Lu, S., Qiao, Y.: Suppressing uncertainties for large-scale facial expression recognition. In: Proceedings of the IEEE/CVF Conference on Computer Vision and Pattern Recognition, pp. 6897–6906 (2020)
22. Wang, Q., Guo, G.: LS-CNN: characterizing local patches at multiple scales for face recognition. IEEE Trans. Inf. Forensics Secur. **15**, 1640–1653 (2020)
23. Yan, S., Xie, J., He, X.: DER: dynamically expandable representation for class incremental learning. In: Proceedings of the IEEE/CVF Conference on Computer Vision and Pattern Recognition, pp. 3014–3023 (2021)

24. Ye, J., He, J., Peng, X., Wu, W., Qiao, Yu.: Attention-driven dynamic graph convolutional network for multi-label image recognition. In: Vedaldi, A., Bischof, H., Brox, T., Frahm, J.-M. (eds.) ECCV 2020. LNCS, vol. 12366, pp. 649–665. Springer, Cham (2020). https://doi.org/10.1007/978-3-030-58589-1_39
25. Zhang, Z., Yi, M., Xu, J., Zhang, R., Shen, J.: Two-stage recognition and beyond for compound facial emotion recognition. In: Proceedings of the IEEE International Conference on Automatic Face & Gesture Recognition, pp. 900–904 (2020)
26. Zhou, D.W., Wang, F.Y., Ye, H.J., Zhan, D.C.: PyCIL: a python toolbox for class-incremental learning. arXiv preprint arXiv:2112.12533 (2021)
27. Zhou, D.W., Wang, Q.W., Qi, Z.H., Ye, H.J., Zhan, D.C., Liu, Z.: Deep class-incremental learning: a survey (2023)
28. Zhou, D.W., Wang, Q.W., Ye, H.J., Zhan, D.C.: A model or 603 exemplars: towards memory-efficient class-incremental learning. In: International Conference on Learning Representations (2023)
29. Zhou, D.W., Ye, H.J., Zhan, D.C.: Co-transport for class-incremental learning. In: Proceedings of the ACM International Conference on Multimedia, pp. 1645–1654 (2021)
30. Zhu, J., Luo, B., Zhao, S., Ying, S., Zhao, X., Gao, Y.: IExpressNet: facial expression recognition with incremental classes. In: Proceedings of the ACM International Conference on Multimedia, pp. 2899–2908 (2020)

RepGCN: A Novel Graph Convolution-Based Model for Gait Recognition with Accompanying Behaviors

Zijie Mei, Zhanyong Mei(✉), He Tong, Sijia Yi, Hui Zeng, and Yingyi Li

College of Computer Science and Cyber Security, Chengdu University of Technology, Chengdu 610059, China
meizhanyong2014@cdut.edu.cn

Abstract. Currently, two challenges exist in the field of gait recognition: (1) there is a lack of gait datasets that include common accompanying behaviors during walking, and (2) it's necessary to improve feature representation in skeleton sequence data for model-based approaches. To address these concerns, we focused on the study of accompanying behavior-based walking conditions and multiple views, and utilizes depth cameras to collect gait data. We presented the *CDUT Gait* dataset to investigate the impact of various accompanying behaviors on gait recognition performance. And we proposed a *RepGCN*, a novel graph convolution networks model with innovative residual strategy in the spatial module, as well as new spatio-temporal feature extraction modules. Experiments demonstrate that *RepGCN* achieves state-of-the-art performance on *CDUT Gait* with minimal model parameters compared to existing model-based approaches. The combination of depth cameras and *RepGCN* has potential applications in access control, smart home, and anti-terrorism areas.

Keywords: Gait recognition · depth cameras · *CDUT Gait* · graph convolution networks

1 Introduction

Gait recognition is a promising research direction in biometrics, because of its advantages and distinctive characteristics, including long-distance recognition capability, non-interactivity, non-invasiveness, and resistance to deception. With the advancements in depth sensing technology, the use of depth cameras has gained traction in the field of gait recognition [1]. The body tracking capabilities of depth cameras allow for the direct acquisition of skeleton sequence data during human walking, making them highly suitable for the development of model-based gait recognition methods. However, the utilization of depth skeleton sequence data for gait recognition currently encounters several challenges that need to be addressed. Firstly, the existing studies often lack comprehensive consideration of accompanying behaviors during walking, as they predominantly focus on cross-view analysis. Secondly, there is a limited application of graph convolution in model-based approaches for extracting feature, highlighting the need for further

Q. Liu et al. (Eds.): PRCV 2023, LNCS 14429, pp. 147–158, 2024.
https://doi.org/10.1007/978-981-99-8469-5_12

advancements in characterizing skeleton sequence data using graph convolution-based methods.

Previous studies in gait recognition using depth cameras have focused on specific walking conditions, e.g. multi-view [2, 3], speed [4], carrying objects [5], and view occlusion [6]. Hofmann et al. created the *TUM GAID* dataset by collecting multimodal data using depth cameras [5]. This dataset includes RGB video, RGB-D, and audio modalities, making it one of the largest datasets available with 305 individuals across three variations i.e., normal walking, backpack, wearing coating shoes. The incorporation of multiple modalities enhances the dataset's potential for various research applications. Li et al. utilized *Azure Kinect DK* to collect multi-modal gait data and built $OG\ RGB + D$ Dataset [6]. $OG\ RGB + D$ is to study the impact of walking conditions on the performance of gait recognition under various visual occlusions. These walking conditions include 7 types: backpack, side small object, side large object, low-occlusion clothing, medium-occlusion clothing, high-occlusion clothing and three people walk together. Then, they proposed Siamese Spatio-Temporal Graph Convolutional Network (Siamese *STGCN*) to solve the case of severe view occlusion.

However, these studies did not fully consider the impact of multiple accompanying behaviors that occur during daily walking. In our study, we aim to explore the impact of different accompanying behaviors during walking (*ABW*) on the performance of a gait recognition model. By considering *ABW* of daily life, we can gain practical insights into the robustness and effectiveness of gait recognition methods in real-world scenarios. Thus, we constructed *CDUT Gait* dataset. The findings of this study have significant practical implications for applications such as access control, smart home, and anti-terrorism, where accurate and reliable gait recognition is crucial.

Gait recognition approach using depth cameras can be classified into two main categories: appearance-based and model-based methods. The appearance-based methods, such as *RGB-D* based approaches, typically achieve higher recognition rates [1, 7]. However, these methods often necessitate an increased number of learnable parameters, resulting in higher storage and computing requirements. On the other hand, model-based methods rely on human skeleton information for gait modeling [3, 6, 8]. These approaches require less storage space and have smaller parameter sizes, making them more resource-efficient. In our research, we adopted model-based methods to model the skeleton sequence data, providing an alternative with reduced resource demands.

Currently, deep learning-based methods have emerged as the mainstream approach in the field of gait recognition, encompassing both model-based and appearance-based methods. Deep learning approaches have proven to be robust in handling various covariate factors within the field of gait recognition [9]. However, popular methods like *CNNs* or *LSTMs* may not be optimal for processing non-Euclidean distance data, such as temporal skeletons [10]. To address this, Graph Convolutional Networks (*GCNs*) [11] have been introduced for gait recognition, effectively handling non-Euclidean distance data and addressing challenges like visual occlusion and cross-view recognition [6, 12]. Teepe et al. proposed *GaitGraph* that combines skeleton poses with *GCN* to obtain a model-based approach for gait recognition. The experimental results demonstrate that *GaitGraph* performs well cross-view compared to existing model-based methods and remains competitive with some appearance-based methods.

While existing *GCN*-based methods have demonstrated their effectiveness, they often lack specific enhancements for spatio-temporal feature extraction in gait recognition. In our work, we propose a novel model-based approach called *RepGCN*. It improves the spatio-temporal feature extraction by incorporating a well-designed residual strategy inspired by *RepVgg* [13]. Our experiments validate the effectiveness of incorporating novel residual connections into the graph convolution module. Overall, the designed residual connection strategy in *RepGCN* plays a crucial role in extracting discriminative spatio-temporal features and enhancing the model's ability to capture relevant patterns in gait sequence data. This could improve performance in gait recognition tasks.

The contributions of our works are as follows:

1. We constructed a new gait dataset with *Azure Kinect DK*: *CDUT Gait*. This dataset is distinct in that it includes more accompanying behaviors during walking and incorporates multiple views.
2. We proposed *RepGCN*, an innovative model-based gait recognition approach that achieves state-of-the-art performance on *CDUT Gait*. By incorporating a novel residual strategy in the spatial and spatio-temporal feature extraction modules, *RepGCN* surpasses existing model-based methods.

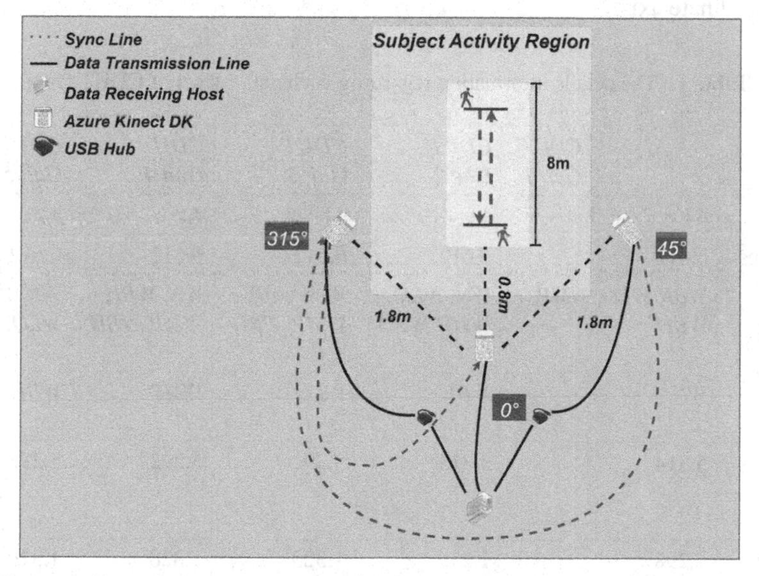

Fig. 1. The experimental scene was set up with three depth cameras positioned at angles of 0°, 315°, and 45°, arranged in a daisy-chain configuration for device synchronization. The subject's walking path is indicated by a light blue dotted line.

2 *CDUT Gait* Datasets

In our experimental design, we aim to investigate the impact of various accompanying behaviors during walking as well as different views on the performance of gait recognition. Figure 1 illustrates our experimental setup where the camera placement is designed to capture a broad range of skeletal motion in the sagittal, coronal, and horizontal planes. This configuration allows us to effectively capture and analyze the dynamics of human movement from multiple views.

For this data collection, 44 graduate students with independent walking ability were invited from *Chengdu University of Technology*. All participants were provided with detailed information regarding the experimental procedure, and they provided informed consent prior to participating in the experiment. The subjects were asked to walk in five different conditions: walking normally (*WN*), walking with a backpack (*WBP*), walking while lifting objects (*WLO*), walking while using a mobile phone (*WMP*) and walking while carrying objects with both hands (*WBH*). They walked on an 8-m-long walkway. Each subject has 150 skeletal gait sequences $\left(= \left[5(accompanyingbehaviorsduringwalking) \times 10(sessions) \times 3(views)\right]\right)$. The collected sequence data has a dimension of $T \times V \times C$, T represents the length of the temporal, V represents the number of human joint points and C represents dimension of the coordinate axis.

Table 1. The details information regarding each data subset of CDUT Gait.

	CDUT Gait 1	*CDUT Gait 2*	*CDUT Gait 3*	*CDUT Gait 4*	*CDUT Gait 5*
Subjects	44	44	44	44	44
Sequences	6,612	6,612	6,612	6,612	6,612
ABW in Training Set	*WBP, WLO, WMP, WBH*	*WN, WLO, WMP, WBH*	*WN, WBP, WMP, WBH*	*WN, WBP, WMP, WBH*	*WN, WBP, WLO, WMP*
ABW in Test Set	*WN*	*WBP*	*WLO*	*WMP*	*WBH*
Sequences in Training Set	5,314	5,313	5,292	5,292	5,292
Sequences in Test Set	1,298	1,299	1,320	1,320	1,320

The above abbreviations as follow:
ABW: accompanying behaviors during walking. *WN*: walking normally; *WBP*: walking with a backpack; *WLO*: walking while lifting objects; *WMP*: walking while using a mobile phone; *WBH*: walking while carrying objects with both hands.

In order to investigate the impact of accompanying behaviors while walking (*ABW*) on gait recognition, we adopt a similar protocol to the Subject-Independent Protocol

(SIP) [14] commonly used in gait recognition datasets i.e., *ABW*-Independent Protocol. (*ABWIP*). *ABWIP* requires that the accompanying behaviors appearing in the train data do not intersect with the ones in the test data. We utilized *Azure Kinect DK* to collect skeletal gait data and created the *CDUT Gait* dataset, which comprises five subsets. Each subset is composed of separate training and test sets. For example, in *CDUT Gait* 1, the test set is constructed using data from *WN*, while the rest of the data is utilized for training purposes. Similar approaches were used to construct the other subsets. Further details about each subset are given in Table 1. The dataset splitting method employed in *CDUT Gait* facilitates thorough and dependable evaluations of model performance in recognizing gait patterns influenced by various accompanying behaviors. In addition, *CDUT Gait* encompasses the cross-view covariate factor commonly found in numerous gait recognition datasets. As a result, *CDUT Gait* is a gait recognition dataset that integrates various intricate covariates, thereby facilitating a more effective evaluation of the robustness and generalization capabilities of gait recognition algorithms. This advancement plays a crucial role in the progression towards more accurate and reliable gait recognition algorithms, especially in real-world applications.

3 Gait Recognition with *RepGCN*

3.1 Preliminary of GCN

The Graph Convolution Network (*GCN*), initially introduced by Kipf et al. [11], pioneered a groundbreaking method that enabled convolution operations on graph-structured data. *GCN* can efficiently extract information between nodes while maintaining low computational cost, even for graph-structured data with arbitrary topology. *CNN* can handle typical computer vision tasks, which is mainly due to properties of the *CNN*, such as rotate invariance and transitional invariance. However, it's difficult for traditional *CNN* to deal with graph-structured data with random topology [15].

Generally, a gait skeleton graph in spatial domain can be defined as:

$$\mathcal{G} = (V, E), \tag{1}$$

where $V = \{v_1, v_2, \ldots, v_N\}$ is the set of vertices, E is the set of edges denotes the connection between vertices. E usually is represented by an adjacency matrix $A \in \mathbb{R}^{N \times N}$, and A can be represented as:

$$A = \begin{cases} 1, & \mathit{if}\ (i,j) \in L \\ 0, & \mathit{else} \end{cases}, \tag{2}$$

where L is the set of node pairs connected in graph. The layer-wise propagation rule of *GCN* as follows:

$$X^{(l+1)} = \sigma\left(\tilde{D}^{-\frac{1}{2}}\tilde{A}\tilde{D}^{-\frac{1}{2}}X^{l}W^{l}\right), \tag{3}$$

where X is feature representation of gait skeleton data. $\tilde{A} = I_n + A$ corresponds to the skeleton graph, where I_n is an identity matrix. $\tilde{D}$ is the diagonal degree matrix of $\tilde{A}$. l represents the number of propagation layers in the network. W^l represents a learnable weight matrix that is utilized throughout the entire propagation process. $\sigma(\cdot)$ is activation function, e.g., *ReLU*, *Sigmoid*.

3.2 RepGCN

In our work, we present *RepGCN*, a novel *GCN*-based network inspired by the *RepVgg* architecture [13]. *GaitGraph* adopts a similar approach to *ResGCN* [16] in constructing graph convolution blocks, specifically by introducing residual connections into the graph convolution. However, the main difference between *RepGCN* and *GaitGraph* lies in their distinct strategies for implementing residual connections. The residual strategy of *GaitGraph* can be expressed as:

$$F(x) = H(x) + g(x) \tag{4}$$

The residual strategy of *RepGCN* can be expressed as:

$$F(x) = H(x) + g(x) + x \tag{5}$$

$H(x)$ represents an underlying mapping that is to be fit by applying a few stacked layers, $g(x)$ indicates the residual connection of non-linear transformation, and x represents identity mapping. In *RepGCN*, the dual-residue connection structure is established by combining the output of $g(x)$ and x. Then, the overall spatio-temporal feature representation in *RepGCN* can be expressed as:

$$Y = \delta\left(Y_{spatial} + Y_{temp}\right) \tag{6}$$

and the counterpart of *GaitGraph* can be expressed as:

$$Y = \delta\left(l(x) + \delta(Y_{spatial} + Y_{temp})\right) \tag{7}$$

where $Y_{spatial}$ and Y_{temp} denote the feature representations of the spatial module and temporal module, respectively. $l(x)$ denotes the long residual connection that spans the feature extraction process within each spatio-temporal module in *GaitGraph*. Notably, we didn't incorporate this kind of long residual connection in *RepGCN*.

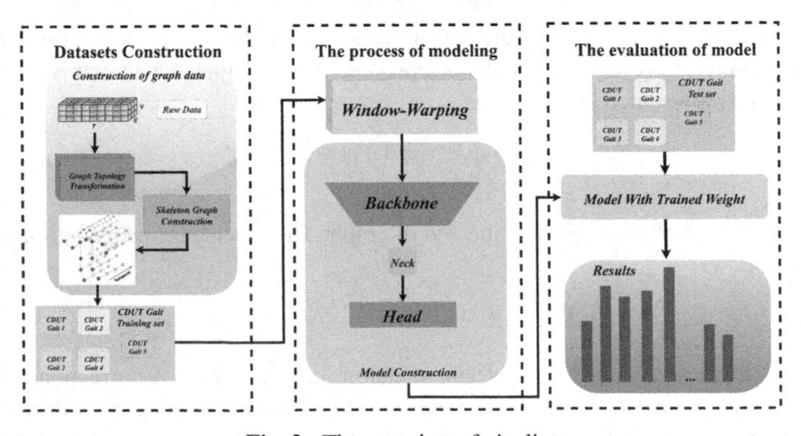

Fig. 2. The overview of pipeline

Figure 2 provides an overview of the pipeline used in our work, which consists of three main parts: dataset construction, model construction, and model evaluation. In the dataset construction part, we have described the process in Sect. 2. For the model construction part, we introduce a *Window-Warping* layer for data augmentation and to address overfitting or underfitting issues. Our model architecture, shown in Fig. 3f, comprises three components: backbone, neck, and head. The backbone includes 5 *RepGCN* blocks, each containing a temporal and spatial feature extraction module as depicted in Fig. 3e. The details of the *spatial module* and *temporal module* are shown in Fig. 3c and Fig. 3d, respectively. The *temporal module* primarily utilizes the Temporal Convolution Network (*TCN*) and used to extract temporal feature, which is implemented using a *1D CNN*. *Spatial module* is used to extract spatial feature. The *up sampling* and *down sampling*, which are included in both the *spatial module* and the *temporal module*, are shown in detail in Fig. 3a and Fig. 3b, respectively. The up-sampling and down-sampling blocks in our model adjust the number of channels in the feature map to facilitate efficient feature extraction and representation. The neck component enhances feature robustness through global average pooling. The head component consists of a linear layer and a softmax layer, working together to generate the final output used for identity recognition. The details of the network architecture are shown in Table 2.

In the *spatial module*, we employ *GCN* as the main feature extractor instead of *CNN*. This choice is motivated by the nature of skeleton-based gait data, where each joint is represented as one node in a graph without a fixed topology. Traditional *CNN* are not well-suited for extracting feature from undirected graph data. In contrast, *GCN*

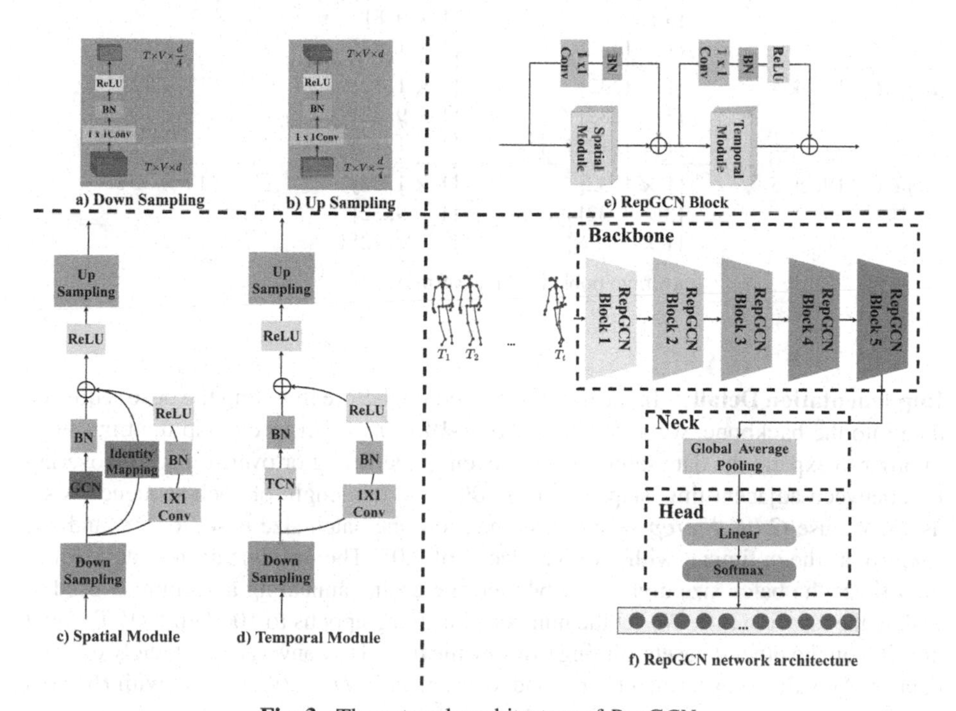

Fig. 3. The network architecture of *RepGCN*

are specifically designed to handle graph-structured data, making them more suitable for extracting feature from gait skeleton data.

4 Experiments

Evaluation Schemes. To comprehensively assess the model's robustness in handling changes in accompanying behaviors, we employ the *ABWIP* for model evaluation. This means that accompanying behaviors present in the training set are intentionally excluded from the test set. This evaluation method was applied to each subset of the *CDUT Gait*, which was specifically constructed in Sect. 2.

Table 2. An overview of the RepGCN network architecture.

layer name	spatial module	temporal module	output size
RepGCN Block 1	[1 × 1, 16] [3 × 3, 16] [1 × 1, 64]	[1 × 1, 16] [1 × 9, 16] [1 × 1, 64]	45 × 32
RepGCN Block 2	[1 × 1, 16] [3 × 3, 16] [1 × 1, 64]	[1 × 1, 16] [1 × 9, 16] [1 × 1, 64]	23 × 32
RepGCN Block 3	[1 × 1, 8] [3 × 3, 8] [1 × 1, 32]	[1 × 1, 8] [1 × 9, 8] [1 × 1, 32]	23 × 32
RepGCN Block 4	[1 × 1, 32] [3 × 3, 32] [1 × 1, 128]	[1 × 1, 32] [1 × 9, 32] [1 × 1, 128]	12 × 32
RepGCN Block 5	[1 × 1, 32] [3 × 3, 32] [1 × 1, 128]	[1 × 1, 32] [1 × 9, 32] [1 × 1, 128]	12 × 32
	average pool, 44-d fc, softmax		1 × 1

Implementation Details. In the training procedure, before inputting the pose sequence data into the backbone, we apply the *Window-Warping* layer to overlap the time steps in order to expand the data samples and prevent underfitting or overfitting. The overlap rate between adjacent time steps is set to 30%, and the length of each sequence is set as 45. We use *Cross Entropy* as the loss function, the batch size is set to 128, and use *AdamW* as the optimizer with a weight decay of 0.05. The initial learning rate is set to 5e-4 times the batch size divided by 64, and the cosine annealing algorithm is used to update the learning rate. We set the number of training epochs to 1000 on *CDUT Gait 4* and 500 on the other datasets, aiming to allow the model to converge completely on each dataset. And all the experiments are conducted on a *NVIDIA 2080ti GPU* with *PyTorch 3.8*.

Comparison with Other Model-based Approach. To ensure a fair comparison of the performance of existing model-based approaches, we implemented the backbone and neck of three model-based approaches, namely *PoseGait*[17], *GaitGraph*[12] and *GaitMixer*[18]. Then, the head part of all models is implemented with the same settings as *RepGCN*. Additionally, we utilized the original hyper parameter for these three compared models.

Ablation Study. We performed ablation experiments specifically targeting the dual-residue connection structure employed in *RepGCN*. Our experiments were primarily designed based on the utilization of *Long Residual Connections* (*LRC*) and *Dual Residual Connections* (*DRC*). The results of the ablation experiment are shown in Table 3. Our findings demonstrate that the absence of the identity mapping in the model leads to a decline in overall performance compared to *RepGCN*. Furthermore, when excluding the employment of *LRC*, there is a significant improvement in accuracy, with an average accuracy enhancement of approximately 1.1%.

Table 3. The ablation experiments of the different residual connection strategy.

LRC	*DRC*	*CDUT Gait*					
		1	2	3	4	5	mean
√	×	*92.55%*	***97.72%***	*96.36%*	*82.97%*	*83.64%*	*90.65%*
×	×	*92.49%*	*96.78%*	*96.07%*	*84.61%*	*88.87%*	*91.76%*
×	√	***93.02%***	*97.58%*	***96.80%***	***85.75%***	***89.81%***	***92.59%***

The above abbreviations as follow:
LRC: long residual connection. *DRC*: dual residual connection.

5 Results and Discussion

We conducted a comprehensive performance comparison between *RepGCN* and existing model-based approaches that utilize human skeleton data. Table 4 provides a comparison of the average performance and the number of parameters between existing model-based methods and *RepGCN* on all subsets of *CDUT Gait*. *RepGCN* achieves the highest recognition performance on *CDUT Gait*, with an accuracy of 92.57%, surpassing existing state-of-the-art methods. Furthermore, *RepGCN* demonstrates efficiency and scalability by having a relatively small number of model parameters compared to other methods.

In summary, graph convolution models, including *RepGCN* and *GaitGraph*, demonstrate the highest effectiveness in gait recognition tasks. *GaitMixer*, which incorporates self-attention mechanisms, also performs well in capturing important feature in gait skeleton sequences. The performance of *PoseGait* is significantly weaker compared to all other models, primarily due to its use of a traditional *CNN* for feature extraction. Traditional *CNNs*, which lack permutation invariance, are not well-suited for handling graph data like gait skeleton sequences. Although *GaitMixer*, a hybrid model combining

Table 4. The comparison of Top-1 average accuracies and the number of backbone parameters with other model-based approaches.

	Top-1 average accuracy	Parameters
PoseGait [17]	16.67%	3,965,344
GaitGraph [12]	90.66%	475,952
GaitMixer [18]	84.84%	165,788
RepGCN (ours)	**92.57%**	**163,224**

CNN and self-attention, exhibits improved performance compared to *PoseGait*, it still does not surpass the performance of the *GCN*-based model on *CDUT Gait*.

Table 5 displays the performance of different gait recognition models, including *RepGCN*, on the *CDUT* Gait dataset. The evaluation is conducted across three different angles: 0°, 45°, and 315°. We observed that the accuracy of these models on the *CDUT Gait 4&5* dataset was lower than remaining datasets. This difference in performance can be attributed to the fact that the gait patterns of the accompanying behaviors in the test set (*CDUT Gait* 4&5) exhibited considerable modification compared to the accompanying behaviors in the training set.

Table 5. The comparison of Top-1 accuracies on CDUT Gait under three view angles with other model-based approaches.

		CDUT Gait 1	*CDUT Gait 2*	*CDUT Gait 3*	*CDUT Gait 4*	*CDUT Gait 5*
0°	*PoseGait* [17]	16.93%	21.58%	19.98%	11.93%	12.77%
	GaitGraph [12]	91.53%	96.08%	96.53%	82.90%	82.14%
	GaitMixer [18]	86.75%	91.66%	92.02%	67.47%	73.18%
	RepGCN (ours)	**93.71%**	**96.16%**	**97.17%**	83.73%	88.24%
45°	*PoseGait* [17]	12.35%	15.30%	15.20%	9.49%	12.60%
	GaitGraph [12]	92.07%	**97.60%**	**96.70%**	83.37%	84.68%
	GaitMixer [18]	88.48%	91.97%	90.82%	75.32%	80.03%
	RepGCN (ours)	**92.82%**	97.10%	96.13%	85.42%	**91.45%**
315°	*PoseGait* [17]	20.55%	26.56%	24.71%	17.17%	13.08%
	GaitGraph [12]	**94.04%**	99.48%	95.86%	82.65%	84.11%
	GaitMixer [18]	89.68%	96.74%	94.97%	73.08%	80.37%
	RepGCN (ours)	92.70%	**99.49%**	96.75%	**88.03%**	**89.71%**

6 Conclusion

In this paper, we verify the impact of accompanying behaviors during walking on gait recognition using the *CDUT Gait* dataset obtained from *Azure Kinect DK*. Our findings show that these accompanying behaviors have a substantial influence on gait recognition performance. These effects are mainly due to the accompanying behaviors of daily life that change the gait patterns. And we propose a model-based gait recognition approach, *RepGCN*. It effectively enhances the capability to extract discriminative feature through the innovative design of residual strategy, and achieves state-of-the-art performance on *CDUT Gait* with minimal parameters size compared to some currently existing model-based gait recognition methods.

Acknowledgements. This work was partially supported by Key R&D support projects 2021-YF05-02175-SN of the Chengdu Science and Technology Bureau, by the Key Project 2023YFG0271 of the Science and Technology Department of Sichuan Province, and by the funding of the China Scholarship Council.

References

1. Zou, Q., Ni, L., Wang, Q., et al.: Robust gait recognition by integrating inertial and RGBD sensors. IEEE Trans. Cybern. **48**, 1136–1150 (2018). https://doi.org/10.1109/TCYB.2017.2682280
2. Yu, S., Wang, Q., Huang, Y.: A large RGB-D gait dataset and the baseline algorithm. In: Sun, Z., Shan, S., Yang, G., Zhou, J., Wang, Y., Yin, Y. (eds.) CCBR 2013. LNCS, vol. 8232, pp. 417–424. Springer, Cham (2013). https://doi.org/10.1007/978-3-319-02961-0_52
3. Wang, Y., Sun, J., Li, J., Zhao, D.: Gait recognition based on 3D skeleton joints captured by kinect. In: 2016 IEEE International Conference on Image Processing (ICIP), pp. 3151–3155. IEEE, Phoenix (2016)
4. Sivapalan, S., Chen, D., Denman, S., et al.: Gait energy volumes and frontal gait recognition using depth images. In: 2011 International Joint Conference on Biometrics (IJCB), pp 1–6. IEEE, Washington, DC (2011)
5. Hofmann, M., Geiger, J., Bachmann, S., et al.: The TUM gait from audio, image and depth (GAID) database: multimodal recognition of subjects and traits. J. Vis. Commun. Image Represent. **25**, 195–206 (2014). https://doi.org/10.1016/j.jvcir.2013.02.006
6. Li, N., Zhao, X.: A multi-modal dataset for gait recognition under occlusion. Appl. Intell. **53**, 1517–1534 (2023). https://doi.org/10.1007/s10489-022-03474-8
7. Marin-Jimenez, M.J., Castro, F.M., Delgado-Escano, R., et al.: UGaitNet: multimodal gait recognition with missing input modalities. IEEE Trans. Inform. Forensic Secur. **16**, 5452–5462 (2021). https://doi.org/10.1109/TIFS.2021.3132579
8. Bari, A.S.M.H., Gavrilova, M.L.: Artificial neural network based gait recognition using kinect sensor. IEEE Access **7**, 162708–162722 (2019). https://doi.org/10.1109/ACCESS.2019.2952065
9. Filipi Gonçalves dos Santos, C., Oliveira, D.D.S., Passos, L.A., et al.: Gait recognition based on deep learning: a survey. ACM Comput. Surv. **55**, 1–34 (2023). https://doi.org/10.1145/3490235
10. Li, G., Muller, M., Thabet, A., Ghanem, B.: DeepGCNs: can GCNs go as deep as CNNs? In: 2019 IEEE/CVF International Conference on Computer Vision (ICCV), pp. 9266–9275. IEEE, Seoul (2019)

11. Kipf, T.N., Welling, M.: Semi-supervised classification with graph convolutional networks. In: 5th International Conference on Learning Representations (ICLR 2017), Conference Track Proceedings, Toulon, 24–26 April 2017 (2017)
12. Teepe, T., Khan, A., Gilg, J., et al.: Gaitgraph: graph convolutional network for skeleton-based gait recognition. In: 2021 IEEE International Conference on Image Processing (ICIP), pp. 2314–2318. IEEE, Anchorage (2021)
13. Ding, X., Zhang, X., Ma, N., et al.: RepVGG: making VGG-style ConvNets great again. In: 2021 IEEE/CVF Conference on Computer Vision and Pattern Recognition (CVPR), pp. 13728–13737. IEEE, Nashville (2021)
14. Sepas-Moghaddam, A., Etemad, A.: Deep gait recognition: a survey. IEEE Trans. Pattern Anal. Mach. Intell. **45**, 264–284 (2023). https://doi.org/10.1109/TPAMI.2022.3151865
15. Defferrard, M., Bresson, X., Vandergheynst, P.: Convolutional neural networks on graphs with fast localized spectral filtering. In: Proceedings of the 30th International Conference on Neural Information Processing Systems, pp. 3844–3852. Curran Associates Inc., Red Hook (2016)
16. Yan, S., Xiong, Y., Lin, D.: Spatial temporal graph convolutional networks for skeleton-based action recognition. AAAI **32** (2018). https://doi.org/10.1609/aaai.v32i1.12328
17. Liao, R., Yu, S., An, W., Huang, Y.: A model-based gait recognition method with body pose and human prior knowledge. Pattern Recogn. **98**, 107069 (2020). https://doi.org/10.1016/j.patcog.2019.107069
18. Pinyoanuntapong, E., Ali, A., Wang, P., et al.: Gaitmixer: Skeleton-based gait representation learning via wide-spectrum multi-axial mixer. In: IEEE International Conference on Acoustics, Speech and Signal Processing (ICASSP 2023–2023), pp. 1–5. IEEE, Rhodes Island (2023)

KDFAS: Multi-stage Knowledge Distillation Vision Transformer for Face Anti-spoofing

Jun Zhang[1], Yunfei Zhang[1], Feixue Shao[1], Xuetao Ma[2], and Daoxiang Zhou[1](✉)

[1] College of Data Science, Taiyuan University of Technology, Taiyuan 030024, China
{zhangjun1337,zhangyunfei1338,shaofeixue1010}@link.tyut.edu.cn, zhoudaoxiang@tyut.edu.cn

[2] School of Artificial Intelligence, Beijing Normal University, Beijing 100875, China
202331081026@mail.bnu.edu.cn

Abstract. With the commercial application of face recognition systems, face anti-spoofing has been studied extensively to enhance security in recent years. In this work, a lightweight network via knowledge distillation for face anti-spoofing is proposed. The main innovations of our approach are threefold: (1) In convolutional neural network based knowledge distillation, the local receptive field of teacher network may be inconsistent with that of student network, which results in misguiding. In our method, vision transformer architecture is leveraged because of its global modeling capabilities. (2) Beyond conventional decision-level knowledge transfer in the classification step via kullback-leibler loss, we present multi-stage feature-level knowledge distillation strategy to guide the feature learning of student network which can transfer richer knowledge from teacher to student network. (3) In contrast to traditional projection head learning, we construct a covariance matrix to solve the embedding dimensionality mismatching problem between teacher and student network in middle layers. Compared to teacher model of 1.28 GB, the memory of student model is only 330.8 MB, which effectively achieves a tradeoff between memory and accuracy. Extensive experiments on three standard benchmarks demonstrate the superiority of our proposed method, which evidently corroborates the significance of multi-stage knowledge distillation for face anti-spoofing.

Keywords: Face Anti-Spoofing · Vision Transformer · Knowledge Distillation

1 Introduction

As a famous and convenient personal identity verification technique, face recognition has become increasingly prevailing in recent years and applied in various

This work was supported by the National Natural Science Foundation of China (62101376) and Natural Science Foundation of Shanxi Province of China (201901D211078).

Q. Liu et al. (Eds.): PRCV 2023, LNCS 14429, pp. 159–171, 2024.
https://doi.org/10.1007/978-981-99-8469-5_13

scenarios. However, as indicated in [26], there are eight points that can be attacked in biometrics authentication system. It is easy for attackers to cheat a biometric identification system by impersonating genuine users through their face image or video [10]. Therefore, face anti-spoofing (FAS) research has attracted a lot of attention and has been studied extensively in the past decades, which aims to determine whether the captured face data is derived from live face or spoofing face, such as print face, video replay, 3D mask and wax figures. FAS will greatly enhance the security and reliability of identity authentication technology [32].

How to learn distinguishable features that can separate genuine faces from attack faces is recognized as a challenging task because they share very similar appearance. Till now, considerable FAS features have been proposed successively in the literature, ranging from the early handcrafted features to latest deep learning features. Handcrafted features rely on experienced domain knowledge, typical ones are local binary pattern (LBP) [5], scale invariant feature transform [23] and histogram of gradient [16]. However, their performance may be dramatically degraded on complex datasets constructed in an unconstrained environment. Over the past decades, advancements in convolutional neural networks (CNNs) have achieved excellent performance. Some works introduced convolutional networks to extract deeper features for FAS, such as 3D-CNN [18] and central difference convolutional network (CDCN) [33]. Although these methods can extract the deeper discriminative features, the parameters and memory of model will grow exponentially as the network layer deepens.

Recently, Vision Transformer (ViT) [9] has achieved great success on many computer vision tasks. Due to the special attention mechanism in ViT, it is able to capture long-range temporal features, which is important for tackling the video replay attack. Many researchers have introduced ViT structures to learn discriminative facial features, such as two-stream vision transformer (TSViT) [24] and vision transformer with depth auxiliary (DE-ViT) [20]. Obviously, ViT obtained better performance compared to traditional convolutional networks due to its global modeling capability. However, they often rely on large-scale labeled data and require more parameters and memory, which prevents them from being deployed on memory-bounded devices such as cell phones. To facilitate the challenges, a series of methods have been proposed to investigate compact deep neural networks, such as network pruning [36] and knowledge distillation (KD) [12].

Latest studies [7,8] indicated that smaller models usually lead to performance degradation. KD is a promising approach for inheriting knowledge from the high-performance teacher to compact student and maintaining strong performance. Some works [19,21] utilized face data from richer and related domains to train powerful teacher network for FAS and obtained the lightweight student network by distillation, but these methods are CNN-based and data-specific. The current common paradigm of KD methods based on ViT [14,28] is to train a student transformer to match soft labels predicted by the pre-trained teacher, but it is not enough to use decision-level knowledge to guide the learning of student. So, it is meaningful to design richer knowledge from other layers to minimize the performance gap between teacher and student.

To this end, we proposed a multi-stage KD based on ViT for FAS (KDFAS). Our proposed method KDFAS is illustrated in Fig. 1, which consists of two parts: pre-trained teacher and student. Considering the local receptive field inconsistency of CNN-based KD, we designed the teacher and student backbone both based on ViT structure. To reduce parameters and memory, we introduced the lightweight ViT with fewer attention heads as student network following DeiT [28]. In addition to traditional decision-level knowledge, we also designed a multi-stage feature-level knowledge to bridge the performance gap between teacher and student. During the transfer of feature-level knowledge, we solved the embedding dimensionality mismatching from middle layers by constructing a covariance matrix. In summary, our work makes the following contributions:

- We proposed a multi-stage knowledge distillation framework for FAS based on ViT structure with powerful global modeling capabilities, which avoids the local receptive field inconsistency of convolutional networks.
- We designed the decision-level and multi-stage feature-level knowledge to guide the learning of student, which could transfer richer knowledge from teacher to student and effectively minimize the performance gap.
- We solved the embedding dimensionality mismatching issue by constructing a covariance matrix, which not only reduces parameters and complexity of network but also avoids the transfer loss of feature-level knowledge.
- We conducted extensive experiments to verify the effectiveness of our proposed method. The results demonstrated that our method achieves a trade-off between memory and accuracy.

2 Related Work

2.1 Face Anti-spoofing

FAS is an active research topic in computer vision and has received an increasing number of publications in recent years. Early methods are designed based on handcrafted features, such as LBP [5] and color texture [1], which need rich task-aware prior knowledge. Recently, CNN-based methods have become mainstream for FAS due to their robust feature extraction and discriminative capabilities. Li et al. [18] designed a deep convolutional network (3D-CNN) to extract high-level features. Chen et al. [4] proposed an attention-based two-stream convolutional network (ATCNN) on RGB and MSR space. However, these methods have large parameters and only focus on the local information of face images. Due to the superiority of attention mechanism for global information extraction, many researchers have attempted to introduce ViT to FAS. Peng et al. [24] proposed two-stream vision transformers (TSViT) for transfer learning in two complementary spaces. Li et al. [20] explored the effectiveness of a vision transformer with depth auxiliary information (DE-ViT). Compared with CNN-based models, the ViT-based methods can achieve better performance but generally need more parameters, making them harder to deploy.

2.2 Knowledge Distillation

KD was first proposed by Hinton et al. [12], which trains a lightweight student model to match soft labels given by a large pre-trained teacher model. Recently, there are a growing number of works that extend KD by novel network architectures or objective functions. Geras et al. [11] proposed to transfer the knowledge from a long short-term memory network to convolutional networks by distillation. Howard et al. [13] and Polino et al. [25] combined knowledge distillation and some specific regularization to compress CNN models. To obtain a lightweight ViT model, Yang et al. [31] proposed a nontrivial way for feature-based ViT distillation, named ViTKD. Huang et al. [14] designed a multi-teacher single-student ViT distillation method (MAMD) with a multi-level attention fusion. However, the backbone of improved KD methods is rarely based on ViT structure and most distillation approaches only use decision-level knowledge. Thus, it is interesting to explore the effectiveness of richer feature-level knowledge for ViT-based distillation methods.

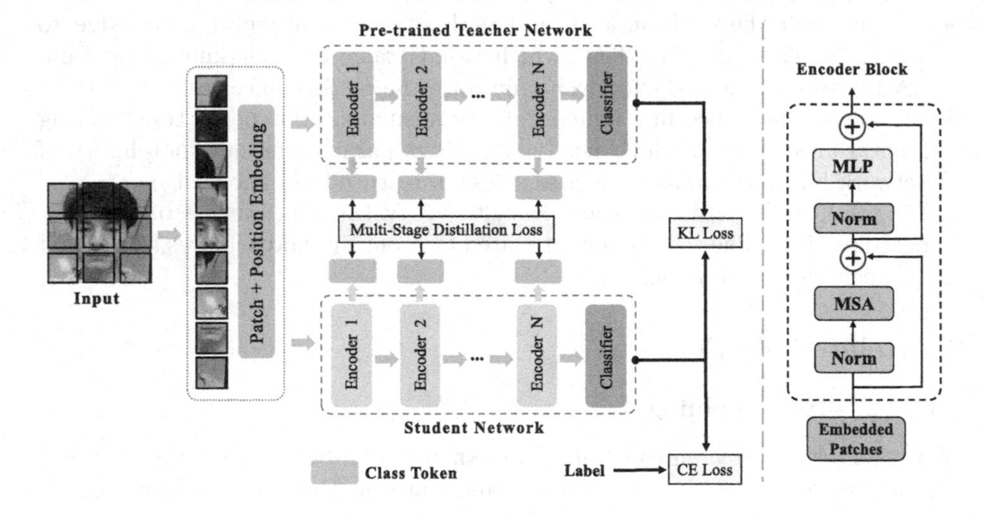

Fig. 1. The architecture of our proposed KDFAS method.

3 Methodology

3.1 Motivation

The motivations for this paper are fourfold: (1) ViT model has achieved excellent performance in FAS due to its powerful global modeling capability, but the training is timeconsuming and it needs large parameters and memory, so devising lightweight ViT is highly desirable. (2) KD as an effective model compression approach is applied in FAS, but most methods are based on convolutional networks and only transfer decision-level knowledge from teacher. (3)

Previous studies [15,22,27] indicated that feature-level knowledge from middle layers is essential to reduce the performance gap between teacher and student, but the shape of corresponding feature maps are different, which will lead to dimensionality mismatching. (4) Most works solve the embedding dimensionality mismatching problem by learning a projection head [3], such strategy may cause loss of feature information when the feature map is projected from high-dimension to low-dimension.

3.2 Feature Encoding

In our work, the backbones of teacher and student network are based on ViT [9]. To encoder facial features, we first split the face image into patches, then project these patches into a D-dimension embedding space and add position embeddings to them. Then, these patch embeddings are fed to encoder blocks, which consist of a multi-head self-attention (MSA) block and a multi-layer perceptron (MLP) block, as shown in Fig. 1. Each encoder block works as follows:

$$\begin{aligned} X_l' &= \text{MSA}(\text{LN}(X_l)) + X_l, \\ X_{l+1} &= \text{MLP}(\text{LN}(X_l')) + X_l' \end{aligned} \tag{1}$$

where X_l denotes patch embeddings from the l-th encoder block and LN denotes the layer normalization.

3.3 Decision-Level and Multi-stage Feature-Level Knowledge

Decision-Level Knowledge. For our proposed method, we first pre-trained a teacher network with abundant face data. Then, we follow the conventional distillation paradigm [12] to use the predictions from teacher to supervise the training of student, as shown in Fig. 1. In particular, given a face image x corresponding to the label y, $f_s(x)$ and $f_t(x)$ represent the prediction of student and teacher, respectively. The loss of decision-level knowledge distillation can be formulated as follows:

$$\mathcal{L}_{KD} = (1-\alpha)\mathcal{L}_{CE}(f_s(x), y) + \alpha\tau^2\mathcal{L}_{KL}(f_s(x)/\tau, f_t(x)/\tau) \tag{2}$$

where $\mathcal{L}_{CE}$ is the cross-entropy loss, $\mathcal{L}_{KL}$ is the Kullback-Leibler divergence loss, τ is a smoothing hyperparameter termed temperature, and α is a balancing hyperparameter.

Multi-stage Feature-Level Knowledge. In this paper, we proposed a multi-stage knowledge distillation strategy for FAS, as shown in Fig. 1. For each pair of manually selected encoder layer, we aim to teach the class token from student to be as close as possible to the one from teacher, since the class token contains rich feature-level knowledge about face.

In particular, given face images of batch size B, we denote the class token from student and teacher encoder layer as $F_S \in R^{B\times D_S}$ and $F_T \in R^{B\times D_T}$, where

D_S and D_T are embedding dimensions. To solve the embedding dimensionality mismatching issue between teacher and student, we first normalize the class token, and construct a covariance matrix as follows:

$$\Psi(F) = \left(\frac{F}{\|F\|}\right)\left(\frac{F}{\|F\|}\right)^T \tag{3}$$

Then, we train student to minimize the gap between $\Psi(F_T)$ and $\Psi(F_S)$ as follows:

$$\mathcal{L}_{MS} = \sum_{l} \|\Psi(F_S) - \Psi(F_T)\|_F^2 \tag{4}$$

where the summation over l means that feature-level knowledge transfer is performed on multi-pair selected encoder layers.

Finally, the total loss function of our proposed multi-stage distillation method KDFAS can be formulated as follows:

$$\mathcal{L}_{total} = \mathcal{L}_{KD} + \beta\mathcal{L}_{MS} = (1-\alpha)\mathcal{L}_{CE} + \alpha\tau^2\mathcal{L}_{KL} + \beta\mathcal{L}_{MS} \tag{5}$$

where $\mathcal{L}_{KD}$ is the decision-level distillation loss, $\mathcal{L}_{MS}$ is the multi-stage feature-level distillation loss, and β is a balancing hyperparameter.

4 Experiments and Analysis

4.1 Experimental Settings

Datasets and Evaluation Metrics. Extensive experiments are performed on three public face databases: CASIA-FASD [35], Replay-Attack [5], and OULU-NPU [2]. Then, we select four evaluation metrics to comprehensively assess the designed model performance following [6], including False Accept Rate (FAR), False Reject Rate(FAR), Half Total Error Rate (HTER), and Equal Error Rate (EER).

Implementation. We implement all networks and training procedures in PyTorch and conduct all experiments on NVIDIA Tesla V100 GPU. Following [34], each frame of videos is aligned into the 224×224-pixel facial image as input. In the pre-training and distillation stage, we select video frames based on the ratio of true and attack videos in each dataset to achieve the data balance. In the testing stage, we select the 21th to 25th frame of each video to evaluate model performance. Before the distillation stage, we first pre-trained a teacher network with the same structure and parameter settings as ViT-B [9]. Then, we need to manually set some parameters based on the results of parameter analysis, such as the scale of student network, temperature factor τ, middle layer selection scheme l, and trade-off weight α, β of the total loss. Other training parameters follow DeiT [28], but we don't use stochastic depth regularization, which will introduce some uncertainty during the training in our experiment. Remarkably, our teacher and student network have the same number of encoder layers. Finally, the teacher network is pre-trained for 500 epochs with batch size 64, and student network is distilled for 300 epochs with the same batch size.

4.2 Parameter Analysis

Some experiments are constructed to evaluate the effect of various parameters on student performance based on protocol 2 of OULU-NPU dataset [2].

Scale of Student Network. To reduce parameters and memory of student, small and tiny ViT models are introduced following DeiT [28], respectively named ViT-S and ViT-Ti, which change the number of attention heads.

The pre-trained teacher is distilled to three different scales of student model, and comparison results are shown in Table 1. It is obvious that the HTER of student model gradually increases as the scale of student decrease, which is related to fewer attention heads. Therefore, considering the trade-off between memory and accuracy, we selected the student network to have the same structure as ViT-S in subsequent experiments.

Table 1. Performance comparison of student networks of different scales by distillation.

Teacher	Student	Params	FAR (%)	FRR (%)	HTER (%)	Memory
ViT-B	ViT-B	86 M	4.19	0.39	2.290	1.28 G
	ViT-S	22 M	4.33	1.00	2.665	330.8 M
	ViT-Ti	5 M	4.94	1.11	3.025	89.49 M

Temperature Factor τ. As stated in [12], temperature factor τ is introduced to control the importance of each soft target. In our experiment, the temperature factor is fine-tuned from 1 to 7 with a stride of 2, and the corresponding HTER are 2.79, 2.67, 2.29, 2.39. When the temperature is small, it is clear that the HTER of student network decreases as τ increases. However, the HTER reaches the lowest value when the temperature is 5. Thus, the temperature factor $\tau = 5$ is applied for distillation.

Middle Layers Selection l. To evaluate the effect of middle layer selection strategy, there are six layer selection strategies being designed and comparison results are shown in Table 2. It can be seen that when fewer layers are used, the model performance is unsatisfying because feature information is insufficient. When adjacent layers are selected, such as 6, 7, 8, the performance is not impressive, a possible reason is that plenty of redundancy features exist in the adjacent layers. Finally, the scheme l = 3, 5, 7, 9 is used to extract multi-stage feature-level knowledge in middle layers.

Weight of the Total Loss Function. Since the total loss is composed of three components, we simply fine-tune the weight α of decision-level distillation loss and weight β of feature-level distillation loss, and comparison results are shown in Table 3. It is clear that increasing the weight α of decision-level distillation loss

Table 2. Effect of different middle layers selection on student performance.

Layers l	FAR (%)	FRR (%)	HTER (%)
4,8	4.33	1.00	2.665
2,3,4	5.33	1.33	3.330
6,7,8	5.78	0.50	3.140
2,4,6,8	4.31	1.11	2.710
3,5,7,9	3.72	0.33	2.025
2,3,4,6,7,8	4.03	0.33	2.180

Table 3. Effect of the weight from total loss function.

α	β	FAR (%)	FRR (%)	HTER (%)
0.3	1	4.28	1.17	2.725
0.5	1	4.33	1.00	2.665
0.7	1	2.86	0.39	1.625
0.9	1	5.06	0.67	2.865
0.7	0.9	4.97	0.28	2.625
0.7	1.1	3.92	0.89	2.405
0.7	0	3.86	0.28	2.070

helps to improve the model performance. Finally, the weights of total loss are set to $\alpha = 0.7$ and $\beta = 1$. Then, we designed ablation experiment to demonstrate the importance of feature-level knowledge, which will be described carefully in ablation study.

4.3 Comparisons with Related Methods

Comparison on CASIA and Replay Database. To compare with early research methods, experiments are performed on CASIA-FASD and Replay-Attack databases and comparison results are shown in Table 4. Specifically, the LBP [5] and Color Texture [1] are based on expert knowledge methods, and another three methods are based on CNN methods. It is obvious that our proposed method acquires the lowest EER and HTER, which proves the effectiveness of multi-stage knowledge distillation method in FAS task.

Table 4. Comparison results on CASIA-FASD and Replay-Attack database. The bold values are the best results for each metric.

Methods	CASIA-FASD	Replay-Attack		Notes
	EER (%)	EER (%)	HTER (%)	
LBP [5]	18.20	13.90	13.80	12 BIOSIG
Color Texture [1]	6.20	0.40	2.90	15 ICIP
CNN [29]	4.64	0.72	1.86	14 arXiv
3D-CNN [18]	1.40	0.30	1.20	18 TIFS
ATCNN [4]	3.14	0.13	0.25	19 TIFS
KDFAS (Ours)	**0.25**	0	**0.075**	–

Comparison on OULU-NPU Database. To compare with recent works, we evaluate the model performance on the challenging OULU-NPU database and tabulate comparison results in Table 5. For example, STASN [30] and CDCN [33] are based on CNN methods, TSViT [24] and DE-ViT [20] are based on

ViT methods, and MAMD [14] is knowledge distillation method based on ViT. It is apparent that our proposed method KDFAS got lower HTER when evaluated on protocol 1 and 3. However, the results of our method are worse on another two protocols, especially on protocol 4. To be specific, TSViT [24] is based on two-stream vision transformers with self-attention fusion, and DE-ViT [20] is based on the vision transformer with depth auxiliary information. However, these methods rely on abundant face data and need large parameters and memory. To address the above problems, MAMD [14] based on the multi-teacher distillation method is proposed. Compared with our method KDFAS, MAMD [14] requires pre-training multiple teacher networks on different databases. In conclusion, although our proposed method obtained unencouraging results on difficult protocols, our method can offer a good trade-off between memory and accuracy.

Table 5. Comparison results on OULU-NPU database. The bold values are the best results for each metric.

Prot.	Methods	FAR (%)	FRR (%)	HTER (%)	Notes
1	STASN [30]	1.2	2.5	1.9	19 CVPR
	CDCN [33]	0.4	1.7	1.0	20 CVPR
	TSViT [24]	1.7	**0.0**	0.9	22 JVCIR
	DE-ViT [20]	0.9	0.1	**0.5**	22 ICONIP
	MAMD [14]	2.0	1.0	1.5	21 BMVC
	KDFAS (Ours)	**0.04**	1.17	0.61	–
2	STASN [30]	4.2	**0.3**	2.2	19 CVPR
	CDCN [33]	1.5	1.4	1.45	20 CVPR
	TSViT [24]	**0.8**	1.3	1.1	22 JVCIR
	DE-ViT [20]	3.0	**0.3**	1.7	22 ICONIP
	MAMD [14]	1.4	**0.3**	**0.85**	21 BMVC
	KDFAS (Ours)	2.53	0.83	1.68	–
3	STASN [30]	4.7 ± 3.9	0.9 ± 1.2	2.8 ± 1.6	19 CVPR
	CDCN [33]	2.4 ± 1.3	2.2 ± 2.0	2.3 ± 1.4	20 CVPR
	TSViT [24]	2.4 ± 2.6	1.4 ± 2.2	1.9 ± 1.3	22 JVCIR
	DE-ViT [20]	1.4 ± 1.0	1.9 ± 3.5	1.7 ± 1.5	22 ICONIP
	MAMD [14]	2.1 ± 1.3	$\mathbf{0.5 \pm 0.4}$	1.3 ± 0.8	21 BMVC
	KDFAS (Ours)	$\mathbf{0.36 \pm 0.45}$	2.22 ± 3.97	$\mathbf{1.29 \pm 1.89}$	–
4	STASN [30]	6.7 ± 10.6	8.3 ± 8.4	7.5 ± 4.7	19 CVPR
	CDCN [33]	$\mathbf{4.6 \pm 4.6}$	9.2 ± 8.0	6.9 ± 2.9	20 CVPR
	TSViT [24]	7.4 ± 5.0	$\mathbf{1.2 \pm 2.2}$	4.3 ± 1.9	22 JVCIR
	DE-ViT [20]	5.7 ± 4.8	1.5 ± 3.2	$\mathbf{3.5 \pm 3.4}$	22 ICONIP
	MAMD [14]	6.6 ± 3.3	2.4 ± 2.8	4.5 ± 2.2	21 BMVC
	KDFAS (Ours)	13.33 ± 10.47	6.33 ± 8.34	9.83 ± 3.98	–

4.4 Ablation Study

Is Knowledge Distillation Necessary? To explore the necessity of distilling smaller ViT models by the large pre-trained ViT, we train a small ViT model from scratch as comparison. Evaluate experiment is performed on protocol 2 of OULU-NPU database, and comparison results are shown in Table 6. It is clear that the small ViT model obtained by distillation gets lower HTER, which can be attributed to the richer knowledge learned by the pre-trained teacher and transferred to student via distillation.

Table 6. Ablation study results of knowledge distillation.

Model	FAR (%)	FRR (%)	HTER (%)
ViT-S(w/ distill)	4.33	1.00	2.665
ViT-S(w/o distill)	8.28	0.28	4.280

Do We Need to Distill the Feature-Level Knowledge? To study the importance of multi-stage feature-level knowledge, we design an ablation experiment and tabulate the comparison results in Table 3. In distillation stage, it is remarkable that combining feature-level with decision-level knowledge ($\alpha = 0.7$, $\beta = 1$) yields a lower HTER than traditional decision-level knowledge ($\alpha = 0.7$, $\beta = 0$), which may be due to feature-level knowledge being a good complement of decision-level knowledge, especially for the training of thinner and deeper networks.

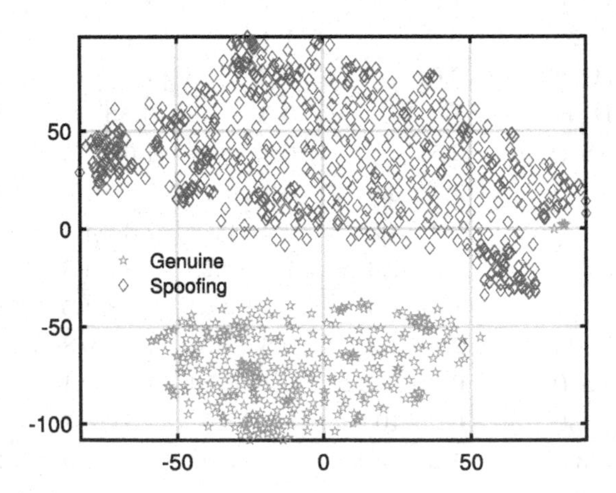

Fig. 2. Feature distribution visualization result via t-SNE.

4.5 Visualization

To visualize the feature distribution learned by our student model, 600 real images and 1200 spoof images from the testing set on protocol 2 of OULU-NPU are randomly chosen, and feature embeddings for the last encode layer are fed into t-SNE [17]. As shown in Fig. 2, it can be observed that the genuine images and spoofing images are very distinguishable, which obviously implies that the features learned by distilling have excellent discriminability.

5 Conclusion

In this paper, we proposed a multi-stage knowledge distillation vision transformer method for face anti-spoofing, which used decision-level knowledge and multi-stage feature-level knowledge from teacher to guide the training of student. Extensive experiments on three face databases demonstrated that our proposed method effectively achieves a trade-off between memory and accuracy. Based on ablation experiments, we got two important observations: (1) Under ViT architecture, knowledge distillation can boost the model performance of student by using a larger pre-trained teacher model. (2) Combining decision-level and feature-level knowledge contributes to minimizing the performance gap between teacher and student. In the future, we will study how to improve the model performance on difficult data by designing richer knowledge or mining difficult samples.

References

1. Boulkenafet, Z., Komulainen, J., Hadid, A.: Face anti-spoofing based on color texture analysis. In: IEEE International Conference on Image Processing, pp. 2636–2640 (2015)
2. Boulkenafet, Z., Komulainen, J., Li, L., Feng, X., Hadid, A.: Oulu-NPU: a mobile face presentation attack database with real-world variations. In: IEEE International Conference on Automatic Face and Gesture Recognition, pp. 612–618 (2017)
3. Chen, D., Mei, J.P., Zhang, H., Wang, C., Feng, Y., Chen, C.: Knowledge distillation with the reused teacher classifier. In: Proceedings of the IEEE Conference on Computer Vision and Pattern Recognition, pp. 11933–11942 (2022)
4. Chen, H., Hu, G., Lei, Z., Chen, Y., Robertson, N.M., Li, S.Z.: Attention-based two-stream convolutional networks for face spoofing detection. IEEE Trans. Inf. Forensics Secur. **15**, 578–593 (2019)
5. Chingovska, I., Anjos, A., Marcel, S.: On the effectiveness of local binary patterns in face anti-spoofing. In: Proceedings of the International Conference of Biometrics Special Interest Group, pp. 1–7 (2012)
6. Chingovska, I., Dos Anjos, A.R., Marcel, S.: Biometrics evaluation under spoofing attacks. IEEE Trans. Inf. Forensics Secur. **9**(12), 2264–2276 (2014)
7. Choudhary, T., Mishra, V., Goswami, A., Sarangapani, J.: A comprehensive survey on model compression and acceleration. Artif. Intell. Rev. **53**, 5113–5155 (2020)
8. Deng, Y.: Deep learning on mobile devices: a review. In: Mobile Multimedia/Image Processing, Security, and Applications 2019, vol. 10993, pp. 52–66 (2019)

9. Dosovitskiy, A., Beyer, L., Kolesnikov, A., Weissenborn, D., Zhai, X., et al.: An image is worth 16x16 words: Transformers for image recognition at scale. In: International Conference on Learning Representations (2021)
10. Galbally, J., Marcel, S., Fierrez, J.: Biometric antispoofing methods: a survey in face recognition. IEEE Access **2**, 1530–1552 (2014)
11. Geras, K.J., et al.: Blending LSTMs into CNNs. In: International Conference on Learning Representations (2016)
12. Hinton, G., Vinyals, O., Dean, J.: Distilling the knowledge in a neural network. In: NIPS Deep Learning and Representation Learning Workshop (2015)
13. Howard, A.G., et al.: Mobilenets: efficient convolutional neural networks for mobile vision applications. arXiv preprint arXiv:1704.04861 (2017)
14. Huang, Y.H., Hsieh, J.W., Chang, M.C., Ke, L., Lyu, S., Santra, A.S.: Multi-teacher single-student visual transformer with multi-level attention for face spoofing detection. In: British Machine Vision Conference, p. 125 (2021)
15. Kim, J., Park, S., Kwak, N.: Paraphrasing complex network: network compression via factor transfer. In: Advances in Neural Information Processing Systems, vol. 31 (2018)
16. Komulainen, J., Hadid, A., Pietikäinen, M.: Context based face anti-spoofing. In: IEEE Sixth International Conference on Biometrics: Theory, Applications and Systems, pp. 1–8 (2013)
17. Van Der Maaten, L., Hinton, G.: Visualizing data using t-SNE. J. Mach. Learn. Res. **9**(2605), 2579–2605 (2008)
18. Li, H., He, P., Wang, S., Rocha, A., Jiang, X., Kot, A.C.: Learning generalized deep feature representation for face anti-spoofing. IEEE Trans. Inf. Forensics Secur. **13**(10), 2639–2652 (2018)
19. Li, H., Wang, S., He, P., Rocha, A.: Face anti-spoofing with deep neural network distillation. IEEE J. Sel. Top. Sig. Process. **14**(5), 933–946 (2020)
20. Li, S., Dong, J., Chen, J., Gao, X., Niu, S.: Vision transformer with depth auxiliary information for face anti-spoofing. In: Neural Information Processing: 29th International Conference, Part III, pp. 335–346 (2023)
21. Li, Z., Cai, R., Li, H., Lam, K.Y., Hu, Y., Kot, A.C.: One-class knowledge distillation for face presentation attack detection. IEEE Trans. Inf. Forensics Secur. **17**, 2137–2150 (2022)
22. Passban, P., Wu, Y., Rezagholizadeh, M., Liu, Q.: Alp-kd: attention-based layer projection for knowledge distillation. In: Proceedings of the AAAI Conference on Artificial Intelligence, vol. 35, pp. 13657–13665 (2021)
23. Patel, K., Han, H., Jain, A.K.: Secure face unlock: spoof detection on smartphones. IEEE Trans. Inf. Forensics Secur. **11**(10), 2268–2283 (2016)
24. Peng, F., Meng, S.H., Long, M.: Presentation attack detection based on two-stream vision transformers with self-attention fusion. J. Vis. Commun. Image Represent. **85**, 103518 (2022)
25. Polino, A., Pascanu, R., Alistarh, D.: Model compression via distillation and quantization. arXiv preprint arXiv:1802.05668 (2018)
26. Ratha, N.K., Connell, J.H., Bolle, R.M.: Enhancing security and privacy in biometrics-based authentication systems. IBM Syst. J. **40**(3), 614–634 (2001)
27. Romero, A., Ballas, N., Kahou, S.E., Chassang, A., Bengio, Y.: Fitnets: hints for thin deep nets. In: International Conference on Learning Representations (2015)
28. Touvron, H., Cord, M., Douze, M., Massa, F., Sablayrolles, A., Jégou, H.: Training data-efficient image transformers & distillation through attention. In: International Conference on Machine Learning, pp. 10347–10357 (2021)

29. Yang, J., Lei, Z., Li, S.Z.: Learn convolutional neural network for face anti-spoofing. arXiv preprint arXiv:1408.5601 (2014)
30. Yang, X., et al.: Face anti-spoofing: model matters, so does data. In: Proceedings of the IEEE Conference on Computer Vision and Pattern Recognition, pp. 3507–3516 (2019)
31. Yang, Z., Li, Z., Zeng, A., Li, Z., Yuan, C., Li, Y.: Vitkd: practical guidelines for ViT feature knowledge distillation. arXiv preprint arXiv:2209.02432 (2022)
32. Yu, Z., Qin, Y., Li, X., Zhao, C., Lei, Z., Zhao, G.: Deep learning for face anti-spoofing: a survey. IEEE Trans. Pattern Anal. Mach. Intell. **45**(5), 5609–5631 (2022)
33. Yu, Z., et al.: Searching central difference convolutional networks for face anti-spoofing. In: Proceedings of the IEEE Conference on Computer Vision and Pattern Recognition, pp. 5295–5305 (2020)
34. Zhang, L.B., Peng, F., Qin, L., Long, M.: Face spoofing detection based on color texture Markov feature and support vector machine recursive feature elimination. J. Vis. Commun. Image Represent. **51**, 56–69 (2018)
35. Zhang, Z., Yan, J., Liu, S., Lei, Z., Yi, D., Li, S.Z.: A face anti-spoofing database with diverse attacks. In: IAPR International Conference on Biometrics, pp. 26–31 (2012)
36. Zhu, M., Tang, Y., Han, K.: Vision transformer pruning. arXiv preprint arXiv:2104.08500 (2021)

Long Short-Term Perception Network for Dynamic Facial Expression Recognition

Chengcheng Lu[1], Yiben Jiang[1], Keren Fu[1,2](✉), Qijun Zhao[1,2], and Hongyu Yang[1,2]

[1] College of Computer Science, Sichuan University, Chengdu, China
fkrsuper@scu.edu.cn
[2] National Key Lab of Fundamental Science on Synthetic Vision, Sichuan University, Chengdu, China

Abstract. Dynamic facial expression recognition (DFER) presents a difficult challenge, and antecedent methodologies leveraging convolutional neural networks (CNNs), recurrent neural networks (RNNs), or Transformers focus on extracting either long-term temporal information or short-term temporal information from facial videos. Unlike prevailing approaches, we design a novel framework named long short-term perception network (LSTPNet). It can easily perceive aforementioned dual temporal cues and bestow notable advantages upon the DFER task. To be specific, a temporal channel excitation (TCE) module is proposed, building upon the previous outstanding efficient channel attention (ECA) module. This extension serves to imbue the backbone network with temporal attention capabilities, thereby facilitating the acquisition of more enriched temporal features. Furthermore, we design a long short-term temporal Transformer (LSTformer) which can capture both short-term and long-term temporal information with efficacy. The empirical findings, as showcased across three benchmark datasets, unequivocally demonstrate the state-of-the-art performance of LSTPNet.

Keywords: Dynamic Facial Expression Recognition · Long Short-Term Perception · Temporal Attention · Transformer

1 Introduction

Diverging from general static facial expression recognition (SFER) [17,18,21,34], dynamic facial expression recognition (DFER) [19,20,24,35] presents a significant challenge within the fields of computer vision and affective computing. Its primary objective lies in the recognition of facial video clips, as opposed to stationary images, necessitating the classification of such clips into distinct fundamental emotional categories such as neutral, happiness, sadness, surprise, fear, anger or disgust. It has all kinds of practical applications, *e.g.*, mental health analyses [8], human-computer interaction [1], and support systems for mentally retarded children [28].

Q. Liu et al. (Eds.): PRCV 2023, LNCS 14429, pp. 172–184, 2024.
https://doi.org/10.1007/978-981-99-8469-5_14

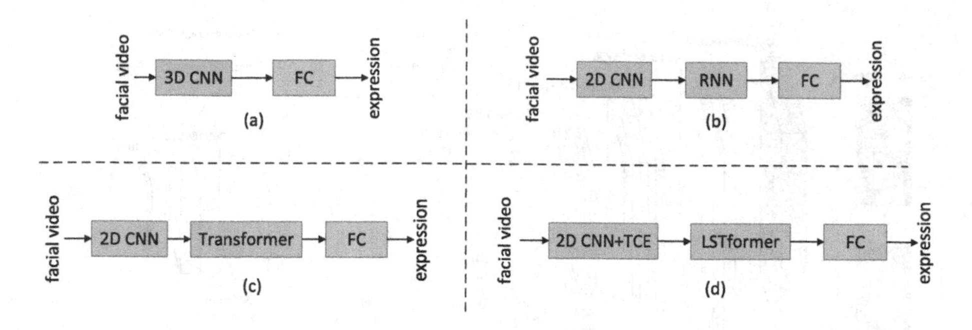

Fig. 1. Feature extraction strategies of existing DFER models ((a) [2,4,14,22], (b) [7,15,29,33] and (c) [19,20,24,35]) as well as our proposed (d).

Recent strides in DFER have been propelled by the evolution of neural network-based methodologies. Noteworthy contributions include models founded on 3D convolutional neural networks-based models [2,4,14,22], recurrent neural networks-based models [7,15,29,33], and Transformer-based models [19,20,24,35], as we sum up in Fig. 1. Despite the progress made by these methods, they are weak in capturing both long-term and short-term temporal features simultaneously. As illustrated in Fig. 1 (a), the 3D CNN is employed to concurrently extract spatial and temporal features. Nonetheless, its principal emphasis lies in the extraction of short-term temporal information due to the inherent constraint of the kernel size. Consequently, the establishment of direct long-range relationships between frames distant in temporal dimension remains challenging. In Fig. 1 (b), initially, the 2D CNN is employed to extract spatial features. After that, the RNN is employed to extract long-term temporal features. Nevertheless, the optimization of RNNs is beset by some challenges, including the emergence of vanishing/exploding gradients which renders them inefficient to train. As depicted in Fig. 1 (c), the 2D CNN is used for the extraction of spatial features. Following that, the Transformer is harnessed to extract global and long-term temporal features, facilitated by its multi-head self-attention mechanism.

In this paper, we think that the strategies delineated above are characterized by a distinct focus on the extraction of either long-term or short-term temporal features, thereby neglecting the importance of jointly emphasizing both facets. Therefore, we propose a novel framework named long short-term perception network (LSTPNet) which is meticulously crafted to realize the concurrent perception of the aforementioned dual temporal features, thereby affording significant advantages to DFER. Concretely, we put forth a unique temporal channel excitation (TCE) module. This module is meticulously devised to undertake attention modeling encompassing both temporal and channel dimensions in order to engender more enriched temporal representations stemming from the backbone network. Besides, an innovative long short-term temporal Transformer (LSTformer) is proposed. It not only perceives long-term temporal dependencies between facial frames through its powerful multi-head self-attention mechanism but also engages in modeling short-term temporal dependencies through the

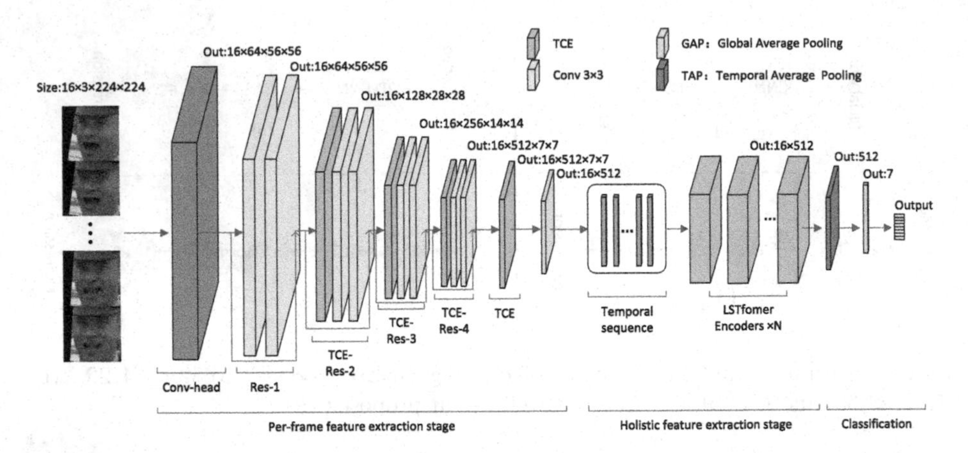

Fig. 2. Block diagram of the proposed LSTPNet, which consists of a per-frame feature extraction stage, a holistic feature extraction stage, and a final classification stage.

integration of one-dimensional (1-D) convolutions which facilitates the acquisition of local temporal induction biases. The overarching framework is shown in Fig. 1 (d), and our principal contributions are succinctly encapsulated as follows:

- We propose an innovative framework for DFER, termed the long short-term perception network (LSTPNet). Remarkably, LSTPNet is the first model within the field of DFER to accentuate both short-term and long-term feature extraction concurrently for improved performance.
- We present a novel temporal channel excitation (TCE) module, designed to imbue intermediate features within the backbone with temporal attention, generating more temporally representative features.
- We also propose the long short-term temporal Transformer, denoted as LSTformer, so as to perceive both short-term and long-term temporal information efficaciously.

2 Proposed Method

2.1 Overview

Our LSTPNet is a hybrid structure, amalgamating elements from both CNN and Transformer. As illustrated in Fig. 2, LSTPNet consists of two handling stages, encompassing the per-frame feature extraction stage and the holistic feature extraction stage. Firstly, a dynamic sampling procedure is employed to extract a set of frames of fixed length from the original facial video, which subsequently serves as the input to the network. Next, to extract fundamental per-frame features from these distinct facial images, a sequence of residual blocks infused with TCE modules is engaged. Then, the generated per-frame features are propagated to a subsequent LSTformer which serves the purpose of perceiving long and short-term temporal dependencies between frames. At last, a global feature embedding from the facial sequence is derived via temporal average pooling (TAP), and it is subsequently subjected to classification, with its categorization into one of seven fundamental expressions being accomplished via a fully connected (FC) layer.

2.2 Temporal Channel Excitation (TCE) Module

In a prior study [30], the efficient channel attention (ECA) module was devised with the intent of capturing interdependencies between feature channels. Distinguished from the conventional squeeze-and-excitation (SE) channel attention module [12], the ECA module offers an advantage by preventing information loss due to the excitation operation within the SE channel attention. Moreover, the ECA module incorporates a local cross-channel interaction mechanism to supplant the global cross-channel interaction mechanism inherent to the SE module. This mechanism yields a marked enhancement in efficiency while upholding the module's effectiveness.

However, while the ECA module exhibits dual traits of efficiency and effectiveness, it regrettably overlooks the temporal associations inherent to the original features. Thus we introduce the 1-D convolution dedicated to temporal modeling. Consequently, we present the TCE module, visually represented in Fig. 3. Given intermediate features denoted as $F \in \mathbb{R}^{T \times C \times H \times W}$ as input, taking inspiration from the work by Woo et al. [32], our approach gathers spatial information from F through the utilization of both global max pooling and global average pooling, yielding two distinct spatial context descriptors, namely F^{max} and F^{avg} as

$$F^{max} = GMP(F), \; F^{avg} = GAP(F), \tag{1}$$

where F^{max} and $F^{avg} \in \mathbb{R}^{T \times C \times 1 \times 1}$, and GMP and GAP are global max pooling and global average pooling operations, respectively. F^{max} and F^{avg} are then permuted into F_1^{max} and $F_1^{avg} \in \mathbb{R}^{T \times 1 \times 1 \times C}$. Subsequently, both F^{max} and F^{avg} are conveyed to a shared network which primarily comprises three 1-D convolutional layers, namely K_1, K_2 and K_3, respectively, to produce a channel attention vector, which can be represented as

$$F_2^{max} = K_3 * P(K_2 * P(K_1 * F_1^{max})), \; F_2^{avg} = K_3 * P(K_2 * P(K_1 * F_1^{avg})) \tag{2}$$

where F_2^{max} and $F_2^{avg} \in \mathbb{R}^{T \times 1 \times 1 \times C}$ are obtained, P is permutation operation, and the size of feature maps before and after P can be observed from Fig. 3. K_1 with kernel size k_c (calculated according to work [30]) outputs channel $\frac{C}{r}$, K_2 with kernel size k^{temp} outputs channel C, and K_3 with kernel size k_c outputs channel 1. We then permute F_2^{max} and F_2^{avg} into F_3^{max} and $F_3^{avg} \in \mathbb{R}^{T \times C \times 1 \times 1}$. Afterwards, F^{max} and F^{avg} are gathered through element-wise summation. The resultant feature is then subjected to a sigmoid activation function σ, producing the channel attention mask denoted as M^c, which can be formulated as

$$M^c = \sigma\left(F_3^{max} + F_3^{avg}\right), \tag{3}$$

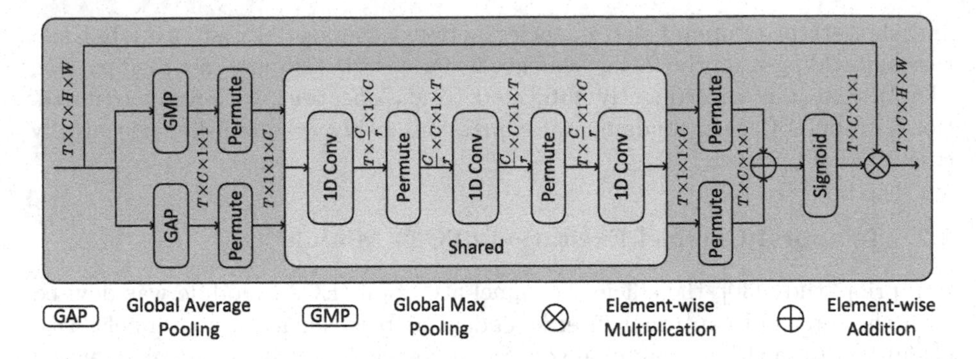

Fig. 3. Detailed structure of the proposed TCE (temporal channel excitation) module.

Ultimately, the output of the TCE module, denoted as F_o, is calculated as

$$F_o = M^c \otimes F, \tag{4}$$

where $F_o \in \mathbb{R}^{T \times C \times H \times W}$, $M^c \in \mathbb{R}^{T \times C \times 1 \times 1}$, and $\otimes$ is broadcast element-wise multiplication.

2.3 Long Short-Term Temporal Transformer

As alluded to previously, the DFER task requests not only the extraction of typical spatial-temporal features from facial videos but also the complete perception of both short-term and long-term dependencies inherent within the facial video data. Hence, a significant majority of recent methodologies [19,20,24,35] lean towards employing pure Transformer architectures to extract temporal features from facial videos due to their inherent capacities for global and long-term modeling. Even so, pure Transformers tend to overlook the significance of locality in visual perception, which has long been proven as helpful for encoding closely correlated visual signals [23]. In light of this, we consider the integration of a 1-D temporal convolutional layer into the pure Transformer. This measure serves the purpose of focusing on the short-term dependencies between neighboring tokens.

Drawing inspiration from [9], we present the LSTformer model, which encompasses both long and short-term temporal modeling abilities. A schematic depiction of the LSTformer is presented in Fig. 4. Specifically, regarding the last output features from the backbone denoted as $Z_0 \in \mathbb{R}^{T \times C \times H \times W}$, where T represents the amount of frames, the spatial information can be gathered through the utilization of global average pooling as

$$Z_0' = GAP\left(Z_0\right), \tag{5}$$

where $Z_0' \in \mathbb{R}^{T \times C \times 1 \times 1}$ is obtained. Afterwards, Z_0' is reshaped into a flattened sequence denoted as $Z^{in} \in \mathbb{R}^{T \times C}$, resulting in the acquisition of T tokens. Each token corresponds to a facial frame and possesses a length of C. We then feed Z^{in}

to the LSTformer which is made up of N transformer-based temporal encoders. During the l_{th} temporal encoder iteration, we first reshape and permute the feature $Z^{l-1} \in \mathbb{R}^{T\times C}$ which is generated by the previous $(l-1)_{th}$ temporal encoder into $Z_*^{l-1} \in \mathbb{R}^{1\times C\times 1\times T}$. Thereafter, a 1-D convolutional layer denoted as K_1^{temp} with a kernel size of k^{temp}, is employed to process the representation Z_*^{l-1} as

$$Z_*^l = Z_*^{l-1} * K_1^{temp} + Z_*^{l-1}, \tag{6}$$

where the output $Z_*^l \in \mathbb{R}^{1\times C\times 1\times T}$. Subsequently, Z_*^l is reshaped and permuted back into $\tilde{Z}^l \in \mathbb{R}^{T\times C}$. The representation $\tilde{Z}^l$ is then input into a typical multi-head self-attention (MHSA) block for further processing as

$$\tilde{Z}_0^l = MHSA\left(LN\left(\tilde{Z}^l\right)\right) + \tilde{Z}^l, \tag{7}$$

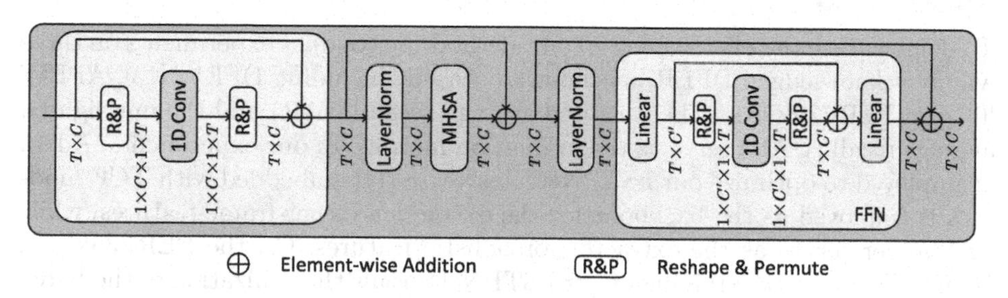

Fig. 4. Detailed structure of the proposed LSTformer (long short-term temporal transformer).

where the output $\tilde{Z}_0^l \in \mathbb{R}^{T\times C}$, and $LN(\cdot)$ is layer normalization. After that, $\tilde{Z}_0^l$ is input to an improved FFN block which is composed of a temporal 1-D convolutional layer and two linear layers. The feature dimension of $\tilde{Z}_0^l$ is initially increased from C to C' via the application of the first linear layer, which can be expressed as

$$\tilde{Z}_1^l = LN\left(\tilde{Z}_0^l\right) W_1^{ffn} + b_1^{ffn}, \tag{8}$$

where $\tilde{Z}_1^l \in \mathbb{R}^{T\times C'}$ represents the output, $W_1^{ffn} \in \mathbb{R}^{C\times C'}$ denotes the linear layer's weights, and $b_1^{ffn} \in \mathbb{R}^{C'}$ is the bias term. We then reshape and permute $\tilde{Z}_1^l$ into $\tilde{Z}_1^{l*} \in \mathbb{R}^{1\times C'\times 1\times T}$. Next, we adopt another 1-D convolutional layer K_2^{temp} with kernel size k^{temp} to deal with $\tilde{Z}_1^{l*}$ as

$$\tilde{Z}_2^l = \tilde{Z}_1^{l*} * K_2^{temp} + \tilde{Z}_1^{l*}, \tag{9}$$

where $\tilde{Z}_2^l \in \mathbb{R}^{1\times C'\times 1\times T}$ is obtained. Thereafter, we reshape and permute $\tilde{Z}_2^l$ back into $\tilde{Z}_2^{l*} \in \mathbb{R}^{T\times C'}$ and then $\tilde{Z}_2^{l*}$ is input to the second linear layer which decreases its feature dimension from C' back to C as

$$Z^l = \tilde{Z}_2^{l*} W_2^{ffn} + b_2^{ffn} + \tilde{Z}_0^l, \tag{10}$$

where $Z^l \in \mathbb{R}^{T\times C}$, $W_2^{ffn} \in \mathbb{R}^{C'\times C}$ and $b_2^{ffn} \in \mathbb{R}^{C}$. The residual connection achieved through the addition of $\tilde{Z}_0^l$ is depicted in Fig. 4. Afterwards, Z^l will be fed to the next temporal encoder if $l < N$. Note that we have $Z^0 = Z^{in}$ when $l = 1$.

In the end, the output $Z^N \in \mathbb{R}^{T\times C}$ which is from the last temporal encoder is processed by a temporal average pooling operation (TAP) and then a fully connected (FC) layer is used for calculating prediction probability p as following

$$p = FC\left(TAP\left(Z^N\right)\right), \tag{11}$$

where $FC \in \mathbb{R}^{C\times G}$ is the FC layer, and G represents the number of facial expression categories.

3 Experiments

To demonstrate the effectiveness of our method, we conduct experiments on three widely acknowledged DFER benchmark datasets, including DFEW [13], AFEW [6], and FERV39k [31]. The weighted average recall (WAR) and the unweighted average recall (UAR) serve as the evaluation metrics. In our experiments, Adam is employed to optimize our LSTPNet. ResNet18 [11] embedded with TCE modules is employed as the backbone for the extraction of per-frame features, while LSTformer serves as the extractor for holistic features. On the FERV39k and DFEW datasets, the training of our LSTPNet entails the utilization of the batch size of 8, coupled with the initial learning rate at 2.56×10^{-5}. As for the AFEW dataset, the training for our LSTPNet involves the batch size of 2, along with the initialization of the learning rate at 6.4×10^{-6}. Each facial clip is divided into $U = 8$ segments and consecutive $V = 2$ frames are stochastically selected from per segment. This process yields $T = 16$ facial frames for training. In the testing phase, a similar procedure is employed, involving the division of each facial clip into $U = 8$ segments. However, the process entails the extraction of successive sets of $V = 2$ frames from the central of each segment, and then a total of $T = 16$ facial frames is generated as well. With regard to the parametric specifics within LSTPNet, $N = 2$, $r = 8$, $C' = 4096$, and $k^{temp} = 3$ are empirically set. The standard cross-entropy loss is adopted to train LSTPNet.

3.1 Ablation Study

Ablation experiments are conducted on both the DFEW and FERV39k datasets, verifying the usefulness of each component integrated within the LSTPNet.

***Basic Evaluations of Key Components*:** In order to demonstrate the effectiveness of our proposed modules, a series of experiments are undertaken wherein the modules are omitted or substituted within the full model. A0 indicates the baseline which consists of the vanilla ResNet18 and pure temporal Transformer. A variant denoted as A1 is established by excluding the TCE modules. Another variant, denoted as A2, is defined by retaining the TCE modules while substituting the LSTformer with the pure temporal Transformer. Results are presented in Table 1. Evidently, these aforementioned variants manifest diminished performance in comparison to the full model, thereby demonstrating the usefulness of the separate components.

***Comparison with Other Channel-Attention Modules*:** In order to demonstrate the usefulness of our proposed TCE modules, We substitute the TCE module with some classical channel attention (CA) modules, namely squeeze-and-excitation (SE) [12], convolutional block attention module (CBAM) [32], efficient channel attention (ECA) [30], along with the recently proposed global convolution-attention block (GCA) [19]. B1, B2, B3, and B4 substitute the TCE module with the SE module, the channel-attention module of the CBAM block, the ECA module, and the GCA block, respectively. As illustrated in Table 2, it is evident that on the two benchmark datasets, our proposed TCE module consistently outperforms the various alternative channel-attention modules.

Table 1. Basic evaluations of the key components in LSTPNet. The best results are highlighted in **bold**.

Setting	Methods		FERV39k (%)		DFEW (%)	
	TCE	LSTformer	WAR	UAR	WAR	UAR
A0			47.34	36.50	67.30	56.97
A1		✓	49.50	39.82	70.75	59.69
A2	✓		49.92	39.61	70.46	59.35
Full	✓	✓	**50.07**	**40.63**	**71.16**	**60.18**

Table 2. Comparison with different channel-attention modules.

Setting	FERV39k (%)		DFEW (%)	
	WAR	UAR	WAR	UAR
B1	49.13	39.14	70.29	58.19
B2	49.05	39.62	70.05	58.66
B3	49.43	39.56	70.23	58.71
B4	49.10	40.26	69.87	59.40

***Effectiveness of the 1-D Convolutions Inside TCE and LSTformer*:** Furthermore, the usefulness of the 1-D convolutions inside TCE and LSTformer is also demonstrated. Regarding the TCE module, we eliminate the second temporal 1-D convolution in B5, while in B6, we retain it but apply it to the channel dimension instead of the temporal dimension. In B7, the second temporal 1-D convolution is retained while the other two 1-D convolutions that operate along the channel dimension are discarded. Besides, in B8, the orientation of these 1-D convolutions is set to the temporal dimension instead of the channel dimension.

Regarding the LSTformer, two 1-D convolutions are integrated into it, as illustrated in Fig. 4. Initially, the first and second 1-D convolutions are omitted, respectively, leading to B9 and B10. Then we alter the orientation of the first and second 1-D convolutions to act on the channel dimension instead of the temporal dimension, respectively, producing B11 and B12. It can be observed, as presented in Table 3 and Table 4, that any form of discarding or altering the original design leads to a deterioration in performance. In other words, the outcomes prove the significance of executing temporal modeling on intermediary features within the backbone network. Furthermore, they emphasize the necessity of enhancing the pure Transformer framework to encompass not only long-term modeling but also the incorporation of short-term temporal dependencies between local tokens.

***Impact of Hyper-Parameters*:** Two hyper-parameters inherent in our method are subjected to experimentation to assess their impact on performance. These include the ratio factor denoted as r within the TCE module, as well as the temporal 1-D convolution's kernel size denoted as k^{temp} in both the TCE module and the LSTformer. As shown in Table 5, a noteworthy observation emerges: a temporal 1-D convolution with a smaller kernel size appears to yield superior performance results. This phenomenon can be attributed to the fact that the larger kernel size in the 1-D convolution tends to emphasize more long-term temporal information, a trait that is already encompassed by the inherent capabilities of the pure temporal Transformer. Additionally, referring to Table 6, it becomes evident that better performance on the two benchmark datasets is attained when the ratio factor r within the TCE module is set to 8. So $r = 8$ is used as the default setting for LSTPNet.

Table 3. Ablation study on the effect of 1-D convolutions in TCE modules.

Setting	FERV39k (%)		DFEW (%)	
	WAR	UAR	WAR	UAR
B5	49.30	40.40	70.07	59.59
B6	48.04	40.17	70.36	57.93
B7	48.89	40.58	70.69	59.71
B8	49.12	40.19	70.10	58.76

Table 4. Ablation study on the effect of 1-D convolutions in LSTformer.

Setting	FERV39k (%)		DFEW (%)	
	WAR	UAR	WAR	UAR
B9	48.87	40.38	70.50	60.02
B10	49.11	40.28	70.92	59.89
B11	49.60	40.30	70.57	58.63
B12	49.50	40.20	70.40	59.32

Table 5. Evaluation results for different 1-D kernel k^{temp}, where the best results are highlighted in **bold**.

Setting	FERV39k (%)		DFEW (%)	
	WAR	UAR	WAR	UAR
$k^{temp} = 5$	48.76	40.64	70.36	60.13
$k^{temp} = 7$	49.91	40.33	69.96	59.65
Ours ($k^{temp} = 3$)	**50.07**	**40.63**	**71.16**	**60.18**

Table 6. Results from using different ratio factor r, where the best are highlighted in **bold**.

Setting	FERV39k (%)		DFEW (%)	
	WAR	UAR	WAR	UAR
r = 2	49.24	39.94	70.61	59.16
r = 4	48.82	40.55	70.28	59.67
r = 16	48.84	40.87	70.58	59.93
Ours (r = 8)	**50.07**	**40.63**	**71.16**	**60.18**

Table 7. Comparison with state-of-the-art methods on DFEW, FERV39k and AFEW. The best results are highlighted in **bold**.

Methods	DFEW (%)		FERV39k (%)		AFEW (%)	
	WAR	UAR	WAR	UAR	WAR	UAR
C3D [26]	53.54	42.74	31.69	22.68	46.72	43.75
P3D [25]	54.47	43.97	33.39	23.20	–	–
3D Resnet18 [10]	54.98	44.73	37.57	26.67	45.67	42.14
R(2+1)D [27]	53.22	42.79	41.28	31.55	46.19	42.89
VGG13+LSTM [31]	–	–	43.37	32.41	–	–
I3D-RGB [3]	54.27	43.40	38.78	30.17	45.41	41.86
VGG11+LSTM [13]	53.70	42.39	–	–	–	–
Two ResNet18+LSTM [31]	–	–	43.20	31.28	–	–
ResNet18+LSTM [13]	53.08	42.86	42.59	30.92	48.82	43.96
VGG16+LSTM [31]	–	–	41.70	30.93	–	–
Resnet18+GRU [35]	64.02	51.68	–	–	49.34	45.12
Two VGG13+LSTM [31]	–	–	44.54	32.79	–	–
EC-STFL [13]	56.51	45.35	–	–	50.66	47.33
Former-DFER [35]	65.70	53.69	46.85	37.20	50.92	47.42
STT [24]	66.45	54.58	–	–	**54.23**	49.11
NR-DFERNet [20]	68.19	54.21	45.97	33.99	53.54	48.37
GCA+IAL [19]	69.24	55.71	48.54	35.82	–	–
LSTPNet (Ours)	**71.16**	**60.18**	**50.07**	**40.63**	53.54	**49.92**

3.2 Comparisons with State-of-the-Art Methods and Visualization

We compare our full LSTPNet with state-of-the-art methods on DFEW, FERV39k and AFEW datasets. As shown in Table 7, our method obtains very competitive performance. Specifically, GCA+IAL [19] currently is the best state-of-the-art model with overall 69.24% WAR and 55.71% UAR on DFEW, while our LSTPNet outperforms GCA+IAL by 1.92% WAR and 4.47% UAR. On FERV39k, our method outperforms Former-DFER [35] by 3.43% UAR and GCA+IAL by 1.53% WAR. For AFEW, our method achieves superior performance against all models on UAR and a comparable result with STT [24] on WAR. It is worth noting that STT employs a facial alignment method called Dlib [16] which distinguishes it from other approaches that adopt another one called RetinaFace [5]. Processed facial frames using this method may achieve better alignments and potentially may be conducive to STT.

In Fig. 5, we visualize attention weights of LSTformer's last encoder. Our LSTPNet emphasizes representative frames (green boxes), improving prediction accuracy compared to the baseline (A0) which pays attention to both representative frames and insignificant frames (blue boxes).

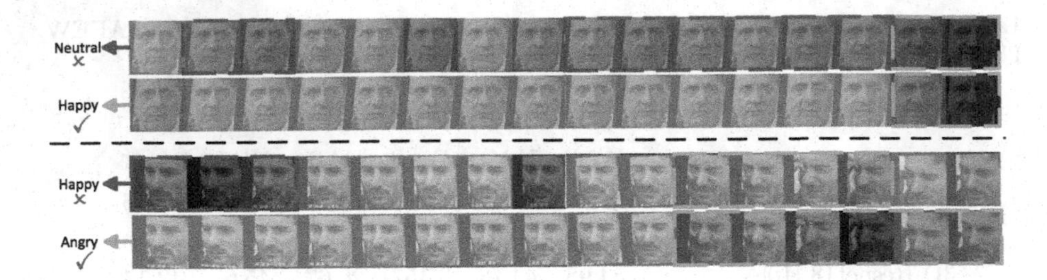

Fig. 5. Visualization of attention weights generated by our model, where each facial video sequence is shown by two rows. The first row shows the weights generated by the baseline (A0) and the second row shows the weights generated by our LSTPNet model. Frames in the green boxes indicate those representative expression frames, while frames in the blue boxes indicate frames with insignificant expressions. (Color figure online)

4 Conclusion

In this paper, we propose a novel long short-term perception network (LSTPNet) for DFER, considering the joint perception of the long-term and short-term temporal features to benefit the DFER task. Two new modules, namely temporal channel excitation (TCE) and long short-term temporal Transformer (LSTformer), are proposed for more temporally representative features and more complete temporal modeling ability. Comprehensive experiments on three benchmark datasets demonstrate the superior performance of LSTPNet over state-of-the-art models, and various ablation experiments validate its key components as well as core inner designs. We hope that more researchers can note the importance of overall temporal modeling for addressing the DFER problem and provide more interesting solutions in the future.

Acknowledgments. This work was supported in part by the NSFC under No. 62176169, and Sichuan Science and Technology Projects (2023ZHCG0007, 2022YFQ0056).

References

1. Abdat, F., Maaoui, C., Pruski, A.: Human-computer interaction using emotion recognition from facial expression. In: 2011 UKSim 5th European Symposium on Computer Modeling and Simulation, pp. 196–201. IEEE (2011)
2. Ayral, T., Pedersoli, M., Bacon, S., Granger, E.: Temporal stochastic softmax for 3D CNNs: an application in facial expression recognition. In: Proceedings of the IEEE/CVF Winter Conference on Applications of Computer Vision, pp. 3029–3038 (2021)
3. Carreira, J., Zisserman, A.: Quo Vadis, action recognition? A new model and the kinetics dataset. In: Proceedings of the IEEE Conference on Computer Vision and Pattern Recognition, pp. 6299–6308 (2017)

4. Chen, W., Zhang, D., Li, M., Lee, D.J.: Stcam: spatial-temporal and channel attention module for dynamic facial expression recognition. IEEE Trans. Affect. Comput. (2020)
5. Deng, J., Guo, J., Ververas, E., Kotsia, I., Zafeiriou, S.: Retinaface: single-shot multi-level face localisation in the wild. In: Proceedings of the IEEE/CVF Conference on Computer Vision and Pattern Recognition, pp. 5203–5212 (2020)
6. Dhall, A., Goecke, R., Lucey, S., Gedeon, T., et al.: Collecting large, richly annotated facial-expression databases from movies. IEEE Multimedia **19**(3), 34 (2012)
7. Fan, Y., Lu, X., Li, D., Liu, Y.: Video-based emotion recognition using CNN-RNN and C3D hybrid networks. In: Proceedings of the 18th ACM International Conference on Multimodal Interaction, pp. 445–450 (2016)
8. Fei, Z., et al.: Deep convolution network based emotion analysis towards mental health care. Neurocomputing **388**, 212–227 (2020)
9. Guo, J., et al.: CMT: convolutional neural networks meet vision transformers. In: Proceedings of the IEEE/CVF Conference on Computer Vision and Pattern Recognition, pp. 12175–12185 (2022)
10. Hara, K., Kataoka, H., Satoh, Y.: Can spatiotemporal 3D CNNs retrace the history of 2D CNNs and imagenet? In: Proceedings of the IEEE Conference on Computer Vision and Pattern Recognition, pp. 6546–6555 (2018)
11. He, K., Zhang, X., Ren, S., Sun, J.: Deep residual learning for image recognition. In: Proceedings of the IEEE Conference on Computer Vision and Pattern Recognition, pp. 770–778 (2016)
12. Hu, J., Shen, L., Sun, G.: Squeeze-and-excitation networks. In: Proceedings of the IEEE Conference on Computer Vision and Pattern Recognition, pp. 7132–7141 (2018)
13. Jiang, X., et al.: Dfew: a large-scale database for recognizing dynamic facial expressions in the wild. In: Proceedings of the 28th ACM International Conference on Multimedia, pp. 2881–2889 (2020)
14. Jung, H., Lee, S., Yim, J., Park, S., Kim, J.: Joint fine-tuning in deep neural networks for facial expression recognition. In: Proceedings of the IEEE International Conference on Computer Vision, pp. 2983–2991 (2015)
15. Kim, D.H., Baddar, W.J., Jang, J., Ro, Y.M.: Multi-objective based spatio-temporal feature representation learning robust to expression intensity variations for facial expression recognition. IEEE Trans. Affect. Comput. **10**(2), 223–236 (2017)
16. King, D.E.: Dlib-ml: a machine learning toolkit. J. Mach. Learn. Res. **10**, 1755–1758 (2009)
17. Li, H., Wang, N., Yang, X., Gao, X.: CRS-CONT: a well-trained general encoder for facial expression analysis. IEEE Trans. Image Process. **31**, 4637–4650 (2022)
18. Li, H., Wang, N., Yang, X., Wang, X., Gao, X.: Towards semi-supervised deep facial expression recognition with an adaptive confidence margin. In: Proceedings of the IEEE/CVF Conference on Computer Vision and Pattern Recognition, pp. 4166–4175 (2022)
19. Li, H., Niu, H., Zhu, Z., Zhao, F.: Intensity-aware loss for dynamic facial expression recognition in the wild. arXiv preprint arXiv:2208.10335 (2022)
20. Li, H., Sui, M., Zhu, Z., et al.: NR-DFERNET: noise-robust network for dynamic facial expression recognition. arXiv preprint arXiv:2206.04975 (2022)
21. Li, S., Deng, W., Du, J.: Reliable crowdsourcing and deep locality-preserving learning for expression recognition in the wild. In: Proceedings of the IEEE Conference on Computer Vision and Pattern Recognition, pp. 2852–2861 (2017)

22. Liu, M., Li, S., Shan, S., Wang, R., Chen, X.: Deeply learning deformable facial action parts model for dynamic expression analysis. In: Cremers, D., Reid, I., Saito, H., Yang, M.-H. (eds.) ACCV 2014. LNCS, vol. 9006, pp. 143–157. Springer, Cham (2015). https://doi.org/10.1007/978-3-319-16817-3_10
23. Lowe, D.G.: Object recognition from local scale-invariant features. In: Proceedings of the Seventh IEEE International Conference on Computer Vision, vol. 2, pp. 1150–1157. IEEE (1999)
24. Ma, F., Sun, B., Li, S.: Spatio-temporal transformer for dynamic facial expression recognition in the wild. arXiv preprint arXiv:2205.04749 (2022)
25. Qiu, Z., Yao, T., Mei, T.: Learning spatio-temporal representation with pseudo-3D residual networks. In: Proceedings of the IEEE International Conference on Computer Vision, pp. 5533–5541 (2017)
26. Tran, D., Bourdev, L., Fergus, R., Torresani, L., Paluri, M.: Learning spatiotemporal features with 3d convolutional networks. In: Proceedings of the IEEE International Conference on Computer Vision, pp. 4489–4497 (2015)
27. Tran, D., Wang, H., Torresani, L., Ray, J., LeCun, Y., Paluri, M.: A closer look at spatiotemporal convolutions for action recognition. In: Proceedings of the IEEE Conference on Computer Vision and Pattern Recognition, pp. 6450–6459 (2018)
28. Udayakumar, N.: Facial expression recognition system for autistic children in virtual reality environment. Int. J. Sci. Res. Publ. **6**(6), 613–622 (2016)
29. Vielzeuf, V., Pateux, S., Jurie, F.: Temporal multimodal fusion for video emotion classification in the wild. In: Proceedings of the 19th ACM International Conference on Multimodal Interaction, pp. 569–576 (2017)
30. Wang, Q., Wu, B., Zhu, P., Li, P., Zuo, W., Hu, Q.: ECA-net: efficient channel attention for deep convolutional neural networks. In: Proceedings of the IEEE/CVF Conference on Computer Vision and Pattern Recognition, pp. 11534–11542 (2020)
31. Wang, Y., et al.: Ferv39k: a large-scale multi-scene dataset for facial expression recognition in videos. In: Proceedings of the IEEE/CVF Conference on Computer Vision and Pattern Recognition, pp. 20922–20931 (2022)
32. Woo, S., Park, J., Lee, J.-Y., Kweon, I.S.: CBAM: convolutional block attention module. In: Ferrari, V., Hebert, M., Sminchisescu, C., Weiss, Y. (eds.) ECCV 2018. LNCS, vol. 11211, pp. 3–19. Springer, Cham (2018). https://doi.org/10.1007/978-3-030-01234-2_1
33. Yu, M., Zheng, H., Peng, Z., Dong, J., Du, H.: Facial expression recognition based on a multi-task global-local network. Pattern Recogn. Lett. **131**, 166–171 (2020)
34. Zhang, Y., Wang, C., Ling, X., Deng, W.: Learn from all: erasing attention consistency for noisy label facial expression recognition. In: Avidan, S., Brostow, G., Cissé, M., Farinella, G.M., Hassner, T. (eds.) ECCV 2022. LNCS, vol. 13686, pp. 418–434. Springer, Cham (2022). https://doi.org/10.1007/978-3-031-19809-0_24
35. Zhao, Z., Liu, Q.: Former-DFER: dynamic facial expression recognition transformer. In: Proceedings of the 29th ACM International Conference on Multimedia, pp. 1553–1561 (2021)

Face Recognition and Pose Recognition

Pyr-HGCN: Pyramid Hybrid Graph Convolutional Network for Gait Emotion Recognition

Li Jing, Guangchao Yang, and Yunfei Yin(✉)

College of Computer Science, Chongqing University, Chongqing 400000, China
yinyunfei@cqu.edu.cn

Abstract. Gait emotion recognition (GER) plays a crucial role in identifying human emotions. Most previous methods apply Spatial-Temporal Graph Convolutional Networks (ST-GCN) to recognize emotions. However, these methods suffer from two serious problems: (1) they ignore the fact that the similarity between emotions with the similar emotional intensity. Consequently, fine-grained information from the low-layer network, which is essential for accurate emotion recognition, is lost. (2) They ignore that the expression of emotion is a continuous process, that is, failing to model the temporal dimension effectively. To address these issues, a novel Pyramid Hybrid Graph Convolutional Network (Pyr-HGCN) is proposed for GER. Specifically, we first introduce and enhance the pyramid structure in GER to compensate for the missing fine-grained information of the ST-GCN structure. Additionally, we design a novel Spatial-Temporal Hybrid Convolution (STHC) block, which can indirectly and simultaneously capture complex spatio-temporal correlations in long-term regions. Extensive experiments and visualizations were performed on several benchmarks, with an accuracy improvement of 0.01 to 0.02 demonstrating the effectiveness of our approach against state-of-the-art competitors.

Keywords: emotion recognition · graph convolutional network · gait

1 Introduction

Emotion refers to a person's attitude towards things and the corresponding behavioral response. Emotion recognition plays an important role in the field of artificial intelligence, with applications including human-computer interaction [18], video surveillance [1]. Emotion recognition can be achieved using different cues, including text [27], facial expression [22], gait [19], etc. Since gait does not require the cooperation of the subject, can be recognized remotely, and emotions are more reliable, GER plays an important role in the field of emotion recognition. GER has various applications, such as emotion-sensing robots [7] or disaster management during evacuation [8], and even mental illness diagnosis.

Q. Liu et al. (Eds.): PRCV 2023, LNCS 14429, pp. 187–198, 2024.
https://doi.org/10.1007/978-981-99-8469-5_15

Current GER methods have mainly used GCN-based methods with 3D skeleton data as input (data obtained from key points of human skeleton are extracted from images through human pose estimation technology [11,17]). These methods typically adopt the stacked network structure of ST-GCN and select feature maps from the top layer network for emotion recognition. However, certain emotions share abstract high-level information, leading to insufficient differentiation. Additionally, these methods tend to emphasize spatial information, inadvertently sidelining the temporal and spatio-temporal dimensions integral to capturing dynamic emotional expressions.

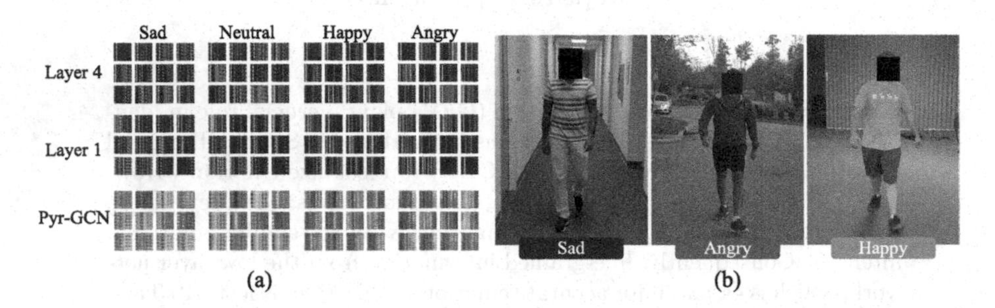

Fig. 1. (a) Visualization of feature maps on 12 channels for low- (Layer 1), high-layer (Layer 4), and Pyr-GCN. (b) Representation of the most pronounced emotions on a frame. The frame that best expresses the emotion is selected to represent this emotion sample. (Image from Ewalk dataset sample [19]).

To illustrate the above problems more clearly, Fig. 1 (a) displays feature maps of different emotions on 12 channels. Interestingly, emotions of varying intensity, encompassing heightened states like happiness and anger, as well as subtler ones like sadness and neutrality, exhibit similarities within high-level feature maps of specific channels. Counterintuitively, these emotions manifest clearer distinctions within low-level feature maps of their respective channels. Furthermore, as shown in Fig. 1 (b), the gait representation of various emotions in a single frame can be ambiguous. Even when the frame that best expresses the emotion is selected to represent the sample, recognizing the emotion accurately can be challenging without incorporating temporal information.

To address these issues, we propose a novel Pyramid Hybrid Graph Convolutional Network (Pyr-HGCN), inspired by the object detection paradigm's Feature Pyramid Network (FPN) [15]. Firstly, we introduce and refine the pyramid structure into the domain of GER. By aggregating low- and high-layer information, our framework aptly modeling distinct gait emotion features. Secondly, we design a novel Spatial-Temporal Hybrid Convolution (STHC). Leveraging the virtues of both GCN and CNN, this convolutional architecture incorporates physical connection priors while enabling direct spatio-temporal feature modeling, effectively curtailing redundancy. Importantly, this approach eliminates the need for prolonged iterations to establish relations between non-adjacent nodes within the graph structure.

In summary, our main contributions are as follows:

- We first introduce and improve a novel pyramid GCN framework for GER. This framework can extract discriminative gait emotion features by aggregating rich detailed and abstract information.
- We propose a novel Spatial-Temporal Hybrid Convolution, which facilitating direct spatio-temporal modeling enriched by physical connection insights, while minimizing redundancy.
- Extensive experiments and visualizations demonstrate the effectiveness of our method against the state-of-the-art competitors.

2 Related Work

2.1 GCN-Based Action Recognition

ST-GCN [28] introduced the graph convolutional network into action recognition (AR), which has since been widely studied. Some scholars have optimized the graph's topology structure together with the network, such as [21] using the two-flow model. MS-G3D [16] improved spatio-temporal topology modeling by constructing a spatio-temporal adjacency matrix. CTR-GCN [5] believed that different channels expressed different information and built a non-shared channel topology. [13] proposed single-oriented pyramid convolutions to capture temporal dynamics at different levels. Like [13], our approach also adopts a pyramid structure, but differs in that we propose a bidirectional interactive pyramid structure in the spatio-temporal domain.

2.2 Skeleton-Based Gait Recognition

Skeleton-based gait recognition (GR) is to recognize a person's identity through the analysis of their skeleton-based gait. [24] adopted the ST-GCN framework of graph convolution to integrate temporal and spatial features. [23] proposed a new GR architecture combining high-order input and residual network. Although skeleton-based GR and gait emotion recognition share the same input data format, the two tasks are fundamentally different. GR focuses on capturing the unique characteristics of each individual's gait, while GER focuses on detecting different emotions expressed in gait.

2.3 Gait Emotion Recognition

In the early stage of GER, [12] applied Fourier transform and principal component analysis to extract gait emotion features. [6] represented human joint motion as a symmetric positive definite matrix and then adopted the nearest neighbor method for sentiment classification. Recently, [4] used a Gated Recurrent Unit to extract features from joint coordinates of a single time step. [3] adopted the graph-based method, uses ST-GCN to extract spatio-temporal features for classification. [20] adopted the attention module and proposed the Attention-enhanced

Time-domain Convolution network (AT-GCN) to capture the distinguishing features in spatially dependent and temporal dynamics for emotion classification. [2] proposed a neural network based on LSTM and multi-layer perceptron to learn features in layers.

Similar to graph-based methods, our method is also ST-GCN based. But the difference is that we take into account the importance of detail information to GER, and make big improvements to the internal block and external framework based on the graph method. Enables the model to extract full of indirect and direct spatial and temporal information.

3 Proposed Method

3.1 Overall Architecture

The overall model architecture of Pyr-HGCN is illustrated in Fig. 2 (a). It consists of three parts: abstraction, concretization and merge. First, the skeleton sequence is gradually sampled by down-sampling to obtain abstract features, and then by up-sampling, where the corresponding features from the down-sampling process are fused to obtain richer information. Then, through the merge stage, the information on the fused features in the up-sampling process is further extracted to promote the fusion. Finally, four emotions are obtained by the fully connected layer classification.

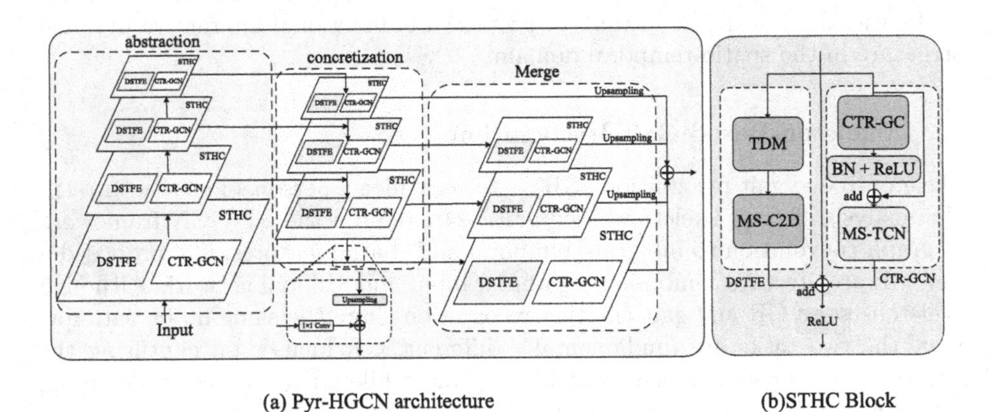

Fig. 2. (a) Architecture Overview. The basic blocks that make up the framework are the Spatial-Temporal Hybrid Convolution (STHC). (b) STHC Block. This diagram shows a simplified structure of this block, which consists of the Direct Spatial-Temporal Feature Extractor (DSTFE) and CTR-GCN [5].

The network is mainly constructed through STHC blocks as shown in Fig. 2 (b). Each STHC block uses two pathways simultaneously to capture complex regional spatial-temporal joint correlations and long-range spatial and temporal dependencies. (1) The CTR-GCN path of the space-time serial structure of the conventional GCN method. (2) The DSTFE path for direct spatial-temporal modeling consisting of Temporal Difference Module (TDM) and Multi-Scale 2D Convolution (MS-C2D). This path avoids information loss caused by the adjacency matrix by directly modeling the space-time information.

The correlation modeling of joints in different frames is limited in ST-GCN [28], as depicted in Fig. 3 (a). Due to architectural constraints, ST-GCN can only capture the relationships between associated nodes on the adjacency matrix, and multiple ST-GCN blocks are required to obtain the adjacency matrix with different node relationships on different frames. However, as the network depth increases, information tends to become consistent in the graph convolutional network [14], resulting in insufficient information obtained through ST-GCN.

3.2 STHC Block

To address these limitations, we propose the STHC inspired by the time-sequence convolution in ST-GCN [28], as depicted in Fig. 3 (b). We extract features through the continuous movement of the convolution kernel, avoiding restrictions imposed by the physical skeleton construction of the adjacency matrix.

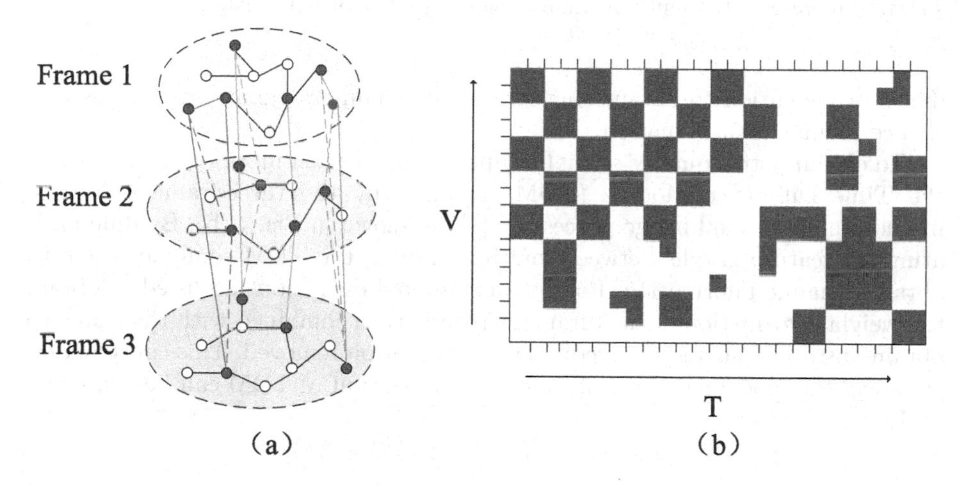

Fig. 3. (a) Variation of node correlations in the three frames. The colored dots represent the nodes that are correlated. The solid lines indicate the association of a node with itself, while the dashed indicate its association with another node within a certain period. (b) Node correlation of V×T plane. Dark blue indicates the correlation between this node and another node. The small red box represents the partial feature map of the information obtained by dilated convolution. (Color figure online)

We also design MS-C2D, which directly models spatial-temporal correlation by discarding the strong constraint brought by the adjacency matrix, as shown in Fig. 4 (a). This method makes up for the deficiency of ST-GCN, which cannot model the correlation of unconnected nodes in different frames. We employ

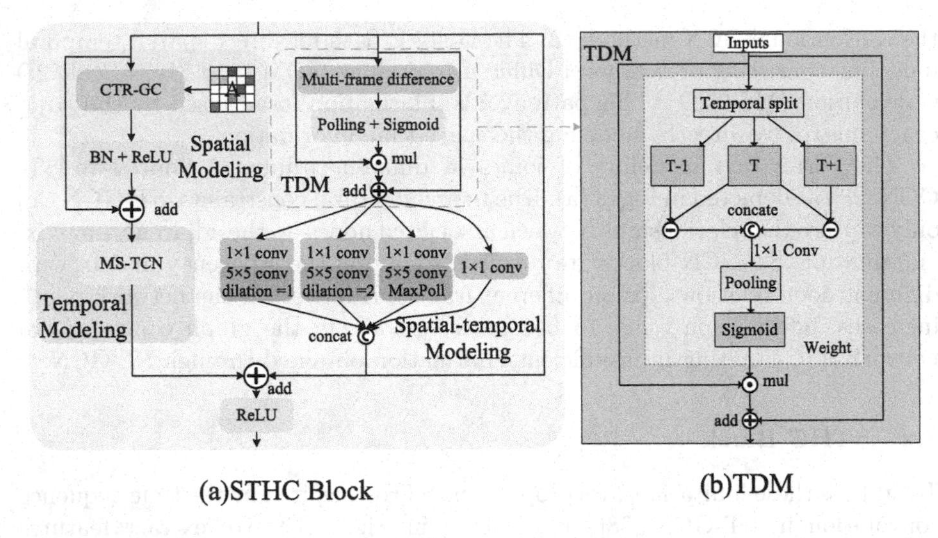

Fig. 4. (a) The structure of STHC. On the left is the CTR-GCN path, and on the right is the DSTFE path combined with TDM (top) and MS-C2D (bottom). The blue Mosaic squares in the upper left corner represent the adjacency matrix. (b) The structure of TDM. ⊙ represents the multiplication of elements. (Color figure online)

dilated convolution to obtain long-term spatio-temporal joint correlations and reduce redundant information.

To obtain better quality spatial-temporal information in MS-C2D, we design the Time Difference Module (TDM) inspired by the three-frame difference method in traditional image processing [9], as shown in Fig. 4 (b). By differentiating the feature graphs between multiple frames, the TDM combines them to obtain dynamic information. Polling and sigmoid operations are used to obtain the weight information of each frame, which is then combined with the frame to obtain abstract features with redundant information removed. This construction enhances the modeling power of time, and the output of TDM can be expressed by Eqs. 1–3.

$$X_d = (X_s^t - X_s^{t-1}) \oplus (X_s^{t+1} - X_s^t), \tag{1}$$

$$X_m = Sigmoid(P(Conv(X_d))) \odot X_s^t, \tag{2}$$

$$TDM(x) = \sum_{T=1}^{T=t}(X_m + x_s^t). \tag{3}$$

where X_s^t represents the output of frame t through CTR-GC, $\oplus$ represents concatenate, $\odot$ representing elements multiplied, $\sum_{T=1}^{T=t} X$ represents the triple frame difference method for frames 1 through t. $TDM(x)$ represents the output of TDM.

4 Experiments

4.1 Experimental Settings

Datasets. Two public benchmarks are adopted: Emotion-Gait-16 [3] and Emotion-Gait-21 [4]. These two datasets are the only publicly available datasets and have been used in several papers. The first dataset has 16 joints and 240 frames, which consists of 2177 ground-truth gait sequences annotated as happy, sad, angry, and neutral. The gait of the second dataset is comprised of a skeleton with 21 joints, and the step of the gait sequence is 48. This dataset contains a total of 1,835 ground-truth gait sequences, each of which has been labeled by 10 annotators.

Implementation Details. All experiments are implemented on a RTX 2080 Ti with the Pytorch framework. 8:1:1 split is used for the training set, validation set, and test set, and the widely-used cross-entropy is applied. For training, we use a batch size of 128 and the Adam optimizer for 70 epochs with an initial learning rate of 0.001. We also use a momentum of 0.9 and a weight decay of 5×10^{-4}.

Table 1. Comparison with state-of-the-art AR, GR and GER on Emotion-Gait-16 dataset. The white lines are AR methods. The light gray lines are the GR method, and the gray lines are the GER method.

Method	Happy	Sad	Angry	Neutral	Accuracy	Precision	Recall	F1
ST-GCN [28]	0.9327	0.8894	0.8942	0.9375	0.9135	0.7520	0.7030	0.7267
MS-G3D [16]	0.9375	0.8942	0.8798	0.9615	0.9183	0.7780	0.7973	0.7875
CTR-GCN [5]	0.9423	0.8798	0.9087	0.9615	0.9231	0.8008	0.7823	0.7914
GaitGraph [24]	0.9219	0.9219	0.8906	0.9375	0.9180	0.7743	0.7668	0.7705
GaitGraph2 [23]	0.9355	0.9274	0.9113	0.9032	0.9194	0.7767	0.7093	0.7415
TEW [4]	0.9327	0.8507	0.9187	0.9402	0.9106	0.7310	0.6893	0.7095
LSTM-MLP [2]	0.9388	0.8798	0.9218	0.9615	0.9255	0.8008	0.8141	0.8074
ProxEmo [18]	0.8173	0.8077	0.8462	0.8990	0.8426	0.5749	0.5683	0.5716
STEP [3]	0.9375	0.8798	0.8990	0.9183	0.9087	0.7257	0.7234	0.7246
G-GCSN [30]	0.9183	0.9135	0.9038	0.9087	0.9111	0.7110	0.6789	0.6946
MSA-GCN [29]	0.9567	0.9038	0.9231	0.9567	0.9351	0.8048	0.8201	0.8124
Pyr-HGCN(ours)	**0.9766**	**0.9141**	**0.9297**	**0.9766**	**0.9493**	**0.8811**	**0.8358**	**0.8579**

4.2 State-of-the-Art Comparisons

Compared to GR and AR, GR has the same action (gait) with us, and AR has the same data form (3D skeleton data) and the same task (classification task) with us. So we perform comparisons with state-of-the-art methods for GER, GR and AR on the above tow datasets. Emotion recognition is a multi-class task, and the commonly used indicators of multi-class tasks, accuracy, precision, recall, and F1-measure, are used as evaluation criteria. The state-of-art methods of GER include the sequential method [2,4], image-based method [18], and graph-based method [3,29,30].

Results on Emotion-Gait-16. Table 1 illustrates the comprehensive experiments on Emotion-Gait-16 dataset using multiple classification metrics. Our proposed method consistently outperforms the SOTAs, particularly in terms of precision and F1 scores.

Results on Emotion-Gait 21. Table 2 illustrates the comprehensive experiments on Emotion-Gait-21 dataset under multiple classification metrics. Our method consistently outperforms the SOTAs, especially on precision, recall, and F1. Notably, CTR-GCN achieves higher accuracy than our method for the angry single category, but lower precision, recall, and F1 scores than ours. This suggests that CTR-GCN simply has a stronger preference for angry samples, resulting in a higher accuracy for this emotion category.

Table 2. Comparison with state-of-the-art AR, GR and GER on Emotion-Gait-21 dataset.

Method	Happy	Sad	Angry	Neutral	Accuracy	Precision	Recall	F1
ST-GCN [28]	0.9511	0.8587	0.9076	0.9348	0.9131	0.7265	0.6943	0.7100
MS-G3D [16]	0.9565	0.9022	0.9348	0.9457	0.9348	0.7859	0.7786	0.7822
CTR-GCN [5]	0.9688	0.9063	**0.9531**	0.9218	0.9375	0.7238	0.6854	0.7041
GaitGraph [24]	0.9545	0.8920	0.9375	0.9432	0.9318	0.7658	0.7871	0.7763
GaitGraph2 [23]	0.9531	0.9063	0.9219	0.9531	0.9336	0.7423	0.6999	0.7205
TEW [4]	0.9622	0.8785	0.9235	0.9505	0.9286	0.7257	0.7110	0.7183
LSTM-MLP [2]	0.9622	0.9297	0.9348	0.9565	0.9458	0.8060	0.7786	0.7921
ProxEmo [18]	0.9545	0.8750	0.8977	0.9205	0.9119	0.6458	0.6302	0.6379
STEP [3]	0.9620	0.8696	0.9185	0.9348	0.9212	0.6108	0.6234	0.6170
G-GCSN [30]	0.9565	0.8641	0.9185	0.9022	0.9103	0.7109	0.7058	0.7083
MSA-GCN [29]	0.9565	0.9185	0.9348	0.9511	0.9402	0.8074	0.7726	0.7896
Pyr-HGCN(ours)	**0.9922**	**0.9375**	0.9453	**0.9688**	**0.9610**	**0.8459**	**0.8300**	**0.8379**

4.3 Ablation Study

Effects of the Proposed Module. The ablation results of the three key components we proposed, namely the Pyramid GCN framework (Pyramid GCN), Spatio-Temporal Hybrid Convolution (STHC), and Time Difference Module (TDM), are shown in Table 3. Each component provides an improvement over the baseline. Among them, Pyramid GCN can bring an improvement of approximately 1.8%. This also demonstrates that while rich abstract information is available, detailed information is also essential.

Study on TDM Module. Table 4 demonstrates that partial channel-containing attention does not perform well. We believe that CTR-GCN itself is innovative and already improved at the channel level, making it difficult for partial channel-inclusive attention to achieve desirable outcomes. However, TDM

Table 3. Ablation experiment of the proposed module.

Model	Accuracy %
CTR-GCN [5]	0.9231
+ STHC	0.9258
+ STHC + TDM	0.9297
+ Pyramid GCN	0.9414
+ Pyramid GCN + STHC	0.9453
+ Pyramid GCN + STHC + TDM	**0.9493**

Table 4. Comparison experiments of the TDM module.

Model	Accuracy %
CTR-GCN [5]	0.9231
+ STHC(with SE [10])	0.9180
+ STHC(with CBAM [26])	0.8789
+ STHC(with ECA [25])	0.9258
+ STHC(with TDM)	**0.9297**

Table 5. Ablation experiment of channel fusion. The first "32" is obtained by down-sampling. The second "32" is obtained by up-sampling.

Method						Accuracy %
CTR-GCN [5] (baseline)						0.9231
Channel	32	32	64	128	256	
With Pyramid GCN merging to 32	✓	✓	✓	✓		**0.9414**
	✓	✓	✓	✓	✓	0.9219
		✓	✓	✓	✓	0.9336
		✓	✓	✓		0.9297
Channel	32	32	64	128	256	
With Pyramid GCN merging to 256	✓	✓	✓	✓		0.9010
	✓	✓	✓	✓	✓	0.9089
		✓	✓	✓	✓	0.9167
		✓	✓	✓		0.9115

performs attention from the temporal dimension. The results suggest that eliminating redundant information and obtaining efficient attention results is effective.

Study on the Pyramid GCN Framework. Table 5 demonstrates the effectiveness of the proposed pyramid GCN framework and the impact of channel dimension fusion. Aggregating all feature channels into 256 channels is less effective than not using the pyramid structure. This may be because in the process of downsampling several distinguishable features are lost and some redundant information is retained. Perhaps, information from the 32 channel overlaps with 256, and the 32 channel has more discriminative information during downsampling. Therefore, including channel 32 in the aggregation produces better results than including channel 256 in the aggregation.

4.4 Visualization and Analysis

To visualize extreme classification difficulty, we selected a few samples and randomly chose five frames from each to represent emotions in Fig. 5. Due to the small number of nodes, the expression forms of different emotions are quite similar. The samples with classification difficulty for anger and neutral, and for happiness and anger, showed similar characteristics. This demands that the model have excellent modeling power in time series and the ability to eliminate redundancy. Our proposed Pyr-HGCN successfully classified such difficult samples, demonstrating the model's effectiveness.

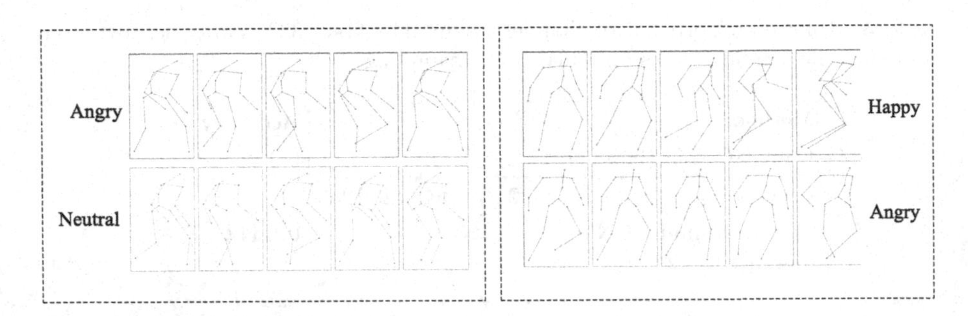

Fig. 5. Visualization of samples that are difficult to classify. Different colors represent different emotional gaits. (Color figure online)

5 Conclusion

In this paper, a novel gait-based Pyr-HGCN network is proposed for human emotion recognition. Considering the similarity of gait-based emotion expression, our method obtains effective emotion feature classification from a fine-grained perspective by proposing a novel pyramid GCN framework and a novel Spatio-Temporal Hybrid Convolution. The proposed framework presents superior performance on two popular benchmarks. There are gait samples collected from different angles in the dataset, and the recognition rate of the side is usually lower than that of the front, because there are more occlusions on the side. In future work, we plan to explore the emotion recognition of the occluded side samples.

References

1. Arunnehru, J., Kalaiselvi Geetha, M.: Automatic human emotion recognition in surveillance video. In: Dey, N., Santhi, V. (eds.) Intelligent Techniques in Signal Processing for Multimedia Security. SCI, vol. 660, pp. 321–342. Springer, Cham (2017). https://doi.org/10.1007/978-3-319-44790-2_15
2. Bhatia, Y., Bari, A.H., Gavrilova, M.: A LSTM-based approach for gait emotion recognition. In: 2021 IEEE 20th International Conference on Cognitive Informatics and Cognitive Computing (ICCI* CC), pp. 214–221. IEEE (2021)

3. Bhattacharya, U., Mittal, T., Chandra, R., Randhavane, T., Bera, A., Manocha, D.: Step: spatial temporal graph convolutional networks for emotion perception from gaits. In: Proceedings of the AAAI Conference on Artificial Intelligence, vol. 34, pp. 1342–1350 (2020)
4. Bhattacharya, U., et al.: Take an emotion walk: perceiving emotions from gaits using hierarchical attention pooling and affective mapping. In: Vedaldi, A., Bischof, H., Brox, T., Frahm, J.-M. (eds.) ECCV 2020. LNCS, vol. 12355, pp. 145–163. Springer, Cham (2020). https://doi.org/10.1007/978-3-030-58607-2_9
5. Chen, Y., Zhang, Z., Yuan, C., Li, B., Deng, Y., Hu, W.: Channel-wise topology refinement graph convolution for skeleton-based action recognition. In: Proceedings of the IEEE/CVF International Conference on Computer Vision, pp. 13359–13368 (2021)
6. Daoudi, M., Berretti, S., Pala, P., Delevoye, Y., Del Bimbo, A.: Emotion recognition by body movement representation on the manifold of symmetric positive definite matrices. In: Battiato, S., Gallo, G., Schettini, R., Stanco, F. (eds.) ICIAP 2017. LNCS, vol. 10484, pp. 550–560. Springer, Cham (2017). https://doi.org/10.1007/978-3-319-68560-1_49
7. Destephe, M., Henning, A., Zecca, M., Hashimoto, K., Takanishi, A.: Perception of emotion and emotional intensity in humanoid robots gait. In: 2013 IEEE International Conference on Robotics and Biomimetics (ROBIO), pp. 1276–1281. IEEE (2013)
8. Gavrilova, M.L., et al.: Multi-modal motion-capture-based biometric systems for emergency response and patient rehabilitation. In: Research Anthology on Rehabilitation Practices and Therapy, pp. 653–678. IGI Global (2021)
9. Han, X., Gao, Y., Lu, Z., Zhang, Z., Niu, D.: Research on moving object detection algorithm based on improved three frame difference method and optical flow. In: 2015 Fifth International Conference on Instrumentation and Measurement, Computer, Communication and Control (IMCCC), pp. 580–584. IEEE (2015)
10. Hu, J., Shen, L., Sun, G.: Squeeze-and-excitation networks. In: Proceedings of the IEEE Conference on Computer Vision and Pattern Recognition, pp. 7132–7141 (2018)
11. Jin, L., et al.: Rethinking the person localization for single-stage multi-person pose estimation. IEEE Trans. Multimedia (2023)
12. Li, B., Zhu, C., Li, S., Zhu, T.: Identifying emotions from non-contact gaits information based on Microsoft Kinects. IEEE Trans. Affect. Comput. **9**(4), 585–591 (2016)
13. Li, F., Zhu, A., Liu, Z., Huo, Y., Xu, Y., Hua, G.: Pyramidal graph convolutional network for skeleton-based human action recognition. IEEE Sens. J. **21**(14), 16183–16191 (2021)
14. Li, Q., Han, Z., Wu, X.M.: Deeper insights into graph convolutional networks for semi-supervised learning. In: Proceedings of the AAAI Conference on Artificial Intelligence, vol. 32 (2018)
15. Lin, T.Y., Dollár, P., Girshick, R., He, K., Hariharan, B., Belongie, S.: Feature pyramid networks for object detection. In: Proceedings of the IEEE Conference on Computer Vision and Pattern Recognition, pp. 2117–2125 (2017)
16. Liu, Z., Zhang, H., Chen, Z., Wang, Z., Ouyang, W.: Disentangling and unifying graph convolutions for skeleton-based action recognition. In: Proceedings of the IEEE/CVF Conference on Computer Vision and Pattern Recognition, pp. 143–152 (2020)

17. Martinez, J., Hossain, R., Romero, J., Little, J.J.: A simple yet effective baseline for 3D human pose estimation. In: Proceedings of the IEEE International Conference on Computer Vision, pp. 2640–2649 (2017)
18. Narayanan, V., Manoghar, B.M., Dorbala, V.S., Manocha, D., Bera, A.: Proxemo: gait-based emotion learning and multi-view proxemic fusion for socially-aware robot navigation. In: 2020 IEEE/RSJ International Conference on Intelligent Robots and Systems (IROS), pp. 8200–8207. IEEE (2020)
19. Randhavane, T., Bhattacharya, U., Kapsaskis, K., Gray, K., Bera, A., Manocha, D.: Learning perceived emotion using affective and deep features for mental health applications. In: 2019 IEEE International Symposium on Mixed and Augmented Reality Adjunct (ISMAR-Adjunct), pp. 395–399. IEEE (2019)
20. Sheng, W., Li, X.: Multi-task learning for gait-based identity recognition and emotion recognition using attention enhanced temporal graph convolutional network. Pattern Recogn. **114**, 107868 (2021)
21. Shi, L., Zhang, Y., Cheng, J., Lu, H.: Two-stream adaptive graph convolutional networks for skeleton-based action recognition. In: Proceedings of the IEEE/CVF Conference on Computer Vision and Pattern Recognition, pp. 12026–12035 (2019)
22. Smith, F.W., Smith, M.L.: Decoding the dynamic representation of facial expressions of emotion in explicit and incidental tasks. Neuroimage **195**, 261–271 (2019)
23. Teepe, T., Gilg, J., Herzog, F., Hörmann, S., Rigoll, G.: Towards a deeper understanding of skeleton-based gait recognition. In: Proceedings of the IEEE/CVF Conference on Computer Vision and Pattern Recognition, pp. 1569–1577 (2022)
24. Teepe, T., Khan, A., Gilg, J., Herzog, F., Hörmann, S., Rigoll, G.: Gaitgraph: graph convolutional network for skeleton-based gait recognition. In: 2021 IEEE International Conference on Image Processing (ICIP), pp. 2314–2318. IEEE (2021)
25. Wang, Q., Wu, B., Zhu, P., Li, P., Zuo, W., Hu, Q.: ECA-net: efficient channel attention for deep convolutional neural networks. In: Conference on Computer Vision and Pattern Recognition (CVPR), pp. 11531–11539 (2020)
26. Woo, S., Park, J., Lee, J.-Y., Kweon, I.S.: CBAM: convolutional block attention module. In: Ferrari, V., Hebert, M., Sminchisescu, C., Weiss, Y. (eds.) ECCV 2018. LNCS, vol. 11211, pp. 3–19. Springer, Cham (2018). https://doi.org/10.1007/978-3-030-01234-2_1
27. Xia, R., Ding, Z.: Emotion-cause pair extraction: a new task to emotion analysis in texts. arXiv preprint arXiv:1906.01267 (2019)
28. Yan, S., Xiong, Y., Lin, D.: Spatial temporal graph convolutional networks for skeleton-based action recognition. In: Thirty-Second AAAI Conference on Artificial Intelligence (2018)
29. Yin, Y., Jing, L., Huang, F., Yang, G., Wang, Z.: MSA-GCN: multiscale adaptive graph convolution network for gait emotion recognition. arXiv preprint arXiv:2209.08988 (2022)
30. Zhuang, Y., Lin, L., Tong, R., Liu, J., Iwamoto, Y., Chen, Y.-W.: G-GCSN: global graph convolution shrinkage network for emotion perception from gait. In: Sato, I., Han, B. (eds.) ACCV 2020. LNCS, vol. 12628, pp. 46–57. Springer, Cham (2021). https://doi.org/10.1007/978-3-030-69756-3_4

A Dual-Path Approach for Gaze Following in Fisheye Meeting Scenes

Long Rao, Xinghao Huang, Shipeng Cai, Bowen Tian, Wei Xu(✉), and Wenqing Cheng

Hubei Key Laboratory of Smart Internet Technology, School of Electronic Information and Communication, Huazhong University of Science and Technology, Wuhan 430074, China

{rao2000,huangxinghao,m202272517,m202072111,xuwei,chengwq}@hust.edu.cn

Abstract. Gaze following plays a crucial role in scene comprehension tasks, as it captures users' visual information from their facial and eye movements, thereby predicting their gaze positions. This technique finds its application in various domains such as human-computer interaction and medical diagnosis. In the domain of multi-party meeting scenes, some studies have utilized fisheye cameras to capture the entire meeting scene. In this work, we focus on gaze following methods that utilize fisheye images for meeting scenes and collect the GazeMeeting dataset that contains 31,915 fisheye samples. We also propose a dual-path feature fusing model for gaze following, which fuses the learned features in the planar and spherical domains by introducing spherical convolutions. The dual-pathway model can learn the distortion information of different positions from scene images, achieving a normalized L2 distance of 0.0657 on our self-built GazeMeeting dataset. This result represents a 22.80% improvement over the current state-of-the-art methods. Additionally, our proposed model achieves a normalized L2 distance of 0.1326 on Gaze-Follow dataset, outperforming the current state-of-the-art methods by 3.35%.

Keywords: Gaze following · Fisheye image · Spherical CNN

1 Introduction

Gaze following is a task that follows the gaze direction of individuals in a scene and infers where they are looking [15]. This technology is widely used in various fields, including human-computer interaction, medical diagnosis, intelligent transportation, and virtual reality, due to its ease of use and good performance [8,18]. Gaze following technology captures visual information of people's facial and eye movements and measures their attention to any object, which helps

This work was supported by the National Key Research and Development Program of China under Grant 2021YFC3340803.

Q. Liu et al. (Eds.): PRCV 2023, LNCS 14429, pp. 199–210, 2024.
https://doi.org/10.1007/978-981-99-8469-5_16

understand their needs. By understanding human gaze behavior, this technology has the potential to enable more intelligent computer systems and improve their interaction and collaboration with humans.

In the field of education, gaze following technology can be used to evaluate students' attention levels to teaching, as well as determining their understanding and interest in particular content, which helps to evaluate the learning outcomes of education. This technology can help teachers adjust their teaching strategies based on students' feedback to increase their engagement and learning outcomes. Additionally, gaze following technology can also be used to detect and correct users' attentional deviations to ensure active participation and effective communication of students. Therefore, gaze following algorithms have significant application value in various fields, especially in the fields of education, and implementing a high-performance gaze following algorithm is of great importance.

However, in multi-party meeting scenes, using gaze following technology to capture participants' attention requires recording their head information and the current scene [11]. A possible approach for detecting the gaze points of multiple users is to equip each user with eye-tracking devices with perspective cameras and then integrate the information obtained from these devices. However, wearable eye trackers are not commonly used and are insufficient to support large-scale participant gaze data collection in environments such as meeting interactions. Wearable eye trackers are also a burden for participants and bring up issues such as calibration, cost, and battery life [9,21]. Another method is to use a fisheye camera [11]. Since people sit around a table, a single fisheye camera can be placed at the focus of attention to record the visual information of the entire meeting scene, including gestures, facial expressions, gaze, and conversation.

In light of these facts, we propose a new task: gaze following in fisheye images. Compared with the application of gaze following in other images, there are two main challenges: (1) Existing deep learning-based gaze following methods are data-driven, but now there is a shortage of sufficiently large dataset based on fisheye images. As shown in Fig. 1, the images in currently available dataset are quite different from the fisheye image. (2) Fisheye images have the characteristic of spherical distortion, which is more pronounced when human objects look into the distance. The mainstream algorithms cannot learn the features of the spherical domain in fisheye images, which poses a challenge for gaze following in fisheye images.

To deal with the first challenge, we used a fisheye camera to collect a large-scale fisheye images dataset in multi-party roundtable meeting scenes, which contains 31,915 samples in total. In addition, to address the second challenge, we proposed a dual-path feature fusion model to learn the distortion feature of different regions. Our model learns the distortion features of different regions in fisheye scene images by constructing two pathways of planar convolution and spherical convolution. The main contributions of this paper are summarized as follows:

- To the best of our knowledge, we are the first to apply gaze following technology to fisheye images.

- We created a dataset using a fisheye camera to collect data, called the GazeMeeting dataset, to address the lack of publicly available datasets for gaze following in fisheye scenes. The dataset contains 31,915 samples in total.
- We proposed a dual-path model for gaze following based on the GazeMeeting dataset. By constructing two pathways of planar convolution and spherical convolution, the model learns the distortion features of different regions in fisheye scene images and achieves significant improvement over the current state-of-the-art models, with a 22.8% improvement over the best public model on the GazeMeeting dataset.

(a) GazeFollow

(b) GazeFollow360

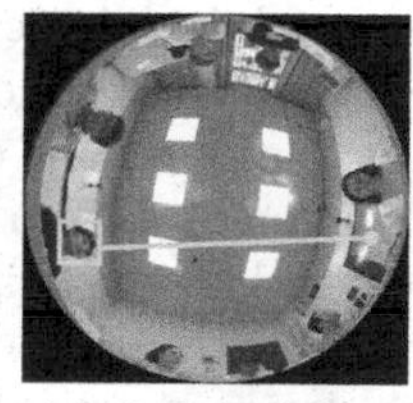

(c) GazeMeeting

Fig. 1. (a) Dataset GazeFollow consists of 2D images. The gaze target of the human subject is out-of-frame or in-frame. (b) Dataset GazeFollow360 consists of 360-degree images, and its distortion characteristics are very different from fisheye image. (c) In the fisheye image, our participants ring sit around the fisheye camera with spherical distortion.

2 Related Works

2.1 Gaze Following

While most of the existing gaze following research has primarily concentrated on real-life scenarios such as socializing [8,15] and shopping [18], our work specifically targets multi-party roundtable meetings. Recasens et al. [15] was the first to define the gaze following problem and constructed the 2D image dataset GazeFollow, which marked a new stage of gaze following research. Cheng et al. [2] predicted gaze direction by combining coarse-grained face features with fine-grained features of the eyes. Lian [13] and Chong [3] introduced the concept of gaze field. Lian [13] used a planar multi-scale gaze direction field to enhance the saliency model's gaze supervision. Chong et al. [3,4] extended gaze target detection to situations where people may look out of the image. Chong et al. [4] attempted to combine head and scene features and employed the generated heatmap of the head features to facilitate the model in learning the most important scene features, ultimately enhancing the model's predictions by identifying whether the gaze direction lies within the frame. Zhuang [22] and Cohen [5] both constructed models to identify the common gaze point of multiple humans in their own work. Li et al. [12] released the GazeFollow360 panoramic image gaze following dataset and combined 2D gaze following technology with 3D gaze estimation [19] to apply gaze following technology to 360-degree panoramic images.

2.2 The Processing of Fisheye Image

In meeting scene, we use fisheye cameras to capture images. Compared with ordinary lenses, fisheye cameras can capture more information but also introduce geometric distortion to the image. One method for handling fisheye image distortion is to improve the adaptability of the algorithm. Su et al. [17] proposed Kernel Transformer Network (KTN). They converted the fisheye images into equidistant rectangular projection images and fed them into KTN, which generates convolution kernels modified by adaptive branches based on the positional information of the equidistant rectangular projection image. However, for different network structures, these parameters need to be adjusted, which means that the original network cannot be applied to other omnidirectional images with fisheye images, and has poor portability and universality. Su et al. [16] designed convolution kernels with different shapes based on different pixel dimensions, while other works [7,20] adjusted convolution kernels on the sphere and resampled features or projected features onto the tangent plane. However, all these methods did not learn features of fisheye images by combining planar convolution and spherical convolution. Cohen et al. [6] extracted rotation-invariant features using spherical convolution and extended convolution to the spectral domain. This provides an idea of combining planar-domain features and spherical-domain features. Therefore, Miao et al. [14] used fisheye images as input for face detection, combined planar-domain features and spherical-domain features by introducing spherical convolution and attention mechanisms, and improved the model's ability to detect faces in fisheye scenes.

3 Method

3.1 Framework

Our architecture comprises of two main stages. A **head feature extraction module** and a **dual-path feature fusion module**. An illustration of the architecture is shown in Fig. 2.

The primary objective of the first stage is to extract corresponding features from the cropped image of the human head and the position of the eye, and then merge them with the scene features in the next stage. The second stage extracts features using both planar convolution and spherical convolution, and then fuses the head features from the first stage. The dual-path module learns distorted features of different regions from the scene image and better learn the mapping relationship of fisheye scenes, thereby improving the performance of the entire gaze following model.

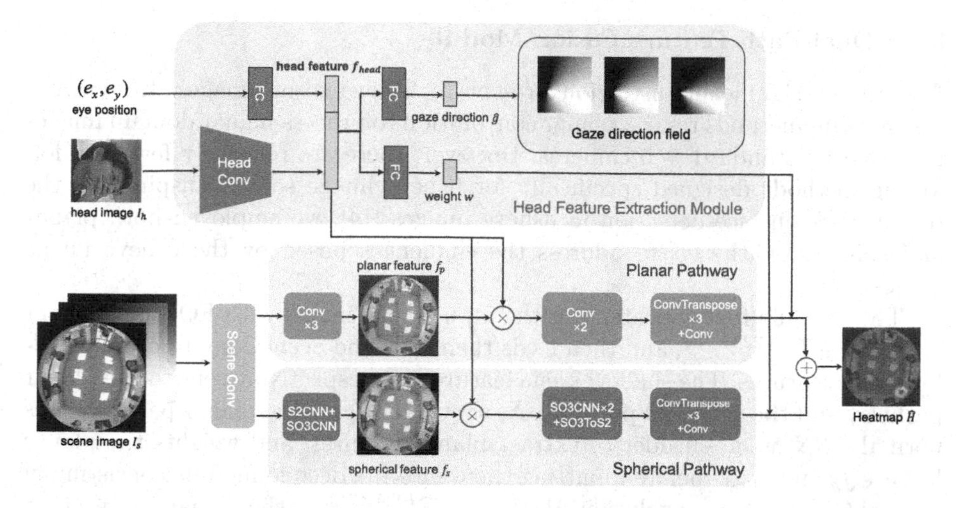

Fig. 2. The overview of our framework. It consists of a head feature extraction module and a dual-path feature fusion module. The head feature extraction module takes eye position and head images as input and outputs a multi-scale gaze field and weighted information. The dual-path feature fusion module takes the meeting scene image as input, and goes through planar pathway and spherical pathway to obtain predict heatmap $\hat{H}$.

3.2 Head Feature Extraction Module

The module first inputs the head image $I_h \in \mathbb{R}^{3\times 224\times 224}$ and the eye center coordinate point (e_x, e_y) of the human object. The head image is sent to the backbone network HeadConv to obtain the feature $f_1 \in \mathbb{R}^{1\times 256}$, which is concatenated with the feature $f_2 \in \mathbb{R}^{1\times 512}$ output by fully connected layer from the 2D eye center position to obtain the final feature $f_{head} \in \mathbb{R}^{1\times 768}$. The feature f_{head} is then passed through two branches to obtain the initial gaze direction $\hat{g}$ and weighted information w.

In fisheye images, the distortion in the sphere domain will greatly degrade the performance of the gaze following methods constructed in the planar domain, because these methods cannot adapt to the distortion caused by the projection from the sphere domain to the plane domain. Based on the characteristic, we set a threshold to distinguish whether the human is looking at the planar domain or the spherical domain, and view w as the weight information for the planar and spherical domains in the second stage.

By combining the predicted gaze direction $\hat{g}$ with the eye center position (e_x, e_y), we further generate the field of view (FOV). The FOV can be viewed as a cone that extends infinitely from the center of the eyes. Its conical sections of different depths are elliptic slices of different sizes, and the projection of these slices onto the camera plane will form sector regions. Based on Lian's work [13], given the eye center point and the predicted gaze direction, we generate three different angles of FOV($\gamma = 5, 2, 1$), for feature extraction in the next stage.

3.3 Dual-Path Feature Fusion Module

Due to the significant distortion present in fisheye scene images, the current mainstream methods rely on planar convolution to process planar domain images captured by standard web cameras. However, there are relatively few gaze following methods designed specifically for fisheye image scenes. Inspired by the recent work on face detection in fisheye images [14], we employed both planar and spherical pathways to address the challenges posed by the fisheye image scenes.

This module first concatenates the input scene image with FOV of different angles $I_s \in \mathbb{R}^{6\times224\times224}$, and then feeds them into the SceneConv to obtain shallow scene features. The shallow scene features are respectively sent to the planar pathway and the spherical pathway. As shown in Fig. 2, the planar pathway uses normal CNN as an encoder to extract planar features, and weights the planar feature f_p and f_{head} before inputting them into the decode module consisting of normal CNN and deconvolution. Here, $f_p \in \mathbb{R}^{1024\times7\times7}$ is the planar feature that perceives the region near the head. The spherical pathway uses S2CNN [6] to map the shallow scene features from the planar coordinate system to the spherical coordinate system while reducing the number of channels of the features. Then, they are fed into SO3CNN [6] to obtain spherical feature $f_s \in \mathbb{R}^{512\times28\times28}$. SO3CNN has the characteristic of rotation invariance, which can learn better semantic features in the spherical domain. After weighting the spherical feature f_s and the head feature f_{head}, it goes through the spherical decode module consisting of SO3CNN and deconvolution. Both the planar pathway and the spherical pathway produce heatmaps of the same size, and the output weights of the first stage extraction module are used to weigh the heatmaps generated by the two branchs to obtain the final prediction result $\hat{H} \in \mathbb{R}^{64\times64}$.

4 Experiments

4.1 Our Dataset—GazeMeeting

Most of the currently available public datasets are formed by selecting images from pre-existing videos or datasets. The image data present in popular datasets like GazeFollow [15] primarily comprises of 2D images. On the other hand, GazeFollow360 [12] is obtained from 360-degree videos sourced from YouTube, which differ significantly from videos captured by fisheye cameras. Given that current deep learning-based gaze following methods are heavily data-driven, there exists a scarcity of publicly available large-scale datasets for gaze following in fisheye images. To address this issue, we developed the GazeMeeting dataset using a fisheye camera to bridge the gap in publicly available datasets for gaze following in fisheye scenes.

The GazeMeeting dataset was created using ezviz fisheye camera to record content from Problem-Based Learning classes and meetings [1] at a medical college in China. A total of 3,573 frames and 31,915 samples were selected from a large number of videos to form the dataset. Eight trained volunteers annotated the human eye center point and gaze point for each human object. A testing set was constructed by holding out approximately 20% of the annotations, ensuring no source-overlap between the train and test splits.

4.2 Experimental Setup

Implementation Details. In the first stage, the output of head feature extraction module is used to predict gaze direction and weight the second stage branch feature. During training, we used a threshold of 300 pixels to distinguish between the planar and spherical regions in the GazeMeeting dataset.

In the second stage, the backbone of SceneConv is ResNet50 [10]. We fused planar and spherical features with the head feature f_{head} and passed them through the planar and spherical decode branches to generate a 64×64 size heatmap. Each pixel in the ground truth heatmap was generated using the following Eq. 1:

$$H(i,j) = \frac{1}{\sqrt{2\pi}\sigma} e^{\frac{(i-g_x)^2+(j-g_y)^2}{2\sigma^2}}, \tag{1}$$

where $g = g(g_x, g_y)$ is the coordinates of point of gaze, and i and j are the horizontal and vertical coordinates of a certain point in the heatmap respectively. σ is the standard deviation of Gaussian kernel, which we set as a rule of thumb to be 3.

During training, we used the Adam optimizer. We first trained head feature extraction module with a batch size of 128 for 30 epochs, with an initial learning rate of 5e-4. The learning rate was decreased by a factor of 0.1 every 15 epochs. We then trained the second stage with a batch size of 96 for 100 epochs, with a learning rate that decreased by a factor of 0.1 at the 70th epoch.

Metric. To evaluate the validity of our method, we adopt these evaluation metrics. ***L2 distance***: we evaluate the Eucludean distance between predict point and the ground truth (lower is better). ***AUC*** refers to the area under the ROC curve (higher is better) [15]. We use the AUC criterion to assess a predicted gaze target heatmap. For fair comparison, all the predicted heatmap are downsample to a 64×64 heatmap and are compared a heatmap of the same size with kernel size 3 to calculate the AUC score.

Loss Function. We train our model in two steps. Firstly, we train the first stage head feature extraction module. Equation 2 represents the loss function of gaze direction in the first stage.

$$\mathcal{L}_g = 1 - \frac{\langle g, \hat{g} \rangle}{\|g\| \, \|\hat{g}\|}. \tag{2}$$

Later, We use binary cross entropy loss as the loss function to train the weight branch of this stage. Finally, the two losses are added with the same weight to joint train the whole module. The loss function of the second stage dual-path feature fusion module is mean squared error loss.

Table 1. Quantitative evaluation on GazeMeeting dataset.

Method	Norm Dist	Pixel Dist	AUC
Recasens [15]	0.1347	180.99	0.8603
Lian [13]	0.0867	116.57	0.8803
Chong [4]	0.0851	114.48	0.8828
Our Method	**0.0657**	**88.36**	**0.8980**

4.3 Comparison with Other Methods

Quantitative Analysis. We compared our proposed method with representative and mainstream gaze following methods, including Recasens [15], Lian [13], Chong [4], and Li [12]. It should be noted that we choose Li's method as a comparison method because they proposed a gaze following method for 360-degree images, which share similarities with fisheye images. However, since the method proposed by Li was not publicly available, we implemented it by ourselves. Li's method was designed for 360-degree images, which may not be suitable for planar and fisheye coordinate systems. Therefore, we only compared Li's method with other methods on the GazeFollow360 dataset. For the GazeFollow dataset, which uses a planar coordinate system, we replaced the spherical CNN in our proposed dual-path feature fusion module with planar CNN during training and testing, while keeping the other structures unchanged. The original model structure was maintained for GazeFollow360 and GazeMeeting dataset.

As shown in Table 1, on our GazeMeeting dataset, the gaze following method proposed by Recasens, which extracts features through parallel gaze estimation and saliency prediction and then fuses them, has the worst performance with a normalized L2 distance of 0.1347. Both Lian and Chong methods use head features to fuse scene features and generate gaze heatmaps using a structure similar to FPN. Therefore, their results are similar, with normalized L2 distance of 0.0867 and 0.0851 respectively. Our proposed method outperforms the other three methods in terms of L2 distance and AUC. Compared to the best-performing method by Chong, our method achieved a 22.80% improvement in normalized L2 distance. These results demonstrate that the dual-path network that combines learning features from the spherical and planar domains can learn distortion features at different regions in the scene image, which better addresses the distortion problem caused by fisheye images.

Table 2. Quantitative evaluation on GazeFollow dataset.

Method	Norm Dist	Pixel Dist	AUC
Recasens [15]	0.1900	108.58	0.8780
Lian [13]	0.1509	76.91	0.9160
Chong [4]	0.1372	70.19	0.9210
Our Method	**0.1326**	**68.29**	**0.9284**

Table 3. Quantitative evaluation on GazeFollow360 dataset.

Method	Norm Dist	Pixel Dist	AUC
Recasens [15]	0.2771	843.48	0.6353
Lian [13]	0.1909	619.72	0.7821
Chong [4]	0.1864	564.30	0.8339
Li [12]	0.1584	443.53	0.7803
Our Method	**0.1240**	**372.20**	**0.8916**

As shown in Table 2, on the GazeFollow dataset, the results show that Recasens has the worst performance, while Chong outperforms Lian. We believe this is because Chong's method fuses head and scene images at the intermediate features, which is better than Lian's approach of concatenating the FOV and scene images along the channel dimension only. Our proposed method also outperforms the other three methods in terms of L2 distance and AUC. Compared to the best-performing method by Chong, our method achieved a 3.35% improvement in normalized pixel distance error. These results demonstrate that our dual-path module with normal CNN can further improve the prediction performance.

Table 3 shows the results of various methods on the GazeFollow360 dataset. For methods using planar CNN, such as Recasens, Lian, and Chong, their performance on GazeFollow360 is not good. Planar CNN are not suitable for 360-degree images, as they cannot learn the spatial mapping relationship of the unfolded panoramic image. Our proposed dual-path network achieved normalized L2 distance of 0.1240 on the GazeFollow360 dataset. Compared to the best-performing comparison method by Li [12], our method achieved a 16.08% improvement in normalized L2 distance. The results in the Table 3 demonstrate that even in 360-degree images, our proposed head feature extraction module can explicitly choose between local region and distant region.

Qualitative Results. To intuitively demonstrate the prediction results and accuracy of mainstream gaze following methods and our proposed method on different datasets. Some qualitative examples of our model is shown in Fig. 3. The first row is the result of GazeMeeting and the second row comes from GazeFollow. In addition, the last row is the result of GazeFollow360. The gaze direction heatmap predicted by our approach is more concentrated and closer to the ground truth than other methods. It is obvious that our method works better than other methods in different dataset.

4.4 Ablation Study

To investigate the role of our proposed model on the gaze following network, we designed ablation experiments on the GazeMeeting dataset. As shown in

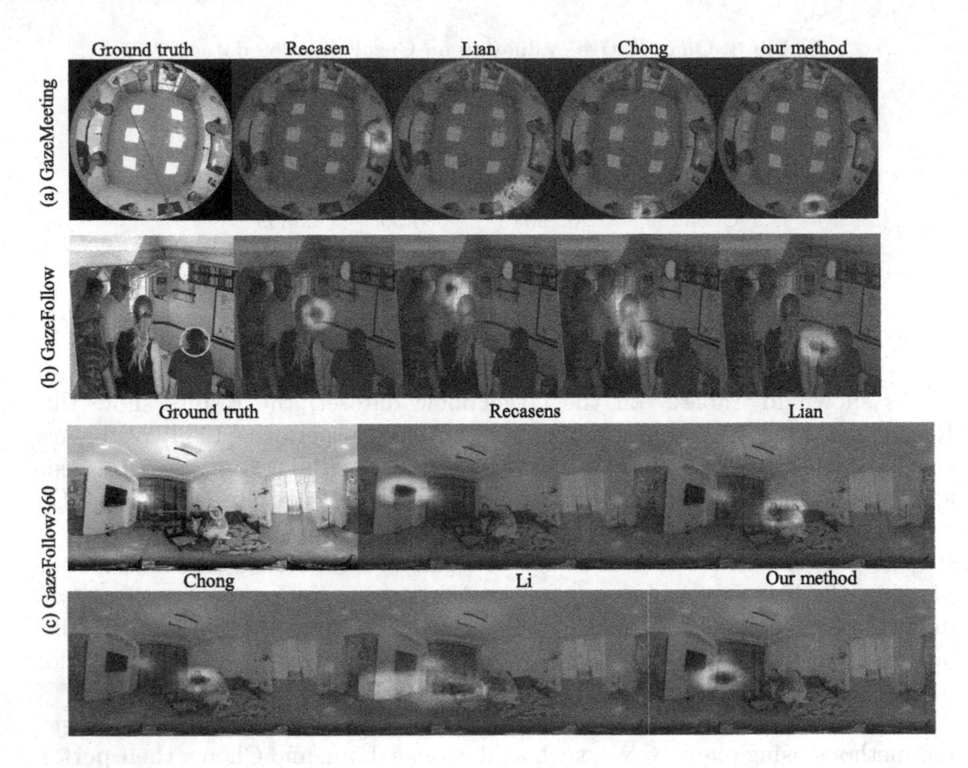

Fig. 3. Qualitative results on GazeMeeting, GazeFollow and GazeFollow360 dataset. The yellow circle indicates the human object, and the arrow and circle indicate the ground truth gaze point.

Table 4, the part of dual-path module has the significant improvement in the prediction performance. Compared to the baseline model, the normalized L2 distance improved by 12.07% on GazeMeeting Dataset. This indicates that the dual-path feature fusion module has big impact on the whole method. Furthermore, we observe that the combination of the dual-path feature fusion module and multi-scale FOV has the most significant improvement in performance, with a 33.94% improvement in normalized L2 distance error compared to the baseline model.

Table 4. Ablation study on the GazeMeeting dataset

Dual-Path	FOV	Norm Dist	Pixel Dist	AUC
		0.099	133.75	0.8668
✓		0.0875	117.61	0.8874
	✓	0.0826	111.06	0.8897
✓	✓	**0.0657**	**88.36**	**0.8980**

5 Conclusion

In this paper, we are the first to apply gaze following technology to fisheye images and collect a large-scale fisheye dataset for multi-party meeting scenes. In addition, this paper proposes a dual-path feature learning gaze following model. By combining visual features from the planar and spherical domains, the entire dual-path network structure can learn the distortion features of different positions in the fisheye image from the scene image to improve the prediction results of gaze following. The experimental results demonstrate that the dual-path network in our model can complement each other to improve the prediction results of entire gaze following model. We believe that our method can effectively assess the attention level of participants or students in educational meeting scenes.

References

1. Barrows, H.S.: Problem-based learning in medicine and beyond: a brief overview. New Dir. Teach. Learn. **1996**(68), 3–12 (1996)
2. Cheng, Y., Huang, S., Wang, F., Qian, C., Lu, F.: A coarse-to-fine adaptive network for appearance-based gaze estimation. In: Proceedings of the AAAI Conference on Artificial Intelligence, vol. 34, pp. 10623–10630 (2020)
3. Chong, E., Ruiz, N., Wang, Y., Zhang, Y., Rozga, A., Rehg, J.M.: Connecting gaze, scene, and attention: generalized attention estimation via joint modeling of gaze and scene saliency. In: Proceedings of the European Conference on Computer Vision (ECCV), pp. 383–398 (2018)
4. Chong, E., Wang, Y., Ruiz, N., Rehg, J.M.: Detecting attended visual targets in video. In: Proceedings of the IEEE/CVF Conference on Computer Vision and Pattern Recognition, pp. 5396–5406 (2020)
5. Cohen, M., Shimshoni, I., Rivlin, E., Adam, A.: Detecting mutual awareness events. IEEE Trans. Pattern Anal. Mach. Intell. **34**(12), 2327–2340 (2012)
6. Cohen, T.S., Geiger, M., Köhler, J., Welling, M.: Spherical CNNs. In: International Conference on Learning Representations, pp. 1–15 (2018)
7. Coors, B., Condurache, A.P., Geiger, A.: Spherenet: learning spherical representations for detection and classification in omnidirectional images. In: Proceedings of the European Conference on Computer Vision (ECCV), pp. 518–533 (2018)
8. Fan, L., Chen, Y., Wei, P., Wang, W., Zhu, S.C.: Inferring shared attention in social scene videos. In: Proceedings of the IEEE Conference on Computer Vision and Pattern Recognition, pp. 6460–6468 (2018)
9. Fathi, A., Li, Y., Rehg, J.M.: Learning to recognize daily actions using gaze. In: Fitzgibbon, A., Lazebnik, S., Perona, P., Sato, Y., Schmid, C. (eds.) ECCV 2012. LNCS, vol. 7572, pp. 314–327. Springer, Heidelberg (2012). https://doi.org/10.1007/978-3-642-33718-5_23
10. He, K., Zhang, X., Ren, S., Sun, J.: Deep residual learning for image recognition. In: Proceedings of the IEEE Conference on Computer Vision and Pattern Recognition, pp. 770–778 (2016)
11. Li, S., Fujii, N.: Estimating gaze points from facial landmarks by a remote spherical camera. In: 2020 25th International Conference on Pattern Recognition (ICPR), pp. 7633–7639. IEEE (2021)

12. Li, Y., Shen, W., Gao, Z., Zhu, Y., Zhai, G., Guo, G.: Looking here or there? Gaze following in 360-degree images. In: Proceedings of the IEEE/CVF International Conference on Computer Vision, pp. 3742–3751 (2021)
13. Lian, D., Yu, Z., Gao, S.: Believe it or not, we know what you are looking at! In: Jawahar, C.V., Li, H., Mori, G., Schindler, K. (eds.) ACCV 2018. LNCS, vol. 11363, pp. 35–50. Springer, Cham (2019). https://doi.org/10.1007/978-3-030-20893-6_3
14. Miao, J., Liu, Y., Liu, J., Argyriou, A., Xu, Z., Han, Y.: Improved face detector on fisheye images via spherical-domain attention. In: 2021 IEEE Symposium on Computers and Communications (ISCC), pp. 1–7. IEEE (2021)
15. Recasens, A., Khosla, A., Vondrick, C., Torralba, A.: Where are they looking? In: Advances in Neural Information Processing Systems, vol. 28 (2015)
16. Su, Y.C., Grauman, K.: Learning spherical convolution for fast features from 360 imagery. In: Advances in Neural Information Processing Systems, vol. 30 (2017)
17. Su, Y.C., Grauman, K.: Kernel transformer networks for compact spherical convolution. In: Proceedings of the IEEE/CVF Conference on Computer Vision and Pattern Recognition, pp. 9442–9451 (2019)
18. Tomas, H., et al.: Goo: a dataset for gaze object prediction in retail environments. In: Proceedings of the IEEE/CVF Conference on Computer Vision and Pattern Recognition, pp. 3125–3133 (2021)
19. Zhang, X., Sugano, Y., Fritz, M., Bulling, A.: Appearance-based gaze estimation in the wild. In: Proceedings of the IEEE Conference on Computer Vision and Pattern Recognition, pp. 4511–4520 (2015)
20. Zhang, Z., Xu, Y., Yu, J., Gao, S.: Saliency detection in 360 videos. In: Proceedings of the European Conference on Computer Vision (ECCV), pp. 488–503 (2018)
21. Zhu, X., Ramanan, D.: Face detection, pose estimation, and landmark localization in the wild. In: 2012 IEEE Conference on Computer Vision and Pattern Recognition, pp. 2879–2886. IEEE (2012)
22. Zhuang, N., et al.: Muggle: multi-stream group gaze learning and estimation. IEEE Trans. Circuits Syst. Video Technol. **30**(10), 3637–3650 (2019)

Mobile-LRPose: Low-Resolution Representation Learning for Human Pose Estimation in Mobile Devices

Xinyu Nan and Chenxing Wang(✉)

School of Automation, Southeast University, Nanjing, China
cxwang@seu.edu.cn

Abstract. Human pose estimation has made great progress in performance due to the development of deep learning. Current methods, including some lightweight networks, usually generate high-resolution heatmaps with rich position information to ensure high accuracy, however, the computational cost is heavy and sometimes unacceptable to mobile devices. In this paper, we construct a network backbone based on the modified MobileNetV2 to only generate low-resolution representations. Then, to enhance the capability of keypoints localization for our model, we also make crucial improvements consisting of bottleneck atrous spatial pyramid, local-space attention, coordinate attention and position embedding. In addition, we design two different network heads for 2D and 3D pose estimation to explore the extensibility of the backbone. Our model achieves superior performance to state-of-the-art lightweight 2D pose estimation models on both COCO and MPII datasets, which achieves 25+ FPS on HUWEI Kirin 9000 and outperforms MoveNet in the same device. Our 3D model also makes nearly 50% and 90% reduction on parameters and FLOPs compared to lightweight alternatives. Code is available at: https://github.com/NanXinyu/Mobile_LRPose.git.

Keywords: Human Pose Estimation · Lightweight · Mobile Application

1 Introduction

Human pose estimation has been widely applied in many fields, such as sports, human-computer interaction, autonomous driving, Virtual Reality (VR) and robotics. It aims to predict the locations of keypoints on every person from single image, which is one of the most important ways for computers to understand human behaviors. Existing human pose estimation methods typically construct network architectures, such as multi-resolution parallel architectures [7,15,26,32] and encoder-decoder architectures [23,31], to generate high-resolution representations that contain explicit position information. Recent researches [14,16] have demonstrated that high-resolution representations may be redundant for lightweight models, however, existing lightweight human pose estimation models

Q. Liu et al. (Eds.): PRCV 2023, LNCS 14429, pp. 211–224, 2024.
https://doi.org/10.1007/978-981-99-8469-5_17

[8,15,29,29] pay little attention to constructing effective network architectures with low-resolution representations while keeping high accuracy.

To address this problem, we propose a novel efficient network architecture based on modified MobileNetV2 [25] to only generate low-resolution representations, which is called *Mobile-LRPose*. Our *Mobile-LRPose* can be used for computation resource-restricted devices, like smartphones, laptops and embedded devices. Furthermore, we find that the efficient modules designed for semantic segmentation can be introduced to our network since both of semantic segmentation and human pose estimation are position-sensitive tasks. Based on this, we propose a Bottleneck Atrous Spatial Pyramid (BASP) module inspired by Atrous Spatial Pyramid Pooling (ASPP) [6] to efficiently integrate multi-scale features at the end of the network. In *Mobile-LRPose*, we also propose a novel attention module called local-space attention, cooperating with coordinate attention [11], which not only improves the position prediction accuracy, but is also beneficial to solve challenges like occlusion and multi-person overlap. We also use absolute-position embedding at the beginning of the network to make our model more position-aware. Our networks can be fitted to different tasks for 2D and 3D human pose estimation respectively.

Our contributions are summarized as follows:

1. We propose an efficient network backbone (See Fig. 1(a)) for human pose estimation consisting of an efficient network architecture with a bottleneck atrous spatial pyramid module, position-aware attention modules and position embedding.
2. We design two different network heads (See Fig. 1(b) and (c)) for 2D and 3D human pose estimation respectively. Both of the final 2D and 3D keypoints representations are efficient low-resolution representations and our models are more efficient than similar alternatives [8,15,32].

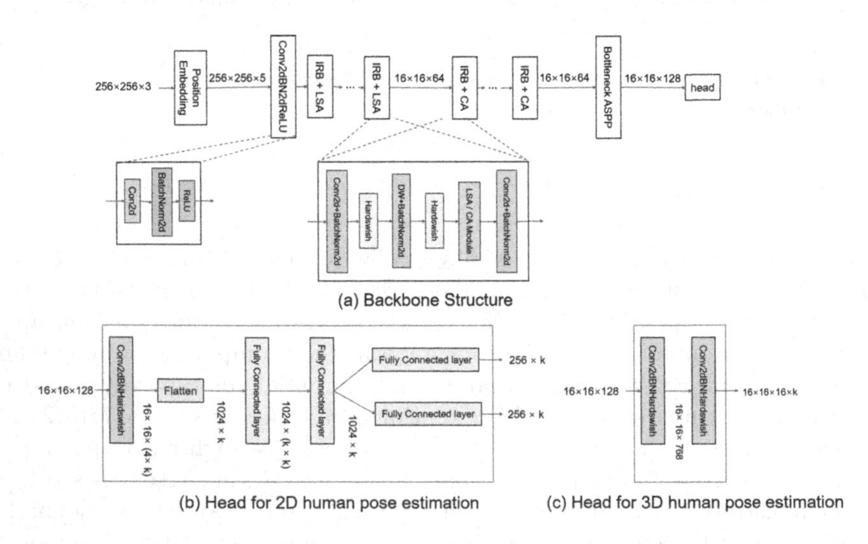

Fig. 1. An illustration of our network. IRB represents Inverted Residual Block.

3. We develop a real-time human pose estimation mobile application based on our 2D model. The inference speed of our model is superior to state-of-the-art real-time human pose estimation model [1] in the same device.

2 Related Work

Human Pose Estimation. Both 2D and 3D human pose estimation methods can be categorized into down-top and top-down approaches. The down-top methods [4,7,10] predict all the candidate keypoints from different bodies in the image and group corresponding ones to each person. In contrast, the top-down methods [23,26] firstly crop the input to several images that only contain one person via a person detector, then the model just predicts keypoints from every single person. The top-down methods usually achieve higher performance.

In this paper, we use the top-down method, which can better balance accuracy and real-time performance for lightweight human pose estimation models.

Keypoints Representation. Regression-based [14,28] and heatmap-based [2,7,26] methods are the two main streams for 2D human keypoints representation. Heatmap-based methods generate the high-resolution heatmap to represent the Gaussian distribution of every keypoint, which is widespread-used for current human pose estimation models.

3D human pose estimation methods can be categorized into two-stage [5,20] and one-stage [21,24,27] methods. Two-stage methods decompose the contents of 3D human pose estimation into 2D human pose estimation and 2D-to-3D lifting, where the keypoints representation is the same as 2D keypoints. One-stage methods predict the 3D keypoints directly from single 2D image. Sun et al. [27] introduced a one-stage regression method based a volumetric heatmap, which represents the weights of probabilities for 3D keypoints.

In this paper, we propose two different models by using the mobile-friendly methods, coordinate classification method [16] and one-stage regression method [27] for 2D and 3D human pose estimation, respectively.

Lightweight Human Pose Estimation Model. There have been various tricks for designing lightweight network architectures, such as replacing standard convolution blocks with efficient convolution blocks [19,25], compressing the existing large models [15,32] and Neural Architecture Search (NAS) [9,12]. Among them, introducing efficient blocks to classical large networks are commonly used for human pose estimation [8,15,29]. Although parameters and FLOPs of the models are dramatically reduced in this way, their structures with heavy computational cost, like high-resolution branches, are redundant for lightweight models.

To handle this problem, we propose a novel network architecture based on MobileNetV2 [25] and some designed modules for human pose estimation.

3 Mobile-LRPose

3.1 The Backbone of Mobile-LRPose

MobileNetV2. MobileNetV2 [25] is one of the most efficient network architectures designed for image classification. In *Mobile-LRPose*, we remove the last down-sampling and following layers of the original MobileNetV2 where crucial position information for human pose estimation is seriously destroyed. By experiments, we find that 1/16 of the input is the minimum representation resolution to predict accurate keypoints, and our *Mobile-LRPose* mainly compute on the feature maps with this low-resolution. In addition, we expand the channels of the previous layers to extract sufficient semantic features in the removed deep layers, and replace ReLU6 in MobileNetV2 with more mobile-friendly activation function hard-swish [12].

Bottleneck Atrous Spatial Pyramid. The modified MobileNetV2 generates low-resolution representations containing essential information of keypoints localization. To integrate the information efficiently, we propose bottleneck Atrous Spatial Pyramid (BASP) inspired by the success of Atrous Spatial Pyramid Pooling (ASPP) [6] in semantic segmentation, which showed that stacking atrous convolutions with different rates is beneficial for integrating multi-scale information, as illustrated in Fig. 2(a). In our *Mobile-LRPose*, down-sampling atrous convolutions and bilinear up-sampling are used to construct a bottleneck structure, which can be seen as an encoder-decoder module that is more efficient than original ASPP, as illustrated in Fig. 2(b).

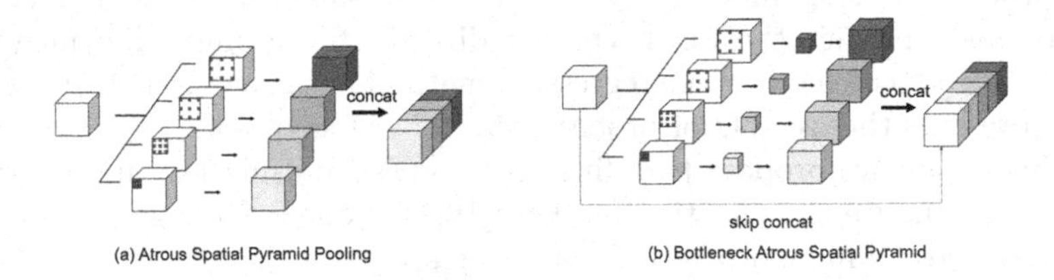

Fig. 2. Two pyramid architectures based on atrous convolution.

Given an input signal $x[m,n]$, the output $\tilde{x}[m,n]$ of the atrous convolution with stride 2, $k \times k$ filter can be defined as:

$$\tilde{x}[m,n] = \sum_{i=1}^{k}\sum_{j=1}^{k} x[m+2+r\cdot i, n+r\cdot j]\,\omega[i,j] \tag{1}$$

where r is the number of filling-zeros between valid values of the kernel. $\tilde{x}[m,n]$ is up-sampled through bilinear-interpolation to the output $y[m,n]$ of BASP, which can be defined as:

$$y[m,n] = 0.5(\tilde{x}[m_1,n_1] + \tilde{x}[m_1,n_2] + \tilde{x}[m_2,n_1] + \tilde{x}[m_2,n_2]) \tag{2}$$

where $[m_1, n_1]$, $[m_1, n_2]$, $[m_2, n_1]$, $[m_2, n_2]$ are four nearest points to the up-sampling target position$[m, n]$ on $\tilde{x}$.

Position Embedding. Position information is crucial to improving the accuracy of human pose estimation and can be expressed explicitly via high-resolution heatmaps. Without generating high-resolution representations, *Mobile-LRPose* use an absolute-position encoding as the follows:

$$PE^h(x, y) = y, PE^w(x, y) = x \tag{3}$$

where (x, y) represents the coordinate of per pixel on the image. The size of each position encoding equals to the input image.

The position encoding is embedded with the RGB image to propagate forward through the network together. Therefore, our model can capture not only image but also position features via position embedding.

3.2 The Attention Modules of Mobile-LRPose

Two different position-aware attention modules are introduced between the depthwise and pointwise convolution layers in inverted residual block, shown as Fig. 1(a), namely the Coordinate Attention (CA) [11], and the designed Local-Space Attention (LSA).

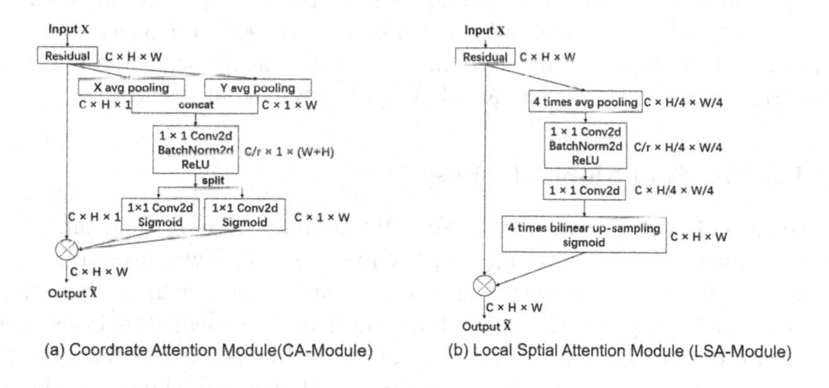

Fig. 3. The schematic conclusion of the attention modules used in our model.

Coordinate Attention. In CA [11], one-dimensional global pooling along the horizontal and vertical directions of an image respectively is performed to generate attention representations, which extracts not only channel-wise but also direction-aware information to accurately locate the positions of the image that are interested. Then two attention representations are encoded and decoded via 1×1 convolutions independently, and re-weight the input residual finally. This process is shown as Fig. 3(a).

Local-Space Attention. By experiments, we find that CA [11] is effective only in deep layers of the network and the 2D relation between horizontal and vertical directions is ignored completely, so we propose novel local-space attention (LSA) to realize resource-redistribution while maintaining the 2D space. In our LSA, an overlapping local large-scale 2D pooling operation is used to integrate continuous local space information, and an attention weights generation operation is used to re-weight every location of the feature map. This process is shown as Fig. 3(b).

Given the input feature map $\mathbf{U} \in \mathbb{R}^{H \times W}$, we use a $k \times k$ (default $k \geq 5$) pooling kernel with the stride s (default 4), to encode the local spatial position information into the attention map $\mathbf{Z} \in \mathbb{R}^{H/s \times W/s}$. This operation is written as:

$$z_c(x, y) = s\frac{1}{k \times k}\sum_{i=0}^{k}\sum_{j=0}^{k} u_c(s \times h + i, s \times w + j) \quad (4)$$

where u_c and z_c represents per channel of $\mathbf{U}$ and $\mathbf{Z}$, respectively, (h, w) represents the corresponding coordinates of $\mathbf{U}$ and $\mathbf{Z}$.

For the feature maps $z \in \mathbb{R}^C$ generated by the overlapping local pooling, the attention weights generation can be defined as follows:

$$s = \sigma(\mathbf{b}(\mathbf{f}^2(\delta(\mathbf{f}^1(\mathbf{Z}))))) \quad (5)$$

where $\mathbf{f}^1 \in \mathbb{R}^{C \times \frac{C}{r}}$ and $\mathbf{f}^2 \in \mathbb{R}^{\frac{C}{r} \times C}$ represent the 1×1 convolution layers respectively, r is the squeeze ratio for encoding, $\mathbf{b}$ is the bilinear up-sampling operation, δ is ReLU activation function and σ is Sigmoid activation function.

The attention modules are used in the entire backbone, while specifically LSA is for high-resolution layers and CA is for low-resolution layers.

3.3 The Heads of Mobile-LRPose

2D Human Pose Estimation. For 2D human pose estimation, we use the simple disentangled coordinate representation [16]. Two disentangled one-dimensional vectors are generated as the network output, which represent the position of each keypoint along the horizontal and vertical directions respectively, so we design the network head of 2D model consisting of flattening and cascading full-connected layers to generate two 1D vectors, shown as Fig. 1(b). The lengths of 1D vectors equal to H and W respectively. The loss function for model training is defined as follows:

$$l = l_x + l_y = \sum_{i=0}^{k}\omega_i \ln\frac{(exp(x_{GT,i})}{\sum_{j=0}^{W} x_{pred,i}(j)} + \sum_{i=0}^{k}\omega_i \ln\frac{(exp(y_{GT,i})}{\sum_{j=0}^{H} y_{pred,i}(j)} \quad (6)$$

where l_x and l_y represent the horizontal and vertical cross entropy loss of the one-dimensional coordinate classification, respectively, k is the number of keypoints and ω_i, $i \in [0, 1, \ldots, k]$ are weights for different keypoints.

3D Human Pose Estimation. For 3D human pose estimation, we use a 3D volumetric heatmap, which can represent the probabilities for keypoints regression through normalization [27]. The loss function for 3D human pose estimation is L1 loss, defined as below:

$$l = \frac{1}{k} \sum_{i=0}^{k} \|X_{pred} - X_{GT}\| \tag{7}$$

where X_{pred} and X_{GT} represent the prediction and ground-truth coordinates, respectively. We generate a low-resolution 3D heatmap (with resolution reduction ratio 1/16) simply via a sequence of 1×1 convolution layers, shown as Fig. 1(c).

4 Experiments

4.1 Implementation Details

Datasets and Evaluation Metrcis. COCO [18] contains over 200K images and 250K 2D person instances, where each person is labeled with 17 keypoints. The evaluation metrics for COCO dataset are the Average Precision (AP) and Average Recall (AR) scores based on Object Keypoint Similarity (OKS). MPII [3] contains about 250K images and 40K 2D person instances, where each person is labeled with 16 keypoints. The evaluation metric for MPII dataset is the head-normalized Probability of Correct Keypoint (PCKh) score. Human3.6M [13] contains 3.6 millions of video frames. There are 15 types of 3D human pose activities constructing by 11 subjects, which are split into two protocols. The Mean per joint position error (MPJPE) and MPJPE calculated after further alignment (PA-MPJPE) are used as the evaluation metrics.

Experiment Setting. *Mobile-LPPose* is trained on one GeForce RTX 3090 GPU with 32 samples and the base learning rate is 1e-3 with the Adam optimizer.

For 2D human pose estimation, the size of training epochs is set as 210 and the learning rate is dropped to 1e-4 and 1e-5 at the 170th and 200th epoch, respectively. The person detector boxes are expanded to be a fixed aspect ratio 4:3, then the original images are cropped to 256×192 for the COCO dataset and 256×256 for the MPII dataset. For 3D human pose estimation, the size of training epochs is set as 25 and the learning rate is dropped to 1e-4 and 1e-5 at the 17th and 21th epoch, respectively. The original images in Human3.6M and MPII dataset are cropped to 256×256.

We use the ground-truth person detector boxes for top-down human pose estimation and perform common data augmentation operations like rotations, random scales and horizontal flipping both on 2D and 3D human pose datasets. The horizontal flipping is also used during the testing procedure.

4.2 Results

COCO Val. Table 1 gives the results of *Mobile-LRPose* and other state-of-the-art models tested on COCO validation set for 2D human pose estimation. Our *Mobile-LRPose* achieves 68.4 AP score and outperforms other lightweight 2D human pose estimation models with less model parameters and GFLOPs. Compared with MobileNetV2 [25], our *Mobile-LRPose* increases with 3.8% on AP, and reduces nearly 85% and 81% parameters and GFLOPs, respectively. Compared with widespread-used Lite-HRNet-30 [32] and newly-proposed Dite-HRNet-30 [15], our *Mobile-LRPose* reduces nearly 17% parameters, and increases with 1.2% and 0.1% on AP respectively. Compared with large models [26,30], our *Mobile-LRPose* has comparable performance on AP^{50} with state-of-the-art large models.

MPII Val. Table 2 gives the results of *Mobile-LRPose* and other state-of-the-art lightweight models tested on MPII validation set for 2D human pose estimation. Our *Mobile-LRPose* increases with 2.1% on AP and reduces about 79% GFLOPs compared with MobileNetV2. With less parameters and GFLOPs, our *Mobile-LRPose* increases with 0.5% on AP compared with Lite-HRNet-30 [32], and achieves comparable AP score with Dite-HRNet-30 [15].

Table 1. Comparison with state-of-the-art models on COCO val. Params. = model parameters. Pretrain = Y means that pretrain the backbone on the ImageNet classification task. GFLOPs and parameters are calculated without person-detectors.

Model	backbone	Pretrain	Params	GFLOPs	AP	AP^{50}	AP^{75}	AR
large models								
SimBa [30]	ResNet-50	Y	34.0M	8.9	70.4	88.6	78.3	76.3
	HRNetV1-W32	Y	28.5M	7.1	74.4	90.5	81.9	79.8
HRNetV1 [26]	HRNetV1-W48	Y	63.6M	14.6	75.1	90.6	82.2	71.5
	HRNetV1-W48	N	66.3M	14.6	75.9	-	-	81.2
SimCC [16]	SimBa-50 [30]	N	25.7M	3.8	70.8	-	-	76.8
	TokenPose-S [17]	N	5.5M	2.2	73.6	-	-	78.9
small models								
MobileNetV2 1 [25]	MobileNetV2	Y	9.6M	1.5	64.6	87.4	72.3	70.7
Lite-HRNet [32]	Lite-HRNet-18	N	1.1M	0.20	64.8	86.7	73.0	71.2
	Lite-HRNet-30	N	1.8M	0.31	67.2	88.0	75.0	73.3
Dite-HRNet [15]	Dite-HRNet-18	N	1.1M	0.2	65.9	87.3	74.0	72.1
	Dite-HRNet-30	N	1.8M	0.3	68.3	88.2	76.2	74.2
Ours	Mobile-LRPose	N	1.5M	0.29	68.4	90.5	76.0	71.8

Table 2. Comparison with state-of-the-art models on MPII val. Params. = model parameters. GFLOPs and parameters are calculated without person-detectors.

Model	backbone	Params.	GFLOPs	PCKh
MobileNetV2 [25]	MobileNetV2 1	9.6M	1.9	85.4
Lite-HRNet [32]	Lite-HRNet-18	1.1M	0.27	86.1
	Lite-HRNet-30	1.8M	0.43	87.0
Dite-HRNet [15]	Dite-HRNet-18	1.1M	0.2	87.0
	Dite-HRNet-30	1.8M	0.4	87.6
Ours	Mobile-LRPose	1.5M	0.39	87.5

Human3.6M. Table 3 and Table 4 give the results of our *Mobile-LRPose* and other state-of-the-art models tested on Human3.6M protocol1 and protocol2 set. Compared with the classical two-stage [20] and one-stage [24] methods, our *Mobile-LRPose* achieves higher accuracy both on protocol1 and protocol2. The lightweight model proposed by Choi et al. [8] has two types of models, small and large models. They both achieve high accuracy and relatively light computational cost. Compared with the small model, our *Mobile-LRPose* can further achieves about 88.5% reduction on GFLOPs and 17.0% reduction on parameters with a little precision sacrifice, about 5.1 mm PA-MPJPE and 3.5 mm MPJPE on protocol1 and protocol2, respectively, which is acceptable in some applications.

Table 3. Comparison with state-of the-art models on Human3.6M protocol1.

Models	Dir	Dis	Eat	Gre	Phon	Pose	Pur	Sit	SitD	Smo	Phot	Wait	Walk	WalkD	WalkP	Avg	Params	GFLOPs
Martinez et al. [20]	39.5	43.2	46.4	47.0	51.0	56.0	41.4	40.6	56.5	69.4	49.2	45.0	49.5	38.0	43.1	47.7	-	-
Pavlakos et al. [24]	-	-	-	-	-	-	-	-	-	-	-	-	-	-	-	51.9	-	-
Moon et al. [22]	31.0	30.0	39.9	35.5	34.8	30.2	32.1	35.0	43.8	35.7	37.6	30.1	24.6	35.7	29.3	34.0	34.3M	-
Choi et al.(S) [8]	30.3	32.9	38.4	35.4	34.9	32.1	32.3	37.6	49.6	38.2	42.2	31.3	26.9	37.8	31.6	35.7	2.24M	3.92
Choi et al.(L) [8]	31.0	32.7	37.5	34.3	35.1	31.4	32.1	37.3	47.9	38.7	40.6	30.6	26.2	37.5	30.6	35.2	4.07M	5.49
Ours	37.1	37.5	46.6	41.7	40.0	37.6	36.0	41.1	53.3	43.4	48.5	36.7	30.0	42.5	36.2	40.8	1.86M	0.45

Table 4. Comparison with state-of-the-art models on Human3.6M protocol2.

Models	Dir	Dis	Eat	Gre	Phon	Pose	Pur	Sit	SitD	Smo	Phot	Wait	Walk	WalkD	WalkP	Avg	Params	GFLOPs
Martinez et al. [20]	51.8	56.2	58.1	59.0	69.5	55.2	58.1	74.0	94.6	62.3	78.4	59.1	49.5	65.1	52.4	62.9	-	-
Pavlakos et al. [24]	67.4	71.9	66.7	69.1	72.0	65.0	68.3	83.7	96.5	71.7	77.0	65.8	59.1	74.9	63.2	71.9	-	-
Moon et al. [22]	50.5	55.7	50.1	51.7	53.9	46.8	50.0	61.9	68.0	52.5	55.9	49.9	41.8	56.1	46.9	53.3	34.3M	-
Choi et al.(S) [8]	51.5	58.7	49.9	53.0	58.3	48.9	53.0	70.9	77.5	58.2	61.0	51.9	42.9	58.6	50.0	56.9	2.24M	3.92
Choi et al.(L) [8]	45.5	51.8	45.9	48.4	52.1	43.7	48.2	63.6	70.2	52.4	56.2	46.2	40.2	54.9	45.4	51.4	4.07M	5.49
Ours	51.5	60.5	57.5	55.8	62.0	52.6	53.9	73.9	87.1	60.9	65.2	54.5	46.0	60.9	53.8	60.42	1.86M	0.45

4.3 Ablation Study and Analysis

Impact of Bottleneck Atrous Spatial Pyramid. Table 5 shows that using the proposed BASP (Y) can achieve higher performance with less GFLOPs than the routine(N) to use single convolution to generate the network output feature map with the same size.

Table 5. Comparison between using BASP (Y) and 1 × 1 convolution (N).

bottleneck atrous spatial pyramid	COCO			MPII		Human3.6M		
	GFLOPs	AP	AR	GFLOPs	PCKh	GFLOPs	MPJPE	PA-MPJPE
Y	0.296	68.4	71.8	0.392	87.5	0.396	65.04	44.16
N	0.303	68.3	71.6	0.404	87.3	0.403	66.70	45.23

Impact of Position Embedding. It is shown in Table 6 that introducing the position embedding into the network can improve the prediction accuracy of both 2D and 3D human pose estimation model with little computational cost.

Table 6. Comparison between using position embedding (Y) or not (N).

position embedding	COCO		MPII	Human3.6M	
	AP	AR	PCKh	MPJPE	PA-MPJPE
Y	68.4	71.8	87.5	65.16	44.16
N	67.7	71.0	87.2	67.77	46.94

Impact of Attention Modules. As shown in Table 7, our attention modules is not only effective for improving the performance but also for solving the challenges like occlusion and multi-person overlap, as demonstrated in Fig. 4.

Table 7. Comparison between using attention modules (Y) or not (N).

attention modules	COCO		MPII	Human3.6M	
	AP	AR	PCKh	MPJPE	PA-MPJPE
Y	68.4	71.7	87.4	66.84	45.13
N	67.0	70.3	86.8	70.08	50.12

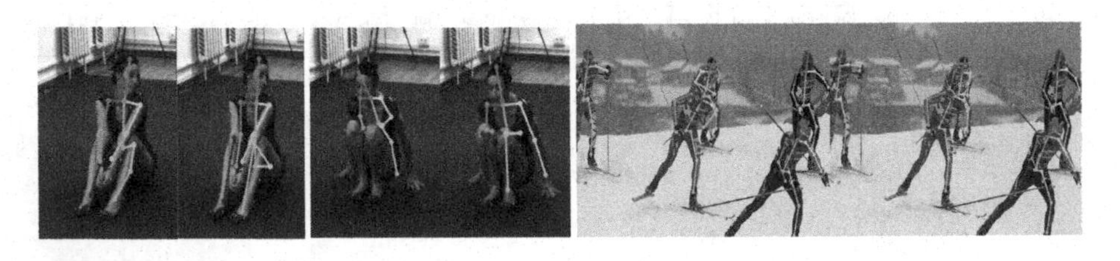

Fig. 4. Comparisons of introducing local-space attention (right) or not (left) under challenging circumstances like occlusion and overlapping multi-person.

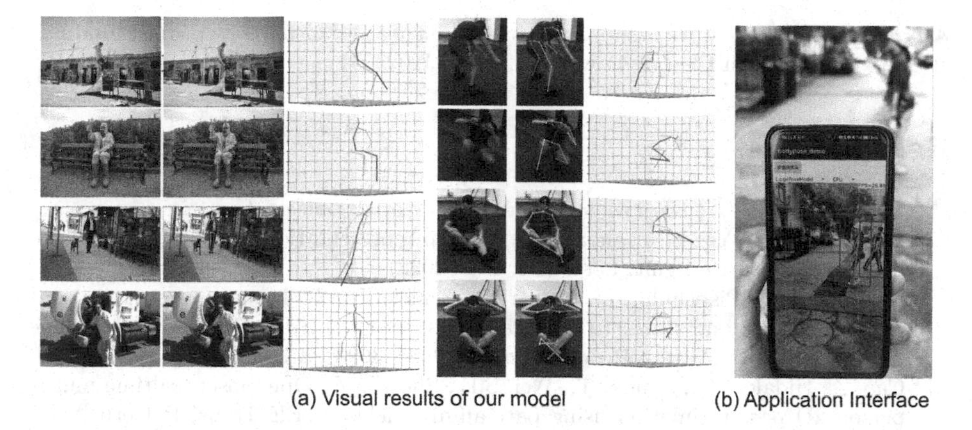

Fig. 5. The visual illustration of *Mobile-LRPose* on the instances of COCO, MPII and Human3.6M datasets and the application interface on HUWEI Mate 40 device.

4.4 Applications

We also applied our method in real applications. The training and testing of *Mobile-LRPose* is based on PyTorch and the mobile deployment of the real-time 2D human pose estimation model is based on mobile-oriented framework, NCNN. The model framework transformation needs to be implemented through ONNX. We develop the application for mobile vision devices with the Android system on HUWEI Mate 40 Pro with HUAWEI Kirin 9000 CPU. The frames per second (fps) of our model can reach 25+ on the smartphone which outperforms MoveNet [1] (fps20+). The visual results of our *Mobile-LRPose* and the application interface are demonstrated as Fig. 5.

5 Conclusion

In this paper, we proposed an efficient network architecture designed for human pose estimation in mobile devices. We studied that there is a lack of lightweight network architectures without generating high-resolution representations that expert in human pose estimation. To handle this problem, we constructed a network backbone based on the modified MobileNetV2 as well as the proposed bottleneck atrous spatial pyramid as the end of the backbone to integrate multi-scale features efficiently. Furthermore, we combine the position-aware attention modules, the coordinate attention and novel local-space attention we designed, and position embedding into the network to enhance the capability of the model to position prediction. Finally, two different heads of the network are designed for 2D and 3D human pose estimation respectively, and the experiments on COCO, MPII and Human3.6M datasets have demonstrated the effectiveness of our network for both 2D and 3D human pose estimation tasks.

Acknowledge. This work is supported by the science and technology project fundings of State Grid Jiangsu Electric Power Co., Ltd. (J2023031).

References

1. Movenet. https://tensorflow.google.cn/hub/tutorials/movenet
2. Yu, H., Du, C., Yu, L.: Scale-aware heatmap representation for human pose estimation. Pattern Recognit. Lett. **154**, 1–6 (2022)
3. Andriluka, M., Pishchulin, L., Gehler, P., Schiele, B.: 2D human pose estimation: new benchmark and state of the art analysis. In: 2014 IEEE Conference on Computer Vision and Pattern Recognition, pp. 3686–3693 (2014)
4. Cao, Z., Hidalgo, G., Simon, T., Wei, S.E., Sheikh, Y.: Openpose: realtime multi-person 2D pose estimation using part affinity fields. IEEE Trans. Pattern Anal. Mach. Intell. **43**(1), 172–186 (2021)
5. Chen, C.H., Ramanan, D.: 3D human pose estimation = 2D pose estimation + matching. In: 2017 IEEE Conference on Computer Vision and Pattern Recognition (CVPR), pp. 5759–5767 (2017)
6. Chen, L.C., Papandreou, G., Kokkinos, I., Murphy, K., Yuille, A.L.: Deeplab: semantic image segmentation with deep convolutional nets, atrous convolution, and fully connected CRFs. IEEE Trans. Pattern Anal. Mach. Intell. **40**(4), 834–848 (2018)
7. Cheng, B., Xiao, B., Wang, J., Shi, H., Huang, T.S., Zhang, L.: Higherhrnet: scale-aware representation learning for bottom-up human pose estimation. In: 2020 IEEE/CVF Conference on Computer Vision and Pattern Recognition (CVPR), pp. 5385–5394 (2020)
8. Choi, S., Choi, S., Kim, C.: Mobilehumanpose: toward real-time 3D human pose estimation in mobile devices. In: 2021 IEEE/CVF Conference on Computer Vision and Pattern Recognition Workshops (CVPRW), pp. 2328–2338 (2021)
9. Elsken, T., Metzen, J.H., Hutter, F.: Neural architecture search: a survey. J. Mach. Learn. Res. **20**(1), 1997–2017 (2019)
10. Geng, Z., Sun, K., Xiao, B., Zhang, Z., Wang, J.: Bottom-up human pose estimation via disentangled keypoint regression. In: 2021 IEEE/CVF Conference on Computer Vision and Pattern Recognition (CVPR), pp. 14671–14681 (2021)
11. Hou, Q., Zhou, D., Feng, J.: Coordinate attention for efficient mobile network design. In: 2021 IEEE/CVF Conference on Computer Vision and Pattern Recognition (CVPR), pp. 13708–13717 (2021)
12. Howard, A., et al.: Searching for MobileNetV3. In: 2019 IEEE/CVF International Conference on Computer Vision (ICCV), pp. 1314–1324 (2019)
13. Ionescu, C., Papava, D., Olaru, V., Sminchisescu, C.: Human3.6m: large scale datasets and predictive methods for 3D human sensing in natural environments. IEEE Trans. Pattern Anal. Mach. Intell. **36**(7), 1325–1339 (2014)
14. Li, J., et al.: Human pose regression with residual log-likelihood estimation. In: 2021 IEEE/CVF International Conference on Computer Vision (ICCV), pp. 11005–11014 (2021)
15. Li, Q., Zhang, Z., Xiao, F., Zhang, F., Bhanu, B.: Dite-HRNet: dynamic lightweight high-resolution network for human pose estimation. In: International Joint Conference on Artificial Intelligence (IJCAI), pp. 1095–1101 (2022)

16. Li, Y., et al.: SimCC: a simple coordinate classification perspective for human pose estimation. In: Avidan, S., Brostow, G., Cissé, M., Farinella, G.M., Hassner, T. (eds.) ECCV 2022. LNCS, vol. 13666, pp. 89–106. Springer, Cham (2022). https://doi.org/10.1007/978-3-031-20068-7_6
17. Li, Y., et al.: Tokenpose: learning keypoint tokens for human pose estimation. In: 2021 IEEE/CVF International Conference on Computer Vision (ICCV), pp. 11293–11302 (2021)
18. Lin, T.Y., et al.: Microsoft coco: common objects in context. In: Fleet, D., Pajdla, T., Schiele, B., Tuytelaars, T. (eds.) ECCV 2014. LNCS, vol. 8693, pp. 740–755. Springer, Cham (2014). https://doi.org/10.1007/978-3-319-10602-1_48
19. Ma, N., Zhang, X., Zheng, H.T., Sun, J.: Shufflenet v2: practical guidelines for efficient CNN architecture design. In: Ferrari, V., Hebert, M., Sminchisescu, C., Weiss, Y. (eds.) ECCV 2018. LNCS, vol. 11218, pp. 122–138. Springer, Cham (2018). https://doi.org/10.1007/978-3-030-01264-9_8
20. Martinez, J., Hossain, R., Romero, J., Little, J.J.: A simple yet effective baseline for 3D human pose estimation. In: 2017 IEEE International Conference on Computer Vision (ICCV), pp. 2659–2668 (2017)
21. Mehta, D., et al.: XNect: real-time multi-person 3D motion capture with a single RGB camera. ACM Trans. Graph. **39**(4) (2020)
22. Moon, G., Chang, J., Lee, K.M.: Camera distance-aware top-down approach for 3D multi-person pose estimation from a single RGB image. In: The IEEE Conference on International Conference on Computer Vision (ICCV) (2019)
23. Newell, A., Yang, K., Deng, J.: Stacked hourglass networks for human pose estimation. In: Leibe, B., Matas, J., Sebe, N., Welling, M. (eds.) ECCV 2016. LNCS, vol. 9912, pp. 483–499. Springer, Cham (2016). https://doi.org/10.1007/978-3-319-46484-8_29
24. Pavlakos, G., Zhou, X., Derpanis, K.G., Daniilidis, K.: Coarse-to-fine volumetric prediction for single-image 3D human pose. In: 2017 IEEE Conference on Computer Vision and Pattern Recognition (CVPR), pp. 1263–1272 (2017)
25. Sandler, M., Howard, A., Zhu, M., Zhmoginov, A., Chen, L.C.: Mobilenetv 2: inverted residuals and linear bottlenecks. In: 2018 IEEE/CVF Conference on Computer Vision and Pattern Recognition, pp. 4510–4520 (2018)
26. Sun, K., Xiao, B., Liu, D., Wang, J.: Deep high-resolution representation learning for human pose estimation. In: 2019 IEEE/CVF Conference on Computer Vision and Pattern Recognition (CVPR), pp. 5686–5696 (2019)
27. Sun, X., Xiao, B., Wei, F., Liang, S., Wei, Y.: Integral human pose regression. In: Ferrari, V., Hebert, M., Sminchisescu, C., Weiss, Y. (eds.) ECCV 2018. LNCS, vol. 11210, pp. 536–553. Springer, Cham (2018). https://doi.org/10.1007/978-3-030-01231-1_33
28. Toshev, A., Szegedy, C.: Deeppose: human pose estimation via deep neural networks. In: 2014 IEEE Conference on Computer Vision and Pattern Recognition, pp. 1653–1660 (2014)
29. Wang, Y., Li, M., Cai, H., Chen, W., Han, S.: Lite pose: efficient architecture design for 2D human pose estimation. In: 2022 IEEE/CVF Conference on Computer Vision and Pattern Recognition (CVPR), pp. 13116–13126 (2022)
30. Xiao, B., Wu, H., Wei, Y.: Simple baselines for human pose estimation and tracking. In: Ferrari, V., Hebert, M., Sminchisescu, C., Weiss, Y. (eds.) ECCV 2018. LNCS, vol. 11210, pp. 472–487. Springer, Cham (2018). https://doi.org/10.1007/978-3-030-01231-1_29

31. Xu, Y., Zhang, J., Zhang, Q., Tao, D.: Vitpose: simple vision transformer baselines for human pose estimation. In: Koyejo, S., Mohamed, S., Agarwal, A., Belgrave, D., Cho, K., Oh, A. (eds.) Advances in Neural Information Processing Systems, vol. 35, pp. 38571–38584. Curran Associates, Inc. (2022)
32. Yu, C., et al.: Lite-HRNet: a lightweight high-resolution network. In: 2021 IEEE/CVF Conference on Computer Vision and Pattern Recognition (CVPR), pp. 10435–10445 (2021)

Attention and Relative Distance Alignment for Low-Resolution Facial Expression Recognition

Liwei An[1,2], Xiao Sun[1,2,3(✉)], Ziyang Zhang[1,2], and Meng Wang[1,2,3]

[1] School of Computer Science and Information Engineering, Hefei University of Technology, Heifei, China
sunx@hfut.edu.cn

[2] Anhui Province Key Laboratory of Affective Computing and Advanced Intelligent Machines, Hefei University of Technology, Heifei, China

[3] Institute of Artificial Intelligence, Hefei Comprehensive National Science Center, Heifei, China

Abstract. In real-world scenarios, facial images obtained by many devices often exhibit low resolution. However, the performance significantly degrades when we apply the existing methods in low-resolution facial expression recognition. Therefore, addressing the problem of low-resolution images in facial expression recognition becomes an important undertaking. Previous attempts to tackle this problem have been limited. For this, we propose a novel Attention and Relative Distance Alignment (ARDA) method by integrating knowledge distillation in low-resolution facial expression recognition. Specifically, the Attention Alignment module guides the student model to focus on the most crucial region of the facial image by enabling the low-resolution student model to learn the attention map of the high-resolution teacher model. The Relative Distance Alignment module utilizes the relative distance between facial image features to transfer differences between different low-resolution facial images from the teacher model to the student model, helping the student model better grasp the differences between expressions. Extensive experiments have shown that the ARDA method effectively transfers knowledge from high-resolution teacher model to low-resolution student model, achieving state-of-the-art performance in synthetic low-resolution facial expression recognition datasets.

Keywords: Low Resolution · Facial Expression Recognition · Attention and Relative Distance Alignment

1 Introduction

Facial Expression Recognition (FER) is one of the current popular research directions in affective computing, computer vision, and pattern recognition. Its primary objective is to enable computers to comprehend human emotional states

Q. Liu et al. (Eds.): PRCV 2023, LNCS 14429, pp. 225–237, 2024.
https://doi.org/10.1007/978-981-99-8469-5_18

and even respond accordingly. Therefore, FER has extensive applications in real-world scenarios, such as intelligent surveillance, deception detection, and human-computer interaction.

Fig. 1. A video frame image obtained in the MOT16, and some facial images in it.

In recent years, remarkable progress has been made in facial expression recognition with the emergence of large-scale datasets such as RAF-DB [7] and FER-Plus [1]. These datasets contain high-resolution facial images, such as those in RAF-DB, which are all 100×100 pixels in size. However, in real-world scenarios, facial images obtained from surveillance camera, smartphone, and other devices are often of low resolution. For example, Fig. 1 shows a video frame extracted from the MOT16 [10] dataset used for multi-object tracking, and it can be evident that even for the two largest faces in the image, it is challenging for humans to discern their facial expressions, let alone the smaller facial images in the scene. Although deep learning has made outstanding achievements in facial expression recognition recently, existing methods have obvious performance degradation in low-resolution facial expression recognition.

To address this problem, we propose a method called Attention and Relative Distance Alignment (ARDA) that leverages knowledge distillation to transfer the capabilities of the high-resolution teacher model to the low-resolution student model. We observe that humans can utilize prior knowledge learned from high-resolution images to approximate the location and regions of interest in low-resolution images. Inspired by this, we introduce Attention Alignment (AA) module that mimics the human visual system. It utilizes attention maps obtained from the high-resolution teacher model to guide the student model to focus on the facial regions relevant to the expression. Furthermore, on the one hand, we recognize that due to substantial disparities between the features of high and low-resolution facial images, direct transference of facial features from the high-resolution teacher model to the low-resolution student model is unfeasible. On the other hand, facial expression features have certain differences, both intra-class and inter-class. Hence, we introduce the Relative Distance Alignment (RDA) module, which quantifies the differences in facial expression features as relative distances between facial features. By transferring the relative distances, the module assists the student model in more effectively discriminating disparities among low-resolution facial images. AA and RDA modules are removed during deployment and incur no extra inference cost. In summary, our contributions are as follows:

1. We propose a novel Attention and Relative Distance Alignment (ARDA) method based on knowledge distillation by utilizing the pre-trained high-resolution teacher model to guide the low-resolution student model, thereby improving the performance of the model in low-resolution facial expression recognition.
2. The Attention Alignment module guides the student model to focus on key regions in facial images that are relevant to facial expression. The Relative Distance Alignment module drives the student model to capture differences between facial expressions. The two work together, enabling the model to distinguish expression even on low-resolution facial images of poor quality.
3. The experimental results show that our proposed ARDA method significantly advances state-of-the-art results on several challenging synthetic low-resolution facial expression recognition datasets.

2 Related Work

2.1 Facial Expression Recognition

In recent years, with the advent of large-scale datasets, deep learning has emerged as the dominant approach for facial expression recognition. Zeng et al. [17] introduced the IPA2LT framework, which trains models to estimate the latent truth by assigning multiple pseudo labels to each sample. Wang et al. [15] proposed the Self-Cure Network (SCN), which suppresses the influence of noisy labels by utilizing a self-attention importance weighting module and a relabeling module. She et al. [14] presented the DMUE, which leverages multiple branches to explore latent distributions and estimates uncertainties based on pairwise relationships within mini-batch. Zhang et al. [19] approached the problem from the relative perspective and proposed the RUL, which uses two branches to compare different facial images and learn relative uncertainties, achieving promising performance. Zhang et al. [20] utilized the flip semantic consistency of facial images and introduced the Erase Attention Consistency (EAC) to suppress noise samples in the datasets.

These methods above are trained on prevalent large-scale datasets with relatively good image quality, while the quality of facial images obtained in real-world scenarios is often worse. Despite this, there are few solutions in facial expression recognition specifically addressing this issue. Nan et al. [11] proposed a feature-level super-resolution approach that transforms facial expression features from low-resolution images into their corresponding high-resolution counterparts. However, super-resolution model significantly increases the computational costs of both training and inference. Moreover, as the image resolution decreases, the loss of spatial information introduces substantial disparities between the features of high-resolution and low-resolution image, making it arduous to restore the genuine details of the image. Our method is based on knowledge distillation, where the teacher model guides the student model to capture important information in low-resolution images. Importantly, this does not incur any additional inference costs.

2.2 Knowledge Distillation

Knowledge distillation is a model compression technique that utilizes a teacher-student network framework for training, aiming to distill knowledge from a complex teacher model into a more compact student model. Hinton et al. [5] pioneered the concept of knowledge distillation, where they achieved knowledge transfer by reducing the difference between the logits distributions of the teacher and student models. Romero et al. [12] introduced a feature-level approach to knowledge distillation, extracting knowledge from intermediate layers of the teacher model to guide the training of the student model. Zagoruyko et al. [16] proposed an attention-level method for knowledge distillation, leveraging channel pooling on feature maps to compute attention maps. However, this method has limitations regarding the precision of the resulting attention maps. Our model employs the Grad-CAM [13] method, which leverages the model's gradient information to generate attention maps with high precision.

3 Method

3.1 Overview

In this section, we provide an overview of the implementation details of our proposed Attention and Relative Distance Alignment (ARDA) method. ResNet [4] is a classic model in the field of deep learning. To facilitate introducing the method proposed in this paper, we adopt ResNet18 as the backbone and integrate it with our proposed ARDA. The pipeline of ARDA is shown in Fig. 2. Unlike traditional knowledge distillation, ARDA employs networks of the same size for teacher and student models. The Attention Alignment module introduces attention alignment loss, which reduces the distance between the attention maps of the teacher and student models. The Relative Distance Alignment module represents the differences among facial images using the relative distances between their feature vectors and utilizes the relative distances obtained from the teacher model to guide the training of the student model. We first train the network on prevalent large-scale datasets to get the well-trained teacher model. Then, we utilize low-resolution images to train the student model. In addition to the commonly used classification loss in facial expression recognition, we introduce attention alignment loss and relative distance alignment loss to guide the training of the student model.

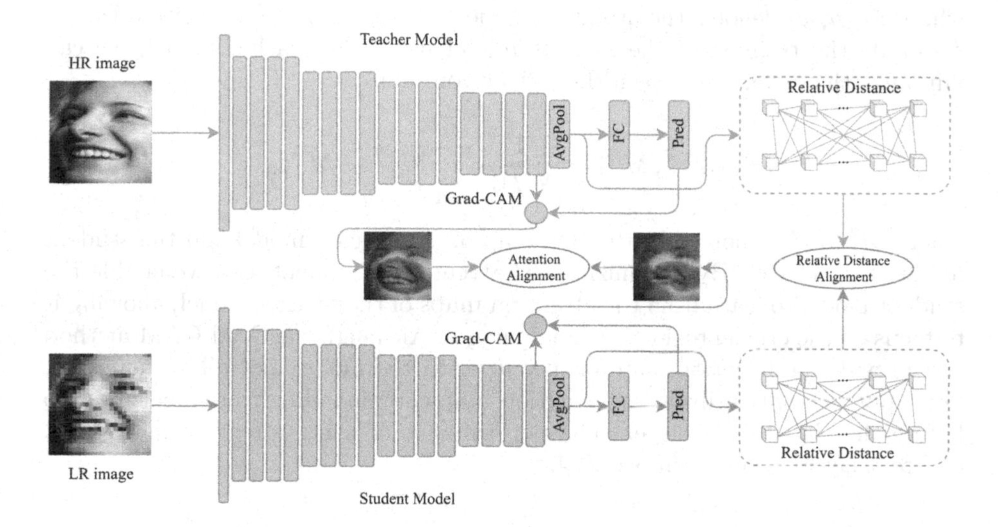

Fig. 2. The overall framework of the Attention and Relative Distance Alignment.

3.2 Attention Alignment

The attention map can clearly show the regions of concern for the model. Various methods have been proposed for generating attention map, and in our approach, we employ the Grad-CAM [13]. This technique leverages the gradient maps obtained through backpropagation during model training to compute a weighted sum of feature maps over the channels, resulting in attention maps with high precision. We define the gradient map of the feature map obtained through backpropagation as follows:

$$G = \frac{\partial}{\partial F} \sum_{i=1}^{N} \sum_{j=1}^{E} \hat{y}_{ij}, \tag{1}$$

where $\hat{y}_{ij}$ is the probability that the i-th image belongs to the j-th category. F, $G \in N \times C \times H \times W$ denote the feature map and its gradient map obtained after backpropagation, respectively. N, C, H, and W denote the number of images, channels, the height and width of the feature map, respectively. E denotes the number of expression categories. Next, we compute the attention map by performing the weighted sum of the feature map F and gradient map G over the channels:

$$M_i(h,w) = \sum_{c=1}^{C} G_i(c,h,w) F_i(c,h,w), \tag{2}$$

where $M_i(h,w)$ denotes the attention value of the i-th image at position (h,w). To guide the training of the student model using the teacher model, we can minimize the distance between their attention maps:

$$l_{AA}(M^T, M^S) = \frac{1}{NHW} \sum_{i=1}^{N} \left\| M_i^T - M_i^S \right\|_2, \tag{3}$$

where M^T, M^S denote the attention map of the teacher model and the student model, respectively. By minimizing the attention alignment loss, we enable the student model to learn from the attention maps of the teacher model, allowing it to focus on the crucial regions in facial images. Although the Grad-CAM method can provide an attention map for any layer in the model, not all are effective. Our approach only utilizes the attention map corresponding to the feature map before the classification layer. The performance of using attention maps from different layers is shown in Sect. 4.4.

3.3 Relative Distance Alignment

To quantify the relative differences, we design binary relative distance and ternary relative distance. The binary relative distance represents the relative difference between two facial expression features, and the ternary relative distance represents the relative difference among three facial expression features.

Binary Relative Distance Alignment. Given a pair of facial expression features, their relative distance can be represented by the Euclidean distance:

$$d_{i_1 i_2} = \left\| f_{i_1} - f_{i_2} \right\|_2, \tag{4}$$

where $f_i \in N \times C$ denotes the feature vector obtained from the feature layer and $d_{i_1 i_2}$ denotes the relative distance between the i_1-th expression feature and the i_2-th expression feature. The binary relative distance alignment loss is defined as

$$l_{2-RDA}(d^T, d^S) = \frac{1}{N^2} \sum_{i_1, i_2 \in [1,N]} \left\| d_{i_1 i_2}^T - d_{i_1 i_2}^S \right\|_2, \tag{5}$$

where d^T and d^S denote the binary relative distance of the teacher model and the student model, respectively. The binary relative distance alignment loss improves the ability of the student model to capture differences between facial images by penalizing the difference between the binary relative distance of the teacher model and the student model.

Ternary Relative Distance Alignment. Given a triplet of expression features, we use the angle they form in the representation space as their relative distance:

$$D_{i_1 i_2 i_3} = \cos\langle f_{i_1} - f_{i_2}, f_{i_3} - f_{i_2}\rangle = \frac{(f_{i_1} - f_{i_2}) \cdot (f_{i_3} - f_{i_2})}{\|f_{i_1} - f_{i_2}\|_2 \|f_{i_3} - f_{i_2}\|_2}. \tag{6}$$

The ternary relative distance alignment loss is defined as

$$l_{3-RDA}(D^T, D^S) = \frac{1}{N^3} \sum_{i_1, i_2, i_3 \in [1,N]} l_s(D^T_{i_1 i_2 i_3}, D^S_{i_1 i_2 i_3}), \tag{7}$$

$$l_s(a,b) = \begin{cases} 0.5(a-b)^2 & \text{, if } |a-b| < 1, \\ |a-b| - 0.5 & \text{, otherwise.} \end{cases} \tag{8}$$

where D^T and D^S denote the ternary relative distance of the teacher model and student model, respectively, and l_s is the Huber loss. The ternary relative distance alignment loss can further improve the ability of the student model to capture differences between facial images. Although higher-order relative distance alignment may further enhance the ability of the student model to capture differences, this significantly increases the computational cost.

3.4 Overall Loss Function

In addition to the attention alignment loss and relative distance alignment loss mentioned above, we also employ the commonly used cross-entropy loss in the classification task. It is defined as

$$l_{cls} = -\sum_{i=1}^{N} (\log \frac{e^{W_{y_i} f_i}}{\sum_j^L e^{W_j f_i}}), \tag{9}$$

where W_{y_i} is the y_i-th weight from the fully connected (FC) layer with y_i as the given label of the i-th image. The total loss function can be summarized as follows

$$l_{total} = l_{cls} + \lambda l_{AA} + \alpha l_{2-RDA} + \beta l_{3-RDA}, \tag{10}$$

where λ, α and β are hyperparameters.

4 Experiment

4.1 Datasets

RAF-DB [7] is a real-world Facial Expression Recognition (FER) dataset. It contains 29,672 real-world facial images annotated by 40 well-trained annotators using basic or compound expressions. For our experiments, we select seven basic expressions, namely neutral, happiness, surprise, sadness, angry, disgust, and fear, including 12,271 images for training and 3,068 images for testing.

FERPlus [1] is an extended version of FER2013 [2] that provides a finer label created by 10 crowd-sourced annotators. It was collected using the Google search engine and contains 28,709 training images, 3,589 validation images, and 3,589 testing images. All the images are grayscale images. Each image is annotated into one of eight classes. The validation set is also used during the training process, and the overall accuracy is reported on the testing set.

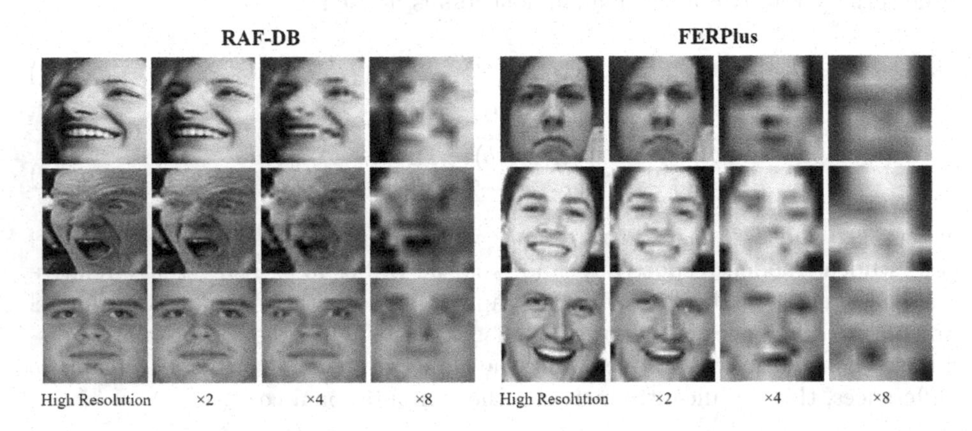

Fig. 3. The samples of high resolution and low resolution images from the datasets.

4.2 Implementation Details

We selected those mentioned above two popular large-scale facial expression recognition datasets to evaluate our method. Facial images are detected and aligned using MT-CNN [18]. Following the common practice in most super-resolution studies [9], we apply bicubic interpolation on the high-resolution images to generate corresponding low-resolution images at downscaling factors of 2, 4, and 8. For RAF-DB, with the original image size of 100×100 pixels, the resulting low-resolution images have sizes of 50×50, 25×25, and 12×12, respectively. For FERPlus, with the original image size of 48×48 pixels, the downsampled low-resolution images have sizes of 24×24, 12×12, and 6×6, respectively. Fig. 3 shows several sample low-resolution images obtained after downsampling. It is evident that the facial images at a downscaling factor of 8 are difficult to discern with the naked eye, revealing only the facial topology.

All experiments are implemented using PyTorch. We utilize a pre-trained ResNet-18 [4] model, trained on MS-Celeb-1M [3], as our baseline model. The input image size is resized to 224×224 pixels, with a batch size of 64 and the epoch size of 80. For the baseline model, we use an ADAM [6] optimizer with the initial learning rate of 0.0001, the weight decay rate of 0.0001, an ExponentialLR [8] learning rate scheduler with the gamma value of 0.9 to decrease the learning rate after each epoch and on RAF-DB, the hyperparameter λ, α, β are set to 10, 0.02, 25. We train our model on NVIDIA 3090 GPU with 24 GB RAM.

4.3 Low-Resolution Facial Expression Recognition Results

Due to the lack of solutions for low-resolution facial expression recognition, we apply our method to state-of-the-art facial expression recognition methods (e.g., SCN, RUL, EAC) and compare their performance. To ensure a fair comparison, we use the official training codes provided by the author, and all experimental settings are kept consistent with the original paper. We use a baseline method, ResNet18, trained with only the classification loss as supervision. As shown in Table 1, the performance of the state-of-the-art facial expression recognition methods significantly drops on low-resolution facial images obtained after downsampling as the downsampling factor increases. This is because these methods don't consider low-resolution facial images, whereas, in real-world scenarios, facial images are often low-resolution. When we apply our proposed ARDA method to these facial expression recognition models, it is evident that all models experience a significant performance improvement on low-resolution images. For the baseline model trained on RAF-DB, the application of ARDA result in an accuracy improvement of 0.13%, 0.59%, and 1.57% for downsampling factors

Table 1. Accuracy (%) comparison with state-of-the-art FER methods.

Resolution	Method	Loss Function	RAF-DB	FERPlus
1×	Baseline	CE	84.29	87.57
	RUL	Add-up	89.37	88.21
	EAC	CE + EAC	89.96	89.23
2×	Baseline	CE	84.16	83.61
	Baseline+ARDA	CE + ARDA	**84.29**	**83.93**
	RUL	Add-up	88.43	84.51
	RUL+ARDA	Add-up + ARDA	**88.59**	**85.08**
	EAC	CE + EAC	89.05	85.69
	EAC+ARDA	CE + EAC + ARDA	**89.34**	**86.04**
4×	Baseline	CE	79.07	70.55
	Baseline+ARDA	CE + ARDA	**79.66**	**71.02**
	RUL	Add-up	83.21	71.57
	RUL+ARDA	Add-up + ARDA	**83.70**	**72.30**
	EAC	CE + EAC	84.16	72.81
	EAC+ARDA	CE + EAC + ARDA	**84.62**	**73.22**
8×	Baseline	CE	67.76	51.35
	Baseline+ARDA	CE + ARDA	**69.33**	**52.28**
	RUL	Add-up	71.38	52.53
	RUL+ARDA	Add-up + ARDA	**73.01**	**53.52**
	EAC	CE + EAC	72.56	54.26
	EAC+ARDA	CE + EAC + ARDA	**73.63**	**55.44**

of 2, 4, and 8, respectively. Even on facial images at a downscaling factor of 8, where it is difficult for the human eye to discern facial expressions, our proposed method could still utilize the remaining facial topology information to recognize the correct facial expression. The experimental results demonstrate that the attention maps and the relative feature differences between images extracted from the teacher and student model can be effectively aligned using the ARDA method, thereby enhancing the model's performance on low-resolution images.

4.4 Ablation Study

Influence of Different Components. To evaluate the impact of each component in ARDA, we conduct an ablation study on the baseline model using RAF-DB with downsampling factor of 8. The study aims to investigate the effects of the AA and RDA module, where the RDA module is further divided into 2-RDA and 3-RDA to represent binary relative distance alignment and ternary relative distance alignment, respectively. The results are in Table 2. Some observations can be summarized. First, all three modules gain different degrees of performance improvement when used alone. Second, when two modules are used, the AA and 2-RDA modules jointly perform the best, which indicates that these two modules have the most collaborative ability. Third, the model performance improve from 67.76% to 69.33% when all three modules are used.

Table 2. Ablation study for each component.

AA	2-RDA	3-RDA	Accuracy(%)
×	×	×	67.76
✓	×	×	67.86
×	✓	×	68.29
×	×	✓	68.32
✓	✓	×	68.81
✓	×	✓	68.55
×	✓	✓	68.77
✓	✓	✓	**69.33**

Table 3. Ablation study for selection of attention maps.

conv2_x	conv3_x	conv4_x	conv5_x	Accuracy(%)
×	×	×	✓	**69.33**
×	×	✓	✓	68.45
×	✓	✓	✓	68.06
✓	✓	✓	✓	67.57

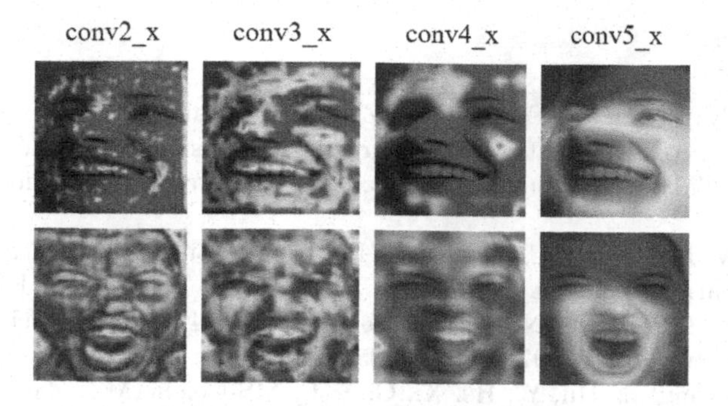

Fig. 4. Attention maps for different layers obtained from the teacher model.

Selection of Attention Maps. We conduct an ablation study on the baseline model to investigate how to select the attention map to align in our proposed Attention Alignment module. The baseline model is based on the ResNet18 architecture, which consists of six parts: conv1, conv2_x, ..., conv5_x, and the classification layer. The outputs of conv2_x, ..., conv5_x are feature maps, and the size of the feature maps decreases gradually. In the ablation study, we analyze the attention maps corresponding to these four feature maps. The results are shown in Table 3. The experimental results indicate that as we increase the shallow-layer attention maps in the attention alignment module, the performance of the model significantly decreases. Figure 4 displays these four attention maps obtained from the teacher model, and it is evident that the shallower layers pay less attention to useful information. This is because shallower layers have smaller receptive fields and can only focus on local information, while deeper layers can better capture spatial and higher-level semantic information. Therefore, adding shallow-layer attention maps in the attention alignment module would degrade performance.

5 Conclusion

In this paper, we propose using knowledge distillation to address the problem of low-resolution facial expression recognition. We introduce a novel and effective method called Attention and Relative Distance Alignment (ARDA) for low-resolution facial expression recognition. The Attention Alignment module guides the student model to focus on the most crucial regions in facial images by leveraging the attention maps of the teacher model. The Relative Distance Alignment module is designed to capture differences between different facial images. Extensive experiments conducted on synthetic low-resolution facial expression recognition datasets show the effectiveness of our proposed ARDA method.

Acknowledgement. This work was supported by the National Key R&D Programme of China (2022YFC3803202), Major Project of Anhui Province under Grant 202203a05020011. This work was done in Anhui Province Key Laboratory of Affective Computing and Advanced Intelligent Machine.

References

1. Barsoum, E., Zhang, C., Ferrer, C.C., Zhang, Z.: Training deep networks for facial expression recognition with crowd-sourced label distribution. In: Proceedings of the 18th ACM International Conference on Multimodal Interaction, pp. 279–283 (2016)
2. Goodfellow, I.J., et al.: Challenges in representation learning: a report on three machine learning contests. In: Lee, M., Hirose, A., Hou, Z.-G., Kil, R.M. (eds.) ICONIP 2013, Part III. LNCS, vol. 8228, pp. 117–124. Springer, Heidelberg (2013). https://doi.org/10.1007/978-3-642-42051-1_16
3. Guo, Y., Zhang, L., Hu, Y., He, X., Gao, J.: MS-Celeb-1M: a dataset and benchmark for large-scale face recognition. In: Leibe, B., Matas, J., Sebe, N., Welling, M. (eds.) ECCV 2016, Part III. LNCS, vol. 9907, pp. 87–102. Springer, Cham (2016). https://doi.org/10.1007/978-3-319-46487-9_6
4. He, K., Zhang, X., Ren, S., Sun, J.: Deep residual learning for image recognition. In: Proceedings of the IEEE Conference on Computer Vision and Pattern Recognition, pp. 770–778 (2016)
5. Hinton, G., Vinyals, O., Dean, J.: Distilling the knowledge in a neural network. arXiv preprint arXiv:1503.02531 (2015)
6. Kingma, D.P., Ba, J.: Adam: a method for stochastic optimization. arXiv preprint arXiv:1412.6980 (2014)
7. Li, S., Deng, W., Du, J.: Reliable crowdsourcing and deep locality-preserving learning for expression recognition in the wild. In: Proceedings of the IEEE Conference on Computer Vision and Pattern Recognition, pp. 2852–2861 (2017)
8. Li, Z., Arora, S.: An exponential learning rate schedule for deep learning. arXiv preprint arXiv:1910.07454 (2019)
9. Ma, C., Jiang, Z., Rao, Y., Lu, J., Zhou, J.: Deep face super-resolution with iterative collaboration between attentive recovery and landmark estimation. In: IEEE/CVF Conference on Computer Vision and Pattern Recognition (CVPR) (2020)
10. Milan, A., Leal-Taixé, L., Reid, I., Roth, S., Schindler, K.: MOT16: a benchmark for multi-object tracking. arXiv preprint arXiv:1603.00831 (2016)
11. Nan, F., et al.: Feature super-resolution based facial expression recognition for multi-scale low-resolution images. Knowl.-Based Syst. **236**, 107678 (2022)
12. Romero, A., Ballas, N., Kahou, S.E., Chassang, A., Gatta, C., Bengio, Y.: Fitnets: hints for thin deep nets. arXiv preprint arXiv:1412.6550 (2014)
13. Selvaraju, R.R., Cogswell, M., Das, A., Vedantam, R., Parikh, D., Batra, D.: Grad-cam: visual explanations from deep networks via gradient-based localization. In: Proceedings of the IEEE International Conference on Computer Vision, pp. 618–626 (2017)
14. She, J., Hu, Y., Shi, H., Wang, J., Shen, Q., Mei, T.: Dive into ambiguity: latent distribution mining and pairwise uncertainty estimation for facial expression recognition. In: Proceedings of the IEEE/CVF Conference on Computer Vision and Pattern Recognition, pp. 6248–6257 (2021)
15. Wang, K., Peng, X., Yang, J., Lu, S., Qiao, Y.: Suppressing uncertainties for large-scale facial expression recognition. In: Proceedings of the IEEE/CVF Conference on Computer Vision and Pattern Recognition, pp. 6897–6906 (2020)
16. Zagoruyko, S., Komodakis, N.: Paying more attention to attention: improving the performance of convolutional neural networks via attention transfer. arXiv preprint arXiv:1612.03928 (2016)

17. Zeng, J., Shan, S., Chen, X.: Facial expression recognition with inconsistently annotated datasets. In: Proceedings of the European Conference on Computer Vision (ECCV), pp. 222–237 (2018)
18. Zhang, K., Zhang, Z., Li, Z., Qiao, Y.: Joint face detection and alignment using multitask cascaded convolutional networks. IEEE Signal Process. Lett. **23**(10), 1499–1503 (2016)
19. Zhang, Y., Wang, C., Deng, W.: Relative uncertainty learning for facial expression recognition. Adv. Neural. Inf. Process. Syst. **34**, 17616–17627 (2021)
20. Zhang, Y., Wang, C., Ling, X., Deng, W.: Learn from all: erasing attention consistency for noisy label facial expression recognition. In: Avidan, S., Brostow, G., Cissé, M., Farinella, G.M., Hassner, T. (eds.) ECCV 2022, Part XXVI. LNCS, vol. 13686, pp. 418–434. Springer, Cham (2022). https://doi.org/10.1007/978-3-031-19809-0_24

Exploring Frequency Attention Learning and Contrastive Learning for Face Forgery Detection

Neng Fang, Bo Xiao(✉), Bo Wang, Chong Li, and Lanxiang Zhou

Beijing University of Posts and Telecommunications, Beijing, China
{fangneng,xiaobo,bobo,lch1203,zhoulanxiang}@bupt.edu.cn

Abstract. Face forgery detection has become a critical security concern due to advances in manipulation techniques. Most methods look for forged clues from the spatial or vanilla frequency domain, leading to serious over-fitting. In this paper, we propose a Frequency Attention Module (FAM) that enhances model generalizability in face forgery detection. We theoretically demonstrate the feasibility of frequency attention learning, which allows the network to automatically refine subtle but discriminative forged features and suppress irrelevant components in the frequency domain without complex manual partitions. Besides, considering that commonly-used cross-entropy loss neglects the intra-class compactness, we design the DeepFake Contrastive Loss (DFCL) to decrease intra-class variances for real faces and enlarge inter-class differences in the feature space. Extensive experiments show that our method significantly outperforms SoTA methods on widely-used benchmarks.

Keywords: Face Forgery Detection · Frequency Attention Learning · Contrastive Learning

1 Introduction

The rapid development of generative techniques, such as Variational Autoencoders (VAE) [27] and Generative Adversarial Networks (GANs) [11], has led to significant progress in face manipulation technologies. With the emergence of deep learning-based forgery methods, *e.g.*, Deepfakes [6] and Face2Face [35], generating highly realistic faces that can deceive human eyes, there is a growing concern over the potential misuse of such manipulated images. The malicious use of these images can result in severe security issues, eroding trust in society. Hence, it is imperative to develop effective techniques for face forgery detection.

Reviewing the literature [5,17,25,41], forgery detection is predominantly approached as a binary supervised classification problem. Early works relied

Supplementary Information The online version contains supplementary material available at https://doi.org/10.1007/978-981-99-8469-5_19.

Q. Liu et al. (Eds.): PRCV 2023, LNCS 14429, pp. 238–251, 2024.
https://doi.org/10.1007/978-981-99-8469-5_19

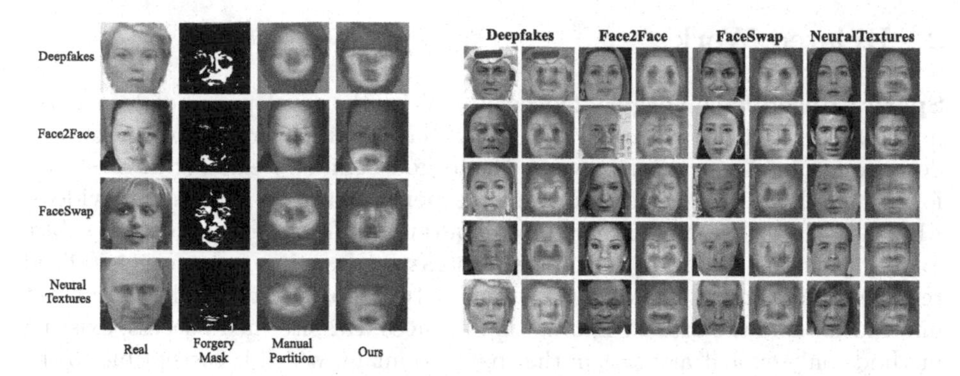

Fig. 1. Grad-CAM [42] **visualization**: The left shows comparison of traditional methods with manual partitions of the frequency domain and our approach. The right illustrates our FACL on different manipulated techniques.

on hand-crafted features. With the advance of convolutional neural networks (CNNs), learning-based forgery detection methods [5,17,25,41] have achieved significant progress. Recently, several works [2,10,22,28] have explored the frequency domain as a promising avenue for detecting face manipulation. However, these methods may introduce bias due to the manual partitions and limit the quality of frequency clues. In this paper, we propose a novel frequency attention module (FAM) for automatic refinement of frequency features in general face forgery detection. Left figure of Fig. 1 compares our FAM to existing methods that utilize manual partitions of the frequency domain and demonstrates its ability to uncover suspicious artifact traces across various manipulation techniques.

Besides extracting discriminative features, we consider representation learning to further improve the generalizability. Inspired by the contrastive learning [12,15], we propose a DeepFake Contrastive Loss (DFCL) to decrease intra-class variances for real faces by pulling real samples closer. For forged images, we only consider real-fake pairs and push them away to enhance inter-class differences.

The main contributions of this paper are:

1. We design a Frequency Attention Module (FAM) to extract discriminative frequency features for face forgery detection. FAM utilizes attention mechanism in the frequency domain to refine subtle but important forgery features and restrain irrelevant coefficients without manual partitions.
2. We propose the DeepFake Contrastive Loss (DFCL) which takes into account the unique characteristics of face forgery detection. DFCL decreases intra-class variance for real faces while increasing inter-class diversities in the feature space, which avoids interference between different manipulation methods.
3. Experiments and visualizations reveal the generalizability and robustness of our method, shows the effectiveness of our method over existing SoTAs.

2 Related Work

Spatial-Based Face Forgery Detection: With the development of deep learning, a wide variety of methods [1,17,31,32] have been proposed for face forgery detection. Face X-ray [17] focuses on the forged boundary existing in most face forgery methods, and achieves remarkable performance in high quality videos. FD^{2}Net [44] decomposes faces into computer graphics views to detect subtle forgery patterns. Dual [32] introduces contrastive learning to learn generalized representations in the spatial domain. DCI [21] proposes to decouple the content information and the artifact information in faces. Despite their success, existing methods only exploit artifacts in the spatial domain, which is susceptible to the data quality and distribution.

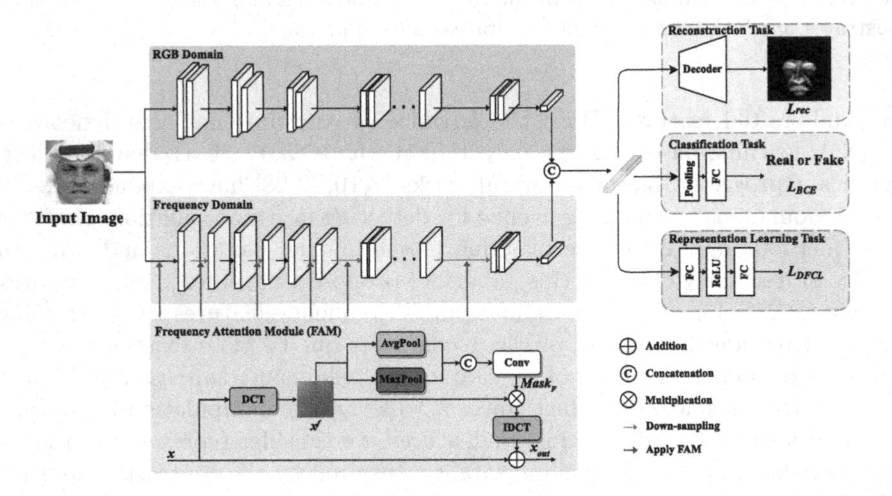

Fig. 2. The architecture of Frequency Attention Learning and Contrastive Learning (FACL). We present a two-branch architecture with multi-tasks to process the image in both the RGB and the frequency domain. We apply our FAM after each down-sampling blocks (red rows) of the origin CNNs, as shown by the blue arrows. In addition to the primary classification task, we introduce auxiliary tasks (*i.e.*, artifact reconstruction and representation learning) to help the model distinguish real/fake faces. (Color figure online)

Frequency-Aware Face Forgery Detection: Recently, several works [2,10,16,23,25,28] have investigated the frequency domain for face forgery detection. Leveraging [10] shows that artifacts in the frequency domain could help distinguish forged faces. GFF [23] argues that high-frequency noises can remove color textures and reveal forgery traces, improving the generalization ability. Two-branch [25] combines information from both the spatial domain and frequency domain using a multi-scale Laplacian of Gaussian operator. F3Net [28] leverages frequency-aware decomposed image components and local frequency statistics to discover the robust forged patterns. Local [2] considers the correlation between local regions to learn generalized features. While these works attempt to mine the implicit forgery artifacts in the frequency domain, they manually slice the whole frequency domain into several bands without discrimination, which is biased due to the manual partitions. In contrast, our proposed FAM allows the network to automatically mine the manipulated clues in the whole frequency domain without manual partitions, leading to better generalizability.

Attention Mechanisms in Face Forgery Detection: Attention mechanisms play an important role in human perception [14,30], and have been widely used in various computer vision tasks. Existing attention mechanisms can be divided into spatial-wise [38], channel-wise [13], or the combination of both [39]. While some recent works have explored attention mechanisms for face forgery detection, they have mostly focused on the spatial domain but neglected the frequency domain. For instance, MAT [41] and RFM [37] use spatial attention to highlight manipulated regions and explore previously overlooked areas in the spatial domain, respectively. FcaNet [29] proposes a multi-spectral channel attention that incorporates more frequency components into the attention mechanism. However, none of these methods directly address the challenge of refining forged features in the frequency domain. To fill this gap, we propose a novel FAM that refines features in the frequency domain by filtering out irrelevant information while highlighting critical forged features.

3 Frequency Attention Learning and Contrastive Learning (FACL)

3.1 A Multi-task Two-Branch Architecture

Figure 2 illustrates the two-branch architecture of our FACL. Considering the complementary property of the spatial and the frequency domain, one branch captures local discriminative features in the spatial domain while the other discovers general forgery patterns in the frequency domain with our FAM. In addition to the intrinsic classification task, we introduce extra tasks to help the model distinguish real/fake faces. Specifically, We use artifact reconstruction to refine the feature space and propose the DeepFake Contrastive Loss (DFCL) to encourage feature discrimination. This multi-task design enables the model to learn more robust and discriminative features for face forgery detection. We will elaborate on each component in turn, followed by the overall loss functions used in FACL.

3.2 Frequency Attention Module

Can Attention Mechanism be Applied to the Frequency Domain? Attention mechanisms give the opportunity to obtain relevant information by re-weighting or refining the knowledge within models, thus enhancing the learning ability. While existing works [4,14,30] mainly focus on attention modules in the spatial domain, face forgery detection requires identifying accurate forgery clues which may not easily discoverable in the spatial domain alone. To address this issue, we explore the feasibility of introducing attention mechanisms in the frequency domain directly to learn more discriminative forgery patterns. We first demonstrate the feasibility of introducing attention mechanisms in the frequency domain. On the basis of the theory of digital image processing, we derive the expression for the Inverse Discrete Cosine Transform (IDCT):

$$\begin{aligned} x(i,j) &= K \sum_{h=0}^{H-1} \sum_{w=0}^{W-1} x^f(h,w) \cos \frac{(2i+1)h\pi}{2H} \cos \frac{(2j+1)w\pi}{2W} \\ &\triangleq \sum_{h=0}^{H-1} \sum_{w=0}^{W-1} \xi(i,j,h,w) x^f(h,w) \end{aligned} \tag{1}$$

Here, $x \in \mathbb{R}^{H\times W\times C}$ denotes the spatial image, where H, W, C represent the height, width, and channel of the image. Similarly, $x^f \in \mathbb{R}^{H\times W\times C}$ represents the frequency feature obtained via the Discrete Cosine Transform (DCT). The coordinates of x and x^f are denoted by (i, j) and (h, w), respectively, while K is a constant coefficient. $\xi(i, j, h, w)$ denotes the correlation coefficients between the spatial and frequency domain. Equation 1 shows that each pixel of the image in the spatial domain is linearly correlated with the coefficient in the frequency domain. This suggests that the refinement of the frequency feature (*i.e.*, frequency attention learning) could further enhance the feature discrimination in the spatial domain from a global perspective, enabling more precise manipulation detection.

Perform Attention Learning in the Frequency Domain. Owing to the linear correlation between the spatial and frequency domains, we can learn attention weights in the frequency domain to extract more discriminative forgery patterns automatically. Specifically, we first obtain the frequency feature of x through DCT:

$$x^f(h,w) = K' \sum_{i=0}^{H-1} \sum_{j=0}^{W-1} x(i,j) \cos \frac{(2i+1)h\pi}{2H} \cos \frac{(2j+1)w\pi}{2W} \tag{2}$$

Here, K' is the constant coefficient. In FAM, we introduce a learning-based attention map generator, denoted as $\mathcal{G}(\cdot, \theta)$, to automatically discover the importance of each frequency component. Specifically, we obtain two frequency views, x^f_{avg} and x^f_{max}, by applying average-pooling and max-pooling to the origin frequency feature x^f. These views are concatenated along the channel dimension and fed through a 1×1 convolution layer followed by a *Sigmoid* non-linearity to generate the attention map in the frequency domain, denoted as Mask_F:

$$\text{Mask}_F = \sigma\left(\text{Conv}_{1\times 1}\left(\text{Concat}\left(x^f_{\text{avg}}, x^f_{\text{max}}\right)\right)\right) \tag{3}$$

Here, σ represents the *Sigmoid* function and $\text{Conv}_{1\times 1}$ denotes the 1×1 convolution operation. The learnable attention map, Mask_F, could automatically re-weight and refine the knowledge in the frequency domain, which encourages the network to focus on the artifacts in the frequency domain generated by the up-sampling procedures. The refined frequency feature is as follows:

$$\begin{aligned} x^f_{\text{ref}} &= x^f \cdot \text{Mask}_F \\ &= \mathcal{D}(x) \cdot \mathcal{G}(\mathcal{D}(x), \theta) \end{aligned} \tag{4}$$

As the frequency domain lacks the shift invariance and local consistency present in natural images, vanilla CNNs are unable to effectively process frequency features. Therefore, we invert the refined frequency feature x^f_{ref} back into the spatial domain through IDCT, allowing the refined information to be preserved in the frequency domain and can be applied to subsequent convolution modules:

$$x_{\text{out}} = \mathcal{D}^{-1}(x^f_{\text{ref}}) + x \tag{5}$$

To preserve the original info of the input, we introduce a residual shortcut at the end of the FAM. Our FAM is specifically designed to capture frequency artifacts that are difficult to detect with spatial-wise and channel-wise attention mechanisms. It is worth noting that FAM can be easily integrated into existing CNN architectures, making it a plug-and-play module that enhances the learning ability of frequency-based clues for detecting face forgery. Detailed experiments are available in *supp*.

3.3 DeepFake Contrastive Learning

In addition to identifying discriminative forgery patterns, we also focus on optimizing the feature space to help the network distinguish real/fake faces. Most existing face forgery detection methods tackle the the problem as a binary classification task and typically utilize cross-entropy loss to supervise training procedure. However, the representation supervised by the cross-entropy loss is not essentially discriminative enough. The main reason lies in the cross-entropy loss assumes all instances in the same category should have similar distribution, disregarding their discrepancies and only focuses on finding a decision boundary to separate different classes. Consequently, inter-class separability and intra-class compactness are not explicitly considered. While some methods have introduced metric learning to optimize the feature space, they have not considered the unique challenges of face forgery detection, where manipulated images can be generated by various techniques with widely varying distribution of representations. Inspired by contrastive learning, we propose a DeepFake Contrastive Learning (DFCL) loss function, which simultaneously pushes away the real-fake pairs and merely pulls close the natural faces:

$$\mathcal{L}_{DFCL} = \sum_{i \in P} -\frac{1}{|P|} \sum_{p \in P} \log \frac{e^{\delta(x_i, x_p)/\tau}}{\sum\limits_{j \in P} e^{\delta(x_i, x_j)/\tau} + \sum\limits_{n \in N} e^{\delta(x_i, x_n)/\tau}} \tag{6}$$

Here, P and N denote real faces and manipulated faces within a batch, respectively; δ means a similarity function; τ is a scalar temperature parameter. Unlike existing supervised contrastive learning methods [15] that maximize the invariance among views of the same category, our $\mathcal{L}_{DFCL}$ only maximizes invariance between real faces and ignores manipulated pairs, which reduces the adverse effect of irrelevant forgery patterns. Further explanation is in *supp*.

3.4 Reconstruction of Manipulated Areas

In recognition of the fact that real faces can be subjected to partial or complete modified, we incorporate a reconstruction task to elucidate the spatial regions on which the model bases its decisions. As depicted in Fig. 2, we utilize a decoder to reconstruct manipulated areas. To create the forgery mask, we pair each forged sample with its corresponding source and calculate the absolute pixel-wise difference in the RGB domain. Then we convert this difference into grayscale and divide it by 255, producing a shadow that captures the degree of alteration. We empirically utilize the threshold of 0.2 to generate the binary ground-truth mask $M \in \mathbb{R}^{H \times W}$, where the value of the real region is 0 and forged region is 1:

$$M(i,j) = \begin{cases} 1 & \text{if } x(i,j) \text{ is manipulated} \\ 0 & \text{if } x(i,j) \text{ is real} \end{cases} \tag{7}$$

Accordingly, we utilize the cross-entropy loss to supervise the predicted mask:

$$\mathcal{L}_{\text{rec}} = \sum_{i,j} - \left[M(i,j) \log \hat{M}(i,j) + (1 - M(i,j)) \log \left(1 - \hat{M}(i,j) \right) \right] \tag{8}$$

Here $\hat{M}$ denotes the output of the decoder, which reflects the probability of a given pixel having been modified.

Table 1. Ablations of the impact of our proposed components. RGB: the RGB stream input, FAM: Frequency Attention Module, DFCL: DeepFake Contrastive Loss, REC: the decoder with the reconstruction task. Red numbers indicate the best performance.

Method	RGB	FAM	DFCL	REC	AUC
Base	✓				91.97%
Base w/ FAM	✓	✓			93.76%
Base w/ DFCL	✓		✓		93.44%
Base w/ REC	✓			✓	92.67%
FACL	✓	✓	✓	✓	94.33%

3.5 Training and Inference

Training: Following the existing approaches, we employ the classical binary cross-entropy loss (BCE) to minimize the divergence between the predictions and their corresponding ground truth (y', y):

$$\mathcal{L}_{BCE} = -[y \log y' + (1-y) \log(1-y')] \tag{9}$$

During training, FACL incorporates auxiliary tasks (*i.e.*, representation learning and artifact reconstruction) to help the model distinguish between real and fake faces, the overall loss can be obtained as:

$$\mathcal{L}_{FACL} = \lambda_1 \mathcal{L}_{BCE} + \lambda_2 \mathcal{L}_{DFCL} + \lambda_3 \mathcal{L}_{\text{rec}} \tag{10}$$

where $\lambda_1, \lambda_2, \lambda_3$ are the hyper-parameters used to balance these loss functions.

Inference: During inference, we rely solely on the classification head to produce the final prediction, where the outputs from other heads can be visualized to demonstrate the effectiveness of our FACL, as illustrated in *supp*.

4 Experiments

4.1 Settings

Datasets: We adopt the challenging **FaceForensics++** (FF++) [31] dataset to train our network. FF++ contains 1,000 real videos, with 720 videos used for training, 140 for validation, and 140 for test. Each video undergoes four manipulation methods [6,9,34,35] to generate four fake videos with different quality levels, *i.e.*, raw (C0), high quality (C23) and low quality (C40). To evaluate the robustness of our method, we also conduct experiments on **Celeb-DF** [20], which is a challenging database to current detection methods. Celeb-DF contains 590 real videos and 5639 fake videos generated through face swapping for each pair of the 59 subjects.

Implementation Details: We apply the advanced RetinaFace [8] to extract and align faces for all datasets, and then randomly select 50 frames from each video for training and testing. We implement our FACL via the open-source PyTorch. Following the convention, we utilized the Xception [3] pretrained on the ImageNet [7] as the backbone to extract image features. To optimize our FACL, we used the Adam optimizer with an initial learning rate 2e-4 with a cosine decay schedule. The input face is resized to 299 × 299, and the batch size is set to 96. We train our FACL for approximately 120k iterations on four GeForce RTX3090s. The τ in Eq. 6 is empirically set to 0.07, and we set the values of $\lambda_1, \lambda_2, \lambda_3$ in Eq. 10 to 1, 10 and 1, respectively.

4.2 Ablation Study

To prove the effectiveness of our approach, we conducted extensive ablations on the FF++ dataset under the low quality setting (C40).

Frequency Attention Module: We observe that **Base w/ FAM** improves the performance by 1.9% (91.97 → 93.76). FAM utilizes the attention mechanism

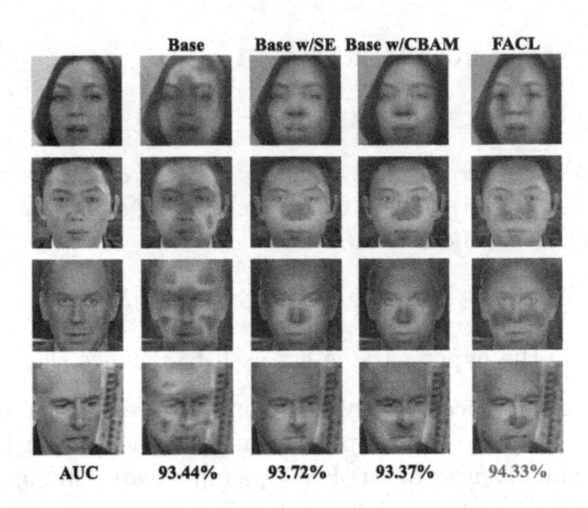

Fig. 3. Comparisons of different attention mechanisms. We visualize the Grad-CAM [42] of existing spatial-aware attention methods and our frequency attention module. We also evaluate the AUC metric on the FF++ dataset. Our FAM could focus on both local attributes (fourth row) and global manipulation areas (second row).

across the entire frequency domain to extract discriminative frequency features naturally, thereby enhancing the generalizability of manipulated image detection. As shown in Fig. 3, our approach achieves 0.6% gains over SE and 1.0% over CBAM. It is worth noting that forged images may replace the whole face, where the petite local features like nose in second row considered by existing spatial attention methods lead to misjudgment. Our FAM overcome this limitation by considering the manipulated clues in the frequency domain, which enables it to pay more attention to the complete manipulated regions, leading to stronger performance and mining specific artifact of different manipulations. **Deepfake Contrastive Loss:** We compare **Base** (with the cross-entropy loss) and **Base w/ DFCL** (our Deepfake Contrastive loss). As shown in Table 1, **Base w/ DFCL** achieves 1.6% (91.97 → 93.44) gains in AUC. To further illustrate the effectiveness of our DFCL, we exploit t-SNE [24] to visualize the feature space. As depicted in Fig. 4, we observe that real samples (blue points) and forgery samples (points in other colors) are pushed out, while real samples are pulled closer. Moreover, samples generated by different forgery techniques still retain their original distributions, which is similar to real-world scenarios.

Benefiting from our novel FAM module, DFCL and the reconstruction decoder, we obtain our full FACL. As shown in Table 1, our **FACL** outperforms **Base** by 1.47%. In conclusion, our FACL consistently improves the performance by incorporating our novel modules and the multi-task two-stream architecture. Further explanation and visualizations about the reconstruction task are in *supp.*

Table 2. Quantitative results in terms of AUC on FF++ dataset with all quality settings, *i.e.*, raw videos without compression (C0), high quality (C23), and low quality (C40). For a fair comparison, the results of other methods are obtained from their papers. Red and blue numbers indicate the best and second best performance.

Method	FF++ (AUC)		
	C0	C23	C40
Xception [3]	–	94.86%	81.76%
Face Xray [17]	98.80%	87.40%	61.60%
SPSL [22]	–	95.32%	82.82%
Two-Branch [25]	–	88.87%	86.59%
MAT [41]	–	98.97%	87.26%
FDFL [16]	99.70%	99.30%	92.40%
F3Net [28]	99.80%	98.10%	93.30%
Dual [32]	–	99.30%	–
Local [2]	99.92%	99.46%	95.21%
FACL (Ours)	99.96%	99.50%	94.33%

4.3 Comparisons

Comparisons with Alternatives Under the Cross-dataset Evaluation: To assess the generalizationablity of our FACL across different data distributions, we follow the evaluation protocol of [41]. Specifically, we train our FACL on the Deepfakes subset of the FF++ dataset under the high quality setting (C23) and evaluate it on the Celeb-DF [20] dataset. We sample 50 frames from each video to compute the AUC scores. As shown in Table 3, our FACL achieves new SoTA performance on the cross-dataset evaluation. Notably, our FACL outperforms DCI [21] by a largin margin (76.91 → 80.69, 4.9%). This significant improvement is mainly attributed to our novel FAM and DFCL, which refine the frequency components of artifact patterns and enhance the differences between real and

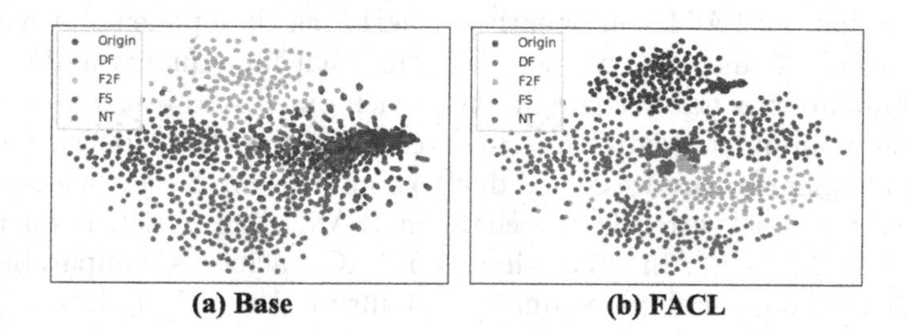

Fig. 4. The t-SNE visualization of (a) Base and (b) FACL on the FF++ dataset. (Color figure online)

Table 3. Cross-dataset evaluation (AUC) on the Celeb-DF dataset. We train our model on the FF++ (DF) under the high quality setting (C23). Red/blue numbers are the best/second best.

Method	FF++ (DF)	Celeb-DF
Two-stream [43]	70.10%	53.80%
Meso4 [1]	84.70%	54.80%
MesoInception4 [1]	83.00%	53.60%
HeadPose [40]	47.30%	54.60%
FWA [19]	80.10%	56.90%
VA-MLP [36]	66.40%	55.00%
Xception-c40 [3]	95.50%	65.50%
Multi-task [26]	76.30%	54.30%
DSP-FWA [19]	93.00%	64.60%
SMIL [18]	96.80%	56.30%
Two-branch [25]	93.20%	73.40%
F3Net [28]	97.97%	65.17%
EfficientNet-B4 [33]	99.70%	64.29%
MAT [41]	99.80%	67.44%
SPSL [22]	96.91%	76.88%
RFM [37]	95.42%	67.21%
DCI (Xception) [21]	96.50%	76.91%
FACL (Ours)	99.96%	80.69%

forged samples. In summary, our FACL demonstrates remarkable transferability compared to other forgery detection techniques.

Comparisons with Alternatives on the Widely-Used Benchmark: Following most existing SoTA alternatives, we compare the performance of our FACL on the FF++ dataset. Our evaluation considers different video quality settings, *i.e.*, C0, C23, and C40. The comparison results are presented in Table 2. We observe that our FACL consistently outperforms all opponents by a considerable increase at C0 and C23. It is worth noting that high quality and raw videos possess abundant frequency components, which our FACL utilizes to efficiently describe the manipulated patterns. However, in low-quality videos, the original frequency components are weakened due to compression. Nevertheless, with the help of our frequency attention mechanism (FAM), our FACL is on par with SoTA method Local [2]. In conclusion, our FACL achieves comparable results with SoTA methods on all video quality settings in the FF++ dataset.

5 Conclusion

This paper presents frequency attention learning and contrastive learning for face forgery detection. Specifically, we propose a Frequency-aware Attention Module (FAM), which can automatically refine subtle but discriminative forged features and suppress irrelevant frequency components without complex manual partition in face forgery detection. Besides, we notice that conventional cross-entropy loss ignores the intra-class compactness and restrains the effectiveness of the model. We design the DeepFake Contrastive Loss (DFCL) to diminish intra-class variances for pristine faces and boost inter-class discrepancies in the feature space. Extensive experiments and visualizations demonstrate the superiority and generalizability of our FACL over widely-used state-of-the-art methods, especially in the challenging cross-manipulated and cross-dataset evaluation scenarios. The future work is to address the limitations mentioned above. Codes and *supp* are available at https://github.com/Jacky-F/FACL.git.

Acknowledgments. Research reported in this paper was supported by the Natural Science Foundation of China under grants 62076031.

References

1. Afchar, D., Nozick, V., Yamagishi, J., Echizen, I.: Mesonet: a compact facial video forgery detection network. In: WIFS (2018)
2. Chen, S., Yao, T., Chen, Y., Ding, S., Li, J., Ji, R.: Local relation learning for face forgery detection. In: AAAI (2021)
3. Chollet, F.: Xception: deep learning with depthwise separable convolutions. In: CVPR (2017)
4. Corbetta, M., Shulman, G.L.: Control of goal-directed and stimulus-driven attention in the brain. Nat. Rev. Neurosci. 201–215 (2002)
5. Dang, H., Liu, F., Stehouwer, J., Liu, X., Jain, A.K.: On the detection of digital face manipulation. In: CVPR (2020)
6. Deepfakes: Deepfakes github (2017). https://www.github.com/deepfakes/faceswap. Accessed 10 Apr 2022
7. Deng, J., Dong, W., Socher, R., Li, L.J., Li, K., Fei-Fei, L.: Imagenet: a large-scale hierarchical image database. In: CVPR (2009)
8. Deng, J., Guo, J., Ververas, E., Kotsia, I., Zafeiriou, S.: Retinaface: single-shot multi-level face localisation in the wild. In: CVPR (2020)
9. FaceSwap: Faceswap github (2016). https://www.github.com/MarekKowalski/FaceSwap. Accessed 10 Apr 2022
10. Frank, J., Eisenhofer, T., Schönherr, L., Fischer, A., Kolossa, D., Holz, T.: Leveraging frequency analysis for deep fake image recognition. In: ICML (2020)
11. Goodfellow, I., et al.: Generative adversarial nets. In: NIPS (2014)
12. He, K., Fan, H., Wu, Y., Xie, S., Girshick, R.: Momentum contrast for unsupervised visual representation learning. In: CVPR (2020)
13. Hu, J., Shen, L., Sun, G.: Squeeze-and-excitation networks. In: CVPR (2018)
14. Itti, L., Koch, C., Niebur, E.: A model of saliency-based visual attention for rapid scene analysis. PAMI, pp. 1254–1259 (1998)
15. Khosla, P., et al.: Supervised contrastive learning. In: NIPS (2020)

16. Li, J., Xie, H., Li, J., Wang, Z., Zhang, Y.: Frequency-aware discriminative feature learning supervised by single-center loss for face forgery detection. In: CVPR (2021)
17. Li, L., et al.: Face x-ray for more general face forgery detection. In: CVPR (2020)
18. Li, X., et al.: Sharp multiple instance learning for deepfake video detection. In: ACMMM (2020)
19. Li, Y., Lyu, S.: Exposing deepfake videos by detecting face warping artifacts. arXiv:1811.00656 (2018)
20. Li, Y., Yang, X., Sun, P., Qi, H., Lyu, S.: Celeb-DF: a large-scale challenging dataset for deepfake forensics. In: CVPR (2020)
21. Liang, J., Shi, H., Deng, W.: Exploring disentangled content information for face forgery detection. In: Avidan, S., Brostow, G., Cissé, M., Farinella, G.M., Hassner, T. (eds.) ECCV 2022. LNCS, vol. 13674, pp. 128–145. Springer, Cham (2022). https://doi.org/10.1007/978-3-031-19781-9_8
22. Liu, H., et al.: Spatial-phase shallow learning: rethinking face forgery detection in frequency domain. In: CVPR (2021)
23. Luo, Y., Zhang, Y., Yan, J., Liu, W.: Generalizing face forgery detection with high-frequency features. In: CVPR (2021)
24. Van der Maaten, L., Hinton, G.: Visualizing data using t-SNE. JMLR, 2579–2605 (2008)
25. Masi, I., Killekar, A., Mascarenhas, R.M., Gurudatt, S.P., AbdAlmageed, W.: Two-branch recurrent network for isolating deepfakes in videos. In: Vedaldi, A., Bischof, H., Brox, T., Frahm, J.-M. (eds.) ECCV 2020. LNCS, vol. 12352, pp. 667–684. Springer, Cham (2020). https://doi.org/10.1007/978-3-030-58571-6_39
26. Nguyen, H.H., Fang, F., Yamagishi, J., Echizen, I.: Multi-task learning for detecting and segmenting manipulated facial images and videos. In: BTAS (2020)
27. Pu, Y., et al.: Variational autoencoder for deep learning of images, labels and captions. In: NIPS (2016)
28. Qian, Y., Yin, G., Sheng, L., Chen, Z., Shao, J.: Thinking in frequency: face forgery detection by mining frequency-aware clues. In: Vedaldi, A., Bischof, H., Brox, T., Frahm, J.-M. (eds.) ECCV 2020. LNCS, vol. 12357, pp. 86–103. Springer, Cham (2020). https://doi.org/10.1007/978-3-030-58610-2_6
29. Qin, Z., Zhang, P., Wu, F., Li, X.: Fcanet: frequency channel attention networks. In: ICCV (2021)
30. Rensink, R.A.: The dynamic representation of scenes. Vis. Cogn. 17–42 (2000)
31. Rossler, A., Cozzolino, D., Verdoliva, L., Riess, C., Thies, J., Nießner, M.: Faceforensics++: learning to detect manipulated facial images. In: ICCV (2019)
32. Sun, K., Yao, T., Chen, S., Ding, S., Ji, R., et al.: Dual contrastive learning for general face forgery detection. arXiv:2112.13522 (2021)
33. Tan, M., Le, Q.: Efficientnet: rethinking model scaling for convolutional neural networks. In: ICML (2019)
34. Thies, J., Zollhöfer, M., Nießner, M.: Deferred neural rendering: image synthesis using neural textures. TOG, 1–12 (2019)
35. Thies, J., Zollhofer, M., Stamminger, M., Theobalt, C., Nießner, M.: Face2face: real-time face capture and reenactment of RGB videos. In: CVPR (2016)
36. WACVW: Exploiting visual artifacts to expose deepfakes and face manipulations (2019)
37. Wang, C., Deng, W.: Representative forgery mining for fake face detection. In: CVPR (2021)
38. Wang, F., et al.: Residual attention network for image classification. In: CVPR (2017)

39. Woo, S., Park, J., Lee, J.-Y., Kweon, I.S.: CBAM: convolutional block attention module. In: Ferrari, V., Hebert, M., Sminchisescu, C., Weiss, Y. (eds.) ECCV 2018. LNCS, vol. 11211, pp. 3–19. Springer, Cham (2018). https://doi.org/10.1007/978-3-030-01234-2_1
40. Yang, X., Li, Y., Lyu, S.: Exposing deep fakes using inconsistent head poses. In: ICASSP (2019)
41. Zhao, H., Zhou, W., Chen, D., Wei, T., Zhang, W., Yu, N.: Multi-attentional deepfake detection. In: CVPR (2021)
42. Zhou, B., Khosla, A., Lapedriza, A., Oliva, A., Torralba, A.: Learning deep features for discriminative localization. In: CVPR (2016)
43. Zhou, P., Han, X., Morariu, V.I., Davis, L.S.: Two-stream neural networks for tampered face detection. In: CVPRW (2017)
44. Zhu, X., Wang, H., Fei, H., Lei, Z., Li, S.Z.: Face forgery detection by 3D decomposition. In: CVPR (2021)

An Automatic Depression Detection Method with Cross-Modal Fusion Network and Multi-head Attention Mechanism

Yutong Li[1], Juan Wang[2(✉)], Zhenyu Liu[1(✉)], Li Zhou[1], Haibo Zhang[1], Cheng Tang[1], Xiping Hu[1(✉)], and Bin Hu[1(✉)]

[1] Gansu Provincial Key Laboratory of Wearable Computing, Lanzhou University, Lanzhou, China
{liuzhenyu,huxp,bh}@lzu.edu.cn

[2] Department of Psychological Medicine, Seventh Medical Center of PLA General Hospital, Beijing, China
imjuan@sina.com

Abstract. Audio-visual based multimodal depression detection has gained significant attention due to its high efficiency and convenience as a computer-aided detection tool, resulting in promising performance. In this paper, we propose a cross-modal fusion network based on multi-head attention and residual structures (CMAFN) for depression recognition. CMAFN consists of three core modules: the Local Temporal Feature Extract Block (LTF), the Cross-Model Fusion Block (CFB), and the Multi-Head Temporal Attention Block (MTB). The LTF module performs feature extraction and encodes temporal information for audio and video modalities separately, while the CFB module facilitates complementary learning between the modalities. The MTB module accounts for the temporal influence of all modalities on each unimodal branch. With the incorporation of the three well-designed modules, CMAFN can refine the inter-modality complementarity and intra-modality temporal dependencies, achieving the interaction between unimodal branches and adaptive balance between modalities. Evaluation results on widely used depression datasets, AVEC2013 and AVEC2014, demonstrate that the proposed CMAFN method outperforms state-of-the-art approaches for depression recognition tasks. The results highlight the potential of CMAFN as an effective tool for the early detection and diagnosis of depression.

Keywords: Depression · Automatic detection · Multi-modal fusion · Multimodal depression detection

1 Introduction

Major depressive disorder (MDD) is one of the major drivers that cause physical and mental disability, leading to severe consequences such as heart attacks

Q. Liu et al. (Eds.): PRCV 2023, LNCS 14429, pp. 252–264, 2024.
https://doi.org/10.1007/978-981-99-8469-5_20

and suicide [1]. At present, the traditional clinical diagnosis of depression is performed by an experienced professional. Therefore, it is necessary to find objective parameter indicators to help improve the accuracy of depression diagnosis.

Machine learning enables automatic depression diagnosis using facial and vocal cues, offering low-cost, non-invasive, adaptable, and non-contact solutions. For instance, [2,3] have demonstrated the effectiveness of acoustic features for depression recognition. At the same time, [4,5] confirm that non-verbal facial behaviours are reliable markers of depression. Consequently, this paper uses facial activities and speech as the biomarkers to analyze the individual depression level, which can be measured through the Beck Depression Inventory-II (BDI-II) score [6].

Currently, there are many methods for the automatic detection of depression based on audio and video features. Most of them consist mainly of hand-crafted methods [7,8] and deep-learning methods [9–11]. However, the above methods only use single modality features, ignoring the information interaction between the two modalities. Moreover, some frameworks [12–15] use fusion methods based on speech and video to estimate depression scores. Feature concatenation and decision weighting are two standard methods. Nonetheless, existing methods have not sufficiently explored the complementarity, redundancy, and interaction between different modalities.

To mitigate the problems mentioned above, we propose a cross-modal fusion network based on multi-head attention and residual structure (CMAFN) for depression recognition. More specifically, as shown in Fig. 1, the CMAFN consists of three components: 1) Local Temporal Feature extract block (LTF), 2) Cross-modal Fusion Block (CFB), and 3) Multi-head Temporal Attention Block (MTB). Concretely, the LTF is designed to encode the local temporal order during the learning process. Next, the CFB obtains the complementary intra- and inter-modal interaction information between the two modalities. At the same time, the MTB further captures the temporal dependency and high-level representations from each modality. Finally, the CMAFN outputs the depression severity.

In summary, the main contributions of this paper can be summarized as follows:

- We propose a cross-modal fusion framework CMAFN, which effectively captures the dynamics features from facial and verbal cues as non-verbal behavior measures for estimating the severity of the depression scale.
- Based on the one-dimensional temporal convolutional and multi-head attention mechanism, we design the LTF and MTB block, which deal with the dynamic feature streams of multi-modal to obtain more comprehensive temporal information.
- To mine more efficient fusion feature representation, we introduce the CFB block to consider the dynamic interactions of different modalities and fully use their complementary information.

- We conducted extensive experiments on AVEC2013 and AVEC2014. The results demonstrate our method's effectiveness and generalization for depression recognition.

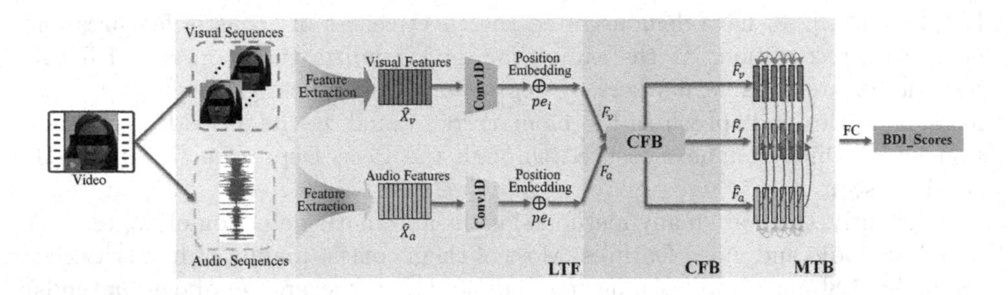

Fig. 1. The framework of CMAFN. Abbreviation: LTF for local temporal feature extract block. CFB for cross-model fusion block. MTB for multi-head temporal attention block.

2 Methodology

2.1 Framework Overview

The multi-modal depression recognition framework CMAFN is depicted in Fig. 1. To effectively learn fusion depression features from facial and vocal expressions, we first encode the local temporal dependency during the learning process using the LTF module. Second, the CFB module models intra- and inter-modality dependencies. Finally, the MTB module obtains each modality's long-range temporal context dependencies and high-level representations. In the following sections, we will describe each component of the CMAFN in detail.

2.2 Local Temporal Feature Extract Block

Multi-modal depression automatic recognition model detects the individual depression level in a video segment by using multi-modal signals. In this paper, our multi-modal framework uses facial expressions and vocal cues as input. Suppose a video segment $X_s = \{x_s^1, x_s^2, \cdots, x_s^{L_s}\}, s \in \{a, v\}$, where x_a^i and x_v^i denote audio and visual frames, respectively. L_s is the sequence length of each modality. For the visual sequences X_v, we use Resnet-50 [16] pre-trained on the FER+ dataset [17] to extract deep appearance spatial features. For the audio sequence X_a, we employ the open source audio processing tool openSmile [18] to extract a set of 25 low-level acoustic descriptors from each clip, including loudness, Mel-frequency cepstral coefficients (MFCCs), spectral flux, and others.

We denote the unimodal feature by $\hat{X}_s \in \mathcal{R}^{L_s \times D_s}$, where $s \in \{a, v\}$, L_s and D_s are the sequence length and feature dimension respectively. Next, we use the

1-D temporal convolution networks [19] to obtain the local temporal information of each modality feature. Considering the time order of the feature vectors, the positional encoding is added to the output of the 1-D temporal convolutional network. Finally, the embedded feature sequence output by the LTF module is defined as:

$$F_s = \mathbf{Conv1D}\left(ReLu\left(Linear\left(\hat{X}_s\right)\right)\right) + pe_i \tag{1}$$

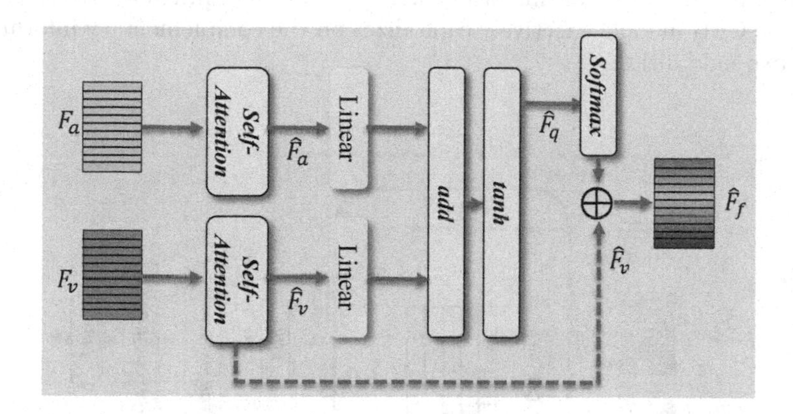

Fig. 2. The detailed illustration of the Cross-model Fusion Block

2.3 Cross-Model Fusion Block

We exploit CFB to compute intra- and inter-modal interaction to obtain complementary information from different modalities. As shown in Fig. 2. Firstly, we use the self-attention mechanism for each modality feature selection. This design enables CFB to focus more on features that significantly impact the result. We then input the learned weight information into the fully connected layer to achieve adaptive learning features. The above process can be described as follows:

$$Attention\,(Q, K, V) = \sigma\left(\frac{QK^{\top}}{\sqrt{d_k}}\right)V \tag{2}$$

$$\hat{F}_s\,(Q, K, V) = \mathbf{LN}\,(Attention\,(Q, K, V)) \tag{3}$$

where $\mathbf{LN}\,(\cdot)$ is the layer normalization, σ stands for the softmax operation. Q, K, and V denote the query matrix, the key matrix and the value matrix respectively. d_k is the key dimensionality.

Secondly, we adopt linear projection $\mathcal{L}$ to derive mapping representations from $\hat{F}_a$ and $\hat{F}_v$. Then, we apply the *add* and *tanh* activation functions to process these two representations. Finally, the $softmax$ is used to obtain the fused features $\hat{F}_f$ and ensure that the information is not lost by employing a residual structure. The $\hat{F}_f$ can be defined as follows:

$$\hat{F}_q = tanh\left(\mathcal{L}\left(\hat{F}_a\right) + \mathcal{L}\left(\hat{F}_v\right)\right) \tag{4}$$

$$\hat{F}_f = \sigma\left(\hat{F}_q\right) \oplus \hat{F}_v \in R^{L_s \times d_f} \tag{5}$$

where d_f and L_s stand for dimension and sequence length, respectively. In summary, the CFB module effectively capitalizes on the complementary information of the two modalities.

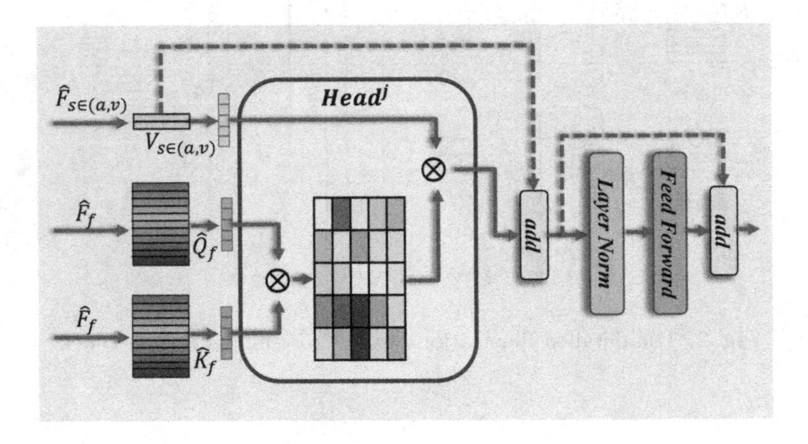

Fig. 3. The detailed illustration of the Multi-head Temporal Attention Block

2.4 Multi-head Temporal Attention Block

In order to account for the temporal impact of all modalities on a single modal branch and achieve adaptive cross-modal balancing, as shown in Fig. 3, we perform MTB on $\hat{F}_f$ and $\hat{F}_s$ to capture the long-range temporal dependency. Specifically, the MTB project the representation of all modalities $\hat{F}_f$ to $\hat{Q}_f$, $\hat{K}_f \in \mathbf{R}^{L_s \times d_f}$ in each unimodal branch and the unimodal representation $\hat{F}_s$ to $\hat{V}_s \in \mathbf{R}^{L_s \times d_f}$, where d_f and L_s stand for dimension and sequence length, respectively. The above process can be described as follows:

$$MultiHead\left(\hat{Q}_f, \hat{K}_f, \hat{V}_s\right) = \mathbf{Concat}\left(head^1, \ldots, head^h\right) \tag{6}$$

$$where \quad head^j = Attention\left(\hat{Q}_f^j, \hat{K}_f^j, \hat{V}_s^j\right) \tag{7}$$

$$\hat{Q}_f = \mathbf{Concat}\left(\hat{Q}_f^1 \ldots \hat{Q}_f^t \ldots \hat{Q}_f^{L_s}\right) W_f^{\hat{Q}} \tag{8}$$

$$\hat{K}_f = \mathbf{Concat}\left(\hat{K}_f^1 \ldots \hat{K}_f^t \ldots \hat{K}_f^{L_s}\right) W_f^{\hat{K}} \tag{9}$$

$$\hat{V}_s = \mathbf{Concat}\left(\hat{V}_s^1 \ldots \hat{V}_s^t \ldots \hat{V}_s^{L_s}\right) W_s^{\hat{V}} \tag{10}$$

where $j \in \{1, h\}$ indicates the j_{th} head of the total number of h heads and time step $t \in \{1, L_s\}$. Finally, the final BDI-II scores is output after a fully connected layer.

3 Experiments

3.1 Experiment Datasets

For the AVEC 2013 depression dataset, there are 150 videos from 82 subjects participating in different human-computer interaction (HCI) tasks. The length of the videos are about 20 to 50 min long (mean = 25 min). The age range for all participants in the dataset is 18 to 63 years old, with an average age is 31.5 years old and a standard deviation of 12.3 years. This depression dataset has been divided into three partitions by the publisher, i.e., training, development, and test set. Each partition has 50 videos, and each video has a label corresponding to its BDI-II score.

The AVEC 2014 depression dataset is a subset of the AVEC 2013 dataset. Only two different HCI tasks are involved, i.e., "FreeForm" and "Northwind", both of each have 150 videos. Specifically, in the "FreeForm" task, the subjects responded to a question about a sad childhood memory. In the "Northwind" task, the subjects were required to read an excerpt audibly from a fable. The same as AVEC 2013, it also has three partitions, i.e., training, development, and test sets. We perform experiments employing training and development sets from both tasks as training data, and the test sets are used to measure the model's performance.

3.2 Experiment Details and Evaluation Metrics

Pre-processing. In the AVEC 2013 and AVEC 2014 datasets, for the video modality, we use the machine learning toolkit DliB [20] to detect faces in all video samples, align by facial landmarks and resize the images to 224 × 224 RGB color channels. As mentioned before, we use Resnet-50 [16] pre-trained on the FER+ dataset [17] to extract frame-level features at a frame rate of 30 frames/second, then average feature of every 3 frames as the input of the model. For the audio modality, we segment each audio clip by the second, then we set the frame size and frame step to 1 s and 0.1 s, respectively. We extract the set of 25 low-level acoustic descriptors for each cropped audio clip and subseqsuently obtained their mean values.

Experiment Setup. The overall framework of CMAFN is shown in Fig. 1. To implement CMAFN, we leveraged the PyTorch framework [21] and trained the model on a local GPU server equipped with a TESLA-A100 GPU featuring 40GB of global memory. The optimizer is Adam with batch size 64 and learning rate is $3e-4$. The weight decay and eps of the Adam optimizer are set to $5e-4$ and $1e-8$, respectively. We use the Mean Absolute Error with the loss function. In addition, for the 1-D CNN used in the LTF module, we stack two layers with a kernel size of 3.

Evaluation Metrics. We offer several evaluation metrics to compare with prior approaches. For the AVEC 2013 and AVEC 2014 datasets, we utilize Mean Absolute Error (MAE) and Root Mean Square Error (RMSE) as the evaluation metrics during testing to ensure fair comparisons, which details are defined as:

$$MAE = \frac{1}{M}\sum_{j=1}^{M} |\hat{y}_j - p_j| \tag{11}$$

$$RMSE = \sqrt{\frac{1}{M}\sum_{j=1}^{M} (\hat{y}_j - p_j)^2} \tag{12}$$

where M is the total number of test samples, p_j and $\hat{y}_j$ are the ground truth and the predicted BDI-II score of the j_{th} subject, respectively.

4 Results and Analysis

4.1 Ablation Study

To verify the effectiveness of each component in CMAFN, we conduct ablation experiments on the AVEC2013 and AVEC2014 datasets. The results in Table 1 indicates that A performs the worst, which is attributed to the fact that A only employs the LTF module and a fusion strategy similar to feature concatenation. The experimental results of B and C indicate that the inclusion of the CFB module slightly improves the performance compared to using only the MTB module, as the CFB module captures complementary information within and across modalities. The combination of all modules achieves the best performance, as it builds on the effective feature representation obtained by the CFB module and considers the temporal influence of all modalities on a single modality branch using the MTB module, thus significantly enhancing the depression assessment capability of CMAFN.

In addition, we investigate the effectiveness of the self-attention mechanism and residual structure. The results from Table 1 shows that their performance is improved compared to the control group. Indicates that the self-attention structure can adaptively select depression-related features and perform multimodal interaction more efficiently. In the meantime, the residual structure can minimize the loss of video features during the interaction process.

Table 1. Ablation study of the individual components on the test set of AVEC2013 and AVEC2014

Combination	AVEC 2013		AVEC2014	
	MAE	RMSE	MAE	RMSE
A :LTF(Conv1D)	9.68	10.79	8.96	9.73
B :LTF+CFB	6.17	7.29	6.05	7.16
C :LTF+MTB	6.85	7.93	6.62	7.33
D :LTF+CFB+MTB (**Ours**)	5.26	6.60	5.16	6.04
w/o self-attention	5.42	6.63	5.37	6.09
w/o residual	5.34	6.57	5.20	6.17

4.2 Comparing Video-Audio and Audio-Video Fusion

In the model design, we fuse the audio modality with the video modality, denoted as A->V. On the contrary, integrating video modality into audio modality is denoted as V->A. The experimental results are shown in Table 2. We find that the performance of V->A is the best, achieving the smallest error in both the AVEC 2013 and AVEC 2014 datasets. We attribute this to the well-designed CFB module, which reduces the redundancy of audio features and produces results more effectively. Additionally, due to the rich spatio-temporal structure of the video modality, the multimodal representation can better capture depression-related information, which also impacts the final output results.

Table 2. Comparative Analysis of Audio-Video and Video-Audio Fusion Methods in a CMAFN

Method	AVEC 2013		AVEC2014	
	MAE	RMSE	MAE	RMSE
A->V	5.38	6.71	5.29	6.17
V->A	5.26	6.60	5.16	6.04

4.3 Compared with the State-of-the-Art Models

To prove the effectiveness of our CMAFN, we present the quantitative performance comparison results on AVEC 2013 and AVEC 2014 datasets in Table 3. We can get the MAE scores of 5.21 and 5.16, and the RMSE scores of 6.60 and 6.04.

Specifically, our approach achieves the best performance compared to single-modality-based methods, indicating that fusion feature representations from different modalities can effectively enhance the model's performance in depression recognition. Our proposed method also stands out among multi-modal fusion methods. On the AVEC 2013 dataset, CMAFN obtains the best precision, which reduces the MAE and RMSE of the best-performing method [22] by 0.12 and 0.41. On the AVEC 2014 dataset, the proposed CMAFN framework still achieves very promising results, with 6.04 for RMSE and 5.16 for MAE. The effectiveness of the proposed method can be seen from the performance of the comparison experiments. This is because CMAFN can fully leverage the complementarity between the two modalities, reduce redundancy in feature information to some extent through attention mechanisms, and thus enhance the model's overall performance.

Table 3. Comparison of our method and state-of-the-art approaches on the test set of the two datasets. "A" and "V" represented audio and video modalities. "$A+V$" is the fusion of audio and video modalities.

Modalities	Methods	AVEC2013		AVEC2014	
		RMSE	MAE	RMSE	MAE
A	Cummins et al. [23]	8.16	/	/	/
	He et al. [2]	10.00	8.20	9.99	8.19
	Niu et al. [24]	9.50	7.14	9.66	8.02
	Niu et. al. [25]	9.79	7.48	9.25	7.86
	Zhao et al. [26]	9.65	7.38	9.57	7.94
V	Zhu et al. [4]	9.82	7.58	9.55	7.74
	Al Jazaery et al. [5]	9.28	7.37	9.20	7.22
	Melo et al. [27]	8.25	6.30	8.23	6.13
	Zhou et al. [28]	8.28	6.20	8.39	6.21
	He et al. [29]	8.39	6.59	8.30	6.51
	Uddin et al. [30]	8.93	7.04	8.78	6.86
	He et al. [31]	9.37	7.02	9.24	6.95
	Liu et al. [32]	7.59	6.08	7.98	6.04
	Li et al. [33]	7.38	6.05	7.60	6.01
$A+V$	Cummins et al. [15]	10.62	/	/	/
	Jan et al. [12]	/	/	7.43	6.14
	Meng et al. [14]	10.96	8.72	10.82	8.99
	Kaya et al. [34]	9.44	7.68	9.61	7.69
	Niu et al. [25]	8.16	6.14	7.03	5.21
	Uddin et al. [22]	6.83	5.38	6.16	5.03
	Ours	**6.60**	**5.26**	**6.04**	**5.16**

4.4 Effectiveness of Different Head Numbers

We further explore the best performance of the CMAFN model by using different numbers of attention heads h. Figure 1 shows the experimental results of CMAFN-h ($h = 1, 2, 4, 6$) on the AVEC2013 and AVEC2014 datasets, where Fig. 4 (a) and (b) represent the experimental results on the AVEC2013 and AVEC2014 datasets, respectively. The results indicate that CMAFN-4 achieved the best performance on both datasets. Specifically, compared with CMAFN-1 with a single attention head, CMAFN-4 reduced the RMSE and MAE by 8.46% and 13.63%, respectively, on the AVEC2013 dataset. Similarly, on the AVEC2014 dataset, CMAFN-4 reduced the RMSE and MAE by 8.07% and 13.71%, respectively, compared with CMAFN-1. Furthermore, we observed that the error increases as h increases, which may be due to overfitting caused by the small dataset.

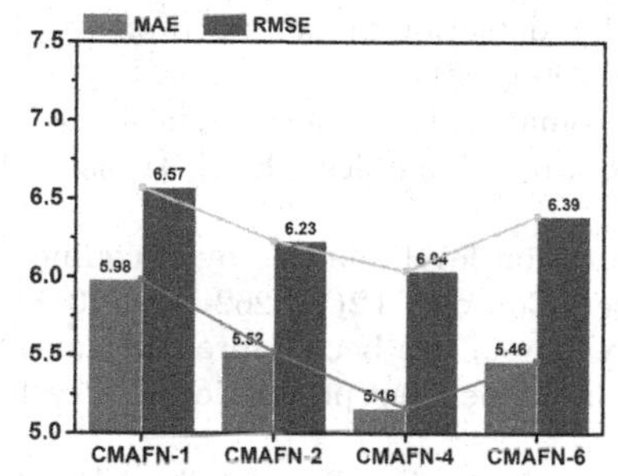

(a) The results on AVEC 2013 dataset.

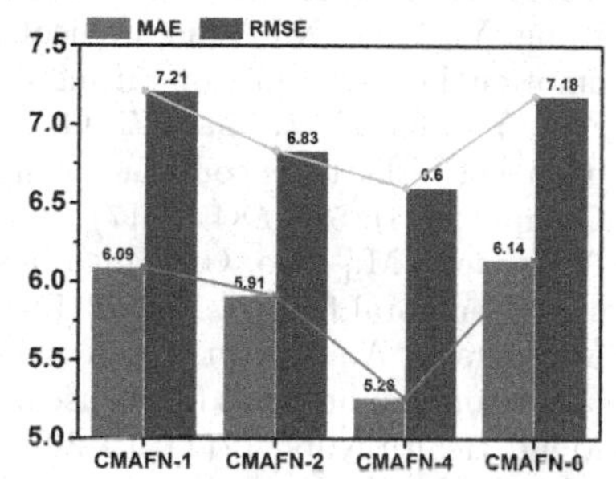

(b) The results on AVEC 2014 dataset.

Fig. 4. Recognition results of different head numbers of the CMAFN model on the AVEC 2013 (a) and AVEC 2014 (b) datasets

5 Conclusion

Physiological studies have revealed that depressive and healthy individuals can simultaneously observe changes in temporal affective state from audio and visual cues. Based on the facts, we propose a visual-audio cross-modal fusion network for automatic depression scale prediction. Firstly, we employ the LTF module to perform feature preprocessing on each modality separately and extract their temporal features. Secondly, our designed CFB module is employed for complementary learning across different modalities. CFB ensures the efficiency and integrity of inter-modal information interaction through self-attention mechanism and residual connection. Finally, we utilize the MTB module to consider the temporal influence of all modalities on each unimodal branch and gradually refine the interplay between modalities. We conduct extensive experiments on two publicly available depression datasets, AVEC 2013 and AVEC 2014, and the results demonstrate the superiority of our method.

In future work, we will focus on introducing more modalities, such as text and body postures, to improve the detection accuracy of depression assessment.

Acknowledgments. This work was supported in part by the National Key Research and Development Program of China (Grant No. 2019YFA0706200), in part by the National Natural Science Foundation of China (Grant No. 62227807, No. 62372217).

References

1. American Psychiatric Association, A., Association, A.P., et al.: Diagnostic and statistical manual of mental disorders: DSM-5, vol. 10. Washington, DC: American psychiatric association (2013)
2. He, L., Cao, C.: Automated depression analysis using convolutional neural networks from speech. J. Biomed. Inform. **83**, 103–111 (2018)
3. Dong, Y., Yang, X.: A hierarchical depression detection model based on vocal and emotional cues. Neurocomputing **441**, 279–290 (2021)
4. Zhu, Y., Shang, Y., Shao, Z., Guo, G.: Automated depression diagnosis based on deep networks to encode facial appearance and dynamics. IEEE Trans. Affect. Comput. **9**(4), 578–584 (2017)
5. Al Jazaery, M., Guo, G.: Video-based depression level analysis by encoding deep spatiotemporal features. IEEE Trans. Affect. Comput. **12**(1), 262–268 (2018)
6. McPherson, A., Martin, C.: A narrative review of the beck depression inventory (BDI) and implications for its use in an alcohol-dependent population. J. Psychiatr. Ment. Health Nurs. **17**(1), 19–30 (2010)
7. Wen, L., Li, X., Guo, G., Zhu, Y.: Automated depression diagnosis based on facial dynamic analysis and sparse coding. IEEE Trans. Inf. Forensics Secur. **10**(7), 1432–1441 (2015)
8. Stasak, B., Joachim, D., Epps, J.: Breaking age barriers with automatic voice-based depression detection. IEEE Pervasive Comput. (2022)
9. He, L., et al.: Deep learning for depression recognition with audiovisual cues: a review. Inf. Fusion **80**, 56–86 (2022)
10. Dubagunta, S.P., Vlasenko, B., Doss, M.M.: Learning voice source related information for depression detection. In: ICASSP 2019–2019 IEEE International Conference on Acoustics, Speech and Signal Processing (ICASSP), pp. 6525–6529. IEEE (2019)
11. Haque, A., Guo, M., Miner, A.S., Fei-Fei, L.: Measuring depression symptom severity from spoken language and 3d facial expressions. arXiv preprint arXiv:1811.08592 (2018)
12. Jan, A., Meng, H., Gaus, Y.F.B.A., Zhang, F.: Artificial intelligent system for automatic depression level analysis through visual and vocal expressions. IEEE Trans. Cogn. Dev. Syst. **10**(3), 668–680 (2017)
13. He, L., Jiang, D., Sahli, H.: Multimodal depression recognition with dynamic visual and audio cues. In: 2015 International Conference on Affective Computing and Intelligent Interaction (ACII), pp. 260–266. IEEE (2015)
14. Meng, H., Huang, D., Wang, H., Yang, H., Ai-Shuraifi, M., Wang, Y.: Depression recognition based on dynamic facial and vocal expression features using partial least square regression. In: Proceedings of the 3rd ACM International Workshop on Audio/Visual Emotion Challenge, pp. 21–30 (2013)

15. Cummins, N., Joshi, J., Dhall, A., Sethu, V., Goecke, R., Epps, J.: Diagnosis of depression by behavioural signals: a multimodal approach. In: Proceedings of the 3rd ACM International Workshop on Audio/Visual Emotion Challenge, pp. 11–20 (2013)
16. He, K., Zhang, X., Ren, S., Sun, J.: Deep residual learning for image recognition. In: Proceedings of the IEEE Conference on Computer Vision and Pattern Recognition, pp. 770–778 (2016)
17. Barsoum, E., Zhang, C., Ferrer, C.C., Zhang, Z.: Training deep networks for facial expression recognition with crowd-sourced label distribution. In: Proceedings of the 18th ACM International Conference on Multimodal Interaction, pp. 279–283 (2016)
18. Eyben, F., et al.: The Geneva minimalistic acoustic parameter set (GeMAPS) for voice research and affective computing. IEEE Trans. Affect. Comput. **7**(2), 190–202 (2015)
19. Bai, S., Kolter, J.Z., Koltun, V.: An empirical evaluation of generic convolutional and recurrent networks for sequence modeling. arXiv preprint arXiv:1803.01271 (2018)
20. King, D.E.: Dlib-ml: a machine learning toolkit. J. Mach. Learn. Res. **10**, 1755–1758 (2009)
21. Stevens, E., Antiga, L., Viehmann, T.: Deep Learning with PyTorch. Manning Publications (2020)
22. Uddin, M.A., Joolee, J.B., Sohn, K.A.: Deep multi-modal network based automated depression severity estimation. IEEE Trans. Affect. Comput. (2022)
23. Cummins, N., Sethu, V., Epps, J., Williamson, J.R., Quatieri, T.F., Krajewski, J.: Generalized two-stage rank regression framework for depression score prediction from speech. IEEE Trans. Affect. Comput. **11**(2), 272–283 (2017)
24. Niu, M., Tao, J., Liu, B., Fan, C.: Automatic depression level detection via lp-Norm pooling. In: Proceedings of the INTERSPEECH, Graz, Austria, pp. 4559–4563 (2019)
25. Niu, M., Tao, J., Liu, B., Huang, J., Lian, Z.: Multimodal spatiotemporal representation for automatic depression level detection. IEEE Trans. Affect. Comput. (2020)
26. Zhao, Z., Li, Q., Cummins, N., Liu, B., Wang, H., Tao, J., Schuller, B.: Hybrid network feature extraction for depression assessment from speech. In: Proceeding of the INTERSPEECH, Shanghai, China, pp. 4956–4960 (2020)
27. De Melo, W.C., Granger, E., Hadid, A.: Depression detection based on deep distribution learning. In: 2019 IEEE International Conference on Image Processing (ICIP), pp. 4544–4548. IEEE (2019)
28. Zhou, X., Jin, K., Shang, Y., Guo, G.: Visually interpretable representation learning for depression recognition from facial images. IEEE Trans. Affect. Comput. **11**(3), 542–552 (2018)
29. He, L., Chan, J.C.W., Wang, Z.: Automatic depression recognition using CNN with attention mechanism from videos. Neurocomputing **422**, 165–175 (2021)
30. Uddin, M.A., Joolee, J.B., Lee, Y.K.: Depression level prediction using deep spatiotemporal features and multilayer Bi-LTSM. IEEE Trans. Affect. Comput. **13**(2), 864–870 (2020)
31. He, L., Tiwari, P., Lv, C., Wu, W., Guo, L.: Reducing noisy annotations for depression estimation from facial images. Neural Netw. **153**, 120–129 (2022)
32. Liu, Z., Yuan, X., Li, Y., Shangguan, Z., Zhou, L., Hu, B.: PRA-Net: part-and-relation attention network for depression recognition from facial expression. Comput. Biol. Med., 106589 (2023)

33. Li, Y., et al.: A facial depression recognition method based on hybrid multi-head cross attention network. Front. Neurosci. **17**, 1188434 (2023)
34. Kaya, H., Çilli, F., Salah, A.A.: Ensemble CCA for continuous emotion prediction. In: Proceedings of the 4th International Workshop on Audio/Visual Emotion Challenge, pp. 19–26 (2014)

AU-Oriented Expression Decomposition Learning for Facial Expression Recognition

Zehao Lin, Jiahui She, and Qiu Shen(✉)

Nanjing University, Nanjing, Jiangsu Province, China
shenqiu@nju.edu.cn

Abstract. Facial Expression Recognition (FER) has received extensive attention in recent years. Due to the strong similarity between expressions, it is urgent to distinguish them meticulously in a finer-grained manner. In this paper, we propose a method, named **A**U-oriented **E**xpression **D**ecomposition **L**earning (AEDL), which aims to decouple expressions into Action Units (AUs) and focuses on subtle facial differences. In particular, AEDL comprises two branches: the AU Auxiliary (AUA) branch and the FER branch. For the former, the generic knowledge of dependencies among AUs is leveraged to supervise AU predictions which are then transformed into new expression predictions with a learnable matrix modeled by the relationship between AUs and expressions. For the latter, fusion features are employed to compensate for the minority classes to ensure adequate feature learning. FER predictions are guided by the AUA branch, mining detailed distinctions between expressions. Importantly, the proposed method is independent of the backbone network and brings no extra burden on inference. We conduct experiments on popular in-the-wild datasets and achieve leading performance, proving the effectiveness of the proposed AEDL.

Keywords: Facial expression recognition · Action units · Expression decomposition

1 Introduction

Facial expression is a crucial nonverbal signal in human communication that reflects underlying emotions. Facial Expression Recognition (FER) has emerged as an important research area, with applications ranging from health treatment to criminal investigation, owing to the authenticity and particularity of facial expressions. In recent years, deep learning representations have replaced traditional handcrafted features as the mainstream choice for FER, and have achieved remarkable results on lab-controlled datasets such as CK+ [12], MMI [17] and Oulu-CASIA [24]. However, there is still room for improvement on in-the-wild datasets (*e.g.,* RAF-DB [10], FERPlus [1] and AffectNet [13]), which pose significant challenges, including occlusion and pose variations.

Supported by the National Natural Science Foundation of China (Grant # 62071216, 62231002 and U1936202.

Q. Liu et al. (Eds.): PRCV 2023, LNCS 14429, pp. 265–277, 2024.
https://doi.org/10.1007/978-981-99-8469-5_21

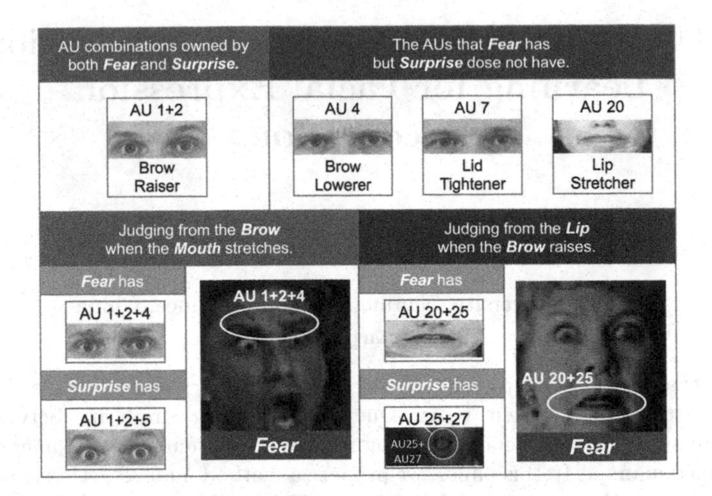

Fig. 1. Common and exclusive AUs for Fear and Surprise. In different cases, they can be accurately distinguished by the different AUs they possess.

Currently, FER encounters a challenge with regard to the strong similarity between expressions. Simply using one-hot labels is insufficient to distinguish similar expressions. Therefore, it is vital to distinguish them in a finer-grained form, and action units (AUs) can act this crucial role. In the Facial Action Coding System (FACS) [4], each expression can be decomposed into a combination of multiple AUs, with each AU representing the muscle movement of a specific part of the face. Based on this, as shown in Fig. 1, in order to distinguish the similar expression pair of fear and surprise, we observe that fear has certain AUs that surprise does not have, such as *Brow Lowerer*, *Lid Tightener* and *Lip Stretcher*. These AUs can be utilized to distinguish the two expressions easily. For instance, we can tell Fear by the *Brow Lowerer* when the mouth stretches in the expression. Besides, we can distinguish Fear by the *Lip Stretcher* when the eyebrows raises in the expression.

To address this issue, several work [3,14] have considered action units as an auxiliary and leverage general knowledge about faces. For instance, in [3], generic knowledge was integrated into the Bayesian network to model the relationship between AUs and expressions. AUE-CRL [14] converted expression labels into AU pseudo labels with the help of prior knowledge for FER. The biggest difference in our work is that we benefit from knowledge distillation [7] and transform AU predictions into FER predictions directly via learned prior knowledge as new supervised information, which is intuitive and effective.

In this paper, we propose a method, named **A**U-oriented **E**xpression **D**ecomposition **L**earning (AEDL), consisting of two branches: the FER branch and the AU auxiliary (AUA) branch. Specifically, in the AUA branch, the prior knowledge of AU positive and negative pairs is utilized to boost the performance of AU detection while the relationship between AUs and expression is modeled as a learnable matrix for the conversion between AU and expression prediction for the further guidance to the FER branch. In the FER branch, in order to compensate for the features of the minority classes, we utilize fusion feature to alleviate the imbalance in the FER datasets. Note that AUA branch is removed during inference as the guidance information has already been fully learned by the FER branch after training, so the proposed method does not bring additional cost on inference. To sum up, the contributions of the proposed method are as follows:

1. We introduce a method named AEDL that can decouple expressions into AUs and convert AU predictions into expression predictions as new supervisory information with a learnable matrix, which looks at the problem of high similarity between expressions from a finer-grained perspective.
2. Our AUA branch can model the generic knowledge between AUs and the relationship between expressions and AUs, while our FER branch can utilize fusion features to compensate for the minority classes.
3. The proposed method achieves leading performance on RAF-DB and AffectNet without bringing extra burden on inference.

2 Related Work

State-of-the-art deep learning-based FER method related to our method can be mainly divided into two branches: expression decoupling [15,19] and face-related task assistance [2,3,14].

Expression Decoupling. Because of the strong similarity between expressions, how to extract fine-grained features to distinguish the small differences between expressions is crucial. FDRL [15] decoupled expression images into a series of action-aware features and assigned weights for intra-feature and inter-feature relation. Then reconstructed them into expression features to predict the label. Besides, TransFER [19] first applied transformer for FER because transformer decoupled the expression images into patches and explored the relationship by self-attention. Different from the aforementioned work, our proposed method decouples expressions into fine-grained AUs to further learn the local features. Then we convert them into expression predictions as supervised information with the learnable transformation matrix with no burden on inference.

Face-Related Task Assistance. Due to the high consistency of face-related tasks, landmark detection [14] and action units detection [3,14] often assist FER. Cui *et al.* [3] encoded the prior knowledge between AUs and expressions into the Bayesian network. AUE-CRL [14] utilized prior knowledge to craft AU pseudo labels and select more useful AU information. JMPL [25] utilized correlation between AUs to increase the sample size and improve the performance with multi-label learning. For the proposed method, we exploit the prior knowledge between AU and expression to transform AU prediction into expression prediction, guiding the training of FER branch. Not only that, the postive and negative pair information is utilized in the AUA branch, while the shared compensation information is inserted in the FER branch.

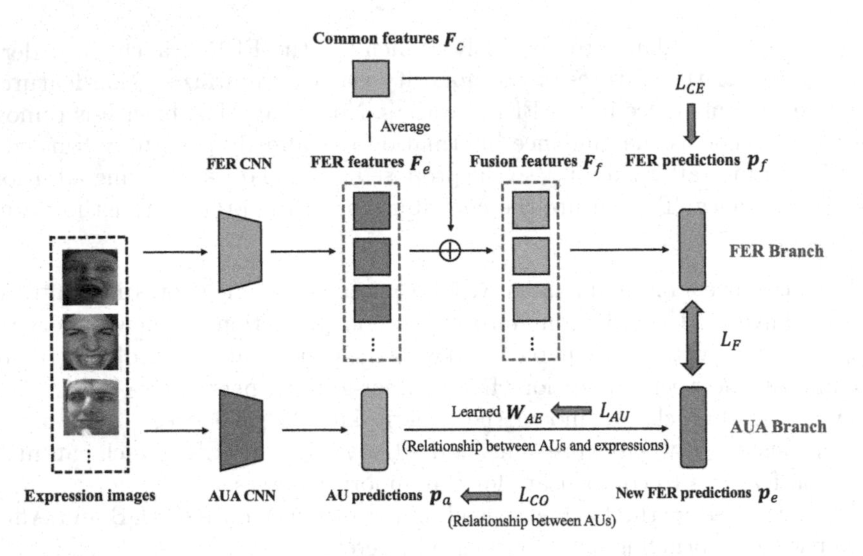

Fig. 2. The overview of AEDL. Firstly, AUA CNN and FER CNN are pre-trained on MS-Celeb1M. Secondly, AUA CNN is fine-tuned on BP4D [22] under the supervision of L_{CO}. Finally, the model is trained with the overall loss function.

3 Methods

3.1 Overview

As shown in Fig. 1, similar expression pairs can be decoupled into corresponding AUs from a finer-grained perspective and then classified more accurately with specific AUs. In order to realize this idea, we proposed a method named AEDL which has two branches: the AUA branch and the FER branch (see in Fig. 2). For the former, AU features are firstly extracted by AUA Convolutional Neural Networks (CNN) and then AU predictions $\mathbf{p}_a$ are obtained under the supervision of L_{CO}. Finally, AU predictions are transformed into FER predictions $\mathbf{p}_e$ with the learnable matrix $\mathbf{W}_{AE}$, so as to provide more professional guidance for FER branch. For the latter, a batch of expression images is fed into the FER CNN and FER features $\mathbf{F}_e$ are obtained. In order to reduce the impact of insufficient feature learning of minority classes, a batch of FER features is averaged as common features $\mathbf{F}_c$, and added them to each FER feature element-wisely to obtain fusion features $\mathbf{F}_f$ for classification. Finally, expression images are classified under the joint supervision of L_{AU} and L_{CE}.

3.2 The AUA Branch

Since facial expressions represent human emotions, previous psychological research is well instructive. As [4,23] argues, there is a strong correlation among AUs, so the relationship between AUs is defined as postive pairs and negative pairs. For example, as show in Fig. 1, *Inner Brow Rasier* and *Outer Brow Rasier*

Table 1. The relationship between expressions and AUs in FACS [4]. A and B represent the primary and secondary AU, respectively.

AU	1	2	4	5	6	7	9	10	12	15	16	17	20	23	24	25	26
anger			A	A		B						B		A	A		
disgust							A	A				B				B	
fear	A	A	A	A		A							A	B		B	
happiness					A	B			A							B	
sadness	A		B		B	A				A		B					
surprise	A	A		A							B					A	A

both indicate the orientation of the eyebrows. They often appear in the expression at the same time, so they are regarded as postive pairs. However, *Lip Corner Puller* and *Lip Corner Depressor* are considered as negative pair owing to the completely opposite direction of lip. To sum up, the probability of positive pairs occurring at the same time is high, while the probability of negative pairs occurring at the same time is extremely small.

For AU i, if p_{a_i} of AU predictions $\mathbf{p}_a$ is greater than 0.5, it means that AU i exists, which can be formulated by $p_{a_{i=1}}$. On the contrary, if AU i dose not exist, it can be expressed by $p_{a_{i=0}}$. For AU i and j, the relationship of positive pair can be written as:

$$p_{a_{i=1|j=1}} > p_{a_{i=0|j=1}} \tag{1}$$

$$p_{a_{i=1|j=1}} > p_{a_{i=1|j=0}} \tag{2}$$

The equivalent formulations are:

$$p_{a_{i=1,j=1}} > p_{a_{i=0,j=1}} \tag{3}$$

$$p_{a_{i=1,j=1}} > p_{a_{i=1}} p_{a_{j=1}} \tag{4}$$

Inspired by [14], the above relationship can be modeled as a regularization term:

$$\begin{aligned} L_P = & \sum_{i,j \in S_P} max(p_{a_{i=1}} p_{a_{j=1}} - p_{a_{i=1,j=1}}, 0) \\ & + \sum_{i,j \in S_P} max(p_{a_{i=1,j=0}} - p_{a_{i=1,j=1}}, 0) \\ & + \sum_{i,j \in S_P} max(p_{a_{i=0,j=1}} - p_{a_{i=1,j=1}}, 0) \end{aligned} \tag{5}$$

where S_P is the set of positive AU pairs. Similarly, the relationship for AU negative pairs can be formulated as:

$$\begin{aligned} L_N = & \sum_{i,j \in S_N} max(p_{a_{i=1,j=1}} - p_{a_{i=1}} p_{a_{j=1}}, 0) \\ & + \sum_{i,j \in S_N} max(p_{a_{i=1,j=1}} - p_{a_{i=1,j=0}}, 0) \\ & + \sum_{i,j \in S_N} max(p_{a_{i=1,j=1}} - p_{a_{i=0,j=1}}, 0) \end{aligned} \tag{6}$$

Table 2. Accuracy(%) on RAF-DB with different backbone architectures.

	ResNet-18	ResNet-50IBN	ShuffleNetV1(group = 3,2.0×)	MobileNetV2
Baseline	86.33	86.57	86.20	86.01
AEDL	89.15	89.78	88.32	88.23

where S_N is the set of negative AU pairs. The loss function of AUA branch is:

$$L_{CO} = L_P + L_N \tag{7}$$

Besides, in order to look at the expressions from a more fine-grained perspective, we followed the idea in [4] and decoupled the expressions into the combination of AUs. The relationship between expressions and AUs is shown in the Table 1. Each expression has its corresponding primary AU and secondary AU. The primary AU represents the most likely AU to appear when annotated with a specific expression, while the secondary AU is the second most likely to appear. Based on this general knowledge, we initialize a matrix $\tilde{\mathbf{W}}_{AE}$ representing the relationship between AUs and expressions. To get more accurate FER predictions, we define a loss function L_{AU} to guide the learning of the matrix $\mathbf{W}_{AE}$.

$$L_{AU} = \left\| \mathbf{W}_{AE} - \tilde{\mathbf{W}}_{AE} \right\|_2^2 \tag{8}$$

The AU predictions are transformed into the new instructive FER predictions $\mathbf{p}_e$ for FER branch by the learned matrix.

$$\mathbf{p}_e = \mathbf{p}_a \mathbf{W}_{AE} \tag{9}$$

3.3 The FER Branch

The phenomenon of class imbalance in FER has been shown in [11]. Inspired by [20], we let the majority classes compensate the features of the minority classes, allowing more sufficient feature learning. Specifically, given a batch of expression images, we fed them into pre-trained FER CNN and obtain FER feature $\mathbf{F}_e$. In order to compensate the minority classes, common feature $\mathbf{F}_c$ is the average of FER feature $\mathbf{F}_e$ and then adds $\mathbf{F}_e$ element-wisely to form fusion feature $\mathbf{F}_f$ for classification.

$$\mathbf{F}_c = \frac{1}{B} \sum_{i=0}^{B} \mathbf{F}_{e_i} \tag{10}$$

$$\mathbf{F}_f = \mathbf{F}_e \oplus \mathbf{F}_c \tag{11}$$

where B denotes the number of a batch, $\oplus$ denotes the element-wise addition. In this way, although it is an expression image of minority class, it also contains some features of the majority class, which can alleviate the impact of class imbalance to a certain extent.

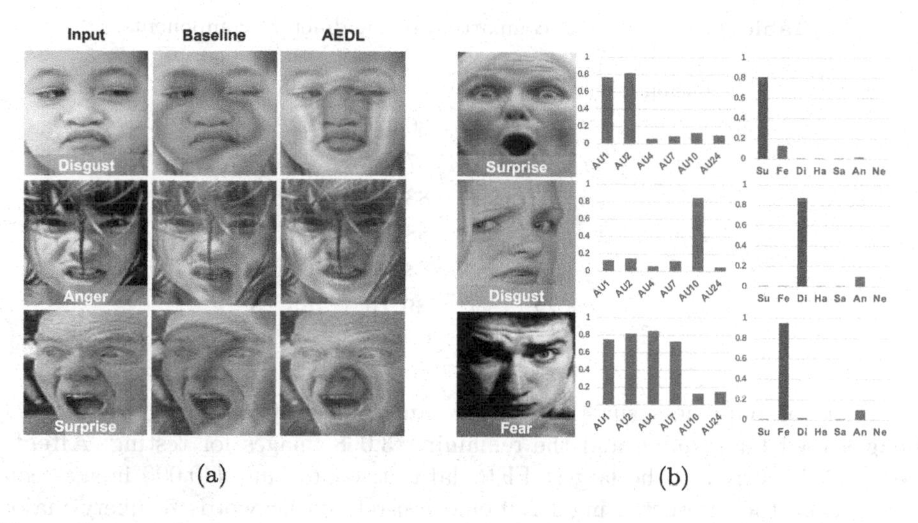

Fig. 3. Text at the bottom of the image is dataset annotation. (a) Attentiton visualization of Baseline and AEDL . (b) The AU and FER predictions of AEDL.

Now that we have obtained the instructive new FER prediction $\mathbf{p}_e$ of AUA branch, inspired by knowledge distillation, we define a loss function L_F as a bridge to connect the two branches:

$$L_F = KL(\mathbf{p}_e^{1/T}, \mathbf{p}_f^{1/T}) \tag{12}$$

where $\mathbf{p}_f$ denotes the prediction of FER branch, KL denotes Kullback-Leibler divergence. T is the temperature factor following with [7].

3.4 Training Strategy

We adopt a three-step training strategy. In the first step, we pre-train FER CNN and AUA CNN on MS-Celeb1M face recognition dataset. In the second step, AUA CNN is fine-tuned on BP4D [22] under the supervision of L_{CO} and binary cross entropy loss function for more detailed AU features extraction. The last step is to train model with the overall loss function of the proposed method:

$$L = L_{CE} + \lambda_1 L_F + \lambda_2 L_{AU} \tag{13}$$

where L_{CE} denotes cross entropy loss function between expression labels and FER prediction $\mathbf{p}_f$, λ_1 and λ_2 both denote the hyper parameters.

Note that the AUA branch is removed during inference cause the FER branch is fully guided by the AUA branch after training, so the proposed AEDL brings no extra burden on inference.

4 Experiments

4.1 Datasets

RAF-DB [10] consists of 30,000 facial images with basic or compound annotations. For our experiment, we selected the images with seven basic expressions

Table 3. Accuracy (%) comparison of the different components.

Fusion feature	L_{AU}	L_F	RAF-DB	AffectNet
-	-	-	86.33	62.34
✓	-	-	87.09	62.76
-	-	✓	88.07	63.52
✓	-	✓	88.43	64.57
-	✓	✓	88.72	64.68
✓	✓	✓	**89.15**	**65.23**

(i.e., neutral, happiness, surprise, sadness, anger, disgust, and fear), with 12,271 images used for training and the remaining 3,068 images for testing. **AffectNet** [13] is currently the largest FER dataset, containing 440,000 images collected from the internet using 1,250 emotion-related keywords to query major search engines. As with previous studies, we chose around 280,000 images for training and 3,500 images for testing. These images all depict one of seven basic expressions. **BP4D** [22] comprises 328 videos of 41 participants (23 females and 18 males) who engaged in eight spontaneous expression sessions. The videos contain a total of 140,000 frames, all of which are annotated with 12 AUs.

4.2 Implementation Details

The default configuration for FER CNN and AUA CNN is to use ResNet18 as the backbone network and we pre-train them on the MS-Celeb-1M face recognition dataset. Besides, AUA CNN is then fine-tuned on BP4D for more detailed AU features extraction following with [14]. Prior to training, facial images are cropped and aligned, and then resized to 256×256 pixels. Random cropping and horizontal flipping are applied to images that are resized to 224×224 pixels. We use a single Nvidia Tesla P40 GPU to train our model, with a batch size of 72. Adam optimization algorithm [9] is deployed with a weight decay of 10^{-4} as the optimizer. Initially, the learning rate is set to 10^{-3}, and then it is divided by 10 after 10 and 20 epochs. The training process lasts for 40 epochs. $\tilde{\mathbf{W}}_{AE}$ is initialized according to Table 1 where A represents 0.9, B represents 0.5 and 0.1 otherwise.

4.3 Performance Evaluation

Evaluation on different backbone networks. In order to verify the feasibility of the proposed AEDL, we apply it in different backbone networks. For ResNet-18, ResNet-50IBN, ShuffleNetV2 and MobileNetV1, our method improves the accuracy by 2.82%, 3.21%, 2.12% and 2.22% respectively compared with the baseline in Table 2. The lightweight MobileNet achieves an impressive accuracy of 88.23%, highlighting the potential of AEDL for practical deployment. Notably, the performance enhancement achieved by AEDL incurs no additional costs on inference and is independent of the backbone network. Our AUA branch can potentially be further embedded into existing methods, providing performance improvements without extra burdens on inference.

Table 4. Accuracy (%) comparison to SOTA methods. † denotes our reproduce results. MAcc. is the abbreviation for mean accuracy.

Method	Backbone	RAF-DB		AffectNet
		Acc.	MAcc.	Acc.
Meta-Face2Exp [21]	ResNet-18	88.54	–	64.23
SCN [18]	ResNet-18	87.03	78.09†	63.40†
DACL [5]	ResNet-18	87.78	80.44	65.20
MA-Net [26]	ResNet-18	88.40	79.73	64.53
DMUE [16]	ResNet-18	88.76	–	–
AUE-CRL [14]	ResNet-101	–	81.00	–
RUL [14]	ResNet-18	88.98	–	- -
FENN [6]	ResNet-18	88.91	–	- -
IPD-FER [8]	ResNet-18	88.89	79.28	62.23
FDRL [15]	ResNet-18	**89.47**	–	–
Ours (AEDL)	ResNet-18	89.15	**81.67**	**65.23**

Attention Visualization of Baseline and AEDL. As shown in Fig. 3a, in comparison to the baseline, AEDL prioritizes local features of critical facial components. For the image labeled as disgust, our method emphasizes AU10 (*Upper Lip Raiser*). Besides, for images labeled as angry, the AEDL primarily focuses on AU4 (*Brow Lowerer*) and AU22 (*Lip Funneler*) while for images labeled as surprise, AU27 (*Mouth Stretch*) receives more attention. These local features correspond to the most likely AUs associated with the respective expressions.

Important AU and FER Predictions of AEDL. We selected several images on RAF-DB, and provided predictions of important AUs and expressions. As shown in Fig. 3b, for the image in the first row, it is challenging to differentiate between surprise and fear. However, by leveraging the distinct AU4 and AU7 predictions, which are different to surprise and fear, the proposed approach successfully predicts the surprise. Similarly, we observe the results of the image in the second row that the AU10 prediction specific to disgust is high. Thus, leveraging the guidance provided by the AUA branch, the FER branch accurately predicts the disgust expression. For the image in the third row, the corresponding AU prediction values are also relatively high so fear is predicted by AEDL.

4.4 Comparison with State-of-the-Arts

In this work, we propose a method that decouples facial expressions into AUs and leverages face-related prior knowledge to enhance the accuracy of expression predictions. To comprehensively demonstrate the superior performance of our AEDL, we use two metrics. Weighted accuracy is most commonly used metric, which is the ratio of correctly classified samples to the total number of samples in the test set while mean accuracy calculates the average accuracy for each category owing to class imbalance in FER datasets.

Table 5. The classification accuracy (%) with different hyper parameters of λ_1, λ_2 and T on RAF-DB.

(a) Influence of λ_1.

λ_1	RAF-DB
0	88.09
0.5	88.73
1.0	**89.15**
1.5	88.62
2.0	88.17

(b) Influence of λ_2.

λ_2	RAF-DB
0	88.36
0.2	**89.15**
0.4	88.85
0.6	88.67
0.8	88.44

(c) Influence of T.

T	RAF-DB
1.0	88.47
1.2	88.65
1.4	88.93
1.6	**89.15**
1.8	88.87

As shown in Table 4, it is worth mentioning that AEDL achieved leading results of 89.15% on RAF-DB and 65.23% on AffectNet. For methods that focus on local changes in face, our method exceeds the result of MA-Net [26] by 0.75% and 0.7% on RAF-DB and AffectNet, respectively. For mean accuracy, our method improves the accuracy by 0.67% over AUE-CRL [14] even though its backbone is ResNet-101 while ours is ResNet-18. For FDRL [15], which also decouples expressions, each module involves the complex process of decomposing the main branch into multiple sub-branches, training them separately, and then merging them back into the main branch. Moreover, these complex operations cannot be eliminated during the inference. In contrast, our method achieves competitive performance results but introduces no additional inference burden, making it possible to achieve lightweight deployment.

4.5 Ablation Studies

Component Analysis. Note that the presence of L_F is necessary in the component analysis. As shown in Table 3, we observed and concluded the following points: (1) When L_F exists in the method as a single component, it can bring 1.74% and 1.18% improvement to RAF-DB and AffectNet, respectively, compared to the baseline. It can prove that L_F is the key to the proposed method and is the bridge connecting AUA branch and FER branch. Besides, decoupling expressions into finer-grained AUs and regarding them as the basis for classification guidance is effective. (2) Although L_{AU} only acts on AUA branch, with the cooperation of L_{AU} and L_F, more accurate AU decoupling results can be obtained, thus bringing more detailed guidance to FER branch. The joint of L_{AU} achieves the improvement by 0.65% and 1.16% on RAF-DB and AffectNet, respectively. (3) The fusion feature is the main trick of the FER branch. It provides more sufficient feature learning for the minority classes which can bring about 0.5% improvement in accuracy.

Trade-off Weight. λ_1 controls the degree to which the AUA branch directs FER branch. Table 5a shows that when λ_1 gradually increases, FER branch can learn finer-grained features. But when it is too large, the effect of AUA branch is too strong and FER branch loses its dominance.

Trade-off Weight. λ_2 is the relevant hyper parameter of learnable matrix $\mathbf{W}_{AE}$, which obtains more accurate expression predictions with fine-grained AU features. Table 5b shows the performance with different λ_2.

Sharpen Temperature. T allows AUA branch to provide smoother guidance to FER branch, which can suppress the sensitivity of the model to incorrect predictions. But when T is too large, the guiding of AUA branch is weakened.

5 Conclusion

To distinguish expressions from finer-grained perspective, we proposed AEDL, a method that decouples expressions into AUs and pays more attention on subtle facial changes. The AEDL approach includes two branches, the AUA branch and the FER branch, that leverage generic knowledge of AUs and expressions to supervise AU predictions and transform them into new expression predictions. Additionally, we use fusion features to compensate for minority classes and ensure adequate feature learning. Our experiments on in-the-wild datasets show the effectiveness of AEDL without adding extra burden to inference.

References

1. Barsoum, E., Zhang, C., Ferrer, C.C., Zhang, Z.: Training deep networks for facial expression recognition with crowd-sourced label distribution. In: Proceedings of the 18th ACM International Conference on Multimodal Interaction, pp. 279–283 (2016)
2. Chen, D., Mei, J.P., Wang, C., Feng, Y., Chen, C.: Online knowledge distillation with diverse peers. In: Proceedings of the AAAI Conference on Artificial Intelligence, vol. 34, pp. 3430–3437 (2020)
3. Cui, Z., Song, T., Wang, Y., Ji, Q.: Knowledge augmented deep neural networks for joint facial expression and action unit recognition. Adv. Neural. Inf. Process. Syst. **33**, 14338–14349 (2020)
4. Ekman, P., Friesen, W.V.: Facial action coding system. Environ. Psychol. Nonverbal Behav. (1978)
5. Farzaneh, A.H., Qi, X.: Facial expression recognition in the wild via deep attentive center loss. In: Proceedings of the IEEE/CVF Winter Conference on Applications of Computer Vision, pp. 2402–2411 (2021)
6. Gu, Y., Yan, H., Zhang, X., Wang, Y., Ji, Y., Ren, F.: Towards facial expression recognition in the wild via noise-tolerant network. IEEE Trans. Circ. Syst. Video Technol. (2022)

7. Hinton, G., Vinyals, O., Dean, J.: Distilling the knowledge in a neural network. arXiv preprint arXiv:1503.02531 (2015)
8. Jiang, J., Deng, W.: Disentangling identity and pose for facial expression recognition. IEEE Trans. Affect. Comput. **13**(4), 1868–1878 (2022)
9. Kingma, D.P., Ba, J.: Adam: a method for stochastic optimization. Comput. Sci. (2014)
10. Li, S., Deng, W., Du, J.: Reliable crowdsourcing and deep locality-preserving learning for expression recognition in the wild. In: 2017 IEEE Conference on Computer Vision and Pattern Recognition (CVPR), pp. 2584–2593. IEEE (2017)
11. Lin, Z., She, J., Shen, Q.: Real emotion seeker: recalibrating annotation for facial expression recognition. Multimedia Syst. **29**(1), 139–151 (2023)
12. Lucey, P., Cohn, J.F., Kanade, T., Saragih, J., Ambadar, Z., Matthews, I.: The extended Cohn-Kanade dataset (CK+): a complete dataset for action unit and emotion-specified expression. In: 2010 IEEE Computer Society Conference on Computer Vision and Pattern Recognition-Workshops, pp. 94–101. IEEE (2010)
13. Mollahosseini, A., Hasani, B., Mahoor, M.H.: AffectNet: a database for facial expression, valence, and arousal computing in the wild. IEEE Trans. Affect. Comput. **10**(1), 18–31 (2017)
14. Pu, T., Chen, T., Xie, Y., Wu, H., Lin, L.: Au-expression knowledge constrained representation learning for facial expression recognition. In: 2021 IEEE International Conference on Robotics and Automation (ICRA), pp. 11154–11161. IEEE (2021)
15. Ruan, D., Yan, Y., Lai, S., Chai, Z., Shen, C., Wang, H.: Feature decomposition and reconstruction learning for effective facial expression recognition. In: Proceedings of the IEEE/CVF Conference on Computer Vision and Pattern Recognition, pp. 7660–7669 (2021)
16. She, J., Hu, Y., Shi, H., Wang, J., Shen, Q., Mei, T.: Dive into ambiguity: latent distribution mining and pairwise uncertainty estimation for facial expression recognition. In: Proceedings of the IEEE/CVF Conference on Computer Vision and Pattern Recognition, pp. 6248–6257 (2021)
17. Valstar, M., Pantic, M.: Induced disgust, happiness and surprise: an addition to the mmi facial expression database. In: Proceedings of the 3rd International Workshop on EMOTION (satellite of LREC): Corpora for Research on Emotion and Affect, p. 65. Paris, France (2010)
18. Wang, K., Peng, X., Yang, J., Lu, S., Qiao, Y.: Suppressing uncertainties for large-scale facial expression recognition. In: Proceedings of the IEEE/CVF Conference on Computer Vision and Pattern Recognition, pp. 6897–6906 (2020)
19. Xue, F., Wang, Q., Guo, G.: Transfer: learning relation-aware facial expression representations with transformers. In: Proceedings of the IEEE/CVF International Conference on Computer Vision, pp. 3601–3610 (2021)
20. Yang, J., Lv, Z., Kuang, K., Yang, S., Xiao, L., Tang, Q.: RASN: using attention and sharing affinity features to address sample imbalance in facial expression recognition. IEEE Access **10**, 103264–103274 (2022)
21. Zeng, D., Lin, Z., Yan, X., Liu, Y., Wang, F., Tang, B.: Face2Exp: combating data biases for facial expression recognition. In: 2022 IEEE/CVF Conference on Computer Vision and Pattern Recognition (CVPR), pp. 20259–20268 (2022). https://doi.org/10.1109/CVPR52688.2022.01965
22. Zhang, X., et al.: BP4D-spontaneous: a high-resolution spontaneous 3D dynamic facial expression database. Image Vis. Comput. **32**(10), 692–706 (2014)

23. Zhang, Y., Dong, W., Hu, B.G., Ji, Q.: Classifier learning with prior probabilities for facial action unit recognition. In: Proceedings of the IEEE Conference on Computer Vision and Pattern Recognition, pp. 5108–5116 (2018)
24. Zhao, G., Huang, X., Taini, M., Li, S.Z., PietikäInen, M.: Facial expression recognition from near-infrared videos. Image Vis. Comput. **29**(9), 607–619 (2011)
25. Zhao, K., Chu, W.S., De la Torre, F., Cohn, J.F., Zhang, H.: Joint patch and multi-label learning for facial action unit detection. In: Proceedings of the IEEE Conference on Computer Vision and Pattern Recognition, pp. 2207–2216 (2015)
26. Zhao, Z., Liu, Q., Wang, S.: Learning deep global multi-scale and local attention features for facial expression recognition in the wild. IEEE Trans. Image Process. **30**, 6544–6556 (2021)

Accurate Facial Landmark Detector via Multi-scale Transformer

Yuyang Sha, Weiyu Meng, Xiaobing Zhai, Can Xie, and Kefeng Li(✉)

Faculty of Applied Sciences, Macao Polytechnic University, Macao SAR, China
kefengl@mpu.edu.mo

Abstract. Facial landmark detection is an essential prerequisite for many face applications, which has attracted much attention and made remarkable progress in recent years. However, some problems still need to be solved urgently, including improving the accuracy of facial landmark detectors in complex scenes, encoding long-range relationships between keypoints and facial components, and optimizing the robustness of methods in unconstrained environments. To address these problems, we propose a novel facial landmark detector via multi-scale transformer (MTLD), which contains three modules: Multi-scale Transformer, Joint Regression, and Structure Loss. The proposed Multi-scale Transformer focuses on capturing long-range information and cross-scale representations from multi-scale feature maps. The Joint Regression takes advantage of both coordinate and heatmap regression, which could boost the inference speed without sacrificing model accuracy. Furthermore, in order to explore the structural dependency between facial landmarks, we design the Structure Loss to fully utilize the geometric information in face images. We evaluate the proposed method through extensive experiments on four benchmark datasets. The results demonstrate that our method outperforms state-of-the-art approaches both in accuracy and efficiency.

Keywords: Facial landmark detection · Vision transformer · Multi-scale feature · Global information

1 Introduction

Facial landmark detection aims to find some pre-defined locations on human face images, which usually have specific semantic meanings, such as the eyebrow or pupil. It has become one of the most fundamental tasks in computer vision and is used for many real-world applications. Thanks to the development of deep learning and computer vision techniques, facial landmark detection algorithms have achieved significant progress in accuracy and efficiency over the past decades.

Since 2012, methods based on deep neural networks have been the dominant solution for many fields in computer vision. Similarly, facial landmark detectors based on deep learning show significant advantages over traditional methods in

Q. Liu et al. (Eds.): PRCV 2023, LNCS 14429, pp. 278–290, 2024.
https://doi.org/10.1007/978-981-99-8469-5_22

terms of accuracy, generalization, and robustness. Recently, several facial landmark detection algorithms [1–3] with excellent performance have been proposed. For instance, Feng *et al.* [2] proposed the Wing-Loss to increase the contribution of the samples with small and medium size errors to the training of the regression framework. The designed Wing-Loss enables coordinate-based methods to achieve promising performance under wild environments. Xia *et al.* [3] leveraged coordinate regression and Transformer to explore the inherent relationships between facial keypoints and achieve impressive results.

In order to achieve excellent performance, existing mainstream methods attempt to utilize a more complex backbone for learning discriminative representations, such as ResNet [4], HRNet [5], *etc.* Other approaches involve complex data augmentation technologies [6], while some methods [1,3] focus on optimizing the regression schemes with the carefully designed detection head or vision transformer. Although these approaches perform well on public benchmark datasets, they are still hard to apply in unconstrained environments and complex scenes. One issue is that most works take deep convolution networks (CNN) as the backbone to extract features for input samples, which may pay more attention to local information but ignore some meaningful global representations and long-range relationships. Additionally, these frameworks often overlook essential prior knowledge of human face images, such as structural information and geometric relationships of different facial components. That may limit the model's performance, especially on occluded and blurred face samples. Moreover, the commonly used approaches struggle to balance accuracy and inference speed.

To address the above issues, we present a novel facial landmark detector via Multi-scale Transformer named MTLD. The proposed method mainly consists of three modules: Multi-scale Transformer, Joint Regression, and Structure Loss. In order to optimize the disadvantages of the facial landmark detector based on CNN, we proposed the Multi-scale Transformer for face alignment by making full use of multi-scale feature maps to capture the global representations and explore long-range relationships between different facial keypoints. The Joint Regression can be regarded as coordination regression, which would generate a group of heatmaps with the output multi-scale feature maps of backbone during the training stage, then apply them as an auxiliary heatmap loss to accelerate convergence. Notable, the heatmap loss is only used in the training stage, so it would not affect the model inference speed. The proposed Joint Regression takes full advantage of both heatmap and coordinate regression, which can improve the accuracy of facial landmark detectors without scarifying the inference speed. Prior knowledge of the human face's structural information can improve the accuracy of facial landmark detection models. However, current methods do not make full use of this information. Therefore, we designed a loss function called Structure Loss to constrain the specific information between facial keypoints. This loss function aims to improve the continuity and consistency of predicted localization, especially in occluded and blurred environments. In summary, the primary contributions of this paper are as follows:

- We propose a Multi-scale Transformer for facial landmark detection to enhance model performance by processing multi-scale feature maps, which can capture global information and long-range relationships between different facial keypoints.
- We introduce the Joint Regression, which applies the auxiliary heatmap loss to accelerate convergence and forces the model to learn more discriminate representations. Additionally, we design the Structure Loss to constrain the structural correlations between keypoints, thus significantly improving the model performance under occlusion, blur, large pose, *etc.*
- We conduct extensive experiments to verify the model effectiveness in four benchmark datasets, including 300W, WFLW, COFW, and AFLW. The results demonstrate that our method obtains competitive results and fast inference speed compared to state-of-the-art methods.

2 Related Work

2.1 Facial Landmark Detection

Facial landmark detection is a crucial technique in numerous applications involving face recognition and emotion estimation. Therefore, optimizing the performance of facial landmark detectors can make excellent benefits for these related tasks. Currently, CNN-based facial landmark detectors in this field primarily fall into coordinate and heatmap regression. Coordinate-based methods directly map the input face samples into 2D coordinates, which usually enjoy a faster inference speed. However, the accuracy of coordinate-based methods still needs to be improved. Therefore, Feng *et al.* [2] introduced a new loss function, termed Wing-Loss, which improved the accuracy of the coordinate regression method, especially for face samples under the occlusion and blur situations. Heatmap-based methods utilize CNNs to encode face images into a group of heatmap representations, each indicating the probability of a landmark localization. Now, most high-performance facial landmark detectors are based on heatmap regression. For instance, HRNet [5] maintained multi-resolution representations in parallel and exchanged information between these streams to obtain high-accuracy prediction results.

2.2 Vision Transformer

Transformer is a deep-learning model originally designed for machine translation. Now, transformer-based models have been shown to significantly enhance the performance of many natural language processing tasks. Inspired by the success of sequence-to-sequence tasks, there is growing interest in exploring the use of Transformer models for various computer vision tasks. For example, DETR [7] proposed a novel object detection system by combining CNN and Transformer, which predicts bounding boxes via bipartite matching. ViT [8] directly extracted representations from flatted image patches with a pure transformer encoder for image classification. In this study, we utilize the transformer to take full advantage of multi-scale feature maps, which can help the model to establish the global information and long-range relationships between different facial keypoints, thus improving the model performance.

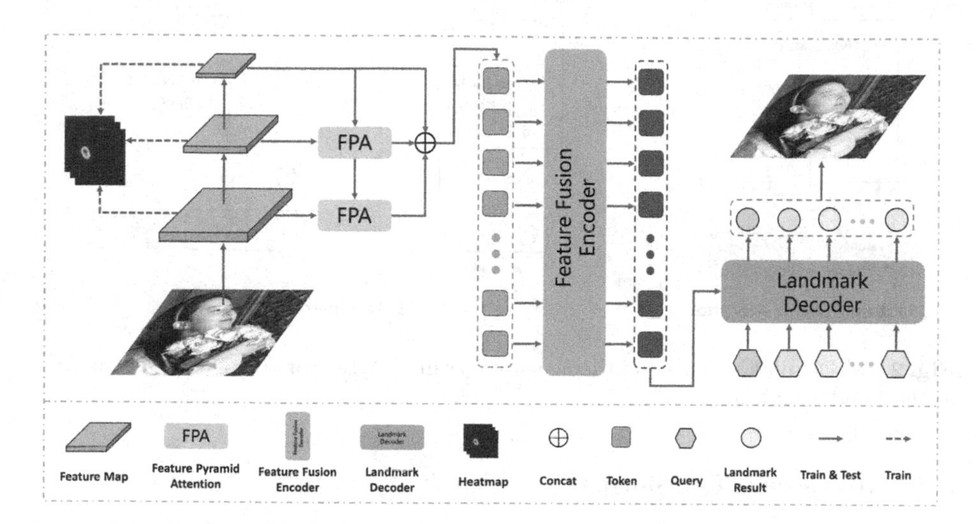

Fig. 1. Overview of the proposed MTLD. Firstly, given a human face image, our framework extracts multi-scale feature maps with the CNN-based backbone. Then, these multi-scale feature maps are processed by two parts simultaneously: one generates heatmaps from the multi-scale feature maps, while the other maps these representations into 2D coordinates. The generated heatmap serves as the auxiliary loss for model training. The proposed multi-scale transformer can fully use cross-level feature maps to extract global information and long-range relationships. Specifically, the heatmap loss is employed in model training and discarded in inference.

3 Method

3.1 Overview

As shown in Fig. 1, we propose a facial landmark detector based on the deep learning module named MTLD. Our method consists of three parts: Multi-scale Transformer, Joint Regression, and Structure Loss. The Multi-scale Transformer can enhance model performance by processing multi-scale feature maps generated from a CNN-based backbone. Joint Regression provides a new scheme for facial landmark detection, which takes advantage of heatmap and coordinate regression. Meanwhile, Structural Loss can constrain the correlation among different facial keypoints, making the model pay more attention to geometric information.

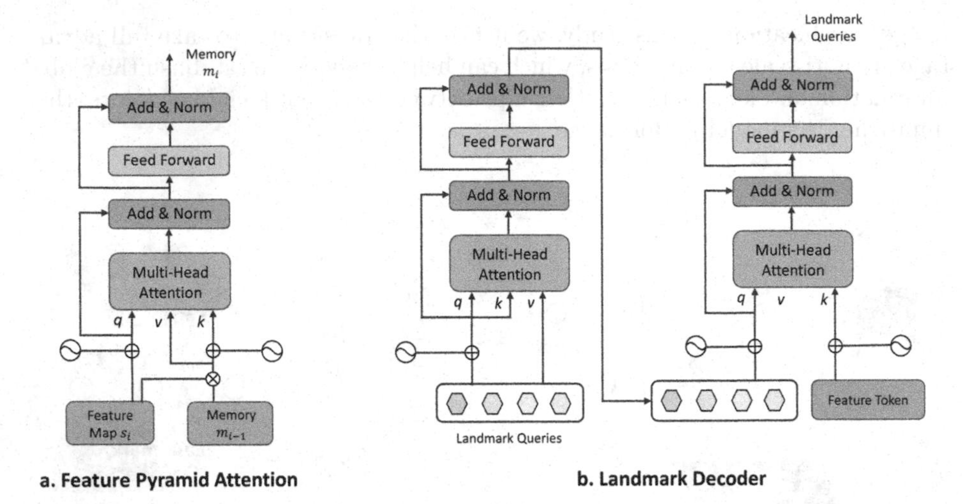

Fig. 2. (a) Detailed structure of the Feature Pyramid Attention. (b) Detailed structure of the Landmark Decoder.

3.2 Multi-scale Transformer

The architecture of our MTLD builds upon the CNN and vision transformer, which is used for exploiting more discriminative representations for facial landmark detection. The introduced network architecture consists of four parts: CNN-based backbone, Feature Pyramid Attention (FPA), Feature Fusion Encoder, and Landmark Decoder. The CNN-based backbone encodes the multi-scale feature maps from input samples, while the FPA obtains cross-scale representations from them. The Feature Fusion Encoder tries to merge and encode the output features from Feature Pyramid Attention to get feature tokens. Then, the Landmark Decoder utilizes landmark queries and feature tokens to predict the coordinates of each facial landmark.

CNN-Based Backbone. In the proposed method, we select the output of the last three stages in the CNN model as multi-scale feature maps. Generally, these feature map contains a large amount of multi-scale semantic and spatial information. Classical CNN models can be directly used in the proposed framework without modification, such as VGG, ResNet, and MobileNet. For instance, we use ResNet-18 as the backbone to illustrate some details. Specifically, given an RGB image $\mathbf{I} \in \mathbb{R}^{H \times W \times 3}$ as input, we can get three feature maps s_3, s_4, and s_5, with the stride of 8, 16, and 32, respectively. Then, these multi-scale feature maps are fed into FPA to get refinement cross-scale representations.

Feature Pyramid Attention (FPA). Encoding multi-scale feature maps can efficiently improve model performance in complex computer vision tasks, such as object detection, instance segmentation, and image generation. Inspired by [9], we introduce a Feature Pyramid Attention formed by the vision transformer to enhance the power of model to utilize multi-level features and capture the long-range relationships between keypoints and facial components. The proposed FPA is shown in Fig. 2-(a). FPA uses the multi-scale feature maps $s_i \in \{s_3, s_4, s_5\}$ generated by CNN-based backbone to output multi-scale feature memory M. In order to make full use of these multi-scale representations, the ith FPA would use the $(i\text{-}1)^{th}$ FPA's output as input.

The input of FPA includes three parties: queries q, keys k, and values v. The q should maintain the relative position by positional embedding p, which is defined as $q_i = s_i + p_i$. We can get the v by Hadamard product with the prior FPA's output m_{i-1}, which can be expressed as $v_i = s_i \circ m_{i-1}$. The k still needs the positional embedding to align semantic meaning as $k_i = v_i + p_i$. Finally, the output of ith FPA can be calculated as:

$$m_i = \mathrm{LN}(t_i + \mathrm{FFN}(t_i)), \tag{1}$$

where, the LN denotes the layer normalization, and FFN is a feed forward network. The t_i is represented by: $t_i = \mathrm{LN}(s_i + \mathrm{MHA}(q_i, k_i, v_i))$, the MHA infers the multi-head attention.

Feature Fusion Encoder. In our method, we employ FPA to encode the global information and long-range relationships from multi-scale feature maps. However, the FPA might pay much attention to high-level representations but ignore some details information contained in these multi-scale feature maps. Therefore, we introduce the Feature Fusion Encoder consisting of multiple transformer encoders to combine and improve these representations from the CNN-based backbone and FPA. The Feature Fusion Encoder can map the input representations into cross-region feature tokens involving rich local and global information.

Landmark Decoder. The proposed MTLD can directly output the coordinates of facial keypoints with Landmark Decoder. The detailed structure is shown in Fig. 2-(b). First, the Landmark Decoder would process the input landmark queries with a self-attention module to make them interact with each other. Then, each landmark query extracts discriminate representations from the input multi-scale feature tokens. Finally, we employ a group of MLPs as the detection head to predict the coordinate results for each facial landmark. In our setting, all the detection heads should output prediction results. Therefore, the first detection head output rough positions, then the subsequent ones can gradually refine the previous results in a coarse-to-fine manner.

3.3 Joint Regression

We propose a novel face alignment scheme named Joint Regression to address the challenging problem of balancing the model accuracy and inference speed. The proposed Joint Regression can be viewed as a combination of heatmap and coordinate regression, which takes full advantage of them. Our framework first employs the CNN-based backbone to encode the input face samples into a group of mulita-scale feature maps. Then, these feature maps would be processed in two parts simultaneously: one generates heatmaps from the multi-scale feature maps, while the other maps the representations into 2D coordinates.

We employ several convolutional layers at the top of the CNN-based backbone for converting the input multi-scale feature maps into a set of heatmap representations $\mathbf{F_H} \in \mathbb{R}^{H_h \times W_h \times N_h}$, where H_h and W_h represent the height and the width. The N_h denotes the number of facial keypoints. The extracted heatmap representation can be seen as the probability of landmark location. Inspired by heatmap regression, we employ the L_2 loss function to compare the ground-truth heatmap $\mathbf{L_H}$ and predict ones $\mathbf{F_H}$. The $\mathcal{L}_{heat}$ is defined as:

$$\mathcal{L}_{heat} = \|\mathbf{L_H} - \mathbf{F_H}\|_2^2. \tag{2}$$

Notable, the heatmap loss can only be used for auxiliary supervision during the training stage and removed when the model testing and deployment. Therefore, it would not affect the inference speed and model efficiency.

At the same time, the selected multi-scale feature maps generated by the CNN-based backbone are also fed to the designed Multi-scale Transformer, which can map them into 2D coordinates of facial landmarks. We adopt L_1 loss to minimize the error $\mathcal{L}_{coord}$ between the predicted results and 2D ground truths:

$$\mathcal{L}_{coord} = \|\mathbf{L_C} - \mathbf{J_C}\|_1, \tag{3}$$

where, the $\mathbf{L_C}$ and $\mathbf{J_C}$ denote the 2D annotations and predicted results, respectively.

3.4 Structure Loss

The human face contains large amounts of geometric information, which are beneficial to improve model performance, especially face under occlusion, blur, lighting, and extreme pose. In order to make full use of these dependencies, we propose Structure Loss to exploit the structural information among facial landmarks effectively. Specifically, the location of facial landmarks is relatively fixed, such as the pupil locates in the center of the iris. Therefore, the structural information can be used to infer the location of adjacent facial keypoints and prevent some abnormal prediction results. The proposed Structure Loss aims to ensure that the distances between predicted keypoints are the same as those calculated from the ground truth. We formulate the Structure Loss as follows:

$$\mathcal{L}_{struc} = \sum_{i \in N} \sum_{j \in C} \| \|\mathbf{J}_i - \mathbf{J}_j\|_2^2 - \|\mathbf{L}_i - \mathbf{L}_j\|_2^2 \|_1, \tag{4}$$

where $\mathbf{J}$ and $\mathbf{L}$ denote the prediction result and ground-truth labels, respectively. The N indicates the total number of facial landmarks, and i represents the ith landmark. C is a collection, which contains M closet landmarks to the ith one, and j denotes the jth landmarks in C. In our setting, the number of adjacent landmarks $Num.$ is 5. Structure Loss is beneficial to enhance the stability and robustness of facial landmark detection approaches.

3.5 Training Objective

We formulate the goals of MTLD and get the overall training objective, which is computed by:

$$\mathcal{L}_{total} = \lambda_h \mathcal{L}_{heat} + \lambda_c \mathcal{L}_{coord} + \lambda_s \mathcal{L}_{struc}, \tag{5}$$

where λ_h, λ_c, and λ_s denote the balancing parameters used to reweight these above loss functions.

Table 1. Facial landmark detection results about NME (%) on 300W, AFLW, and COFW. Lower is better. Red denotes the best, and **blue** indicates the second best.

Method	Backbone	300W			AFLW		COFW
		Full	Comm.	Chal.	Full	Fron.	
LAB [6]	ResNet-18	3.49	2.98	5.19	1.85	1.62	3.92
Wing [2]	ResNet-50	4.04	3.27	7.18	1.47	-	5.07
ODN [10]	ResNet-18	4.17	3.56	6.67	1.63	1.38	-
HRNet [5]	HRNet-W18	3.32	2.87	5.15	1.56	1.46	3.45
AWing [11]	Hourglass	**3.07**	**2.72**	**4.52**	-	-	-
PIPNet [1]	ResNet-101	3.19	2.78	4.89	1.42	-	3.08
SDFL [12]	ResNet-18	3.28	2.88	4.93	-	-	3.63
SLPT [3]	ResNet-34	3.20	2.78	4.93	-	-	4.11
MTLD	ResNet-18	3.28	2.81	4.96	1.42	1.31	3.25
MTLD	ResNet-50	3.20	2.75	4.94	**1.40**	**1.30**	**3.06**
MTLD	ResNet-101	**3.15**	**2.74**	**4.85**	**1.39**	**1.28**	**3.04**

4 Experiments

4.1 Implementation Details and Datasets

In the training phase, all input images need to be cropped by bounding boxes, then resized to the size of 256 × 256. The data augmentations are applied for model training, including random rotation, occlusion, scaling, horizontal flipping, and blurring. We adopt pre-trained ResNet-18 as the default CNN-based backbone. In order to get more accurate results, we also conduct experiments based on ResNet-50 and ResNet-101. The total epochs are 150, and the mini-batch size is 64. We choose Adam as the optimizer with an initial learning rate is 3.0×10^{-4}, and then decay by 10 at 70th and 120th separately. The implementation of our method is based on PyTorch with one NVIDIA Tesla A100 GPU.

Table 2. Facial landmark detection results about NME (%) on WFLW test and 6 subsets: pose, expression (expr.), illumination (illu.), make-up (mu.), occlusion (occu.) and blur. For the NME and FR, lower is better.

Method	backbone	Test	Pose	Expr.	Illu.	Mu.	Occl.	Blur
LAB [6]	ResNet-18	5.27	10.24	5.51	5.23	5.15	6.79	6.32
Wing [2]	ResNet-50	5.11	8.75	5.36	4.93	5.41	6.37	5.81
HRNet [5]	HRNet-W18	4.60	7.94	4.85	4.55	4.29	7.33	6.88
Awing [11]	Hourglass	4.36	7.38	4.58	4.32	4.27	**5.19**	4.96
PIPNet [1]	ResNet-101	4.31	7.51	**4.44**	4.19	4.02	5.36	5.02
SDFL [12]	ResNet-18	4.35	7.42	4.63	4.29	4.22	**5.19**	5.08
SLPT [3]	ResNet-34	4.20	7.18	4.52	4.07	4.17	5.01	4.85
MTLD	ResNet-18	4.47	7.80	4.54	4.40	4.31	5.52	5.23
MTLD	ResNet-50	4.39	7.70	4.41	4.22	4.15	5.43	5.12
MTLD	ResNet-101	**4.25**	**7.29**	4.37	**4.10**	**4.03**	5.31	**4.91**

In order to evaluate the performance of our proposed method, we conduct extensive experiments on four benchmark datasets: 300W [13], COFW [14], AFLW [15], and WFLW [6]. Most of the experiment setting about datasets follow [5]. We adopt the normalized mean error (NME) to evaluate the performance of our approach on the benchmark dataset. Specifically, the inter-ocular distance is used as the normalization distance for 300W, COFW, and WFLW, while using the face bounding box as the normalization distance in AFLW.

4.2 Main Results

We compare our proposed method with several state-of-the-art approaches on four benchmark datasets in terms of NME. To further explore the effectiveness of the backbone, we conduct experiments with different CNN modules, including ResNet-18, ResNet-50, and ResNet-101. Some visualization results of our proposed MTLD on 300W and WFLW are shown in Fig. 3.

300W. We compare the proposed MTLD with other state-of-the-art methods on 300W and its subsets. Table 1 shows that MTLD with ResNet-18 obtains comparable results with existing approaches. We can find that our method with ResNet-101 achieves the second best detection accuracy on 300W-Full, Common, and Challenging sets, which is only slightly behind AWing [11].

AFLW. The AFLW dataset is a challenging benchmark for evaluating facial landmark detectors. We compare the proposed MTLD with the existing methods, and the results are shown in Table 1. Obviously, our framework with ResNet-18 gets 1.42% NME, which is comparable with SOTA methods. Furthermore, MTLD with ResNet-50 or ResNet-101 outperforms all the existing methods on AFLW-Full and AFLW-Frontal datasets.

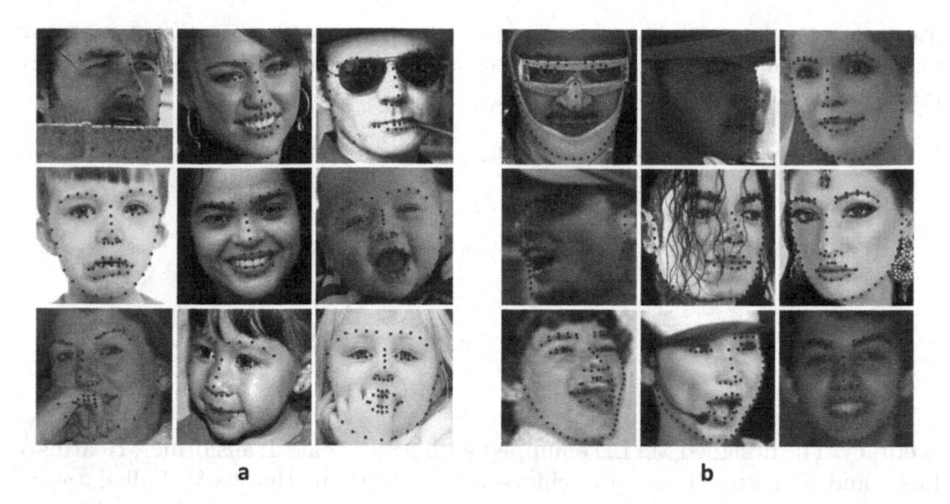

Fig. 3. Visualization results of our method. (a) Results on 300W. (b) Results on WFLW.

COFW. In Table 1, we report the comparison results with existing SOTA methods on the COFW dataset. The results indicate that MTLD with ResNet-18 obtains 3.25% NME, which is slightly higher than Wing-Loss [2] and HRNet [5]. Furthermore, the MTLD with ResNet-101 gets 3.04% NME and outperforms the previous methods by a significant margin.

WFLW. The WFLW dataset is more challenging than 300W and COFW, which provides many face images with various senses. We conduct experiments on WFLW and six subsets. Table 2 demonstrates NME results about the SOTA methods and MTLD with different backbones. We observe that MTLD with ResNet-18 achieves comparable performance to the SOTA methods equipped with more complex models, such as Hourglass and HRNet. Furthermore, MTLD with ResNet-101 achieves SOTA performance on one subset and second best results on four subsets.

4.3 Ablation Study

In this section, we conduct several experiments to verify the effectiveness of the proposed module. Then, we evaluate the model size, computational cost, and inference speed of MTLD. Besides, we also design experiments to explore the appropriate number of adjacent landmarks in Structure Loss.

Effectiveness of Proposed Modules. In this section, we investigate the effectiveness of these modules and conduct experiments on the 300W-Full dataset with NME. For easy comparison, we make the coordinate regression framework with ResNet-18 as the baseline model. Then, add different modules proposed in this paper and analyze their impact on the results. The results are shown in Table 3. We observe that each proposed module is beneficial to improve model

Table 3. The NME (%) of different modules on 300W-Full dataset, including: Baseline (Base.), Multi-scale Transformer (MST.), Heatmap Loss (Heat.), Structure Loss (Sturc.). Lower is better.

Base.	MST.	Heat.	Struc.	NME
✓				4.02
✓	✓			3.39
✓	✓	✓		3.34
✓	✓		✓	3.30
✓	✓	✓	✓	**3.25**

accuracy. The designed MTLD equipped with Multi-scale Transformer, Heatmap Loss, and Structure Loss can achieve 3.25% NME in the 300W-Full dataset. Specifically, the Multi-scale Transformer can significantly boost the model performance.

Table 4. The comparison of different approaches in backbone, model size (Param) computational cost (GFLOPs), and inference speed (fps) on CPU and GPU.

Method	Backbone	Param	GFLOPs	CPU	GPU
LAB [6]	ResNet-18	24.1M	26.7G	2.1	16.7
Wing [2]	ResNet-50	91.0M	5.5G	8.0	30.0
HRNet [5]	HRNet-W18	**9.7M**	4.8G	4.4	11.7
PIPNet [1]	ResNet-18	12.0M	**2.4G**	35.7	200
MTLD	ResNet-18	12.6M	2.7G	**45.8**	**213.5**
MTLD	ResNet-50	27.3M	5.8G	13.5	112.2
MTLD	ResNet-101	46.0M	10.7G	7.6	66.5

Model Size and Speed Analysis. To further evaluate the model's effectiveness, we compare the model size (Params), computational cost (FLOPs), and inference speed (FPS) of our MTLD with SOTA methods. Specifically, the input samples are resized to 256×256, and models are implemented with PyTorch. To compare the inference speed, we evaluate these frameworks on CPU (Intel i7-9700@3.00GHz) and GPU (Nvidia Tesla A100), respectively. Results are shown in Table 4, which indicates that our proposed method with ResNet-18 gets 45.8 FPS and 213.5 FPS on CPU and GPU, respectively. Compared with existing methods, MTLD obtains comparable performance while maintaining a fast inference speed.

Number of Adjacent Points. To explore the appropriate collection number of adjacent points in Structure Loss, we conduct experiments on the 300W-Full dataset in terms of NME. The results are shown in Table 5. It can be observed that when the **Num.** is set to **5**, our method can deliver the best performance.

Table 5. The NME (%) results of our method with different number of adjacent points on 300W-Full dataset. Lower is better.

Num	0	1	3	**5**	8	10	15	20	30
NME	3.29	3.28	3.26	**3.25**	3.27	3.27	3.30	3.36	3.45

5 Conclusion

In this paper, we propose a facial landmark detector named MTLD, which includes three modules: Multi-scale Transformer, Joint Regression, and Structure Loss. Specifically, the carefully designed Multi-scale Transformer enables the model to capture the global dependencies between keypoints and facial components from multi-scale feature maps. The Joint Regression takes advantage of heatmap and coordinate regression, which can achieve superior accuracy results and faster inference speed compared with existing methods. In order to make full use of geometric information contained in the human face, the proposed Structure Loss can force the model to pay more attention to the correlation between landmarks. Additionally, we validate the efficiency and effectiveness of different modules in this paper. Extensive experiments on several benchmark datasets show that MTLD can outperform previous works and offer a better trade-off between accuracy and efficiency.

Acknowledgements. This research was supported by the Macao Polytechnic University (RP/FCSD-02/2022).

References

1. Jin, H., Liao, S., Shao, L.: Pixel-in-pixel net: towards efficient facial landmark detection in the wild. IJCV **129**(12), 3174–3194 (2021)
2. Feng, Z.H., Kittler, J., Awais, M., Huber, P., Wu, X.J.: Wing loss for robust facial landmark localisation with convolutional neural networks. In: CVPR, pp. 2235–2245 (2018)
3. Xia, J., Qu, W., Huang, W., Zhang, J., Wang, X., Xu, M.: Sparse local patch transformer for robust face alignment and landmarks inherent relation learning. In: CVPR, pp. 4052–4061 (2022)
4. He, K., Zhang, X., Ren, S., Sun, J.: Deep residual learning for image recognition. In: CVPR, pp. 770–778 (2016)
5. Wang, J., et al.: Deep high-resolution representation learning for visual recognition. TPAMI **43**(10), 3349–3364 (2020)

6. Wu, W., Qian, C., Yang, S., Wang, Q., Cai, Y., Zhou, Q.: Look at boundary: a boundary-aware face alignment algorithm. In: CVPR, pp. 2129–2138 (2018)
7. Carion, N., Massa, F., Synnaeve, G., Usunier, N., Kirillov, A., Zagoruyko, S.: End-to-end object detection with transformers. In: Vedaldi, A., Bischof, H., Brox, T., Frahm, J.-M. (eds.) ECCV 2020. LNCS, vol. 12346, pp. 213–229. Springer, Cham (2020). https://doi.org/10.1007/978-3-030-58452-8_13
8. Dosovitskiy, A., et al.: An image is worth 16x16 words: transformers for image recognition at scale, arXiv preprint arXiv:2010.11929 (2020)
9. Chen, C.-F., Fan, Q., Panda, R.: Crossvit: cross-attention multi-scale vision transformer for image classification. In: CVPR, pp. 357–366 (2021)
10. Zhu, M., Shi, D., Zheng, M., Sadiq, M.: Robust facial landmark detection via occlusion-adaptive deep networks. In: CVPR, pp. 3486–3496 (2019)
11. Wang, X., Bo, L., Fuxin, L.: Adaptive wing loss for robust face alignment via heatmap regression. In: CVPR, pp. 6971–6981 (2019)
12. Lin, C., et al.: Structure-coherent deep feature learning for robust face alignment. TIP **30**, 5313–5326 (2021)
13. Sagonas, C., Tzimiropoulos, G., Zafeiriou, S., Pantic, M.: 300 faces in-the-wild challenge: the first facial landmark localization challenge. In: ICCV Workshops, pp. 397–403 (2013)
14. Burgos-Artizzu, X.P., Perona, P., Dollár, P.: Robust face landmark estimation under occlusion. In: ICCV, pp. 1513–1520 (2013)
15. Koestinger, M., Wohlhart, P., Roth, P.M., Bischof, H.: Annotated facial landmarks in the wild: a large-scale, real-world database for facial landmark localization. In: ICCV Workshops, pp. 2144–2151. IEEE (2011)

ASM: Adaptive Sample Mining for In-The-Wild Facial Expression Recognition

Ziyang Zhang[1,2], Xiao Sun[1,2,3(✉)], Liuwei An[1,2], and Meng Wang[1,2,3]

[1] School of Computer Science and Information Engineering, Hefei University of Technology, Heifei, China
[2] Anhui Province Key Laboratory of Affective Computing and Advanced Intelligent Machines, Hefei University of Technology, Heifei, China
[3] Institute of Artificial Intelligence, Hefei Comprehensive National Science Center, Heifei, China
sunx@hfut.edu.cn

Abstract. Given the similarity between facial expression categories, the presence of compound facial expressions, and the subjectivity of annotators, facial expression recognition (FER) datasets often suffer from ambiguity and noisy labels. Ambiguous expressions are challenging to differentiate from expressions with noisy labels, which hurt the robustness of FER models. Furthermore, the difficulty of recognition varies across different expression categories, rendering a uniform approach unfair for all expressions. In this paper, we introduce a novel approach called **A**daptive **S**ample **M**ining (ASM) to dynamically address ambiguity and noise within each expression category. First, the Adaptive Threshold Learning module generates two thresholds, namely the clean and noisy thresholds, for each category. These thresholds are based on the mean class probabilities at each training epoch. Next, the Sample Mining module partitions the dataset into three subsets: clean, ambiguity, and noise, by comparing the sample confidence with the clean and noisy thresholds. Finally, the Tri-Regularization module employs a mutual learning strategy for the ambiguity subset to enhance discrimination ability, and an unsupervised learning strategy for the noise subset to mitigate the impact of noisy labels. Extensive experiments prove that our method can effectively mine both ambiguity and noise, and outperform SOTA methods on both synthetic noisy and original datasets. The supplement material is available at https://github.com/zzzzzyang/ASM.

Keywords: Facial Expression Recognition · Noisy Label Learning · Adaptive Threshold Mining

1 Introduction

Facial expressions have important social functions and can convey emotions, cognition, and attitudes in an intangible way. Facial expression recognition (FER)

Q. Liu et al. (Eds.): PRCV 2023, LNCS 14429, pp. 291–302, 2024.
https://doi.org/10.1007/978-981-99-8469-5_23

has been widely applied in various fields, such as media analysis, academic research, etc. It can be used to evaluate and guide psychological counseling, assist decision-makers in making decisions, or analyze the trajectory of emotions, etc. In recent years, with the emergence of large-scale in-the-wild datasets, such as RAF-DB [12], FERPlus [2], and AffectNet [15], deep learning-based FER researches [18,19,28] have made remarkable progress.

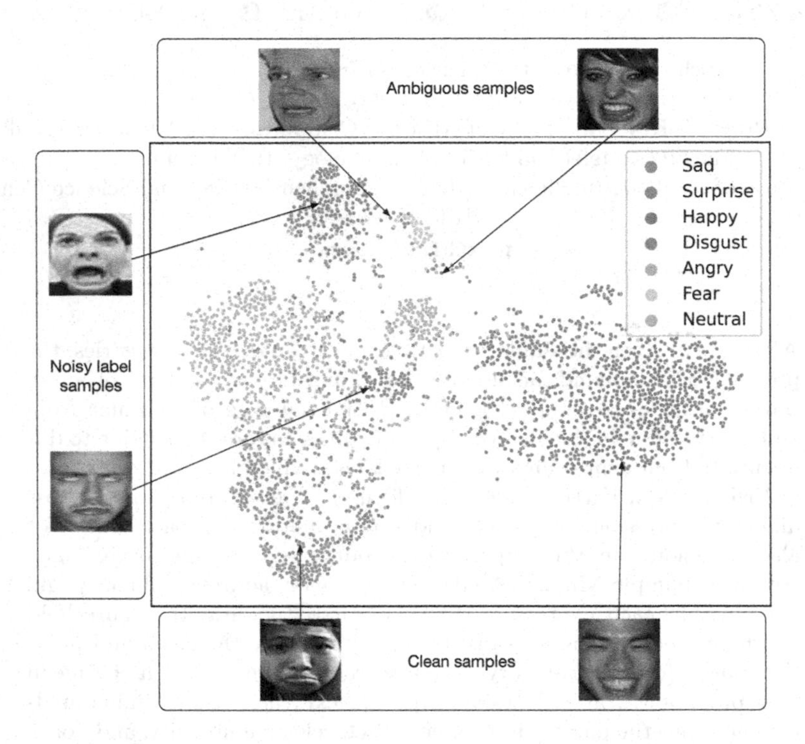

Fig. 1. t-SNE [13] visualizations of facial expression features obtained by ResNet-18 [10] on RAF-DB test set. Clean samples have similar features that cluster together, while the features of ambiguous samples are distributed near the decision boundary. As for noisy label samples, the information provided by the given label is irrelevant with their features. **Best viewed in color. Zoom in for better view.**

However, due to the ambiguity of facial expressions, the subjectivity of annotators, there are a large number of noisy labels in large-scale in-the-wild FER datasets, which can lead to overfitting of models in the supervised learning paradigm and seriously affect the robustness of FER models. Many existing methods use noisy label learning [9,11] to address this issue, treating large-loss samples as noisy label samples. Although these methods achieve significant progress, they still have two problems: (1) They cannot distinguish between ambiguous expressions and noisy expressions. As shown in Fig. 1, ambiguous expressions usually have complexity, and their features are usually distributed near the decision boundary, and they will have large losses regardless of whether their labels are correct or incorrect. Confusing ambiguous expressions with noisy expressions may bias the model towards learning easy samples, making it difficult to learn hard samples and limiting the generalization ability. (2) The presence of intra-class differences and inter-class similarities in facial expressions, coupled with the varying distribution of sample numbers across different expression categories in diverse datasets, introduces variations in the difficulty of recognizing each category. In addition, as the training process progresses, the model's recognition ability also dynamically improves, so using a fixed loss to select noisy samples is not accurate enough. To address the first question, we propose dividing the dataset into three subsets: clean, ambiguity, and noisy. In the field of noisy label learning, co-training has proven to be an exceptionally effective approach. It leverages the utilization of two networks, each providing a unique perspective, to better combat noise. This method is also beneficial for mining both noisy and ambiguous samples. Regarding the second question, we believe that setting different thresholds dynamically for each category based on the recognition difficulty is crucial.

In this paper, we propose a novel method called **A**daptive **S**ample **M**ining (ASM). ASM consists of three key modules: adaptive threshold learning, sample mining, and tri-regularization. For a FER dataset, the adaptive threshold learning module first dynamically updates the clean threshold t_c and the noisy threshold t_n for each category based on their learning difficulties. Then, the sample mining module divides all samples into three subsets according to their confidence scores and the category-specific thresholds: (1) clean samples whose confidence scores are higher than T_c; (2) noisy samples whose confidence scores are lower than T_n; and (3) ambiguous samples whose confidence scores are between T_n and T_c. Finally, the tri-regularization module employs different learning strategies for these three subsets. For clean samples, which are typically simple and easy to learn, we use supervised learning. For ambiguous samples, whose features are usually located around the decision boundary and difficult to distinguish. We design a sophisticated mutual learning strategy. Specifically, mutual learning is guided by the mutuality loss, which comprises a supervised loss and a contrastive loss. The former fits clean expressions in the early stage, and the latter maximizes the consistency between the two networks in the later stage to avoid memorizing the samples with noisy labels. For noisy label samples, we adopt an unsupervised consistency learning strategy to enhance the discriminative ability of the models without using noisy labels. In summary, our contributions are as follows:

1. We innovatively investigate the difference between ambiguous and noisy label expressions in the FER datasets and proposed a novel end-to-end approach for adaptive sample mining.
2. We elaborately design a category-related dynamic threshold learning module and adopt it as a reference to mine noisy and ambiguous samples.
3. Extensive experiments on synthetic and real-world datasets demonstrate that ASM can effectively distinguish between ambiguous and noisy label expressions and achieves state-of-the-art performance.

2 Related Work

2.1 Facial Expression Recognition

In recent years, many researchers have focused on the problem of noisy labels and uncertainty in the FER datasets. SCN [21] adopts a self-attention importance weighting module to learn an import weight for each image. Low weight samples are treated as noisy and relabeled if the maximum prediction probability is higher than the given label with a margin threshold. DMUE [19] uses several branches to mine latent distribution and estimates the uncertainty by the pairwise relationship of semantic features between samples in a mini-batch. RUL [28] adopt two branch to learn facial features and uncertainty values simultaneously, and then the mix-up strategy [26] is used to mix features according to their uncertainty values. EAC [29] utilize the flip and earse semantic consistency strategy to prevent the model from focusing on a part of the features.

2.2 Co-training

Co-training [3,9,16] is a popular approach in noisy label learning that leverages the idea of training multiple classifiers simultaneously on the same dataset. It aims to reduce the impact of noisy labels and improve the overall model performance. The key idea behind co-training is that the two classifiers learn complementary information from different views of the data, and by exchanging and updating their predictions, they can correct each other's mistakes and improve the overall performance. Due to the nonconvex nature of DNNs, even if the network and optimization method are same, different initializations can lead to different local optimum. Thus, following Co-teaching [9], Decoupling [14] and JoCoR [24], we also take two networks with the same architecture but different initializations as two classifiers which can provide different views. The detailed theoretical proof can be found in Decoupling.

3 Method

3.1 The Overall Framework

To distinguish between ambiguous and noisy facial expressions in FER, as well as dynamically set thresholds for each class, we propose an Adaptive Threshold Sample Mining (ASM) method. As shown in Fig. 2, ASM consists of three modules: i) Adaptive Threshold Learning, ii) Sample Mining, and iii) Tri-Regularization.

At the beginning of each epoch, we first make predictions on facial images using two networks and obtain the average probabilities. The adaptive threshold learning module selects samples that are predicted correctly (matching the labels) and calculates the average probabilities per class to obtain the clean threshold T_c. Based on [6,7,23], T_c can reflect the models' recognition ability for

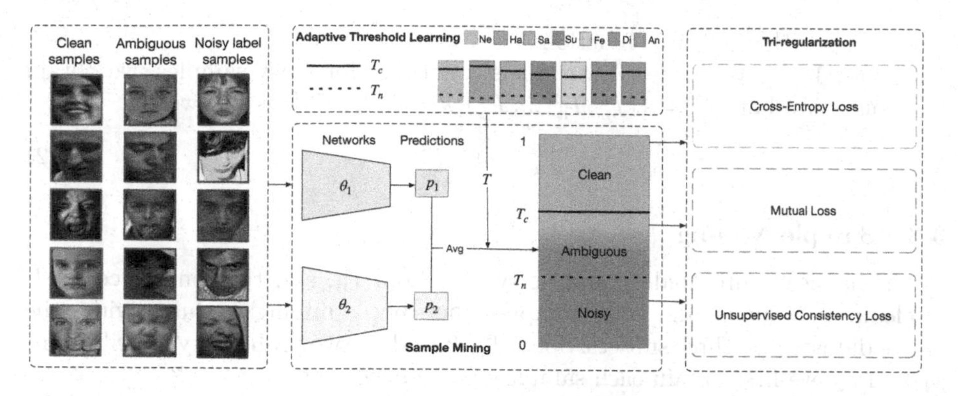

Fig. 2. The pipeline of our ASM. p_1 and p_2 are the predicted probabilities from network's softmax layer. T_c and T_n denote the clean threshold and noisy threshold, respectively.

each class, and conversely, $1 - T_c$ reflects the models' tolerance for noisy samples. Therefore, we consider $1 - T_c$ as the noisy threshold T_n. As for ambiguous expressions, they often share certain features with two or more classes, and they distribute very close to the decision boundaries. Their probability distribution is more chaotic (higher entropy). Based on our observation, the confidence score (average probabilities) of ambiguous expression is between T_c and T_n. Therefore, the sample mining module compares the confidence score of samples with their corresponding labels to T_c and T_n, and divide the dataset into three subsets: clean, ambiguous, and noisy. Finally, the tri-regularization module applies different optimization strategies to each subset. Clean samples are simple and easy to learn, undergo supervised learning. Ambiguous samples, which are more hard to learn, are subjected to mutual learning. Noisy samples, with erroneous labels, we adopt unsupervised consistency learning to improve the models' robustness without using labels.

3.2 Adaptive Threshold Learning

We introduce the adaptive threshold learning module to explore the clean and noisy threshold for different facial expressions. Specifically, given a dataset $S = \{(x_i, y_i), i = 1, 2, ..., N\}$ in which each image x belongs to one of K classes, and y denotes the corresponding label, we first obtain the predictions of all samples and determine predicted labels. Compared to the ground truth $y_i \in \{1, 2, ..., K\}$, we select the correctly-predicted samples $S' = \{(S'_1, S'_2, ..., S'_k), k = 1, 2, ...K\}$, and $S'_k = \{(x'_i, y'_i, s_i), i = 1, 2, ..., N_{sk}\}$ where s_i denote the maximum value in the average probability distribution and N_{sk} denote the number of samples that labeled with the k-th class in S'. We obtain the clean threshold $T_c = \{(T_c^1, T_c^2, ..., T_c^k), k = 1, 2, ...K\}$ as follows

$$T_c^k = \frac{1}{N_{sk}} \sum_{i=1}^{N_{sk}} s_i \tag{1}$$

where $\mathbb{1}(\cdot)$ is the indicator function.

As for $1 - T_c$ can reflect the models' tolerance for noisy samples, we obtain the noisy threshold $T_n = \{(T_n^1, T_n^2, ..., T_n^k), k = 1, 2, ...K\}$ as follows:

$$T_n^k = 1 - T_c^k \tag{2}$$

3.3 Sample Mining

With the clean threshold T_c and noisy threshold T_n, and the confidence levels of clean, ambiguous, and noisy samples, the sample mining module divides the entire dataset into three subsets. Specially, based on two probability distributions p_1 and p_2, we first obtain each sample's confidence score:

$$s = max(\frac{1}{2}(p_1 + p_2)) \tag{3}$$

Then we compare the confidence score with two threshold values associated with the class of samples to dynamically divide them.

Clean samples are easy to fit and their confidence score is the highest. Therefore, we classify samples with confidence score greater than T_c as clean samples.

$$S_{clean} = \{(x_i, y_i, s_i) \mid s_i > T_c^{y_i}\} \tag{4}$$

Ambiguous samples have feature distributions near the decision boundary and are related to two or more classes. Based on our observations, their confidence should be between clean samples and noisy samples. Therefore, we classify samples with confidence between T_n and T_c as ambiguous samples.

$$S_{ambiguous} = \{(x_i, y_i, s_i) \mid T_n^{y_i} <= s_i <= T_c^{y_i}\} \tag{5}$$

Noisy samples have features that are unrelated to their labels, so their confidence is the lowest. Therefore, we classify samples with confidence less than T_n as noisy samples.

$$S_{noisy} = \{(x_i, y_i, s_i) \mid s_i < T_n^{y_i}\} \tag{6}$$

3.4 Tri-regularization

Based on the different characteristics of clean samples, ambiguous samples and noisy samples, tri-regularization module employs different training strategies respectively.

Supervised Learning. Clean samples are highly correlated with their labels and features. Networks can easily fit these samples in the early stages of training, our ASM adopts supervised learning strategy to clean samples, which would consider the classification losses from both two networks.

$$L_{sup} = L_{CE}(p_1, y) + L_{CE}(p_2, y) \tag{7}$$

where L_{CE} denotes the standard cross-entropy loss and y denotes the ground truth.

Mutual Learning. Ambiguous samples contains both samples with clean and noisy labels. As mentioned by [1], DNNs tend to prioritize learning simple patterns first while memorizing noisy samples as training progresses, which will eventually deteriorate the generalization ability. Inspired by this and the view of agreement maximization principles [20], we design a mutual learning strategy which consists two components:

$$L_{mut} = (1 - \lambda) \cdot L_{CE} + \lambda \cdot L_{Con} \tag{8}$$

The former L_{CE} is used to guide networks to fit ambiguous with clean labels in the early stage. The latter is the regularization from peer networks helps maximize the agreement between them, which is expected to provide better generalization performance. In ASM, we adopt the contrastive loss to make the networks guide each other.

$$L_{con} = D_{KL}(p_1 \parallel p_2) + D_{KL}(p_2 \parallel p_1) \tag{9}$$

where

$$D_{KL}(p_1 \parallel p_2) = \sum_{k=1}^{K} p_1^k \log \frac{p_1^k}{p_2^k}$$

The symmetric KL divergence has two advantages: on the one hand, it allows two networks to guide each other to reduce confirmation bias, and on the other hand, it can compensate for the lack of semantic information caused by one-hot labels. In addition, inspired by [17], we adopt a dynamic balancing scheme which gradually increases the weight of the contrastive loss while decreasing the weight of the supervision loss. The dynamic shift is based on a sigmoid ramp-up function, which can be formulated as:

$$\lambda = \lambda_{max} \ * \ e^{-\beta \ *\left(1-\frac{e}{e_r}\right)^2} \tag{10}$$

where λ_{max} is the maximum lambda value, e is the current epoch, e_r is the epoch threshold at which λ gets the maximum value and β controls the shape of the function.

Unsupervised Consistency Learning. The noisy samples have erroneous labels which are irrelevant with the features. To fully leverage noisy samples, the distributions of weak-augmented and strongly-augmented images are aligned using MSE loss, which can be formulated as:

$$L_{usc} = MSE\left(p_1^w, p_1^s\right) + MSE\left(p_2^w, p_2^s\right) \tag{11}$$

where p_1^w and p_2^w denote the probability distribution of weak-augmented images, and p_1^s and p_2^s denote the probability distribution of strongly-augmented images.

Overall Objective Function.

$$L_{total} = L_{sup} + \omega L_{mut} + \gamma L_{usc} \tag{12}$$

where ω and γ are the hyper-parameters, and the corresponding ablation study is provided in the supplementary material.

4 Experiments

4.1 Datasets

RAF-DB [12] comprises over 29,670 facial images annotated with basic or compound expressions by 40 trained annotators. For our experiments, we focus on the seven basic expressions: neutral, happiness, surprise, sadness, anger, disgust, and fear. The dataset is divided into a training set of 12,271 images and a testing set of 3,068 images.

FERPlus [2] is an extension of FER2013 [5] and consists of 28,709 training images, 3,589 validation images, and 3,589 testing images. The dataset was collected using the Google search engine. Each image is resized to 48 × 48 pixels and annotated with one of eight classes. The validation set is also utilized during the training process.

AffectNet [15] is currently the largest FER dataset, containing over one million images collected from the Internet using 1,250 expression-related keywords. Approximately half of the images are manually annotated with eight expression classes. The dataset consists of around 280,000 training images and 4,000 testing images.

4.2 Implementation Details

In our ASM, facial images are detected and aligned using MT-CNN [27]. Subsequently, the images are resized to 224 × 224 pixels and subjected to data augmentation techniques such as random horizontal flipping and random erasing. As a default configuration, we employ ResNet-18 as the backbone network, pre-trained on the MS-Celeb-1M dataset [8]. All experiments are conducted using Pytorch on a single RTX 3090 GPU. The training process spans 100 epochs with a batch size of 128. A warm-up epoch of 10 is implemented. We utilize an Adam optimizer with a weight decay of 1e−4. The initial learning rate is set to 0.001 and exponentially decayed by a gamma value of 0.9 after each epoch. The initial T_c and T_n is set to $\{0.8\}^K$ and $\{0.2\}^K$ based on the ablation study. Following the guidelines of NCT [17], we set the hyper-parameters λ_{max}, β and e_r to 0.9, 0.65 and 90, respectively.

4.3 Evaluation on Synthetic Noise

Follow [21,28,29], we evaluate our proposed ASM with three level of noisy label including the ratio of 10%, 20% and 30% on RAF-DB, FERPlus and AffectNet. As shown in Table 1, our method outperforms the baseline and all previous state-of-the-art methods under all circumstances by a large margin. For example, ASM outperforms SCN under 30% label noise by 8.21%, 4.49%, 5.15% on RAF-DB, FERPlus, AffectNet respectively. This can be attributed to ASM's ability to dynamically differentiate between ambiguous and noisy samples, and mitigate the harmful effects of synthetic noise through mutual learning and unsupervised consistency learning.

Table 1. Evaluation of ASM on noisy FER datasets. Results are computed as the mean of the accuracy of the last 5 epochs.

Method	Noisy(%)	RAF-DB(%)	FERPlus(%)	AffectNet(%)
Baseline	10	81.01	83.29	57.24
SCN (CVPR20)	10	82.15	84.99	58.60
RUL (NeurIPS21)	10	86.17	86.93	60.54
EAC (ECCV22)	10	88.02	87.03	61.11
ASM (Ours)	10	88.75	88.51	61.21
Baseline	20	77.98	82.34	55.89
SCN (CVPR20)	20	79.79	83.35	57.51
RUL (NeurIPS21)	20	84.32	85.05	59.01
EAC (ECCV22)	20	86.05	86.07	60.29
ASM (Ours)	20	87.75	87.41	60.52
Baseline	30	75.50	79.77	52.16
SCN (CVPR20)	30	77.45	82.20	54.60
RUL (NeurIPS21)	30	82.06	83.90	56.93
EAC (ECCV22)	30	84.42	85.44	58.91
ASM (Ours)	30	85.66	86.69	59.75

4.4 Ablation Study

Please note that due to page limitation, we place the ablation experiments with hyper-parameters, ablation experiments with fixed and adaptive thresholds, and feature visualization in the supplementary material.

Effectiveness of Each Component in ASM. We evaluate the three key modules of the proposed ASM to find why ASM works well under label noise. The experiment results are shown in Table 2. Several observations can be concluded in the following. First, excluding L_{mut} and L_{usc} and only adding an adaptive threshold (2nd row) is equivalent to adding a co-training strategy on top of the baseline (1st row), resulting in a significant improvement, which highlights the advantage of model ensembling. Second, when adding L_{mut} (3rd row) or L_{usc} (4th row), we achieve higher accuracy, which can be attributed to these two carefully designed loss functions. The former addresses the information insufficiency issue caused by one-hot labels and promotes consensus learning to better resist the influence of noise. The latter enhances the model's discriminative ability through contrastive learning, unaffected by noise. Finally, by integrating all modules, we achieve the highest accuracy on RAF-DB, raising the baseline from 87.25% to 90.58%.

Table 2. Evaluation of the three modules in ASM. Note that the exclusion of L_{mut} and L_{usc} implies the replacement with the Cross-Entropy (CE) loss.

AT	L_{mut}	L_{usc}	RAF-DB(%)
×	×	×	87.25
✓	×	×	88.63
✓	✓	×	89.47
✓	×	✓	89.72
✓	✓	✓	90.58

Table 3. Comparison with other state-of-the-art results on different FER datasets. † denotes training with both AffectNet and RAF-DB datasets. * denotes test with 7 classes on AffectNet. We report the results for both AffectNet8 and AffectNet7.

RAF-DB		FERPlus		AffectNet	
Methods	Acc. (%)	Methods	Acc. (%)	Methods	Acc. (%)
IPA2LT† [25]	86.77	IPA2LT† [25]	–	IPA2LT† [25]	57.31
RAN [22]	86.90	RAN [22]	88.55	RAN [22]	59.50
SCN [21]	87.03	SCN [21]	88.01	SCN [21]	60.23
DACL [4]	87.78	DACL [4]	–	DACL* [4]	65.20
RUL [28]	88.98	RUL [28]	88.75	RUL [28]	61.43
EAC [29]	89.99	EAC [29]	89.64	EAC* [29]	65.32
ASM (Ours)	90.58	ASM (Ours)	90.21	ASM (Ours)	62.36 | 65.68

4.5 Comparison with the State-of-the-art

We compare our ASM with several state-of-the-art methods on three popular benchmarks. The results are shown in Table 3. RAN [22] is designed to address the occlusion and head pose problem. DACL [4] proposes a Deep Attentive Center Loss method to adaptively select a subset of significant feature elements for enhanced discrimination. SCN [21], RUL [28] and EAC [29] are noise-tolerant methods. The first two have an uncertainty estimation module to reflect the uncertainty of each sample, and the third adpots erasing attention consistency method to prevent the model from remembering noisy labels. Our ASM outperforms these state-of-the-art methods with 90.58%, 90.21%, 62.36% and 65.68% on RAF-DB, FERPlus, AffectNet8 and AffectNet7, respectively.

5 Conclusion

In this paper, we highlight the significance of distinguishing between ambiguous and noisy expressions in FER datasets and dynamically handling each expression category. We propose an adaptive threshold learning module that dynamically generates distinct clean and noisy thresholds tailored to each expression class.

These thresholds serve as references for effectively identifying clean, noisy and ambiguous samples. To enhance generalizability, we employ distinct optimization strategies for the three subsets. Extensive experiments verify that ASM outperforms other state-of-the-art noisy label FER methods on both real-world and noisy datasets.

Acknowledgments. This work was supported by the National Key R&D Programme of China (2022YFC3803202), Major Project of Anhui Province under Grant 202203a05020011. This work was done in Anhui Province Key Laboratory of Affective Computing and Advanced Intelligent Machine.

References

1. Arpit, D., et al.: A closer look at memorization in deep networks. In: International Conference on Machine Learning, pp. 233–242. PMLR (2017)
2. Barsoum, E., Zhang, C., Ferrer, C.C., Zhang, Z.: Training deep networks for facial expression recognition with crowd-sourced label distribution. In: Proceedings of the 18th ACM International Conference on Multimodal Interaction, pp. 279–283 (2016)
3. Blum, A., Mitchell, T.: Combining labeled and unlabeled data with co-training. In: Proceedings of the Eleventh Annual Conference on Computational Learning Theory, pp. 92–100 (1998)
4. Farzaneh, A.H., Qi, X.: Facial expression recognition in the wild via deep attentive center loss. In: Proceedings of the IEEE/CVF Winter Conference on Applications of Computer Vision, pp. 2402–2411 (2021)
5. Goodfellow, I.J., et al.: Challenges in representation learning: a report on three machine learning contests. In: Lee, M., Hirose, A., Hou, Z.-G., Kil, R.M. (eds.) ICONIP 2013. LNCS, vol. 8228, pp. 117–124. Springer, Heidelberg (2013). https://doi.org/10.1007/978-3-642-42051-1_16
6. Guo, L.Z., Li, Y.F.: Class-imbalanced semi-supervised learning with adaptive thresholding. In: Proceedings of the 39th International Conference on Machine Learning, pp. 8082–8094 (2022)
7. Guo, L.Z., Zhang, Y.G., Wu, Z.F., Shao, J.J., Li, Y.F.: Robust semi-supervised learning when not all classes have labels. In: Advanced Neural Information Processing Systems, vol. 35, pp. 3305–3317 (2022)
8. Guo, Y., Zhang, L., Hu, Y., He, X., Gao, J.: MS-Celeb-1M: a dataset and benchmark for large-scale face recognition. In: Leibe, B., Matas, J., Sebe, N., Welling, M. (eds.) ECCV 2016. LNCS, vol. 9907, pp. 87–102. Springer, Cham (2016). https://doi.org/10.1007/978-3-319-46487-9_6
9. Han, B., et al.: Co-teaching: robust training of deep neural networks with extremely noisy labels. In: Advances in Neural Information Processing Systems 31 (2018)
10. He, K., Zhang, X., Ren, S., Sun, J.: Deep residual learning for image recognition. In: Proceedings of the IEEE Conference on Computer Vision and Pattern Recognition (CVPR), June 2016
11. Li, J., Socher, R., Hoi, S.C.: DIVIDEMIX: learning with noisy labels as semi-supervised learning. arXiv preprint arXiv:2002.07394 (2020)
12. Li, S., Deng, W., Du, J.: Reliable crowdsourcing and deep locality-preserving learning for expression recognition in the wild. In: Proceedings of the IEEE Conference on Computer Vision and Pattern Recognition, pp. 2852–2861 (2017)

13. Van der Maaten, L., Hinton, G.: Visualizing data using t-SNE. J. Mach. Learn. Res. **9**(11), 2579–2605 (2008)
14. Malach, E., Shalev-Shwartz, S.: Decoupling "when to update" from "how to update". In: Advances in Neural Information Processing Systems 30 (2017)
15. Mollahosseini, A., Hasani, B., Mahoor, M.H.: AffectNet: a database for facial expression, valence, and arousal computing in the wild. IEEE Trans. Affect. Comput. **10**(1), 18–31 (2017)
16. Nigam, K., Ghani, R.: Analyzing the effectiveness and applicability of co-training. In: Proceedings of the Ninth International Conference on Information and Knowledge Management, pp. 86–93 (2000)
17. Sarfraz, F., Arani, E., Zonooz, B.: Noisy concurrent training for efficient learning under label noise. In: Proceedings of the IEEE/CVF Winter Conference on Applications of Computer Vision, pp. 3159–3168 (2021)
18. Shao, J., Wu, Z., Luo, Y., Huang, S., Pu, X., Ren, Y.: Self-paced label distribution learning for in-the-wild facial expression recognition. In: Proceedings of the 30th ACM International Conference on Multimedia, pp. 161–169 (2022)
19. She, J., Hu, Y., Shi, H., Wang, J., Shen, Q., Mei, T.: Dive into ambiguity: latent distribution mining and pairwise uncertainty estimation for facial expression recognition. In: Proceedings of the IEEE/CVF Conference on Computer Vision and Pattern Recognition, pp. 6248–6257 (2021)
20. Sindhwani, V., Niyogi, P., Belkin, M.: A co-regularization approach to semi-supervised learning with multiple views. In: Proceedings of ICML Workshop on Learning with Multiple Views, vol. 2005, pp. 74–79. Citeseer (2005)
21. Wang, K., Peng, X., Yang, J., Lu, S., Qiao, Y.: Suppressing uncertainties for large-scale facial expression recognition. In: Proceedings of the IEEE/CVF Conference on Computer Vision and Pattern Recognition, pp. 6897–6906 (2020)
22. Wang, K., Peng, X., Yang, J., Meng, D., Qiao, Y.: Region attention networks for pose and occlusion robust facial expression recognition. IEEE Trans. Image Process. **29**, 4057–4069 (2020)
23. Wang, Y., et al.: FreeMatch: self-adaptive thresholding for semi-supervised learning. arXiv preprint arXiv:2205.07246 (2022)
24. Wei, H., Feng, L., Chen, X., An, B.: Combating noisy labels by agreement: a joint training method with co-regularization. In: Proceedings of the IEEE/CVF Conference on Computer Vision and Pattern Recognition, pp. 13726–13735 (2020)
25. Zeng, J., Shan, S., Chen, X.: Facial expression recognition with inconsistently annotated datasets. In: Ferrari, V., Hebert, M., Sminchisescu, C., Weiss, Y. (eds.) ECCV 2018. LNCS, vol. 11217, pp. 227–243. Springer, Cham (2018). https://doi.org/10.1007/978-3-030-01261-8_14
26. Zhang, H., Cisse, M., Dauphin, Y.N., Lopez-Paz, D.: mixup: beyond empirical risk minimization. arXiv preprint arXiv:1710.09412 (2017)
27. Zhang, K., Zhang, Z., Li, Z., Qiao, Y.: Joint face detection and alignment using multitask cascaded convolutional networks. IEEE Signal Process. Lett. **23**(10), 1499–1503 (2016)
28. Zhang, Y., Wang, C., Deng, W.: Relative uncertainty learning for facial expression recognition. In: Advances in Neural Information Processing Systems 34 (2021)
29. Zhang, Y., Wang, C., Ling, X., Deng, W.: Learn from all: Erasing attention consistency for noisy label facial expression recognition. In: Avidan, S., Brostow, G., Cissé, M., Farinella, G.M., Hassner, T. (eds.) Computer Vision-ECCV 2022: 17th European Conference, Tel Aviv, Israel, October 23–27, 2022, Proceedings, Part XXVI. pp. 418–434. Springer, Cham (2022). https://doi.org/10.1007/978-3-031-19809-0_24

Deep Face Recognition with Cosine Boundary Softmax Loss

Chen Zheng[1,2], Yuncheng Chen[1,2], Jingying Li[1,2](✉), Yongxia Wang[1,2], and Leiguang Wang[3]

[1] School of Mathematics and Statistics, Henan University, Kaifeng 475004, China
10475221083@henu.edu.cn

[2] Henan Engineering Research Center for Artificial Intelligence Theory and Algorithms, Institute of Applied Mathematics, Henan University, Kaifeng 475004, China

[3] Institutes of Big Data and Artificial Intelligence, Southwest Forestry University, Kunming 650224, China

Abstract. To improve the accuracy of face recognition when there are wrong-labeled samples, a new deep face recognition model with cosine boundary loss is proposed in this paper. First, the proposed model uses the cosine similarity to determine the boundary that divides training samples into easy samples, semi-hard samples and harder samples, which play different roles during the training process. Then, an adaptive weighted piecewise loss function is developed to emphasize semi-hard samples and suppress wrong-labeled samples in harder samples by assigning different weights to related types of samples during different training stages. Compared with the state-of-the-art face recognition methods, i.e., CosFace, CurricularFace, and EnhanceFace, experimental results on CFP_FF, CFP_FP, AgeDB, LFW, CALFW, CPLFW, VGG2_FP datasets demonstrate that the proposed method can effectively reduce the impact of the wrong-labeled samples and provide a better accuracy.

Keywords: Face Recognition · Deep Learning · Loss Function · Cosine Similarity

1 Introduction

Face recognition (FR) is a prominent research area in artificial intelligence. In recent years, the advancements in deep learning technology have significantly improved the accuracy of FR and facilitated its large-scale application in real-world scenarios [1, 2]. A deep FR model consists of three essential modules: training data, network structure and loss function. Among these modules, the loss function plays an important role as it determines the optimization direction of model training [3]. Therefore, the design and optimization of the loss function are crucial for achieving superior FR performance.

Loss functions in FR can be divided into two categories: loss functions based on Euclidean distance and softmax loss and its variants. The former aims to find an appropriate distance measurement function to reduce the intra-class variance and increase the

Q. Liu et al. (Eds.): PRCV 2023, LNCS 14429, pp. 303–314, 2024.
https://doi.org/10.1007/978-981-99-8469-5_24

inter-class variance by projecting the face sample into Euclidean space, such as triplet loss [4], center loss [5] and so on. The latter is widely used in deep learning, especially in FR, where it effectively enhances the discriminative ability of network features by incorporating the angle interval within the loss function. It has gained significant research attention in recent years. In SphereFace, Liu et al. [6] introduced the Softmax loss based on the angle interval as an extension of the classic Softmax loss function. This formulation utilizes vector inner product and introduces the parameter 'm' to control the decision boundary of the loss function, effectively reducing intra-class variance and improving inter-class distance. In CosFace [7], the multiplicative parameter 'm' in SphereFace for the angle θ is adjusted to an additive parameter in the cosine space. This adjustment aims to improve face recognition ability when the decision boundary is inconsistent due to variations in the angle θ between faces. To further enhance the separability between different faces, the ArcFace [8] proposed an additive angular margin loss utilizing inverse trigonometric functions.

In FR, the collected face data varies in lighting, scene, age, partial occlusion, or pose angles. These samples are vital for accurate decision boundaries in face distinction but can impede initial model training. Consequently, Optimizing FR requires assigning appropriate weights to samples based on difficulty during different training stages. However, the softmax loss function based on the angle interval is deterministically invariant during training. To address this issue, CurricularFace [9] proposed an adaptive curriculum learning loss function based on measuring the difficulty of training data. It emphasizes easy samples, such as fronts, with adaptive weights in initial stages, focusing on harder samples, such as sides, in later stages, effectively improves accuracy. FR databases contain not only hard samples like side shots and lighting variations but also noise samples with mislabel, as shown in Fig. 1. Some databases have noise samples comprising nearly 30% [10]. Therefore, during training process, it is crucial to consider both the variation in sample importance and the impact of noise samples on the model. EnhanceFace [11] proposed a partial noise sample suppression method by leveraging cosine being negative and using a 90° angle interval as the boundary. However, there are numerous noise samples in practice that exhibit features close to correctly labeled samples, with angle intervals less than 90° and cosine values greater than zero.

This paper proposes a deep FR model based on the cosine boundary loss function to address this issue. The main contributions of this model are as follows:

1. A decision boundary based on the cosine values within angle intervals is introduced to classify samples into three types: easy, semi-hard, and harder. This classification helps to effectively differentiate between samples captured from frontal views, samples with side angles or lighting variations, and noise samples.
2. A novel adaptive segmented loss function is proposed to assign adaptive weights to samples based on difficulty during different training stages. This adaptive weighting scheme highlights the varying impact of different types of samples throughout the training process.

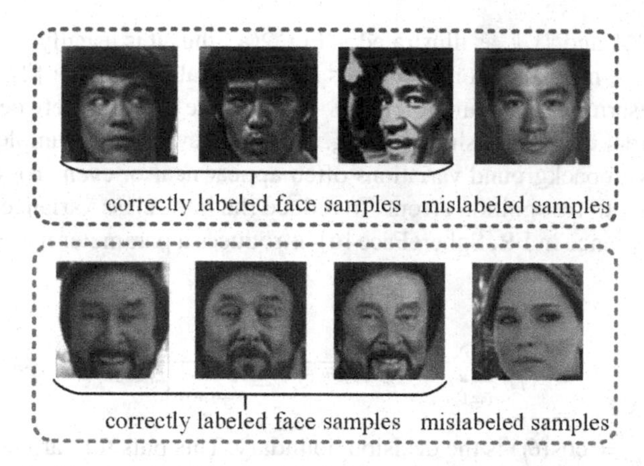

Fig. 1. Illustrations of wrong-labeled face data within the face dataset

2 Method

The deep FR model based on the cosine boundary loss function proposed in this paper aims to use the cosine decision boundary to distinguish the difficulty of different types of training samples, and accordingly design a new adaptive loss function to suppress noise samples, such as annotation errors, and improve recognition accuracy. This section first briefly reviews the existing angle interval softmax loss function, then discusses the loss function based on the cosine boundary, and finally gives the corresponding deep FR model and algorithm flow.

2.1 The Softmax Loss Function Based on Angle Intervals

In FR, the formulation of the classic softmax loss function is:

$$L = \frac{1}{N}\sum_{i=1}^{N} - \ln \frac{e^{W_{y_i}^T x_i + b_{y_i}}}{\sum_{j=1}^{n} e^{W_j^T x_j + b_j}} \tag{1}$$

where $x_i \in R^d$ represents the feature of ith sample belonging to y_i^{th} class. In this paper, W_{y_i} represents y_i^{th} column of the weight matrix $\mathbf{W} \in R^{d\times n}$, b_{y_i} is the corresponding bias term, $d = 512$. N is the number of training sample for each batch, n is the number of total classes. In practical, all bias items are usually set to 0, i.e. $b_{y_i}, b_j = 0$. According to the properties of vector inner product, $W_j^T x_i$ in Eq. 1 is equal to $\| W_j \| \cdot \|x_i\| \cdot \cos\theta_j$, where θ_j is the angle between the weight vector W_j and the sample features x_i. The softmax function can be represented as:

$$L = \frac{1}{N}\sum_{i=1}^{N} - \ln \frac{e^{\| W_{y_i} \| \cdot \|x_i\| \cdot \cos\theta_{y_i}}}{\sum_{j=1}^{n} e^{\| W_j \| \cdot \|x_i\| \cdot \cos\theta_j}} \tag{2}$$

Under the loss function based on the angle interval, the recognition of the sample mainly depends on the size of the angle θ_j between the vectors and is less affected by the sample norm $\|W_j\|$ and $\|x_i\|$, as illustrated in Fig. 2(a), thus it is usually set to $\|W_j\| = 1$ and $\|x_i\| = s$ by L_2 normalization, where s is rescaled scale parameter [2]. The decision boundary for discriminating sample classes is the angle bisector between the weight vectors of different classes, as shown in Fig. 2(b). However, some samples taken from side angles or with background variations often appear near or even cross the decision boundary, leading to recognition errors. To reduce the intra-class variance and increase the inter-class variance in FR, SphereFace [6] introduces a parameter m ($\geq$2) in the loss function, i.e.

$$L = \frac{1}{N}\sum_{i=1}^{N} - \ln \frac{e^{s\cdot\cos(m\cdot\theta_{y_i})}}{e^{s\cdot\cos(m\cdot\theta_{y_i})} + \sum_{j\neq y_i} e^{s\cdot\cos\theta_j}} \tag{3}$$

Here, $\cos(m \cdot \theta_{y_i}) = \cos(\theta_j)$ is the decision boundary. This puts forward higher requirements for correctly identifying the sample class, that is, not only needs to be satisfied $\cos(\theta_{y_i}) > \cos(\theta_j)$, but also needs to be satisfied $\cos(m \cdot \theta_{y_i}) > \cos(\theta_j)$. The aforementioned conditions compress the range of angle values for each facial class, transforming the decision boundaries between classes into decision margins, as shown in Fig. 2(c). This achieves a reduction in intra-class variance and an increase in inter-class variance. In Eq. (3), different cosine functions are used to distinguish the true class and other classes, denoted respectively as positive cosine similarity and negative cosine similarity, i.e., $T(\cos(\theta_{y_i})) = \cos(m \cdot \theta_{y_i})$ and $N(\cos(\theta_j)) = \cos(\theta_j)$ [9]. The positive cosine similarity, $T(\cos(\theta_{y_i})) = \cos(\theta_{y_i}) - m$ and $T(\cos(\theta_{y_i})) = \cos(\theta_{y_i} + m)$, was proposed successively in CosFace [7] and ArcFace [8].

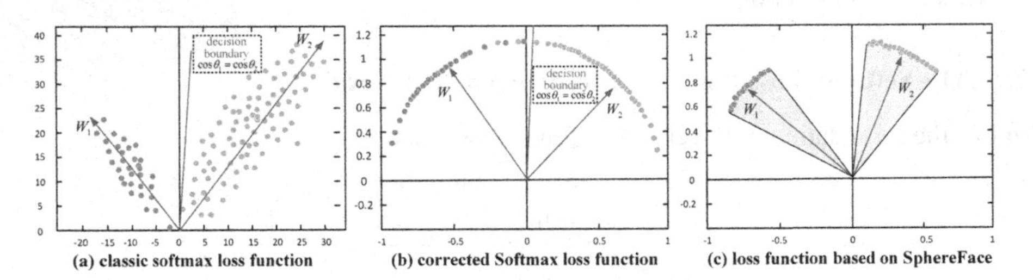

Fig. 2. Demonstration of different types of loss functions

To deal with hard samples such as side-face shots, environmental variations, and lighting changes, CurricularFace [9] further introduces a novel segmented negative cosine similarity function

$$N(t, \cos(\theta_j)) = \begin{cases} \cos(\theta_j) & T(\cos(\theta_{y_i})) \geq \cos(\theta_j) \\ \cos(\theta_j)(t + \cos(\theta_j)) & T(\cos(\theta_{y_i})) < \cos(\theta_j) \end{cases} \tag{4}$$

where $t > 0$ is a hyper-parameter that characterizes the different training stages of the model. In practical, the hard samples consist not only correctly labeled samples but also mislabeled samples. Among them, correctly labeled hard samples hold significant value in determining the decision boundary for face recognition and need to be emphasized.

On the other hand, mislabeled hard samples will lead to boundary confusion, making the iterative convergence of the model training process difficult and slow, thereby affecting the final accuracy. Therefore, EnhanceFace [11] is further divided into semi-hard samples $T(\cos(\theta_{y_i})) < \cos(\theta_j)$ and harder samples $0 \leq T(\cos(\theta_{y_i})) < \cos(\theta_j)$ in Eq. (4), and assigns different weights.

2.2 Cosine Function Based Decision Boundary

Despite existing research distinguishing different hard samples based on whether the cosine similarity is greater than 0, its decision boundary does not apply to all noise samples, as illustrated by the red circles in Fig. 3, which represent mislabeled noise samples with cosine similarity greater than 0. In other words, the angle between the feature vector of a noise sample and the weight vector of its labeled class is not always greater than 90°. To address this issue and better differentiate noise samples within hard samples, this paper proposes a new decision boundary based on the cosine function to assess the difficulty level of samples. Specifically, the decision boundaries for easy samples, semi-hard samples, and harder samples are defined as follows:

$$\begin{cases} \text{easy samples}, & T(\cos(\theta_{y_i})) \geq \cos(\theta_j) \\ \text{semi - hard samples}, & \cos(\theta_{y_i}) > \cos(\theta_j) > T(\cos(\theta_{y_i})) \\ \text{hard samples}, & \cos(\theta_j) > \cos(\theta_{y_i}) > T(\cos(\theta_{y_i})) \end{cases} \tag{5}$$

where $T(\cos(\theta_{y_i})) = \cos(\theta_{y_i} + m), m > 0$. In Eq. (5), the positive cosine similarity $T(\cos(\theta_{y_i}))$ is used to reduce the intra-class variance, so that the simple samples with correct labels are more concentrated near the weight vector W_{y_i} of class y_i. Simultaneously, the decision boundary between semi-hard and harder samples is defined as $\cos(\theta_{y_i}) = \cos(\theta_j)$, as illustrated in Fig. 4. Compared to existing loss functions, the proposed decision boundary in this paper achieves adaptive partitioning of semi-hard and harder samples by comparing the values of positive and negative cosine functions. The example samples in Fig. 4 also demonstrate that the mislabeled samples with cosine values greater than 0 are correctly classified as harder samples under the new decision boundary.

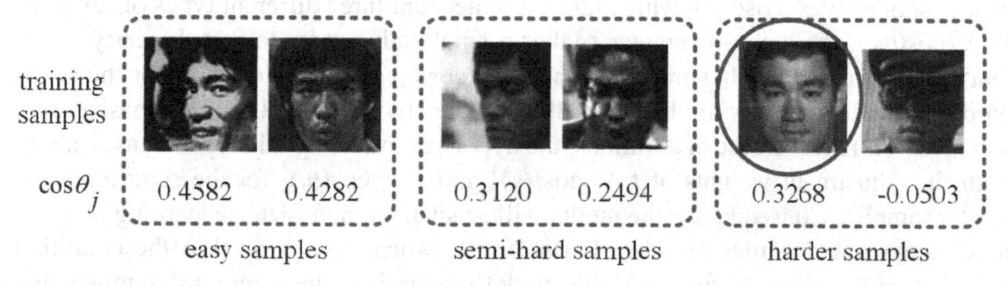

Fig. 3. Different samples for the same person and their cosine likelihood values

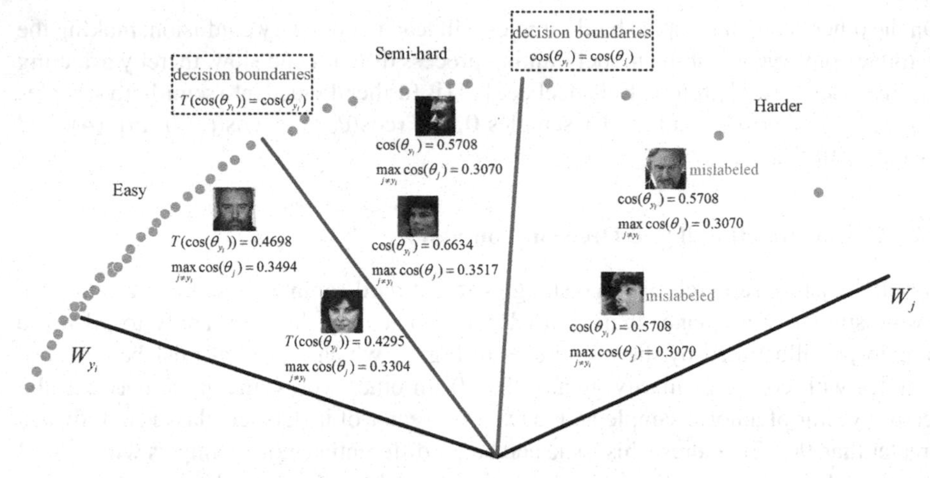

Fig. 4. Illustration of cosine decision boundary

2.3 The Loss Function Based on Cosine Decision Boundary

Based on the different types of training samples divided in the previous section, this paper proposes a deep FR model based on cosine boundary loss. In this model, the loss function is defined as follows:

$$L = \frac{1}{N}\sum_{i=1}^{N} - \ln \frac{e^{s \cdot T(\cos(\theta_{y_i}))}}{e^{s \cdot T(\cos(\theta_{y_i}))} + \sum_{j \neq y_i} e^{s \cdot N(\cos \theta_j)}} \tag{6}$$

and

$$T(\cos(\theta_{y_i})) = \cos(\theta_{y_i} + m), m > 0$$

$$N(\cos(\theta_j)) = \begin{cases} \cos(\theta_j), & T(\cos(\theta_{y_i})) \geq \cos(\theta_j) \\ (t + |\cos(\theta_j)|) \cdot \cos(\theta_j), & \cos(\theta_{y_i}) > \cos(\theta_j) > T(\cos(\theta_{y_i})) \\ (t - |\cos(\theta_j)|) \cdot \cos(\theta_j), & \cos(\theta_j) > \cos(\theta_{y_i}) > T(\cos(\theta_{y_i})) \end{cases}$$

In the proposed loss function, the model incorporates an adaptive segmented negative cosine similarity $N(\cos(\theta_j))$, which takes into account three different types of samples. In $N(\cos(\theta_j))$, the hyper-parameter t takes a small value close to 0 at the early stage. During this stage, the adjustment weights $t + |\cos(\theta_j)|$ and $t - |\cos(\theta_j)|$ for the semi-hard and harder samples are both less than 1, thereby the model will emphasize easy samples. With the increase of iterations, the hyper-parameter t gradually increases, along with the adjustment weights of $t + |\cos(\theta_j)|$ and $t - |\cos(\theta_j)|$ for the semi-hard and harder samples. Consequently, the model will gradually emphasize the learning of semi-hard samples and harder samples. Furthermore, when $\cos(\theta_j) > 0$ in the condition $(t + |\cos(\theta_j)|) > (t - |\cos(\theta_j)|)$, the model emphasizes the semi-hard samples and suppresses the weights of harder samples, thereby reducing the impact of noisy samples. Conversely, when $\cos(\theta_j) < 0$, indicating the presence of outliers, $N(\cos(\theta_j))$ assigns smaller weights to semi-hard samples compared to harder samples, reducing the impact of outliers.

2.4 Algorithm

The optimization of the loss function based on cosine decision boundary can be realized by the stochastic gradient descent (SGD). The exponential part of the cosine similarity of jth class in the loss function is f_j, i.e.,

$$f_j = \begin{cases} sT(\cos(\theta_{y_i})) & j = y_i \\ s\cos(\theta_j) & \text{easy}, j \neq y_i \\ s(t + \left|\cos(\theta_j)\right|) \cdot \cos(\theta_j) & \text{semi - hard}, j \neq y_i \\ s(t - \left|\cos(\theta_j)\right|) \cdot \cos(\theta_j) & \text{harder}, j \neq y_i \end{cases} \tag{7}$$

Hence, x_i and W_j needs to discuss and calculate in four different situations during the backward propagation process, when $\cos(\theta_j) > 0 \cos(\theta_j) > 0$

$$\frac{\partial L}{\partial x_i} = \begin{cases} \frac{\partial L}{\partial f_{y_i}}(s\frac{\sin(\theta_{y_i}+m)}{\sin(\theta_{y_i})})W_{y_i}, & j = y_i \\ \frac{\partial L}{\partial f_j} sW_j, & \text{easy}, j \neq y_i \\ \frac{\partial L}{\partial f_j} s(t + 2\cos(\theta_j))W_j, & \text{semi - hard}, j \neq y_i \\ \frac{\partial L}{\partial f_j} s(t - 2\cos(\theta_j))W_j, & \text{harder}, j \neq y_i \end{cases} \tag{8}$$

$$\frac{\partial L}{\partial W_j} = \begin{cases} \frac{\partial L}{\partial f_{y_i}}(s\frac{\sin(\theta_{y_i}+m)}{\sin(\theta_{y_i})})x_{y_i}, & j = y_i \\ \frac{\partial L}{\partial f_j} sx_j, & \text{easy}, j \neq y_i \\ \frac{\partial L}{\partial f_j} s(t + 2\cos(\theta_j))x_j, & \text{semi - hard}, j \neq y_i \\ \frac{\partial L}{\partial f_j} s(t - 2\cos(\theta_j))x_j, & \text{harder}, j \neq y_i \end{cases} \tag{9}$$

When $\cos(\theta_j) < 0 \cos(\theta_j) < 0$

$$\frac{\partial L}{\partial x_i} = \begin{cases} \frac{\partial L}{\partial f_{y_i}}(s\frac{\sin(\theta_{y_i}+m)}{\sin(\theta_{y_i})})W_{y_i}, & j = y_i \\ \frac{\partial L}{\partial f_j} sW_j, & \text{easy}, j \neq y_i \\ \frac{\partial L}{\partial f_j} s(t - 2\cos(\theta_j))W_j, & \text{semi - hard}, j \neq y_i \\ \frac{\partial L}{\partial f_j} s(t + 2\cos(\theta_j))W_j, & \text{harder}, j \neq y_i \end{cases} \tag{10}$$

$$\frac{\partial L}{\partial W_j} = \begin{cases} \frac{\partial L}{\partial f_{y_i}}(s\frac{\sin(\theta_{y_i}+m)}{\sin(\theta_{y_i})})x_{y_i}, & j = y_i \\ \frac{\partial L}{\partial f_j} sx_j, & \text{easy}, j \neq y_i \\ \frac{\partial L}{\partial f_j} s(t - 2\cos(\theta_j))x_j, & \text{semi - hard}, j \neq y_i \\ \frac{\partial L}{\partial f_j} s(t + 2\cos(\theta_j))x_j, & \text{harder}, j \neq y_i \end{cases} \tag{11}$$

During the above model optimization process, two parameters are involved, i.e., s and t. In this paper, the scale parameter s is set to 64, and hype-parameter t is set to an adaptive value related to the number of iterations

$$t^{(k)} = \alpha\gamma^{(k)} + (1 - \alpha)t^{(k-1)} \tag{12}$$

where k is the iteration number, $\alpha = 0.99$, $\gamma^{(k)} = \sum_i \cos(\theta_{y_i})$, and the initial $t^{(0)}$ set to 0. The algorithm of this method is given in Algorithm 1.

Algorithm 1: The loss function based on cosine decision boundary

Step 1(Input):

The sample feature x_i belongs to the class y_i, the weight matrix of the fully connected layer $\boldsymbol{W}^{(0)}$, cosine similarity $\cos(\theta_j)$ of x_i and $W_j^{(0)}$, embedding model parameters $\Theta^{(0)}$, learning rate λ, and margin m.

Step 2(Initialization)

Begin with $k=0, t=0, m=0.5$

Step 3 (Iterative training stage)

While not converged **do**

if $\cos(\theta_{y_i}+m) \geq \cos(\theta_j)$ **then**

$$N(\cos(\theta_j)) = \cos(\theta_j)$$

else if $\cos(\theta_{y_i}) > \cos(\theta_j) > \cos(\theta_{y_i}+m)$ **then**

$$N(\cos(\theta_j)) = (t + |\cos(\theta_j)|) \cdot \cos(\theta_j)$$

else

$$N(\cos(\theta_j)) = (t - |\cos(\theta_j)|) \cdot \cos(\theta_j)$$

end

$$T(\cos(\theta_{y_i})) = \cos(\theta_{y_i}+m)$$

Compute loss function L by Eq. 6,

Compute the gradient of x_i and W_j by Eq.8,9,10,11,

Update weight matrix $\boldsymbol{W}$ and embedding model parameters $\Theta^{(0)}$:

$$\boldsymbol{W}^{(k+1)} = \boldsymbol{W}^{(k)} - \lambda^{(k)} \frac{\partial L}{\partial \boldsymbol{W}}, \quad \Theta^{(k+1)} = \Theta^{(k)} - \lambda^{(k)} \frac{\partial L}{\partial x_i} \cdot \frac{\partial x_i}{\partial \Theta^{(k)}}, \text{ let } k = k+1;$$

Update the parameter t by Eq.12

end while

Step 4(Output): weight matrix $\boldsymbol{W}$ and embedding model parameters $\Theta^{(0)}$

3 Experimental Results

3.1 Implementation Details

- Datasets: In our experiments, nine datasets are employed with two datasets used for training and seven datasets used for testing. Specifically, to validate the robustness of the proposed method, the training datasets consisted of the WebFace [12] and UMDFace [13] datasets. WebFace contains 494,414 images from 10,575 distinct individuals, while UMDFace contains 367,920 images from 8,501 distinct individuals. To evaluate the generalization ability of the method, seven different face databases, namely LFW [14], CFP_FF [15], CFP_FP [15], AgeDB [16], CALFW [17], CPLFW [18], and VGG2_FP [19], were used for testing.

- Experimental Setting: The experiments are implemented with PyTorch framework, with the SGD algorithm used for optimizing the model. The IR_SE_50 network [20] is used as the base network to train the model. The size of samples is set to 112 × 112. The experiments are performed on an RTX5000 GPU environment, the number of epoch is 50. The initial learning rate is set to 0.1 and is divided by 10 at the 28th, 38th, and 46th epoch. The batch size is set to 256, the momentum is set to 0.9, and the weight decay is set to 5e-4.

3.2 Evaluation Results

- Experimental results based on Webface
  The trained model on the Webface dataset is tested on five different face datasets: CFP_FF, CFP_FP, AgeDB, CALFW, and CPLFW. CFP dataset comprises 7,000 facial images from 500 individuals, with 10 frontal face images and 4 profile face images per person, which CFP_FF contains only frontal face images and CFP_FP includes both frontal and profile face images. AgeDB consists of 16,488 labeled images from 568 individuals, with information on person identity, age, and gender. The age range in this database is 1 to 101 years, and the images have real-world backgrounds. CALFW and CPLFW are extensions of the LFW database, incorporating age and pose variations. These datasets encompass various types of challenging samples for face recognition, including profile shots, background variations, age variations, and noise samples.
  Meanwhile, the proposed method is compared to three state-of-the-art angular/cosine-margin-based methods: CosFace, CurricularFace, and EnhanceFace. Experimental accuracies of these methods are shown in Table 1. It can be seen that in CFP_FF dataset consisting of simple samples, the proposed method achieves comparable accuracy to existing methods. However, on the remaining four datasets that focus on semi-hard and harder samples, particularly the CFP_FP, CALFW, and CPLFW with significant amount of noise samples, the proposed method outperforms the existing methods. This highlights the effectiveness of the proposed method, especially when dealing with mislabeled training data. Moreover, the experiment demonstrates the good transferability and generalization of the proposed method by training and testing on different datasets.

Table 1. Verification results (%) of different loss function on five various verification datasets by training on Webface. The highest accuracies are in bold.

Dataset / method	CFP_FF	CFP_FP	AgeDB	CALFW	CPLFW
CosFace	99. 51	97. 16	94. 55	93. 75	90. 72
CurricularFace	99. 51	96. 81	94. 83	93. 43	90. 47
EnhanceFace	99. 44	97. 36	94. 60	93. 67	90. 00
the proposed method	**99. 51**	**97. 41**	**94. 83**	**93. 83**	**90. 82**

- **Experimental results based on UMDFace**

  To validate the robustness of the proposed method, the proposed method is trained on UMDFace dataset, and tested on CFP_FF, CFP_FP, CALFW, LFW and VGG2_FP datasets. CFP_FF, CFP_FP, and CALFW are used to test the recognition stability of the model on the same datasets, while LFW and VGG2_FP are used to test the generalization ability of the model. LFW consists of 13,233 face images from 5,749 different individuals, and VGG2_FP contains 3.31 million face images from 9,131 different individuals. To further verify the effectiveness of the method, the proposed method is also compared with ArcFace, and the experimental accuracy is shown in Table 2.

  The experimental results in Table 2 show that the proposed method robustly achieves better recognition accuracy across different datasets. The LFW further validate the good generalization ability of this method. And the accuracy results in VGG2_FP indicate that the proposed method outperforms even on large-scale datasets with millions of samples. Above experiments indicate that classifying the difficulty level of sample types and using segmented adaptive weighting can effectively emphasize semi-hard samples while mitigating the impact of mislabeled and other noise samples on accuracy.

Table 2. Verification results (%) of different loss function on five various verification datasets by training on UMDFace. The highest accuracies are in bold.

Dataset / method	CFP_FF	CFP_FP	CALFW	LFW	VGG2_FP
CosFace	99. 46	96.20	94. 37	99.45	93.80
ArcFace	99. 41	95.86	94. 23	99.40	93.60
CurricularFace	99. 41	95.60	94. 20	99.47	93.46
EnhanceFace	99.39	95.64	94.25	99.45	93.64
the proposed method	**99. 51**	**96.26**	**94. 57**	**99.53**	**94.01**

4 Conclusion

This paper proposes a deep face recognition model based on the cosine boundary loss function, which has two main characteristics. Firstly, a novel cosine decision boundary is proposed to address mislabeled and noise samples in FR datasets. It categorizes into easy, semi-hard, and harder samples based on angular intervals. Secondly, an adaptive segmented loss function is introduced, considering the roles of samples during training. It emphasizes easy samples in early stages, focuses on semi-hard samples while suppressing harder samples, effectively controlling the impact of noise samples. Experimental results on different FR datasets demonstrate the robustness and generalization of the proposed method, even in scenarios involving millions of samples.

Acknowledgements. This study was supported by National Natural Science Foundation of China under Grant 41771375, Grant 31860182, and Grant 41961053, Natural Science Foundation of Henan under Grant 232300421071, Scientific and Technological Innovation Talent in Universities of Henan Province under Grant 22HASTIT015, and Youth key Teacher of Henan under Grant 2020GGJS030.

References

1. Shan, X., Lu, Y., Li, Q., Wen, Y.: Model-based transfer learning and sparse coding for partial face recognition. IEEE Trans. Circuits Syst. Video Technol. **31**(11), 4347–4356 (2021)
2. Wang, M., Deng, W.: Deep face recognition: a survey. Neurocomputing **429**, 215–244 (2021)
3. Deng, J., Zhou, Y., Zafeiriou, S.: Marginal loss for deep face recognition. In: 2017 IEEE Conference on Computer Vision and Pattern Recognition Workshops (CVPRW), Hawaii, pp. 2006–2014 (2017)
4. Schroff, F., Kalenichenko, D., Philbin, J.: Facenet: a unified embedding for face recognition and clustering. In: 2015 IEEE Conference on Computer Vision and Pattern Recognition (CVPR), Boston, pp. 815–823 (2015)
5. Wu, Y., Liu, H., Li, J., Fu, Y.: Deep face recognition with center invariant loss. In: Proceedings of the on Thematic Workshops of ACM Multimedia 2017 (Thematic Workshops 2017), pp. 408–414. Association for Computing Machinery, New York (2017)
6. Liu, W., Wen, Y., Yu, Z., Li, M., Raj, B., Song, L.: Sphereface: deep hypersphere embedding for face recognition. In: 2017 IEEE Conference on Computer Vision and Pattern Recognition (CVPR), Boston, pp. 6738–6746 (2017)
7. Wang, H., et al.: Cosface: large margin cosine loss for deep face recognition. In: 2018 IEEE/CVF Conference on Computer Vision and Pattern Recognition (CVPR), Salt Lake City, pp. 5265–5274 (2018)
8. Deng, J., Guo, J., Yang, J., Xue, N., Kotsia, I., Zafeiriou, S.: Arcface: additive angular margin loss for deep face recognition. In: Proceedings of the IEEE Conference on Computer Vision and Pattern Recognition, Long Beach, pp. 4690–4699 (2019)
9. Huang, Y., et al.: Curricularface: adaptive curriculum learning loss for deep face recognition. In: 2020 IEEE/CVF Conference on Computer Vision and Pattern Recognition, Seattle, pp. 5900–5909 (2020)
10. Hu, W., Huang, Y., Zhang, F., Li, R.: Noise-tolerant paradigm for training face recognition CNNs. In: 2019 IEEE/CVF Conference on Computer Vision and Pattern Recognition (CVPR), Long Beach, pp. 11879–11888 (2019)
11. Wang, J., Zheng, C., Yang, X., Yang, L.: Enhanceface: adaptive weighted softmax loss for deep face recognition. IEEE Signal Process. Lett. **29**, 65–69 (2022)
12. Yi, D., Lei, Z., Liao, S., Li, S.Z.: Learning face representation from scratch (2014). https://arxiv.org/pdf/1411.7923.pdf
13. Bansal, A., Nanduri, A., Castillo, C.D., Ranjan, R., Chellappa, R.: Umdfaces: an annotated face dataset for training deep networks. In: 2017 IEEE International Joint Conference on Biometrics (IJCB), pp. 464–473 (2017)
14. Huang, G., Mattar, M., Berg, T., Learned-Miller, E.: Labeled faces in the wild: a database for studying face recognition in unconstrained environments. Tech. rep. University of Massachusetts (2007)
15. Sengupta, S., Chen, J.C., Castillo, C., Patel, V.M., Chellappa, R., Jacobs, D.W.: Frontal to profile face verification in the wild. In: 2016 IEEE Winter Conference on Applications of Computer Vision (WACV), Lake Placid, pp. 1–9 (2016)

16. Moschoglou, S., Papaioannou, A., Sagonas, C., Deng, J., Kotsia, I., Zafeiriou, S.: Agedb: the first manually collected, in-the-wild age database. In: 2017 IEEE Conference on Computer Vision and Pattern Recognition Workshops, Honolulu, pp. 1997–2005 (2017)
17. Zheng, T., Deng, W.: Cross-pose IFW: a database for studying cross-pose face recognition in unconstrained environments. arXiv preprint arXiv:1708.08197(2017)
18. Zheng, T., Deng, W., Hu, J.: Cross-age IFW: a database for studying cross-age face recognition in unconstrained environments. Tech. rep. Beijing University of Posts and Telecommunications (2018)
19. Cao, Q., Shen, L., Xie, W., Parkhi, O.M., Zisserman, A.: Vggface2: a dataset for recognising faces across pose and age. In: IEEE International Conference on Automatic Face and Gesture Recognition, Xi'an, pp. 67–74 (2018)
20. Zhang, K., Zhang, Z., Li, Z., Qiao, Y.: Joint face detection and alignment using multitask cascaded convolutional networks. IEEE Signal Process. Lett. **23**(10), 1499–1503 (2016)

A Mix Fusion Spatial-Temporal Network for Facial Expression Recognition

Chang Shu(✉) and Feng Xue

University of Electronic Science and Technology of China, Chengdu, Sichuan, China
changshu@uestc.edu.cn

Abstract. Facial expression is a powerful, natural and universal signal for human beings to convey their emotional states and intentions. In this paper, we propose a new spatial-temporal facial expression recognition network which outperforms many state-of-the-art methods. Our model is composed by two networks, a temporal feature extraction network based on facial landmarks and a spatial feature extraction network based on densely connected network. Image preprocessing method is optimized according to the features of the expression image to reduce network's overfitting on small datasets. In addition, we propose a mix fusion strategy to better combine spatial and temporal features. Finally, experiments on public datasets are conducted to verify the effectiveness of each module and the improvement of expression recognition accuracy of the spatial-temporal fusion network. The accuracies on OULU-CASIA and CK + datasets reach 90.21% and 99.82% respectively.

Keywords: Facial Expression Recognition · Facial Landmarks · Model Fusion · Spatial-Temporal Network

1 Introduction

Facial expression is one or more motions or positions of the muscles beneath the skin of the face. These movements can convey the emotional state of an individual to observers. With the development of technologies like human-computer interaction and image processing, facial expression recognition (FER) has become a hot research field. Research results of FER have been widely used in driver fatigue surveillance, online teaching, medical treatment and other fields [1].

FER can be divided into two categories: still frame based and image sequence based. Still frame based methods extract spatial features from a single image. Traditional methods depend on handcrafted features like local binary patterns (LBP) and histogram of oriented gradients (HOG). With relatively sufficient data collected by various competitions like FER2013 [2] and emergence of more deliberately designed Convolutional Neural Networks (CNN), many deep learning methods were brought into this field. Image sequence based methods focus more on extracting temporal features from consecutive frames and try to capture dynamic information. LBP-TOP [4] uses time as the

This work has been supported by the Fundamental Research Funds for the Central Universities.

Q. Liu et al. (Eds.): PRCV 2023, LNCS 14429, pp. 315–326, 2024.
https://doi.org/10.1007/978-981-99-8469-5_25

third orthogonal plane, STSGN [5] uses graph network, PHRNN [6] uses bidirectional recurrent neural network (BRNN).

While neural networks significantly improve FER performance, small datasets have become a big impediment to apply more powerful networks. There are mainly two ways to overcome this problem. Firstly, data augmentation can be applied. There are many common augmentation technics in the field of computer vision such as rotation, translation, random cropping and adding noises. The second way is to apply networks with small amount of layers and parameters to alleviate overfitting caused by small datasets [6, 7].

In this paper, we are interested in improving recognition accuracy for image sequences on small datasets. We construct two networks to extract different features and combine them as a spatial-temporal network. One network takes facial landmarks gained from image sequences as input and tries to capture dynamic information form consecutive frames. The other network is trained on static frames with densely connected network. To make full use of spatial and temporal features, we design a mix fusion strategy to combine information gained by two networks.

Three main contributions of this paper can be summarized as follows:

1. Targeted augmentation method for landmarks is designed according to the characteristics of FER.
2. Feature reusing strategy is applied to alleviate overfitting problem on deep networks.
3. A mix fusion strategy is presented to compose information extracted from two different networks. Compared with traditional fusion strategies like middle fusion and late fusion, better expression recognition accuracy is achieved by our strategy.

2 Related Work

2.1 Facial Expression Recognition

The development of facial expression recognition (FER) methods can be divided into four stages. The first stage is to classify images with handcrafted features extracted by methods like Gabor texture [8], LBP and HOG. Although handcrafted features suffered the problem of hard to design, it is a compromise of lacking proper dataset.

With relatively sufficient data collected by various competitions like FER2013 [2] and rapidly increasing computational ability, CNNs were brought into this field of study. Tang [3], winner of FER2013 uses a simple CNN with support vector machine(SVM) as classifier and outperforms other methods by 2%.

The third stage is to combine spatial and temporal information. Image sequence based methods using temporal features generally achieve better results than still image based methods [12]. Thus, models composed by a temporal network extracting features from image sequences and a spatial network extracting appearance features are put forward. And various classifiers are used, DTGN [7] uses fully connect network, PHRNN [6] uses bidirectional recurrent neural network (BRNN).

The fourth stage is to optimize neural networks. Common datasets like CK + [9] and OULU-CASIA [10] only have hundreds of samples and that impedes using larger and more powerful networks. For example, DTAN [7] only has two convolutional layers and MSCNN [6] only has four. To get better performance, many deliberately designed networks are proposed. Compact-CNN [11] uses PReLU and larger convolution filters to improve the DTAN model. STSGN [5] uses sematic facial graph to encode both spatial and temporal information and constructs a novel graph neural network. Zhu [12] uses cascade attention blocks to improve feature aggregation results.

2.2 Model Fusion

Multimodal data fusion has long been investigated in medical imaging since images with multimodalities are wildly used in computed tomography (CT) and magnetic resonance imaging (MRI) [13]. The underlying motivation for model fusion is to extract complementary features from different modalities and thus improve decision making. There are three main fusion strategies: early fusion, joint fusion and late fusion. Early fusion is also known as feature level fusion. Features extracted from different modalities are concatenated or added as a new vector before feeding into classifier. Joint fusion is similar to early fusion but loss can be back propagated from classifier to feature extraction models. Therefore, better features can be extracted through the training process. Late fusion also known as decision level fusion, makes final decision by aggregation function to combine decisions from each model. Common aggregation functions include averaging, weighted sum and majority voting.

3 Our Model

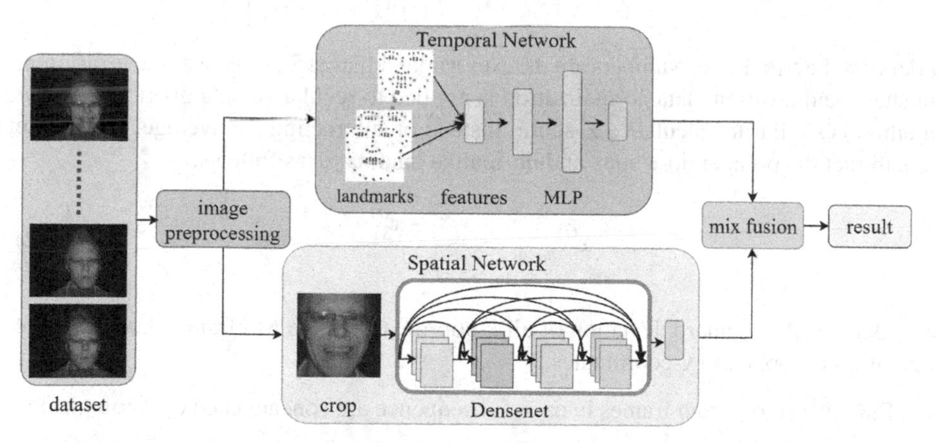

Fig. 1. Spatial-temporal network for FER

Figure 1 demonstrates our proposed FER model which contains two networks: a landmark based network to extract temporal features from image sequences and a densely connected convolutional network to extract spatial features from static images. Then we design a mix fusion strategy to combine two kinds of features and improve the recognition accuracy. We will describe details of our method in the following section.

3.1 Temporal Network

In temporal network, we follow steps in Fig. 2 to extract basic temporal features.

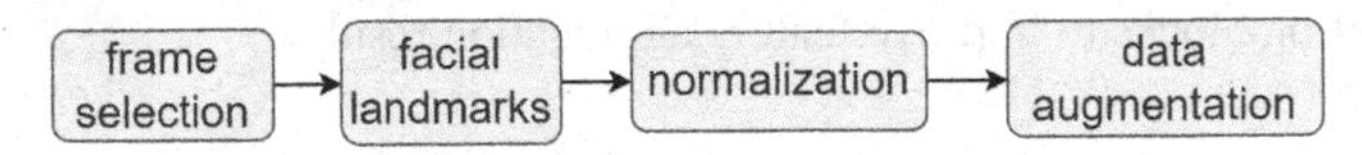

Fig. 2. Image preprocessing for temporal network

Frame Selection. Neural networks generally need inputs with the same dimension. To solve the problem that each image sequence in the dataset has different length, for each sequence we select a subsequence which contains five frames with equal intervals. Each subsequence contains the first frame with neutral expression and the last frame with the peak expression. Frame selection can retain dynamic temporal information while reducing data needs to be processed. Therefore, it can also accelerate the whole network.

Facial Landmarks. To get facial landmarks, we use Haar Cascade [14] to detect bounding boxes for faces. Then coordinates of landmarks are localized by Deep Alignment Network (DAN) [15]. DAN can extract 68 points for each image and shows the contour of eyes, nose, mouth, eyebrows and check.

Normalization. We reform the landmarks coordinates of each frame into a vector as:

$$v^{(t)} = \left[x_1^{(t)}, x_2^{(t)}, \cdots, x_{68}^{(t)}, y_1^{(t)}, y_2^{(t)}, \cdots, y_{68}^{(t)}\right]^T \tag{1}$$

t denotes the t-th frame. Numbers are indexes for landmarks. Since every face is different in shape and position, data normalization is applied to regularize data distribution. Our method is similar to calculate a Z-score. Instead of subtracting the average, we choose to subtract the point at nose apex and normalize landmarks as follows:

$$x_{i-norm}^{(t)} = \frac{x_i^{(t)} - x_{31}^{(t)}}{\sigma_x^{(t)}} \tag{2}$$

$\sigma_x^{(t)}$ denotes the standard deviation of all x-coordinates at the t-th frame. And the same operation is applied to y-coordinates.

Finally, vectors from frames in each subsequence are concatenated chronologically.

$$v = \left[x_{1-norm}^{(1)}, \cdots, x_{68-norm}^{(1)}, y_{1-norm}^{(1)}, \cdots, y_{68-norm}^{(1)}, \cdots, y_{1-norm}^{(5)}, \cdots, y_{68-norm}^{(5)}\right]^T \tag{3}$$

And v is the input feature for this sample.

Data Augmentation. We utilize data augmentation technics to expand vectors gained from former step by 22 times. We use three technics to augment the data: rotation, flipping and stretching. They can be achieved by Eqs. 4, 5 and 6. Rotation and flipping can alleviate the influence of position and asymmetry of faces. Stretching is to mitigate different aspect ratio of faces.

Firstly, we horizontally flip landmarks. Secondly, we rotate original and flipped landmarks by degrees in $\{-15^{\circ}, -10^{\circ}, -5^{\circ}, 5^{\circ}, 10^{\circ}, 15^{\circ}\}$. Rotation can be achieved by multiplying a 2x2 rotation matrix.

$$\left(x_i^{(t)}, y_i^{(t)}\right)_{rotated}^T = \begin{bmatrix} \cos\theta & \sin\theta \\ \sin\theta & \cos\theta \end{bmatrix} \left(x_{i-norm}^{(t)}, y_{i-norm}^{(t)}\right)^T \tag{4}$$

$$\left(x_i^{(t)}, y_i^{(t)}\right)_{stretched}^T = \left(\beta_1 x_{i-norm}^{(t)}, \beta_2 y_{i-norm}^{(t)}\right)^T \tag{5}$$

$$\left(x_i^{(t)}, y_i^{(t)}\right)_{flipped}^T = \left(-x_i^{(t)}, y_i^{(t)}\right)^T \tag{6}$$

Finally, we rescale x-coordinates and y-coordinates by ratio of $\{0.5, 0.8, 1.5, 2\}$ respectively. Thus, we get original(1) and flipped(1) landmarks, their rotated versions(2x6) and stretched versions(2x4).

Network Architecture

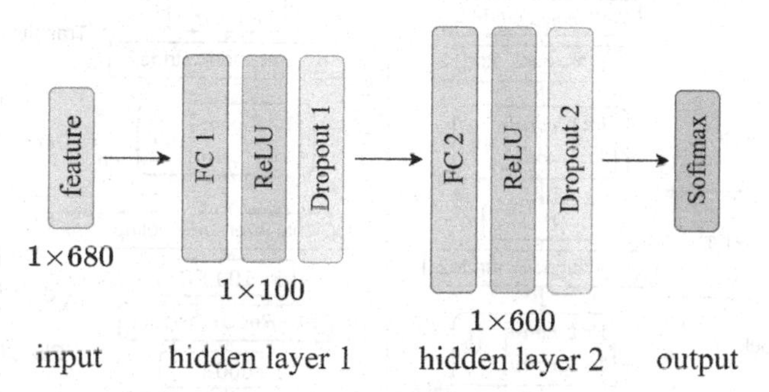

Fig. 3. Temporal network structure

We use a similar two-layer multilayer perceptron in [7]. It has two fully connected layers and a Softmax layer. To avoid overfitting, dropout is adopted after every hidden layer and L2 regularization is introduced. L2 regularization can be described as following function (Fig. 3):

$$Loss = -\frac{1}{N}\sum_{i}\sum_{c=1}^{M} y_{ic} ln(p_{ic}) + \alpha||w||_2^2 \tag{7}$$

M is the number of total classes, y_{ic} is a sign function, it is 1 when sample i belongs to class c, p means probability, w denotes weights in the second fully connected layer.

ReLU is chosen to be activation function. For training, we use Stochastic gradient descend with momentum and decay, and use cross-entropy as loss function.

3.2 Spatial Network

While landmark-based network is a light weight network with high efficiency, it does not use spatial information in every frame sufficiently. For example, landmarks of both sadness and disgust have downward corners of lips and similar contours of faces. The main difference between them is the amount of wrinkles. To detect such difference we use a densely connected CNN.

Preprocessing. We choose the last frame from each sequence and use Haar Cascade [14] to detect faces. Then all images are reshaped to 190x190. Data augmentation is also applied with slight difference from landmark-based network. We use online data augmentation method provided by TensorFlow. It takes original dataset as input and generates a random batch of images in real-time. Parameters controlling rotation, shifting, flipping and etc. can be easily set up. We set specs as follows: rotation range from $[-20^{\circ}, 20^{\circ}]$, horizontal and vertical shifting and zoom are set to 0.2, random horizontal flipping is also used.

Network Architecture

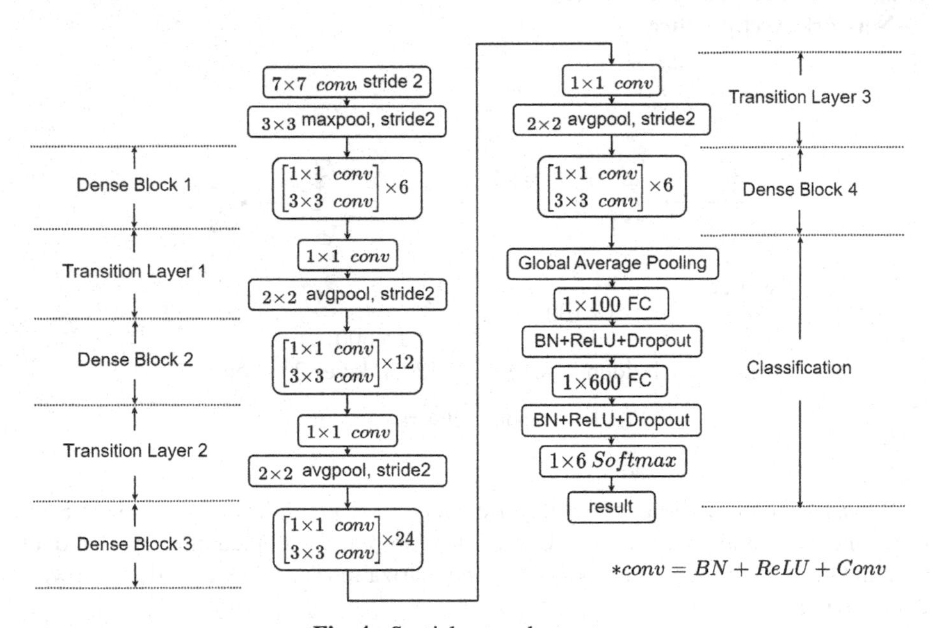

Fig. 4. Spatial network structure

Generally, CNNs used in FER are shallow networks because they are easier to prevent overfitting problems on small datasets. Although deeper networks can extract more complex representations and benefit classification accuracy, they are hard to train on small datasets. Densely connected network (DenseNet) [16] reduces network's parameters by bottleneck layers and increases feature reuse ability by dense connections. Therefore, we use DenseNet-121 with 4 dense blocks and 2 fully connected layers shown in Fig. 4 as our model.

3.3 Mix Fusion Strategy

In general, combination of different networks can improve classification performance. And the reason is obvious, different network focus on different features and the combination can provide more useful information. Many model fusion strategies have been used in FER. We evaluate their performance and design a mix-fusion strategy which achieves better result.

Early Fusion, Joint Fusion and Late Fusion. Early fusion is a very challenging task because it usually requires both inputs have the same dimension and type. Therefore, early fusion turns to use simple networks to extract low level features and combine them at feature level. On this basis, joint fusion can fine tune the whole model by back propagation. Late fusion is to combine outputs of different networks by aggregation functions. The choice of aggregation function varies with specific problem, but usually relies on subjective experience.

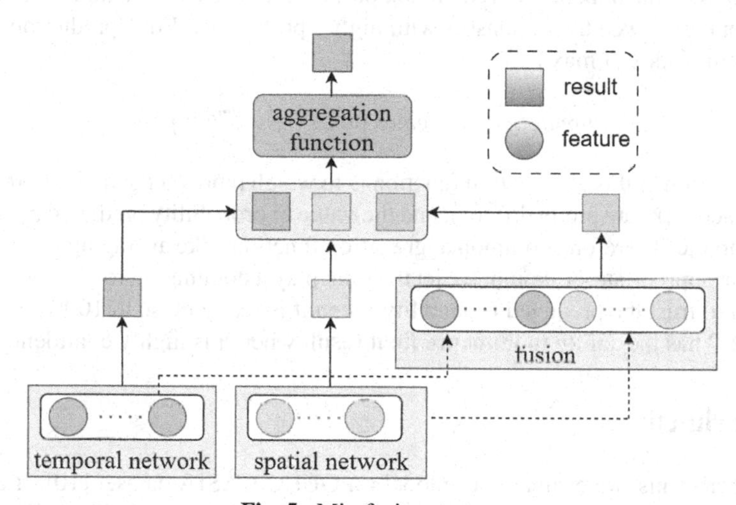

Fig. 5. Mix fusion strategy

Mix Fusion. Early fusion makes use of extracted features but ignores logic of prediction while late fusion is exactly contrary. To make full use of information got by two networks, we proposed a mix-fusion strategy in Fig. 5. It is similar to late fusion, but adds early fusion network as the third network. The third network is aimed to combine feature-wise information.

In the third network, preprocessed data are sent into temporal network and spatial network respectively. After independent feature extraction, two networks are combined before fully connected layers to form a fusion model. We concatenate features from the last fully connected layers before Softmax of two networks. To balance their contribution, two fully connected layers are designed to have same number of neurons. Parameters in two feature extraction models are frozen during training. Only layers after concatenation are trained.

Then we use the following aggregation function to get final classification result from three networks.

$$v^j = \sum_{i=1}^{2} \alpha_i \left[R_i^j + \beta_i P_i^j \right] \quad (8)$$

α and β are tuning parameters. α is used to balance trust on two networks. β is to adjust trust on rank and probability. j denotes j-th class, i denotes $i - th$ network. v^j is the pseudo Softmax output for j-th class. P_i^j is the probability or original Softmax output generates by network i for j-th class. R is rank points given by function 9:

$$R_i^j = num_{class} - rank_i(j) \quad (9)$$

Rank points are generated by subtracting probability rank from number of expression classes. For example, if there are 6 classes and happy expression has the highest probability, then the rank of it is 1, and rank point for it is 6. The function of rank points is to assign higher weights for classes with higher probability. Final prediction is made by getting the index of max v:

$$prediction = argmax\left(v^1, \cdots, v^{num_{class}}\right) \quad (10)$$

The intuition of this aggregation function is to weigh more on the rank of probability since two networks are pretty different and the value of probability predicted by them are not comparable. Therefore, common aggregation functions like averaging and weighted sum are not appropriate. Although we let the rank play a dominating role in the function, it is worth to trust the predicted probability when it is really close to 100%. So we use β to make P has the ability to influence final result when it is highly confident.

4 Experiments

In our experiments, we evaluate our model on OULU-CASIA dataset [10]. It contains image sequences captured by near-infrared and visible light under different illumination condition. In each sequence, the first frame is neutral and the last frame has peak expression. There are six expressions in the dataset: anger, disgust, fear, happiness, sadness and surprise. We use 480 image sequences captured by visible light under normal illumination condition. To better evaluate robustness of our model, 10-fold cross validation strategy is adopted in all experiments. This strategy is to equally divide the whole dataset into ten parts. And nine parts are used as training set, the rest one as validation set. Each part will be used as validation set once and final result is the average of ten validation accuracy.

4.1 Frame Selection

As mentioned earlier in this paper, select fixed length of frames can align input dimension and reduce workload of data processing. Therefore, we evaluate some frame selection strategies (Table 1).

Table 1. Influence of frame selection strategy on temporal network

Selection strategy	Accuracy
8 frames, equal interval	77.71
5 frames, from the last, stride 2	74.58
5 frames, first and last 4	76.25
5 frames, equal interval	**79.79**

An intuitive strategy is to pick more frames similar to peak expressions. But without neutral frames, dynamic information is hard to extract. And more inconspicuous expressions will be chosen when we select more frames. Therefore, a common strategy in many researches is to pick last 3 frames and the first neutral frame [1]. We pick last 4 frames to make subsequences have the same length. To further improve dynamic information extraction ability, we spread selected frames with equal intervals and get a gain in accuracy of 3%. Theoretically, selecting more frames brings more information to the network. However, degradation is found when we increase frames to 8. This may be caused by facial landmark localization algorithm. There are different perturbations between localized landmarks and ground truth among frames in each subsequence. And such perturbations bring negative impact to the final results.

4.2 Augmentation Method

Table 2. Influence of augmentation method on temporal network

Method	Accuracy
flipping*1, rotation*6	79.79
flipping*1, rotation*12,	80.41
flipping*1, rotation*12, stretching*8	**82.29**

Neural networks have huge amount of parameters to be determined, and big amount of data is needed. Different augmentation method has quite diverse performance. We evaluate some of them and design a better augmentation method. Traditional methods use flipping and rotation to alleviate influence of different face positions. However, aspect ratios of faces are ignored. Different face aspect ratios also lead to different distances between eyes and between mouth and nose. To reduce impact of such irrelevant aspect, we stretch landmarks horizontally and vertically. With our method, over 2% of improvement is achieved (Table 2).

4.3 Regularization

Regularization is a common technic to generate better results by adding extra terms to the loss function. L1-norm is used to generate sparse results, but some features will be deprecated. Therefore, it is not appropriate for small datasets in our task. We use L2-norm to alleviate overfitting problem and that brings promotion in recognition accuracy (Table 3).

Table 3. Influence of regularization on temporal network

Method	Accuracy
none	82.29
$l_1 \alpha = 0.1$	80.41
$l_2 \alpha = 0.01$	81.04
$l_2 \alpha = 0.05$	82.08
$\boldsymbol{l_2 \alpha = 0.1}$	**83.33**
$l_2 \alpha = 0.2$	81.04

4.4 Fusion Strategy

We evaluate common model fusion strategies and compare them with our mix fusion.

Table 4. Accuracy of our method compared with other FER models

Model	OULU-CASIA	CK +
DTGN [7]	74.17	92.35
DTAGN [7]	81.46	97.25
PHRNN [6]	78.96	96.36
PHRNN-MSCNN [6]	86.25	98.50
STSGN [5]	87.23	98.63
Cascade Attention [12]	89.29	99.23
Our temporal network	83.33	95.11
Our spatial network	89.17	99.39
Early fusion	89.17	99.08
Late fusion	87.08	97.55
Joint fusion	89.17	98.47
Our method	**90.21**	**99.82**

Result in Table 4 shows that all three common fusion strategies cannot improve recognition accuracy in this circumstance. Moreover, the performance of late fusion is even worse than a single DenseNet. We think the difference of accuracy in two networks impede fusion models to achieve better performance. In early fusion and joint fusion, fusion model tends to use more information from the network with higher accuracy and gives low weight to the landmark network. This may explain why they generate similar results with DenseNet. In late fusion, final decision is made by aggregating outputs of two networks and may be degraded by the lower accuracy network.

But joint fusion is a little bit different. Although final result does not change, weights are different. This means joint fusion model recognizes expressions from a new perspective and may complement our two networks. We evaluate our model on OULU-CASIA and CK + datasets, experiment result in Table 4 proves our method and mix fusion model achieves the best performance. Figure 6 shows confusion matrices on both datasets.

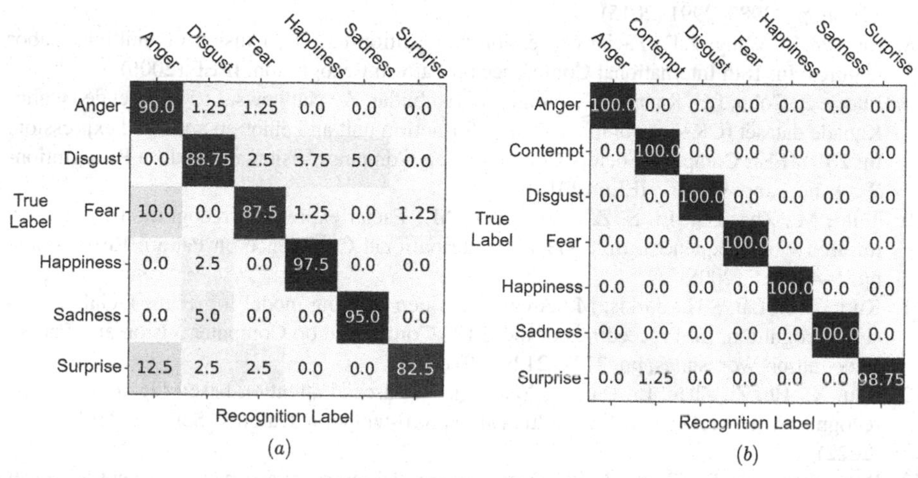

Fig. 6. Confusion matrix of our model, (a) is on OULU-CASIA dataset, (b) is on CK + dataset.

5 Conclusion

In this paper, we proposed a new spatial-temporal FER model and a mix fusion strategy. We demonstrate that specific image preprocessing method, regularization and feature reusing can improve model's performance on small datasets. We also show our mix fusion strategy outperforms common fusion strategies in experiment. With the help of mix fusion strategy, our model provides the state-of-the-art accuracies on OULU-CASIA and CK + datasets.

References

1. Li, S., Deng, W.: Deep facial expression recognition: a survey. IEEE Trans. Affective Comput. **13**, 1195-1215 (2022)
2. Goodfellow, I.J., et al.: Challenges in representation learning: a report on three machine learning contests. In: Lee, M., Hirose, A., Hou, ZG., Kil, R.M. (eds.) Neural Information Processing. ICONIP 2013. LNCS, vol. 8228. Springer, Heidelberg (2013). https://doi.org/10.1007/978-3-642-42051-1_16

3. Tang, Y.: Deep learning using linear support vector machines (2013). arXiv preprint arXiv: 1306.0239
4. Zhao, G., Pietikainen, M.: Dynamic texture recognition using local binary patterns with an application to facial expressions. IEEE Trans. Pattern Anal. Mach. Intell. **29**(6), 915–928 (2007)
5. Zhou, J., Zhang, X., Liu, Y., Lan, X.: Facial expression recognition using spatial-temporal semantic graph network. In: 2020 IEEE International Conference on Image Processing (ICIP), pp. 1961–1965. IEEE (2020)
6. Zhang, K., Huang, Y., Du, Y., Wang, L.: Facial expression recognition based on deep evolutional spatial-temporal networks. IEEE Trans. Image Process. **26**(9), 4193–4203 (2017)
7. Jung, H., Lee, S., Yim, J., Park, S., Kim, J.: Joint fine-tuning in deep neural networks for facial expression recognition. In: Proceedings of the IEEE International Conference on Computer Vision, pp. 2983–2991 (2015)
8. Liu, W. F., Wang, Z.F.: Facial expression recognition based on fusion of multiple Gabor features. In: 18th International Conference on Pattern Recognition. IEEE (2006)
9. Lucey, P., Cohn, J.F., Kanade, T., Saragih, J., Ambadar, Z., Matthews, I.: The extended Cohn-Kanade dataset (CK+): a complete dataset for action unit and emotion-specified expression. In: 2010 IEEE Computer Society Conference on Computer Vision and Pattern Recognition-Workshops, pp. 94–101. IEEE (2010)
10. Taini, M., Zhao, G., Li, S. Z., Pietikainen, M.: Facial expression recognition from near-infrared video sequences. In: 2008 19th International Conference on Pattern Recognition, pp. 1–4. IEEE (2008)
11. Kuo, C.M., Lai, S.H., Sarkis, M.: A compact deep learning model for robust facial expression recognition. In: Proceedings of the IEEE Conference on Computer Vision and Pattern Recognition Workshops, pp. 2121–2129 (2018)
12. Zhu, X., He, Z., Zhao, L., Dai, Z., Yang, Q.: A cascade attention based facial expression recognition network by fusing multi-scale spatio-temporal features. Sensors **22**(4), 1350 (2022)
13. Ramachandram, D., Taylor, G.W.: Deep multimodal learning: a survey on recent advances and trends. IEEE Signal Process. Mag. **34**(6), 96–108 (2017)
14. Viola, P., Jones, M.: Rapid object detection using a boosted cascade of simple features. In: Proceedings of the 2001 IEEE Computer Society Conference on Computer Vision and Pattern Recognition. CVPR 2001, vol. 1, pp. I-I. IEEE (2001)
15. Kowalski, M., Naruniec, J., Trzcinski, T.: Deep alignment network: a convolutional neural network for robust face alignment. In: Proceedings of the IEEE Conference on Computer Vision and Pattern Recognition Workshops, pp. 88–97 (2017)
16. Huang, G., Liu, Z., Van Der Maaten, L., Weinberger, K.Q.: Densely connected convolutional networks. In: Proceedings of the IEEE Conference on Computer Vision and Pattern Recognition, pp. 4700–4708 (2017)

CTHPose: An Efficient and Effective CNN-Transformer Hybrid Network for Human Pose Estimation

Danya Chen, Lijun Wu(✉), Zhicong Chen, and Xufeng Lin

College of Physics and Information Engineering, Fuzhou University, Fuzhou, China
lijun.wu@fzu.edu.cn

Abstract. Recently, CNN-Transformer hybrid network has been proposed to resolve either the heavy computational burden of CNN or the difficulty encountered during training the Transformer-based networks. In this work, we design an efficient and effective CNN-Transformer hybrid network for human pose estimation, namely CTHPose. Specifically, Polarized CNN Module is employed to extract the feature with plentiful visual semantic clues, which is beneficial for the convergence of the subsequent Transformer encoders. Pyramid Transformer Module is utilized to build the long-term relationship between human body parts with lightweight structure and less computational complexity. To establish long-term relationship, large field of view is necessary in Transformer, which leads to a large computational workload. Hence, instead of the entire feature map, we introduced a reorganized small sliding window to provide the required large field of view. Finally, Heatmap Generator is designed to reconstruct the 2D heatmaps from the 1D keypoint representation, which balances parameters and FLOPs while obtaining accurate prediction. According to quantitative comparison experiments with CNN estimators, CTHPose significantly reduces the number of network parameters and GFLOPs, while also providing better detection accuracy. Compared with mainstream pure Transformer networks and state-of-the-art CNN-Transformer hybrid networks, this network also has competitive performance, and is more robust to the clothing pattern interference and overlapping limbs from the visual perspective.

Keywords: Human pose estimation · Transformer · Long-range dependency

1 Introduction

Based on the locations of human body keypoints, the human pose that contains rich features of skeletons and limbs, can be predicted. It plays an important role in several visual applications and researches, such as behavior analysis [5], human-machine interaction [17], etc.

This work was supported by the National Natural Science Foundation of China (Grant Nos. 51508105 and 61601127), and the Fujian Provincial Department of Science and Technology of China (Grant Nos. 2019H0006, 2021J01580 and 2022H0008).

Q. Liu et al. (Eds.): PRCV 2023, LNCS 14429, pp. 327–339, 2024.
https://doi.org/10.1007/978-981-99-8469-5_26

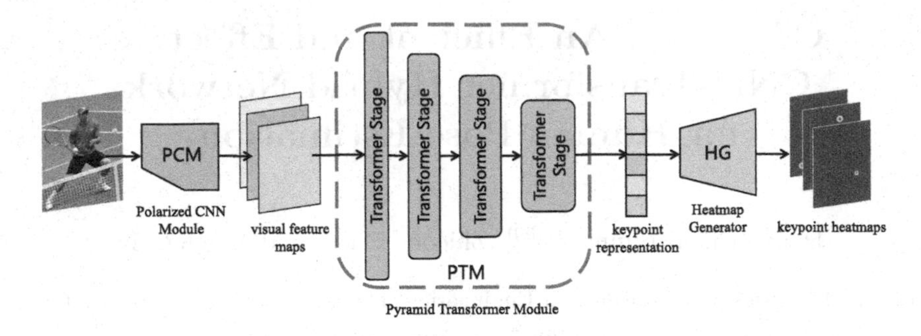

Fig. 1. Schematic illustration of CTHPose

The convolutional networks have been widely used in human pose estimators [2,18,20,26] by virtue of the powerful semantic extraction ability. As known, accurate extraction of keypoints usually requires a large receptive field to provide better visual clues. For this reason, the CNN network is constantly expanded to improve the detection accuracy [18,20,26], which brings a heavy computational burden to the network.

In the past few years, Transformer has achieved great success in computer vision tasks with its powerful global dependency modeling capability [1,4,24,25]. In Transformer, the input image is divided into patches and global features can be well extracted through self-attention and feed-forward layers. However, compared to convolution models, the pure Transformer architecture usually needs more training resources to converge. For example, each self-attention layer will occupy large memory space and more computational complexity that is proportional to the square of the input feature size [11,16].

Recently, the CNN-Transformer hybrid estimators are proposed [10,12,27, 28], which usually extract low-level features through CNN first, and then build the lone-range relationship by Transformer. They utilizes both local and global dependencies to obtain accurate results. However, the existing methods usually pays more attention to study the Transformer part, and seldom explore the contribution of CNN. The improper strategies of heatmap generation also brings more computing burden [12,27].

In this work, we design an efficient and effective CNN-Transformer hybrid human pose estimator (CTHPose), which fully utilizes the powerful visual semantics extraction ability of CNN and the long-range dependencies modeling capability of Transformer to accurately locate the keypoints. As depicted in Fig. 1, CTHPose is composed of three submodules: Polarized CNN Module (PCM), Pyramid Transformer Module (PTM) and Heatmap Generator (HG).

In PCM, parallel branches together with polarized blocks [15] that is inspired by the Polaroid filter in camera, is utilized to better extract the semantic features. Better semantic features will be beneficial to the convergence of the subsequent Transformer encoders. To reduce the computing complexity, a shallow PTM is then employed to model the long-range dependencies of feature maps. Moreover, a shifted window is introduced to reduce the input feature size of the self-attention module, which can effectively reduce the computation burden. At last, HG is proposed to recover the 1D representation into 2D heatmaps with target resolution. Compared with the other hybrid networks [12,27], the adoption of HG can balance the required resource and the accuracy. After the above efforts, CTHPose provides a comparable performance while it can be trained in only one GPU with about $\frac{1}{4}$ batch size of the general setting. In summary, our main contributions are as follows:

- We design PCM to better extract the semantic features, and prove that it is beneficial to the convergence of the subsequent Transformer.
- To reduce the computation cost, we design PTM to build long-range dependencies and propose HG to reconstruct heatmaps. The combined use of them maintains high accuracy with small model scale.
- Extensive experiments show that CTHPose achieves competitive performance with less computational burden, fewer parameters and less training resources compared with mainstream CNN, pure Transformer and hybrid estimators.

2 Related Work

Human pose estimators based on CNN [2,18,20,26] have been widely proposed due to its powerful feature extraction ability. Chen et al. [2] design a two-stage pyramid network that can detect keypoints from coarse to fine. Newell et al. [18] stack hourglass blocks to gain multi-scale heatmaps. Bin et al. [26] use deep residual network to build a simple baseline for 2D human pose estimation (HPE). Ke et al. [20] propose a strong backbone which maintains high resolution representation throughout the whole network. In order to achieve a better prediction effect, these estimators usually expand receptive field by increasing the depth [26], the width [20] or the number of stacked blocks [18]. Although this method compensates the weakness of CNN in modeling global dependencies to some extend, it leads to the increased parameters and computational burden.

Since Dosovitskiy et al. [4] introduced Transformer into computer visual field, the deep neural networks based on Transformer have gained success in many tasks [1,24,25]. Recently, researchers begin to introduce Transformer into HPE as well. TokenPose [12] apparently represents keypoints as token embeddings to get keypoint constraints directly. TransPose [27] utilizes self-attention mechanism to predict the position of keypoints by the principle of activation maximization. PRTR [10] uses cascade Transformer structure to do HPE end to end. HRFormer [28] utilizes Transformer to learn high-resolution representations for dense prediction tasks. Although these estimators can build the long-range interaction between features, they usually need more training resources, such as more GPUs and extra pretraining.

In this work, our hybrid estimator maximizes the advantages of CNN and Transformer, and make up for their shortcomings mentioned above. It is an efficient and effective network with less training resources.

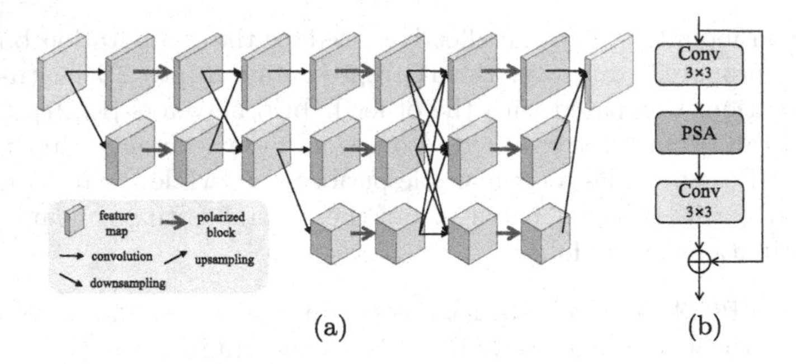

Fig. 2. The architecture of Polarized CNN Module (PCM) (a) The main body of PCM (b) The structure of polarized block

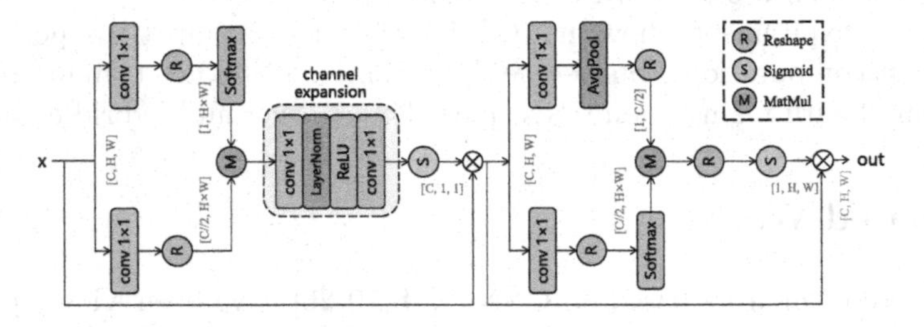

Fig. 3. The structure of polarized self-attention module (PSA)

3 CTHPose

3.1 Polarized CNN Module

The convolutional network is good at extracting low-level visual semantics [9,19]. Compared with raw images, these feature maps extracted are beneficial to the training and convergence of Transformer [12]. Thus, we utilize CNN as the first stage of CTHPose. In order to extract high quality feature maps, we mainly refer to [15,20] to design Polarized CNN Module (PCM), which can adjust weights dynamically while keeping high resolution of features.

As shown in Fig. 2, PCM consists of bridged polarized blocks with multi-scale feature fusion. It totally consists of three parallel branches which keeps $1\times$, $0.5\times$ and $0.25\times$ resolution respectively. Because high resolution is helpful to improve detection accuracy in dense prediction, the first branch is used to generate target feature maps, while the others mainly make feature fusion after up- or down-sampling to obtain multi-scale reception fields.

The polarized block is a residual structure with two 3×3 convolution layers and a polarized self-attention module (PSA) [15] which is a two-stage two-branch residual block shown in Fig. 3. To meet the demand for high resolution of dense prediction, PSA keeps the spatial resolution in the lower branch and halves the channels to balance computational complexity. Moreover, similar to the idea that Polaroid filters light to improve the contrast of photos, the upper branch collapses the spatial or channel dimension in two stages respectively, then keeps and acts on the corresponding orthogonal dimension to adjust weights dynamically. Finally, the outputs of the two branches are fused by matrix multiplication, and the original feature maps are weighted after channel expansion and reshaping.

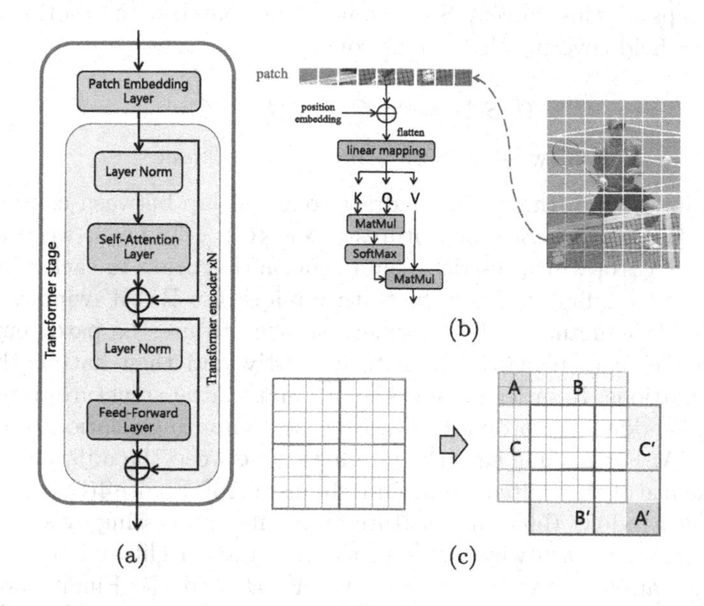

Fig. 4. The architecture and principle of the Transformer stage (a) The structure of the Transformer stage (b) The principle of window self-attention (c) The principle of shifted window mechanism

3.2 Pyramid Transformer Module

For further achieving the target of effectiveness and efficiency, we design Pyramid Transformer Module (PTM) to build the relationship between human body parts. Due to the long-range receptive field of Transformer, PTM can obtain accurate detection with only a few stacked layers. Therefore, the model parameters are obviously decreased.

The whole PTM consists of four Transformer stages, as Fig. 4(a) shows, each stage is composed of a patch embedding layer and N_i (i = 1,2,3,4) Transformer encoders. The main body of the encoder is two residual blocks with self-attention layer and feed-forward layer. In every stage, the feature representations are halved the size and doubled the channels after processed by the patch embedding layer. They form a pyramid structure which is proved to be beneficial to dense prediction tasks [13].

The natural advantage of Transformer in modeling long-range dependencies is mainly comes from the self-attention mechanism (SA) [21]. Its principle can be described as Eq. (1):

$$Attention\left(Q,K,V\right)=softmax\left(\frac{QK^{T}}{\sqrt{d_{k}}}\right)V \tag{1}$$

The self-attention layer linearly maps the input features into query (Q), key (K) and value (V) and then matching Q with K and reacting with V to obtain the attention maps. In this process, SA matches every patch with all others to form the receptive field covering the action scope.

$$\Omega\left(SA\right)=4hwC^{2}+2\left(hw\right)^{2}C \tag{2}$$

$$\Omega\left(\text{window } SA\right)=4hwC^{2}+2W^{2}hwC \tag{3}$$

As Eq. (2) illustrated, there is a quadratic relationship between computational complexity (Ω) and the size of features ($h\times w\times C$) [16]. Thus, we take use of the window SA [16] which restricts self-attention operation to each window. As Fig. 4(b) shows, we firstly divide the feature maps into $W\times W$ windows. Owing to the all-MLP structure of Transformer, we need to use the patch embedding layer to divide feature maps into patches firstly and then flatten them into 1D representations. In order to keep the original spatial structure, the position embedding is added. According to Eq. (3), the computing relationship is linear in window SA. For making sure the interaction between the adjacent windows, we also take use of the shifted window mechanism [16]. Figure 4(c) is an example of 4 adjacent windows (blue) in a feature map, after processing by self-attention operation, we move each window $\frac{W}{2}$ to the down and right and then move the A, B and C areas of the feature map to A', B' and C'. Finally, we do the self-attention again in the new windows (red).

3.3 Heatmap Generator

The output of Transformer encoder is a 1D representation, so we need to recover 2D keypoint heatmaps from it. According to the existing hybrid estimators [12, 27], there are two main methods. For TransPose-H [27], features keep the target resolution in Transformer, which leads to high computational complexity due to the quadratic relationship between FLOPs and feature size. TokenPose [12] upsamples keypoint tokens with the MLP head which brings more parameters. To overcome the above shortcomings, we design a heatmap recovery module, namely Heatmap Generator (HG), which balances parameters and FLOPs while obtaining accurate prediction.

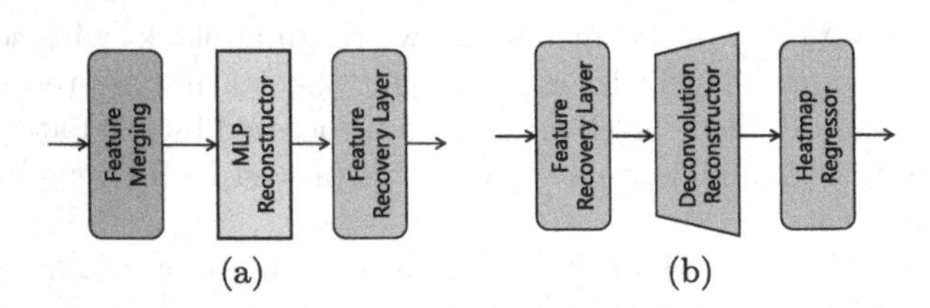

Fig. 5. The design of Heatmap Generator (HG) (a) The linear HG (b) The spatial HG

In order to explore the appropriate architecture, we have 2 designs as Fig. 5 shows. The linear HG is mainly composed of MLP layers. It firstly does feature merging operation designed by hand to make the most use of each feature point of the keypoint representation. Then, it uses the MLP reconstructor to get every pixel making up the heatmaps and recovers the 2D spatial structure. The main body of the spatial HG is deconvolution. In contrast to the linear design, it firstly transforms the keypoint representation to 2D structure and then achieves the reconstruction goal directly in spatial dimension via the deconvolution reconstructor. Finally, the heatmaps are obtained by using the heatmap regressor. Experiments show that, compared with the MLP structure, the spatial one performs better which mainly thanks to the spatial recovery ability of deconvolution.

4 Experiments

4.1 Model Setting

For PCM, we set the dimension of the first branch as 32 and the number of Transformer blocks as $N_i = \{2,2,6,2\}$ $(i = 1, 2, 3, 4)$. In addition, the window size of the window SA is set as W=8. For spatial HG, there are three deconvolution layers utilized. Specially, only PCM loads the weights pretrained in ImageNet partially, PTM and HG are just trained from scratch.

4.2 Implementation Details

Dataset and Evaluation Metric. COCO [14] is one of the standard benchmarks of HPE. It includes more than 20K images and 25K human instances. COCO is divided into three parts: train, val and test-dev set which include 57K, 5K and 20K images respectively. For each human instance, there are at most 17 keypoints labeled.

The standard evaluation metric of COCO is calculated according to Object Keypoint Similarity (OKS):

$$OKS = \frac{\sum_i exp\left(-d_i^2/2\, s^2 k_i^2\right) \delta\left(v_i > 0\right)}{\sum_i \delta\left(v_i > 0\right)} \tag{4}$$

where v_i stands for the visibility of the i-th keypoint, d_i is used to measure the distance between the prediction and corresponding GT, s means the scale of current human body and k_i is a constant factor that controls falloff. We mainly report average precision (AP) and recall (AR) score to evaluate the model performance. Specifically, AP and AR present the mean scores at OKS = 0.5, 0.55, ..., 0.95; AP^{50} and AP^{75} stand for the scores at OKS = 0.5 and 0.75; AP^M and AP^L are used to evaluate the medium and large human instances.

Technical Setting. We train and evaluate CTHPose in an NVIDIA Geforce 2080Ti and the experiments follow the top-down paradigm of HPE. For fair comparison, we use the person detection result comes from [26]. Before inputting the human samples detected, we resize them to 256 × 192 and make data augmentation the same as [20]. We totally train 300 epochs with AdamW optimizer and the heatmap decoding strategy [29]. The initial learning rate is set as 1e−3, and changed to 1e−4 and 1e−5 at 200-th and 260-th epoch respectively. We evaluate the performance of our work on COCO2017 validation set and test-dev set.

5 Results

The performance of CTHPose on COCO validation set and test-dev set are as shown in Table 1 and Table 2, respectively.

Table 1. Comparisons on COCO2017 validation set. Pretrain means the corresponding part has been initialized by the weights pretrained in ImageNet. Trans. denotes Transformer.

Method	Pretrain		Input size	#Params	GFLOPs	gtbbox	AP	AP^{50}	AP^{75}	AP^{M}	AP^{L}	AR
	CNN	Trans.				AP						
Hourglass [18]	N	–	256×192	25.1M	14.3	–	66.9	–	–	–	–	–
SimpleBaseline-R50 [26]	Y	–	256×192	34.0M	8.9	72.4	70.4	88.6	78.3	67.1	77.2	76.3
SimpleBaseline-R101 [26]	Y	–	256×192	53.0M	12.4	–	71.4	89.3	79.3	68.1	78.1	77.1
SimpleBaseline-R152 [26]	Y	–	256×192	68.6M	15.7	74.3	72.0	89.3	79.8	68.7	78.9	77.8
TransPose-R-A4 [27]	Y	N	256×192	6.0M	8.9	75.1	72.6	89.1	79.9	68.8	79.8	78.0
TransPose-H-S [27]	Y	N	256×192	8.0M	10.2	76.1	74.2	89.6	80.8	70.6	81.0	78.0
HRNet-W32 [20]	Y	–	256×192	28.5M	7.1	76.5	74.4	90.5	81.9	70.8	81.0	79.8
CTHPose	Y	N	256×192	11.0M	6.4	77.3	74.7	89.8	81.3	71.1	81.7	79.9

On the validation set, CTHPose outperforms the convolution-based models with lightweight structure and without extra pretraining. Compared to the counterpart CNN method HRNet-W32, the AP of CTHPose improves by 0.3 with only 38.6% parameters and 9.9% fewer GFLOPs. Moreover, compared to hybrid networks, TransPose-R-A4 and TransPose-H-S, the prediction accuracy improves 2.1 and 0.5 with the reduction of computing complexity of 2.5×10^9 and 3.8×10^9 FLOPs respectively.

On test-dev set, compared with various state-of-the-art methods, our CTHPose further demonstrates its efficiency and effectiveness. Specifically, compared to the larger scale CNN model HRNet-W48, CTHPose achieves higher prediction accuracy 74.4 AP with significant improvement in both model parameters (↓82.7%) and GFLOPs (↓56.2%). Meanwhile, CTHPose has advantage in detecting large scale human instances, its AP^L is 80.2 which improves 0.5. It is also competitive compared with hybrid networks [12,27]. For TokenPose-S-v2 and TransPose-H-S, it outperforms 1.3 and 1.0 AP with only almost 50% of GFLOPs. And for TokenPose-B, it improves by 0.4 AP with 2.5×10^9 fewer parameters.

Table 2. Comparisons on COCO2017 test-dev set with sate-of-the-art methods

Method	Input size	#Params	GFLOPs	AP	AP^{50}	AP^{75}	AP^{M}	AP^{L}	AR
HigherHRNet [3]	640 × 640	63.8M	154.3	70.5	89.3	77.2	66.6	75.8	74.9
DEKR [7]	640 × 640	65.7	141.5	71.0	89.2	78.0	67.1	76.9	76.7
PRTR [10]	512 × 384	57.2M	37.8	72.1	90.4	79.6	68.1	79.0	79.4
CPN [2]	384 × 288	–	–	72.1	91.4	80.0	68.7	77.2	78.5
RMPE [6]	320 × 256	28.1M	26.7	72.3	89.2	79.1	68.0	78.6	–
TokenPose-S-v2 [12]	256 × 192	6.2M	11.6	73.1	91.4	80.7	69.7	79.0	78.3
TransPose-H-S [27]	256 × 192	8.0M	10.2	73.4	91.6	81.1	70.1	79.3	–
SimpleBaseline-R152 [26]	384 × 288	68.6M	35.6	73.7	91.9	81.1	70.3	80.0	79.0
TokenPose-B [12]	256 × 192	13.5M	5.7	74.0	91.9	81.5	70.6	79.8	79.1
HRNet-W48 [20]	256 × 192	63.6M	14.6	74.2	92.4	82.4	70.9	79.7	79.5
CTHPose	256 × 192	11.0M	6.4	74.4	92.0	81.9	71.1	80.2	79.5

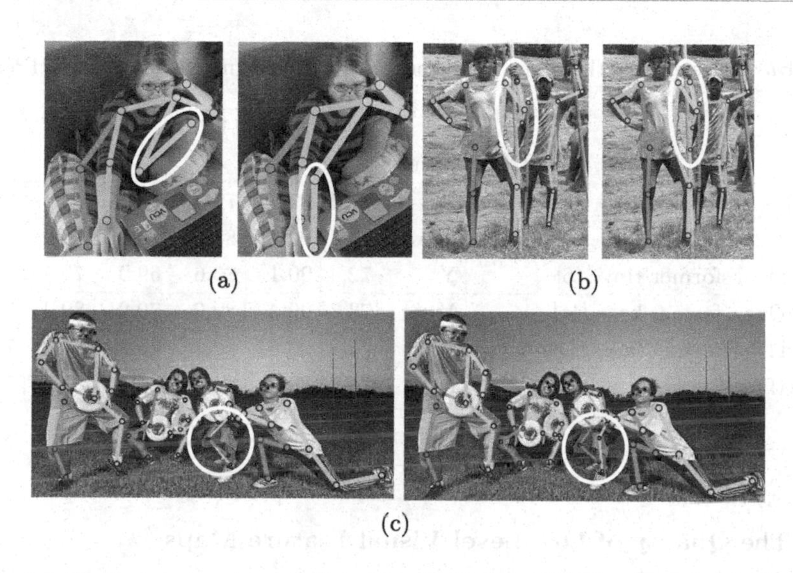

Fig. 6. Qualitative comparison of TransPose(left) and CTHPose(right) (a) The influence of clothing pattern (b)The influence of neighbouring limbs (c) The influence of self-symmetrical limbs

In the above comparisons, CTHPose shows its efficiency which mainly thanks to Transformer part. The parameters of PTM is only 1.73M in our estimator. Because of the high quality feature maps output by PCM and SA's strong ability of modeling long-range relationship, PTM can ensure the effectiveness with small scale. And its FLOPs is only 0.3G which mainly benefits from window self-attention mechanism. What's more, our design of Heatmap Generator makes balance between parameters and FLOPs compared with other hybrid networks.

In addition, we also show the comparison of qualitative results of CTHPose and TransPose on COCO2017 dataset in Fig. 6. Our hybrid estimator is more robust to the influences of clothing color, neighbors' and self-symmetrical limbs.

6 Ablation Studies and Analyses

6.1 The Pure Transformer and Hybrid Architecture

The pure Transformer architecture [1,16,24,25] achieves good performance in CV tasks, however, it also faces huge challenges of more training cost and inflexibility to resolution changes which usually requires re-pretraining. [16,22] have tried overcoming it by interpolation, yet have not achieved satisfactory results.

The comparison of pure Transformer networks and CTHPose is shown in Table 3. CTHPose outperforms pretrained PVT, PVTv2 and Swin Transformer, which proves that the low-level features extracted by CNN are helpful to the convergence and performance of Transformer encoders with less training cost. What's more, because PTM is trained from scratch, it can flexibly cope with different resolutions without extra re-pretraining or processing.

Table 3. Results of the mainstream pure Transformer networks and CTHPose

Method	Pretrain		AP	AP^{50}	AP^{75}	AP^{M}	AP^{L}	AR
	CNN	Trans.						
PVT-s [22]	–	Y	71.4	89.6	79.4	67.7	78.4	77.3
PVTv2-b2 [23]	–	Y	73.7	90.5	81.2	70.0	80.6	79.1
SwinTransformer tiny [16]	–	Y	72.4	90.1	80.6	69.0	79.1	78.2
SwinTransformer base [16]	–	Y	73.7	90.4	82.0	70.2	80.4	79.4
SwinTransformer large [16]	–	Y	74.3	90.6	82.1	70.6	81.2	79.8
CTHPose	Y	N	74.7	89.8	81.3	71.1	81.7	79.9

6.2 The Quality of Low-Level Visual Feature Maps

In order to better construct long-range relationship between human body parts, it is beneficial to get the high quality feature maps as the input of PTM. To evaluate it, we compare the performance of PCM with basic blocks [8] and polarized blocks respectively. The results are shown in Table 4.

Since the feature maps keep high resolution and make multi-scale feature fusion in PCM, we can obtain abundant low-level semantics with human instances of different sizes. Moreover, with the help of PSA [15], CTHPose may determine the initial regions of interest for keypoints in the orthogonal dimensions. Due to both of which, Transformer can be trained from scratch directly and achieves fine performance in CTHPose rather than pretraining in ImageNet as general setting.

Table 4. The comparison of PCM composed of basic blocks and polarized blocks

Block	AP	AP^{50}	AP^{75}	AP^{M}	AP^{L}	AR
Basic block	73.8	89.4	80.4	70.4	80.5	79.0
Polarized block	74.8	89.7	81.4	71.1	81.7	79.9

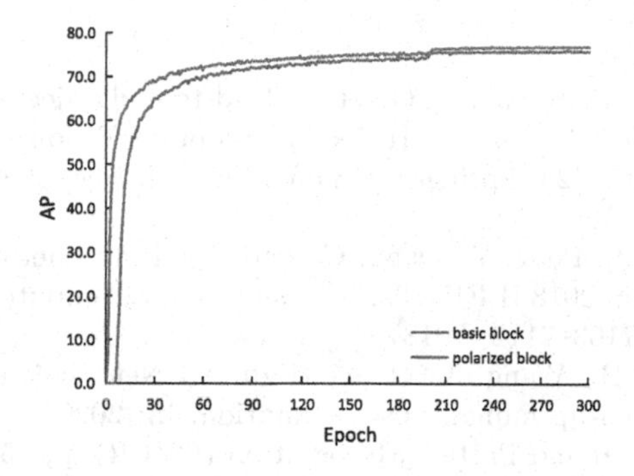

Fig. 7. The convergence of PCM with basic blocks and polarized blocks

In short, the high quality feature maps enable the shifted window self-attention to locate the keypoints' scope and then converge more easily as shows in Fig. 7.

6.3 The Design of Heatmap Generator

According to the evaluation results in Table 5, the spatial HG generally performs better considering the balance of parameters, FLOPs and prediction accuracy. The main reason is deconvolution does well in recovering spatial information which is heatmap prediction needs.

In other words, the spatial positioning bias of the refined HG is smaller than the linear one, though MLP structure is used more widely in Transformer-related networks.

Table 5. The comparison of Heatmap Generators

Heatmap Generator	#Prams	GFLOPs	AP	AP^{50}	AP^{75}	AP^{M}	AP^{L}	AR
Linear HG	11.9M	5.3	74.1	91.7	81.4	70.7	80.0	79.2
Spatial HG	11.0M	6.4	74.4	92.0	81.9	71.1	80.2	79.5

7 Conclusion

We design an efficient and effective CNN-Transformer hybrid network, CTH-Pose, for 2D human pose estimation. This estimator utilizes both visual clues and long-range spatial dependencies to locate keypoints. It achieves good performance with lightweight architecture and less training cost. On one hand, our work has proved that the high quality feature maps CNN extracted are beneficial to the convergence of subsequent Transformer. On the other, Pyramid Transformer Module and Heatmap Generator designed significantly reduce the model parameters and computing burden while ensuring accuracy.

References

1. Carion, N., Massa, F., Synnaeve, G., et al.: End-to-end object detection with transformers. In: Vedaldi, A., Bischof, H. (eds.) Computer Vision – ECCV 2020. LNCS, vol. 12346, pp. 213–229. Springer, Cham (2020). https://doi.org/10.1007/978-3-030-58452-8_13
2. Chen, Y., Wang, Z., Peng, Y., et al.: Cascaded pyramid network for multi-person pose estimation. In: 2018 IEEE/CVF Conference on Computer Vision and Pattern Recognition, pp. 7103–7112 (2018)
3. Cheng, B., Xiao, B., Wang, J., et al.: HigherHRNet: scale-aware representation learning for bottom-up human pose estimation. In: 2020 IEEE/CVF Conference on Computer Vision and Pattern Recognition (CVPR), pp. 5385–5394 (2020)
4. Dosovitskiy, A., Beyer, L., Kolesnikov, A., et al.: An image is worth 16 × 16 words: transformers for image recognition at scale. In: 9th International Conference on Learning Representations (ICLR), pp. 1–21 (2021)
5. Duan, H., Zhao, Y., Chen, K., et al.: Revisiting skeleton-based action recognition. In: 2022 IEEE/CVF Conference on Computer Vision and Pattern Recognition (CVPR), pp. 2959–2968 (2022)
6. Fang, H.S., Xie, S., Tai, Y.W., et al.: RMPE: regional multi-person pose estimation. In: 2017 IEEE International Conference on Computer Vision (ICCV), pp. 2353–2362 (2017)
7. Geng, Z., Sun, K., Xiao, B., et al.: Bottom-up human pose estimation via disentangled keypoint regression. In: 2021 IEEE/CVF Conference on Computer Vision and Pattern Recognition (CVPR), pp. 14671–14681 (2021)
8. He, K., Zhang, X., Ren, S., et al.: Deep residual learning for image recognition. In: 2016 IEEE Conference on Computer Vision and Pattern Recognition (CVPR), pp. 770–778 (2016)
9. Krizhevsky, A., Sutskever, I., Hinton, G.E.: ImageNet classification with deep convolutional neural networks. Commun. ACM **60**(6), 84–90 (2017)
10. Li, K., Wang, S., Zhang, X., et al.: Pose recognition with cascade transformers. In: 2021 IEEE/CVF Conference on Computer Vision and Pattern Recognition (CVPR), pp. 1944–1953 (2021)
11. Li, Y., Xie, S., Chen, X., et al.: Benchmarking detection transfer learning with vision transformers. CoRR (2021). https://arxiv.org/abs/2111.11429
12. Li, Y., Zhang, S., Wang, Z., et al.: TokenPose: learning keypoint tokens for human pose estimation. In: 2021 IEEE/CVF International Conference on Computer Vision (ICCV), pp. 11293–11302 (2021)
13. Lin, T.Y., Dollár, P., Girshick, R., et al.: Feature pyramid networks for object detection. In: 2017 IEEE Conference on Computer Vision and Pattern Recognition (CVPR), pp. 936–944 (2017)
14. Lin, T.Y., Maire, M., Belongie, S., et al.: Microsoft COCO: common objects in context. In: Fleet, D., Pajdla, T. (eds.) Computer Vision – ECCV 2014. LNCS, vol. 8693, pp. 740–755. Springer, Cham (2014). https://doi.org/10.1007/978-3-319-10602-1_48

15. Liu, H., Liu, F., Fan, X., et al.: Polarized self-attention: towards high-quality pixel-wise mapping. Neurocomputing **506**, 158–167 (2022)
16. Liu, Z., Lin, Y., Cao, Y., et al.: Swin transformer: hierarchical vision transformer using shifted windows. In: 2021 IEEE/CVF International Conference on Computer Vision (ICCV), pp. 9992–10002 (2021)
17. Madrigal, F., Lerasle, F.: Robust head pose estimation based on key frames for human-machine interaction. EURASIP J. Image Video Process. **2020**, 13 (2020)
18. Newell, A., Yang, K., Deng, J.: Stacked hourglass networks for human pose estimation. In: Ferrari, V., Hebert, M. (eds.) Computer Vision – ECCV 2016. LNCS, vol. 9912, pp. 483–499. Springer, Cham (2016). https://doi.org/10.1007/978-3-319-46484-8_29
19. Ramachandran, P., Parmar, N., Vaswani, A., et al.: Stand-alone self-attention in vision models. In: Advances in Neural Information Processing Systems (2019)
20. Sun, K., Xiao, B., Liu, D., et al.: Deep high-resolution representation learning for human pose estimation. In: 2019 IEEE/CVF Conference on Computer Vision and Pattern Recognition (CVPR), pp. 5686–5696 (2019)
21. Vaswani, A., Shazeer, N., Parmar, N., et al.: Attention is all you need. In: Proceedings of the 31st International Conference on Neural Information Processing Systems, pp. 6000–6010. Curran Associates Inc., Red Hook, NY, USA (2017)
22. Wang, W., Xie, E., Li, X., et al.: Pyramid vision transformer: A versatile backbone for dense prediction without convolutions. In: 2021 IEEE/CVF International Conference on Computer Vision (ICCV), pp. 548–558 (2021)
23. Wang, W., Xie, E., Li, X., et al.: PVT v2: improved baselines with pyramid vision transformer. Comput. Vis. Media **8**(3), 415–424 (2022)
24. Wang, Y., Xu, Z., Wang, X., et al.: End-to-end video instance segmentation with transformers. In: 2021 IEEE/CVF Conference on Computer Vision and Pattern Recognition (CVPR), pp. 8737–8746 (2021)
25. Wang, Z., Cun, X., Bao, J., et al.: Uformer: a general u-shaped transformer for image restoration. In: 2022 IEEE/CVF Conference on Computer Vision and Pattern Recognition (CVPR), pp. 17662–17672 (2022)
26. Xiao, B., Wu, H., Wei, Y.: Simple baselines for human pose estimation and tracking. In: Ferrari, V., Hebert, M. (eds.) Computer Vision – ECCV 2018. LNCS, vol. 11210, pp. 472–487. Springer, Cham (2018). https://doi.org/10.1007/978-3-030-01231-1_29
27. Yang, S., Quan, Z., Nie, M., et al.: Transpose: keypoint localization via transformer. In: 2021 IEEE/CVF International Conference on Computer Vision (ICCV), pp. 11782–11792 (2021)
28. YUAN, Y., Fu, R., Huang, L., et al.: Hrformer: high-resolution vision transformer for dense predict. In: Advances in Neural Information Processing Systems, pp. 7281–7293. Curran Associates, Inc. (2021)
29. Zhang, F., Zhu, X., Dai, H., et al.: Distribution-aware coordinate representation for human pose estimation. In: 2020 IEEE/CVF Conference on Computer Vision and Pattern Recognition (CVPR), pp. 7091–7100 (2020)

Structural Pattern Recognition

DuAT: Dual-Aggregation Transformer Network for Medical Image Segmentation

Feilong Tang, Zhongxing Xu, Qiming Huang, Jinfeng Wang, Xianxu Hou, Jionglong Su(✉), and Jingxin Liu(✉)

School of AI and Advanced Computing, Xi'an Jiaotong-Liverpool University, Suzhou, China
{Jionglong.Su,Jingxin.Liu}@xjtlu.edu.cn

Abstract. Transformer-based models have been widely demonstrated to be successful in computer vision tasks by modeling long-range dependencies and capturing global representations. However, they are often dominated by features of large patterns leading to the loss of local details (*e.g.*, boundaries and small objects), which are critical in medical image segmentation. To alleviate this problem, we propose a Dual-Aggregation Transformer Network called DuAT, which is characterized by two innovative designs, namely, the **G**lobal-to-**L**ocal **S**patial **A**ggregation (**GLSA**) and **S**elective **B**oundary **A**ggregation (**SBA**) modules. The GLSA has the ability to aggregate and represent both global and local spatial features, which are beneficial for locating large and small objects, respectively. The SBA module aggregates the boundary characteristic from low-level features and semantic information from high-level features for better-preserving boundary details and locating the re-calibration objects. Extensive experiments in six benchmark datasets demonstrate that our proposed model outperforms state-of-the-art methods in the segmentation of skin lesion images and polyps in colonoscopy images. In addition, our approach is more robust than existing methods in various challenging situations, such as small object segmentation and ambiguous object boundaries. The project is available at https://github.com/Barrett-python/DuAT.

Keywords: Polyp segmentation · Dual decoder · Vision Transformers

1 Introduction

Medical image segmentation is a computer-aided automatic procedure for extracting the region of interest, *e.g.*, tissues, lesions, and body organs. It can assist clinicians in improving diagnostic and treatment processes more efficient and precise. For instance, colonoscopy is the gold standard for detecting colorectal lesions, and accurately locating early polyps is crucial for the clinical prevention of rectal cancer [1]. Likewise, melanoma skin cancer is one of the most rapidly increasing cancers worldwide. The segmentation of skin lesions from dermoscopic images is a critical step in skin cancer diagnosis and treatment

Q. Liu et al. (Eds.): PRCV 2023, LNCS 14429, pp. 343–356, 2024.
https://doi.org/10.1007/978-981-99-8469-5_27

planning [2]. However, manually annotating these structures in clinical practice is impractical due to the tedious, time-consuming, and error-prone nature of the process.

The advance in deep learning has led to the emergence of numerous deep convolutional neural networks (DCNNs) [3–5] for medical image segmentation. However, the limited receptive field of DCNNs poses challenges in effectively capturing global representation. To address this issue, attention models [6–8] have been developed to enhance the capture of long-range context information. These alternatives achieve promising results in semantic segmentation. More recently, Transformer-based methods [9,10] have been proposed and achieved comparable performance to state-of-the-art results. For instance, Vision Transformer (ViT) [11] divides images into patches and applies a **m**ulti-head **s**elf-attention (MHSA) and the multi-layer perceptron (MLP) structure. However, this approach may overlook local features, resulting in overly smoothed predictions for small objects and blurred boundaries between objects.

The challenge lies in developing a model that effectively combines both local and global features. Researchers have explored solutions such as incorporating local context in aggregated long-range relationships [12] and hybrid architectures that combine the strengths of transformers and CNNs [13–15]. However, directly feeding local information into the transformer may not handle local context relationships, resulting in the local information being overwhelmed by the dominant global context. Ultimately it leads to inferior results in the medical image segmentation of small objects.

In this paper, we propose a novel approach called Dual-Aggregation Transformer Network (DuAT) for medical image segmentation. DuAT incorporates the **G**lobal-to-**L**ocal **S**patial **A**ggregation (*GLSA*) module to combine local and global features, as well as the **S**elective **B**oundary **A**ggregation (*SBA*) module to enhance the boundary information and improve object localization. Our approach recognizes the importance of both global and local spatial features, with the aim of accurately identifying objects of different sizes. By aggregating boundary information, our model refines object boundaries and recalibrates coarse predictions. Specifically, the *GLSA* module is responsible for extracting and fusing both local and global spatial features from the backbone. We separate the channels, one for global representation extracted by Global context (GC) block [16], and the other for local information extracted by multiple depthwise convolutions. This separation allows for a comprehensive understanding of the image at both global and local scales. The *SBA* module aims to simulate the biological visual perception process, distinguishing objects from the background. Specifically, it incorporates shallow- and deep-level features to establish the relationship between body areas and boundaries, thereby enhancing boundary characteristics.

In summary, the main contributions of this paper are three-fold.

- We propose a novel framework named Dual-Aggregation Transformer Network (**DuAT**) that leverages the pyramid vision transformer as an encoder.

This choice enables DuAT to extract more robust features than the existing CNN-based methods for medical images.

- We propose dual aggregation modules, Global-to-Local Spatial Aggregation (**GLSA**) module and Selective Boundary Aggregation (**SBA**) module. The GLSA module addresses the challenge of integrating both local spatial detail information and global spatial semantic information, which reduces incorrect information in the high-level features. The SBA module focuses on fine-tuning object boundaries, effectively tackling the issue of the "ambiguous" problem of boundaries.
- Extensive experiments on five polyp datasets (ETIS [17], CVC-ClinicDB [18], CVC-ColonDB [19], EndoScene-CVC300 [20], Kvasir [21]), skin lesion dataset (ISIC-2018 [22]) and 2018 Data Science Bowl [23] demonstrate that the proposed DuAT methods advances the state-of-the-art (SOTA) performance.

2 Related Work

2.1 Vision Transformer

Transformer has dominated the field of NLP with its MHSA layer to capture the pure attention structure of long-range dependencies. Different from the convolutional layer, the MHSA layer has dynamic weight and a global receptive field, which makes it more flexible and effective. Dosovitskiy *et al.* propose a vision transformer (ViT) [11], which is an end-to-end model using the Transformer structure for image recognition tasks. In addition, previous work has proved that the pyramid structure in convolutional networks is also suitable for Transformer and various downstream tasks, such as PVT [10], TransUNet [13], Segformer [24], etc. PVT requires less computation than ViT and adopts the classical Semantic-FPN to deploy the task of semantic segmentation. In medical image segmentation, TransUNet [13] demonstrates that the Transformer can be used as powerful encoders for medical image segmentation. TransFuse [25] is proposed to improve efficiency for global context modeling by fusing transformers and CNNs. Furthermore, to train the model effectively on medical images, Polyp-PVT [26] introduces Similarity Aggregation Module based on graph convolution domain. Inspired by these approaches, we propose a new transformer-based medical segmentation framework that can accurately locate small objects.

2.2 Image Boundary Segmentation

Recently, learning additional boundary information has shown superior performance in many image segmentation tasks. In the early research on FCN-based semantic segmentation, Chen *et al.* [27] uses boundaries for post-starting to refine the result at the end of the network. Recently, several approaches explicitly model boundary detection as an independent sub-task in parallel with semantic segmentation for sharper results. Ma *et al.* [28] explicitly exploits the boundary information for context aggregation to further enhance the semantic representation of the model. Ji *et al.* [29] fuse the low-level edge-aware features and constraint it with explicit edge supervision. Unlike the above works, we propose a novel aggregation method to achieve more accurate localization and boundary delineation of objects.

3 Methodology

3.1 Transformer Encoder

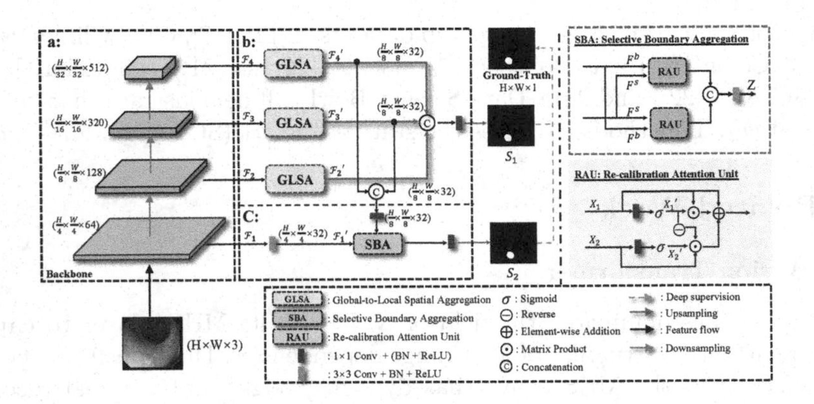

Fig. 1. The overall architecture of Dual-Aggregation Transformer Network (DuAT). The entire model is divided into three parts: (a) pyramid vision transformer (PVT) as backbone; (b) pyramid Global-to-Local Spatial Aggregation (GLSA) Module; (c) Selective Boundary Aggregation (SBA) module and it shown on the red box. (Color figure online)

Some recent studies [24,30] report that vision transformers [10,11] have stronger performance and robustness to input disturbances (*e.g.*, noise) than CNNs. Inspired by this, we use the Transformer based on a pyramid structure as the encoder. Specifically, the pyramid vision transformer (PVT) [10] is utilized as the encoder module for multi-level feature maps $\{\mathcal{F}_i | i \in (1, 2, 3, 4)\}$ extraction. Among these feature maps, $\mathcal{F}_1$ gives detailed boundary information of target, and $\mathcal{F}_2$, $\mathcal{F}_3$ and $\mathcal{F}_4$ provide high-level features.

3.2 Selective Boundary Aggregation

As observed in [31], shallow- and deep-layer features complement each other. The shallow layer has less semantics but is rich in details, with more distinct boundaries and less distortion. Furthermore, the deep level contains rich semantic information. Therefore, directly fusing low-level features with high-level ones may result in redundancy and inconsistency. To address this, we propose the SBA module, which selectively aggregates the boundary information and semantic information to depict a more fine-grained contour of objects and the location of re-calibrated objects.

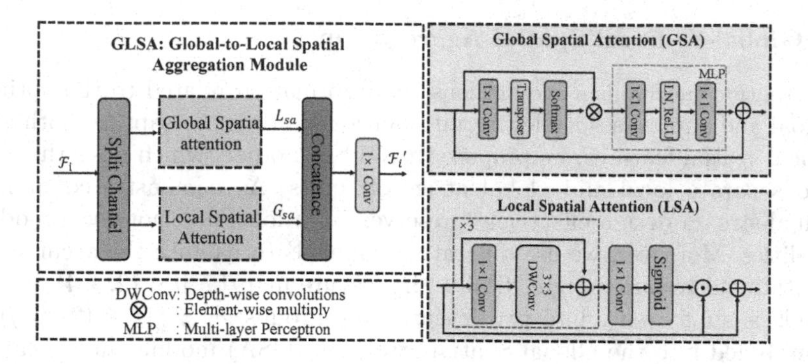

Fig. 2. Overview of the Global-to-Local Spatial Aggregation Module *GLSA*, it is composed of global spatial attention (*GSA*) and local spatial attention *(LSA)*.

Different from previous fusion methods, we design a novel Re-calibration attention unit (RAU) block that adaptively picks up mutual representations from two inputs (F^s, F^b) before fusion. As given in Fig. 1, the shallow- and deep-level information is fed into the two RAU blocks in different ways to make up for the missing spatial boundary information of the high-level semantic features and the missing semantic information of low-level features. Finally, the outputs of two RAU blocks are concatenated after a 3×3 convolution. This aggregation strategy realizes the robust combination of different features and refines the rough features. The RAU block function $PAU(\cdot, \cdot)$ process can be expressed as:

$$T_1' = W_\theta(T_1), T_2' = W_\phi(T_2) \tag{1}$$

$$PAU(T_1, T_2) = T_1' \odot T_1 + T_2' \odot T_2 \odot (\ominus(T_1')) + T_1, \tag{2}$$

where T_1, T_2 are the input features, two linear mapping and sigmoid functions $W_\theta(\cdot)$, $W_\phi(\cdot)$ are applied to the input features to reduce the channel dimension to 32 and obtain feature maps T_1' and T_2'. $\odot$ is Point-wise multiplication. $\ominus(\cdot)$ is the reverse operation by subtracting the feature T_1', refining the imprecise and coarse estimation into an accurate and complete prediction map [3]. We take a convolutional operation with a kernel size of 1×1 as the linear mapping process. As a result, the process of SBA can be formulated as:

$$Z = C_{3\times 3}(\text{Concat}(PAU(F^s, F^b), PAU(F^b, F^s))), \tag{3}$$

where $C_{3\times 3}(\cdot)$ is a 3×3 convolution with a batch normalization and a ReLU activation layer. $F^s \in \mathbb{R}^{\frac{H}{8} \times \frac{W}{8} \times 32}$ contains deep-level semantic information after fusing the third and fourth layers from the encoder, $F^b \in \mathbb{R}^{\frac{H}{4} \times \frac{W}{4} \times 32}$ is the first layer with rich boundary details from the backbone. $Concat(\cdot)$ is the concatenation operation along the channel dimension. $Z \in \mathbb{R}^{\frac{H}{4} \times \frac{W}{4} \times 32}$ is the output of the SBA module.

3.3 Global-to-Local Spatial Aggregation

The attention mechanism strengthens the information related to the optimization goal and suppresses irrelevant information. In order to capture both global and local spatial features, we propose the GLSA module, which fuses the results of two separate local and global attention units. As demonstrated in Fig. 2, this dual-stream design effectively preserves both local and non-local modeling capabilities. Moreover, we use separating channels to balance the accuracy and computational resources. Specifically, the feature map $\{\mathcal{F}_i | i \in (2, 3, 4)\}$ with 64 channels is split evenly into two feature map groups $\mathcal{F}_i^1, \mathcal{F}_i^2 (i \in (2, 3, 4))$ and separately fed into the Global Spatial attention (GSA) module and Local Spatial attention (LSA) module. The outputs of those two attention units are finally concatenated, followed by a 1×1 convolution layer. We formulate such a process as

$$\mathcal{F}_i^1, \mathcal{F}_i^2 = \text{Split}(\mathcal{F}_\text{i}) \tag{4}$$

$$\mathcal{F}_i^{'} = C_{1\times1}(\text{Concat}(G_{sa}(\mathcal{F}_i^1), L_{sa}(\mathcal{F}_i^2))). \tag{5}$$

where G_{sa} denotes the global spatial attention and L_{sa} denotes the local spatial attention. $\mathcal{F}_i^{'} \in \mathbb{R}^{\frac{H}{8} \times \frac{W}{8} \times 32}$ is the output features. We will introduce LSA and GSA module in detail in the following.

(1) GSA module: The GSA emphasizes the long-range relationship of each pixel in the spatial space and can be used as a supplement to local spatial attention. Many efforts [16,32] claim that the long-range interaction can make the feature more powerful. Inspired by the manners of extracting long-range interaction in [32], we simply generate global spatial attention map ($G_{sa} \in \mathbb{R}^{\frac{H}{8} \times \frac{W}{8} \times 32}$) and $\mathcal{F}_i^1$ as input as following:

$$Att_G(\mathcal{F}_i^1) = Softmax(Transpose(C_{1\times1}(\mathcal{F}_i^1))), \tag{6}$$

$$G_{sa}(\mathcal{F}_i^1) = MLP(Att_G(\mathcal{F}_i^1) \otimes \mathcal{F}_i^1) + \mathcal{F}_i^1. \tag{7}$$

where $Att_G(\cdot)$ is the attention operation, $C_{1\times1}$ means 1×1 convolution. $\otimes$ denotes matrix multiplication. $MLP(\cdot)$ consists of two fully-connection layers with a ReLU non-linearity and normalization layer. The first layer of MLP transforms its input to a higher-dimensional space which the expansion ratio is two, while the second layer restores the dimension to be the same as the input.

(2) LSA module: The LSA module extracts the local features of the region of interest effectively in the spatial dimension of the given feature map, such as small objects. In short, we compute local spatial attention response ($L_{sa} \in \mathbb{R}^{\frac{H}{8} \times \frac{W}{8} \times 32}$) and $\mathcal{F}_i^2$ as input as follow:

$$Att_L(\mathcal{F}_i^2) = \sigma(C_{1\times1}(\mathcal{F}_c(\mathcal{F}_i^2)) + \mathcal{F}_i^2)), \tag{8}$$

$$L_{sa} = Att_L(\mathcal{F}_i^2) \odot \mathcal{F}_i^2 + \mathcal{F}_i^2. \tag{9}$$

where $\mathcal{F}_c(\cdot)$ denotes cascading three 1×1 convolution layers and 3×3 depth-wise convolution layers. The number of channels is adjusted to 32 in the $\mathcal{F}_c$.

$Att_L(\cdot)$ is the local attention operation, $\sigma(\cdot)$ is the sigmoid function, $\odot$ is point-wise multiplication. This structural design can efficiently aggregate local spatial information using fewer parameters.

3.4 Loss Function

[33] reports that combining multiple loss functions with adaptive weights at different levels can improve the performance of the network with better convergence speed. Therefore, we use binary cross-entropy loss ($\mathcal{L}^{\omega}_{BCE}(\cdot)$) and the weighted IoU loss ($\mathcal{L}^{\omega}_{Iou}(\cdot)$) for supervision. Our loss function is formulated in Eq. 10, where S is the two side-outputs ($i,e., S_1, S_2$) and G is the ground truth, respectively. λ_1 and λ_2 are the weighting coefficients.

$$\mathcal{L}(S,G) = \lambda_1 \mathcal{L}^{\omega}_{IoU}(S,G) + \lambda_2 \mathcal{L}^{\omega}_{BCE}(S,G) \tag{10}$$

Therefore, the total loss $\mathcal{L}_{total}$ for the proposed DuAT can be formulated as:

$$\mathcal{L}_{total} = \mathcal{L}(S_1,G) + \mathcal{L}(S_2,G). \tag{11}$$

4 Experiments

4.1 Datasets

In the experiment, we evaluate our proposed model on three different kinds of medical image sets: colonoscopy images, dermoscopic images, and microscopy images, so as to assess the learning ability and generalization capability of our model.

Colonoscopy Polyp Images: Experiments are conducted on five polyp segmentation datasets (ETIS [17], CVC-ClinicDB (ClinicDB) [18], CVC-ColonDB (ColonDB) [19], EndoScene-CVC300 (EndoScene) [20], Kvasir-SEG (Kvasir) [21]). We follow the same training/testing protocols in [3,26], *i.e.*, the images from the Kvasir and ClinicDB are randomly split into 80% for training, 10% for validation, and 10% for testing (seen data). And test on the out-of-distribution datasets, which are ColonDB, EndoScene, and ETIS (unseen data).

ISIC-2018 Dataset: The dataset comes from ISIC-2018 challenge [22] [34] and is useful for skin lesion analysis. It includes 2596 images and the corresponding annotations, which are resized to 512 × 384 resolution. The images are randomly split into 80% for training, 10% for validation, and 10% for testing.

2018 Data Science Bowl (2018-DSB): The dataset comes from 2018 Data Science Bowl challenge [23] and is used to find the nuclei in divergent images, including 670 images and the corresponding annotations, which are resized to 256 × 256 resolution.

4.2 Evaluation Metrics and Implementation Details

Evaluation Metrics. We employ three widely-used metrics *i.e.*, mean Dice (mDice), mean IoU (mIoU), and mean absolute error (MAE), to evaluate the model performances. Mean Dice and IoU are the most commonly used metrics and mainly emphasize the internal consistency of segmentation results. MAE evaluates the pixel-level accuracy representing the average absolute error between the prediction and true values.

Implementation Details. We use rotation and horizontal flip for data augmentation. Considering the differences in the sizes and color of each polyp image, we adopt a multi-scale training [3,35] and the color exchange [36]. The network is trained end-to-end by AdamW [37] optimizer. The learning rate is set to 1e-4 and the weight decay is adjusted to 1e-4 too. The batch size is set at 16. We use the PyTorch framework for implementation with an NVIDIA RTX 3090 GPU. We will provide the source code after the paper is published.

4.3 Results

Learning Ability. We first evaluate our proposed DuAT model for its segmentation performance on seen datasets. As summarized in Table 1, our model is compared to six recently published models: U-Net [4], UNet++ [38], PraNet [3], CaraNet [39], TransUNet [13], TransFuse [25], UCTransNet [40] and Polyp-PVT [26]. It can be observed that our DuAT model outperforms all other models, and achieving 0.924 and 0.948 mean Dice scores on Kvasir and ClinicDB segmentation respectively. For the ISIC dataset, our DuAT model achieves a 1.0% improvement in terms of mDice and 1.5% of mIoU over SOTA method. For 2018-DSB, DuAT achieves a mIoU of 0.87, mDice of 0.926 and 0.027 of MAE, which are 1.1%, 1.0%, 0.03% higher than the best performing Polyp-PVT. These results demonstrate that our model can effectively segment polyps.

Generalization Capabilities. We further evaluate the generalization capability of our model on unseen datasets (ETIS, ColonDB, EndoScene). These three datasets have their own specific challenges and properties. For example, ColonDB is a small-scale database that contains 380 images from 15 short colonoscopy sequences. ETIS consists of 196 polyp images for early diagnosis of colorectal cancer. EndoScene is a re-annotated branch with an associated polyp and background (mucosa and lumen). As seen in Table 2, our model outperforms the existing medical segmentation baselines on all unseen datasets for all metrics. Moreover, our DuAT is able to achieve an average dice of 82.2 % on the most challenging ETIS dataset, 3.5% higher than Polyp-PVT.

Visual Results. We also demonstrate qualitatively the performance of our model on five benchmarks, as given in Fig. 3. On ETIS (the first and second row), DuAT is able to accurately capture the target object's boundary and detect a

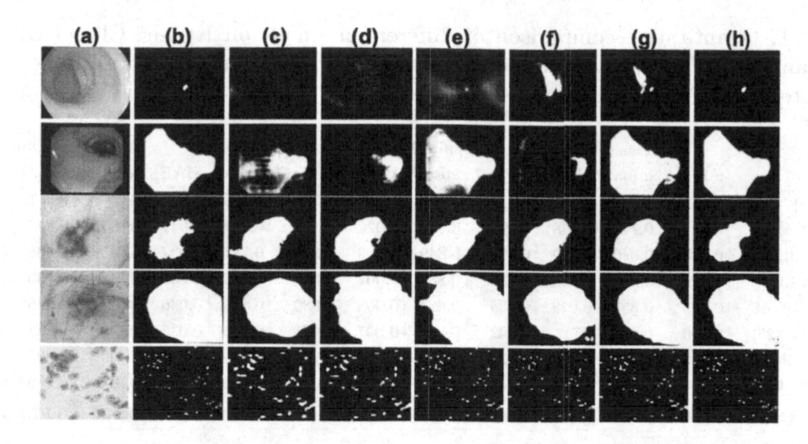

Fig. 3. Qualitative results of different methods. (a) Inputs images, (b) GT, which stands for the ground truths, (c) U-Net [4], (d) U-Net++ [5], (e) PraNet [3], (f) TransUnet [13], (g) Polyp-PVT [26], (h) Our method (DuAT).

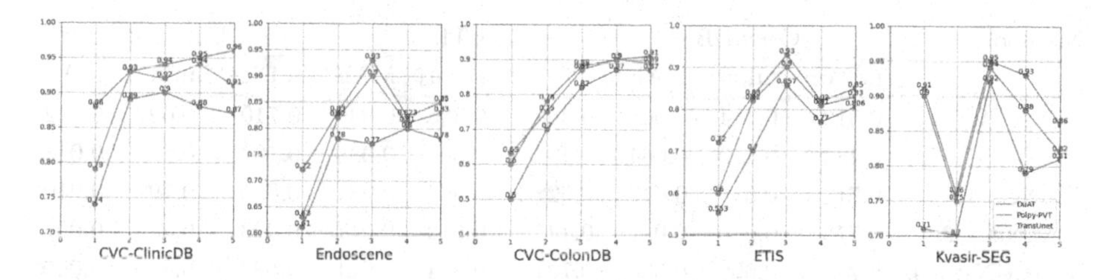

Fig. 4. Performance vs. Size on the five polyp datasets. The x-axis is the proportion size % of the polyp and the y-axis is the averaged mDice coefficient. Blue is for our DuAT, orange is for the Polyp-PVT, and green is for the TransUnet. (Color figure online)

small polyp while other methods fail to detect it. On ISIC-2018 (third row), all methods are able to segment the lesion skin, but our method shows the most similar results compared to the ground truth. On 2018-DSB (the fourth row), we can observe that our DuAT can better capture the presence of nuclei and obtain better segmentation predictions. More qualitative results can be found in the supplementary material.

Small Object Segmentation Analysis. To demonstrate the detection ability of our model for small objects, the ratio of the number of pixels in the object to the number of pixels in the entire image is used to account for the size of the object. We then evaluate the performance of the segmentation model based on the size of the object. We set the area with a proportion less than 5%. For the segmentation model, we first obtain the mean Dice coefficient of the five polyp datasets. Similar to computing the histogram, we calculate the average mean Dice of test data whose size values fall into each interval. For the small object

Table 1. Quantitative comparison of different methods on Kvasir, ClinicDB, ISIC-2018, and 2018-DSB datasets (seen datasets) to validate our model's learning ability. ↑ denotes higher the better and ↓ denotes lower the better.

Methods	Kvasir			ClinicDB			ISIC-2018			2018-DSB		
	mDice↑	mIou↑	MAE↓	mDice↑	mIou↑	MAE↓	mDice↑	mIou↑	MAE↓	mDice↑	mIou↑	MAE↓
U-Net [4]	0.818	0.746	0.055	0.823	0.755	0.019	0.855	0.785	0.045	0.908	0.831	0.040
UNet++ [38]	0.821	0.743	0.048	0.794	0.729	0.022	0.809	0.729	0.041	0.911	0.837	0.039
PraNet [3]	0.898	0.840	0.030	0.899	0.849	0.009	0.875	0.787	0.037	0.912	0.838	0.036
CaraNet [39]	0.918	0.865	0.023	0.936	0.887	0.007	0.870	0.782	0.038	0.910	0.835	0.037
TransUNet [13]	0.913	0.857	0.028	0.935	0.887	0.008	0.880	0.809	0.036	0.915	0.845	0.033
TransFuse [25]	0.920	0.870	0.023	0.942	0.897	0.007	0.901	0.840	0.035	0.916	0.855	0.033
UCTransNet [40]	0.918	0.860	0.023	0.933	0.860	0.008	0.905	0.83	0.035	0.911	0.835	0.035
Polyp-PVT [26]	0.917	0.864	0.023	0.937	0.889	0.006	0.913	0.852	0.032	0.917	0.859	0.030
DuAT (Ours)	**0.924**	**0.876**	**0.023**	**0.948**	**0.906**	**0.006**	**0.923**	**0.867**	**0.029**	**0.926**	**0.870**	**0.027**

Table 2. Quantitative comparison of different methods on ColonDB, ETIS, and EndoScene datasets (unseen datasets) to validate the generalization capability of our model.

Methods	ColonDB			ETIS			EndoScene		
	mDice↑	mIou↑	MAE↓	mDice↑	mIou↑	MAE↓	mDice↑	mIou↑	MAE↓
U-Net [4]	0.512	0.444	0.061	0.398	0.335	0.036	0.710	0.627	0.022
UNet++ [38]	0.483	0.410	0.064	0.401	0.344	0.035	0.707	0.624	0.018
PraNet [3]	0.712	0.640	0.043	0.628	0.567	0.031	0.851	0.797	0.010
CaraNet [39]	0.773	0.689	0.042	0.747	0.672	0.017	0.903	0.838	0.007
TransUNet [13]	0.781	0.699	0.036	0.731	0.824	0.021	0.893	0.660	0.009
TransFuse [25]	0.781	0.706	0.035	0.737	0.826	0.020	0.894	0.654	0.009
SSformer [15]	0.772	0.697	0.036	0.767	0.698	0.016	0.887	0.821	0.007
Polyp-PVT [26]	0.808	0.727	0.031	0.787	0.706	0.013	0.900	0.833	0.007
DuAT (Ours)	**0.819**	**0.737**	**0.026**	**0.822**	**0.746**	**0.013**	**0.901**	**0.840**	**0.005**

segmentation analysis, we compare our DuAT with Polyp-PVT and TransUnet, and the results are given in Fig. 4. The overall accuracy of DuAT is higher than TransUnet [13] and Polyp-PVT [10] on samples with small size polyps.

4.4 Ablation Study

We further conduct ablation study to demonstrate the necessity and effectiveness of each component of our proposed model on three datasets.

Effectiveness of SBA and GLSA. We conduct an experiment to evaluate DuAT without SBA module "(w/o SBA)". The performance without the SBA drops sharply on all three datasets are shown in Table 3. In particular, the mDice is reduced from 0.822 to 0.814 on ETIS. Moreover, we further investigate the contribution of the Global-to-Local Spatial Aggregation by removing it from

Table 3. Ablation study for DuAT on the Kvasir, ETIS, and ISIC-2018 datasets.

Methods	Kvasir-SEG (seen)			ETIS (unseen)			ISIC-2018 (seen)		
	mDice↑	mIou↑	MAE↓	mDice↑	mIou↑	MAE↓	mDice↑	mIou↑	MAE↓
Baseline	0.910	0.856	0.030	0.759	0.668	0.035	0.877	0.783	0.040
+ GSA	0.912	0.860	0.029	0.772	0.675	0.030	0.887	0.803	0.038
+ LSA	0.916	0.863	0.028	0.785	0.690	0.027	0.900	0.839	0.035
+ GSA + LSA (Serial)	0.914	0.863	0.028	0.786	0.695	0.025	0.909	0.845	0.034
+ LSA + GSA (Serial)	0.910	0.860	0.029	0.799	0.713	0.021	0.910	0.852	0.033
+GLSA	0.917	0.864	0.025	0.814	0.723	0.016	0.916	0.816	0.031
w/o SBA	0.917	0.864	0.025	0.814	0.723	0.016	0.916	0.816	0.031
w/o GLSA	0.915	0.863	0.026	0.790	0.696	0.023	0.901	0.800	0.033
SBA + GLSA (Ours)	**0.924**	**0.876**	**0.023**	**0.822**	**0.746**	**0.013**	**0.923**	**0.867**	**0.029**

the overall DuAT and replacing it with a convolution operation with a kernel size of 3, which is denoted as "(w/o GLSA)". The performance of the complete DuAT shows an improvement of 2.2 % and 6.7% in terms of mDice and mIoU, respectively, on ISIC-2018. After using the two modules (SBA + GLSA), the model's performance is improved again. These results demonstrate that these modules enable our model to distinguish polyp and lesion tissues effectively.

Arrangements of GSA and LSA. *GSA* and *LSA* represent the global spatial attention module and local spatial attention module, respectively. We further study the effectiveness and different arrangements of *GSA* and *LSA*. The results tested on Kavsir, ETIS, and ISIC-2018 datasets are shown in Table 3 (the second and sixth row), and all the methods are using the same backbone PVTv2 [10]. *GSA + LSA(Serial)* means first performing *GSA* then *LSA*, while *LSA + GSA(Serial)* is the opposite. Overall, all improve the baseline, and our GLSA group achieves more accurate and reliable results. The GLSA module outperforms the *GSA*, *LSA*, *GSA + LSA (Serial)*, *LSA + GSA (Serial)* by 4.2%, 2.9%, 2.8%, 1.5% in term of mean *Dice* on the ETIS dataset.

5 Conclusions

In this work, we propose DuAT address the issues related to medical image segmentation. Two components, the **G**lobal-to-**L**ocal **S**patial **A**ggregation (**GLSA**) and **S**elective **B**oundary **A**ggregation (**SBA**) modules are proposed. Specifically, the GLSA module extracts the global and local spatial features from the encoder and is beneficial for locating the large and small objects. The SBA module alleviates the unclear boundary of high-level features and further improves its performance. As a result, DuAT can achieve strong learning, generalization ability, and lightweight segmentation efficiency. Both qualitative and quantitative results demonstrate the superiority of our DuAT over other competing methods.

Acknowledgments. This work was jointly supported by the National Natural Science Foundation of China (62201474 and 62206180), the Natural Science Foundation of the Jiangsu Higher Education Institutions of China (22KJB520009), XJTLU Research Development Fund (RDF-21-02-084) and the Key Program Special Fund in XJTLU (KSF-A-22).

References

1. Favoriti, P., Carbone, G., Greco, M., Pirozzi, F., Pirozzi, R.E.M., Corcione, F.: Worldwide burden of colorectal cancer: a review. Updat. Surg. **68**(1), 7–11 (2016). https://doi.org/10.1007/s13304-016-0359-y
2. Mathur, P., et al.: Cancer statistics, 2020: report from national cancer registry Programme, India. JCO Glob. Oncol. **6**, 1063–1075 (2020)
3. Fan, D.-P., et al.: PraNet: parallel reverse attention network for polyp segmentation. In: Martel, A.L., et al. (eds.) MICCAI 2020. LNCS, vol. 12266, pp. 263–273. Springer, Cham (2020). https://doi.org/10.1007/978-3-030-59725-2_26
4. Ronneberger, O., Fischer, P., Brox, T.: U-Net: convolutional networks for biomedical image segmentation. In: Navab, N., Hornegger, J., Wells, W.M., Frangi, A.F. (eds.) MICCAI 2015. LNCS, vol. 9351, pp. 234–241. Springer, Cham (2015). https://doi.org/10.1007/978-3-319-24574-4_28
5. Zhou, Z., Rahman Siddiquee, M.M., Tajbakhsh, N., Liang, J.: UNet++: a nested U-Net architecture for medical image segmentation. In: Stoyanov, D., et al. (eds.) DLMIA/ML-CDS -2018. LNCS, vol. 11045, pp. 3–11. Springer, Cham (2018). https://doi.org/10.1007/978-3-030-00889-5_1
6. Wang, X., Girshick, R., Gupta, A., He, K.: Non-local neural networks. In: Proceedings of the IEEE Conference on Computer Vision and Pattern Recognition (CVPR) (2018)
7. Huang, Z., Wang, X., Huang, L., Huang, C., Wei, Y., Liu, W.: CCNet: criss-cross attention for semantic segmentation. In: 2019 IEEE/CVF International Conference on Computer Vision (ICCV) (2019)
8. Zhang, L., Li, X., Arnab, A., Yang, K., Tong, Y., Torr, P.H.: Dual graph convolutional network for semantic segmentation (2019). arXiv preprint arXiv:1909.06121
9. Zheng, S., et al.: Rethinking semantic segmentation from a sequence-to-sequence perspective with transformers. In: Proceedings of the IEEE/CVF Conference on Computer Vision and Pattern Recognition (CVPR) (2021)
10. Wang, W., et al.: PVT v2: improved baselines with pyramid vision transformer. Comput. Vis. Media **8**, 415–424 (2022)
11. Dosovitskiy, A., et al.: An image is worth 16x16 words: Transformers for image recognition at scale (2020). arXiv preprint arXiv:2010.11929
12. Li, X., Zhang, L., You, A., Yang, M., Yang, K., Tong, Y.: Global aggregation then local distribution in fully convolutional networks (2019). arXiv preprint arXiv:1909.07229
13. Chen, J., et al.: TransUNet: Transformers make strong encoders for medical image segmentation (2021). arXiv preprint arXiv:2102.04306
14. Ranftl, R., Bochkovskiy, A., Koltun, V.: Vision transformers for dense prediction. In: Conference on Computer Vision and Pattern Recognition (CVPR) (2021)
15. Wang, J., Huang, Q., Tang, F., Meng, J., Su, J., Song, S.: Stepwise feature fusion: local guides global. In: Wang, L., Dou, Q., Fletcher, P.T., Speidel, S., Li, S. (eds.) Medical Image Computing and Computer Assisted Intervention – MICCAI 2022.

MICCAI 2022. LNCS, vol. 13433. Springer, Cham (2022). https://doi.org/10.1007/978-3-031-16437-8_11
16. Cao, Y., Xu, J., Lin, S., Wei, F., Hu, H.: GCNet: non-local networks meet squeeze-excitation networks and beyond. In: Conference on Computer Vision and Pattern Recognition (CVPR) (2019)
17. Vázquez, D., et al.: A benchmark for endoluminal scene segmentation of colonoscopy images. J. Healthc. Eng. **2017**, 4037190 (2017)
18. Silva, J., Histace, A., Romain, O., Dray, X., Granado, B.: Toward embedded detection of polyps in WCE images for early diagnosis of colorectal cancer. Int. J. Comput. Assist. Radiol. Surg. **9**(2), 283–293 (2014)
19. Bernal, J., Sánchez, F.J., Fernández-Esparrach, G., Gil, D., Rodríguez, C., Vilariño, F.: WM-DOVA maps for accurate polyp highlighting in colonoscopy: validation vs. saliency maps from physicians. Comput. Med. Imaging Graph. **43**, 99–111 (2015)
20. Tajbakhsh, N., Gurudu, S.R., Liang, J.: Automated polyp detection in colonoscopy videos using shape and context information. IEEE Trans. Med. Imaging **35**(2), 630–644 (2015)
21. Jha, D., et al.: Kvasir-SEG: a segmented polyp dataset. In: Ro, Y.M., et al. (eds.) MMM 2020. LNCS, vol. 11962, pp. 451–462. Springer, Cham (2020). https://doi.org/10.1007/978-3-030-37734-2_37
22. Codella, N., et al.: Skin lesion analysis toward melanoma detection 2018: A challenge hosted by the international skin imaging collaboration (ISIC) (2019). arXiv preprint arXiv:1902.03368
23. Caicedo, J.C., et al.: Nucleus segmentation across imaging experiments: the 2018 data science bowl. Nat. Methods **16**(12), 1247–1253 (2019)
24. Xie, E., Wang, W., Yu, Z., Anandkumar, A., Alvarez, J.M., Luo, P.: SegFormer: simple and efficient design for semantic segmentation with transformers. Adv. Neural Inf. Process. Syst. **34**, 12077–12090 (2021)
25. Zhang, Y., Liu, H., Hu, Q.: TransFuse: fusing transformers and CNNs for medical image segmentation. In: Medical Image Computing and Computer Assisted Intervention (MICCAI) (2021)
26. Dong, B., Wang, W., Fan, D.P., Li, J., Fu, H., Shao, L.: Polyp-PVT: Polyp segmentation with pyramid vision transformers (2021). arXiv preprint arXiv:2108.06932
27. Chen, L.C., Papandreou, G., Kokkinos, I., Murphy, K., Yuille, A.L.: DeepLab: semantic image segmentation with deep convolutional nets, Atrous convolution, and fully connected CRFs. IEEE Trans. Pattern Anal. Mach. Intell. **40**(4), 834–848 (2017)
28. Ma, H., Yang, H., Huang, D.: Boundary guided context aggregation for semantic segmentation (2021). arXiv preprint arXiv:2110.14587
29. Ji, G.P., Zhu, L., Zhuge, M., Fu, K.: Fast camouflaged object detection via edge-based reversible re-calibration network. Pattern Recogn. **123**, 108414 (2022)
30. Bhojanapalli, S., Chakrabarti, A., Glasner, D., Li, D., Unterthiner, T., Veit, A.: Understanding robustness of transformers for image classification. In: Conference on Computer Vision and Pattern Recognition (CVPR) (2021)
31. Li, X., et al.: Improving semantic segmentation via decoupled body and edge supervision. In: European Conference on Computer Vision (ECCV) (2020)
32. Bello, I., Zoph, B., Vaswani, A., Shlens, J., Le, Q.V.: Attention augmented convolutional networks. In: Conference on Computer Vision and Pattern Recognition (CVPR) (2019)

33. Wei, J., Wang, S., Huang, Q.: F^3net: fusion, feedback and focus for salient object detection. In: Proceedings of the AAAI Conference on Artificial Intelligence (AAAI) (2020)
34. Tschandl, P., Rosendahl, C., Kittler, H.: The ham10000 dataset, a large collection of multi-source Dermatoscopic images of common pigmented skin lesions. Sci. Data **5**(1), 1–9 (2018)
35. Huang, C.H., Wu, H.Y., Lin, Y.L.: HarDNet-MSEG: a simple encoder-decoder polyp segmentation neural network that achieves over 0.9 mean dice and 86 FPS (2021). arXiv preprint arXiv:2101.07172
36. Wei, J., Hu, Y., Zhang, R., Li, Z., Zhou, S.K., Cui, S.: Shallow attention network for polyp segmentation. In: de Bruijne, M., et al. (eds.) MICCAI 2021. LNCS, vol. 12901, pp. 699–708. Springer, Cham (2021). https://doi.org/10.1007/978-3-030-87193-2_66
37. Loshchilov, I., Hutter, F.: Decoupled weight decay regularization (2017). arXiv preprint arXiv:1711.05101
38. Jha, D., et al.: ResUNet++: an advanced architecture for medical image segmentation. In: 2019 IEEE International Symposium on Multimedia (ISM), pp. 225–2255. IEEE (2019)
39. Lou, A., Guan, S., Ko, H., Loew, M.H.: CaraNet: context axial reverse attention network for segmentation of small medical objects. In: Medical Imaging 2022: Image Processing. SPIE (2022)
40. Wang, H., Cao, P., Wang, J., Zaiane, O.R.: UCTransNet: rethinking the skip connections in U-Net from a channel-wise perspective with transformer. In: Proceedings of the AAAI Conference on Artificial Intelligence (AAAI) (2022)

MS UX-Net: A Multi-scale Depth-Wise Convolution Network for Medical Image Segmentation

Mingkun Zhang[1], Zhijun Xu[1], Qiuxia Yang[2], and Dongyu Zhang[1](✉)

[1] School of Computer Science and Engineering, Sun Yat-sen University, Guangzhou, China
zhangdy27@mail.sysu.edu.cn

[2] Department of Medical Imaging Center, Sun Yat-sen University Cancer Center, State Key Laboratory of Oncology in South China, Guangzhou, China

Abstract. Semantic segmentation of 3D medical images plays an important role in assisting physicians in diagnosing and successively studying the progression of the disease. In recent years, transformer-based models have achieved state-of-the-art performances on several 3D medical image segmentation tasks. However, these methods still suffer from huge model sizes and high complexity. On the other hand, large-kernel depth-wise convolution networks have shown great potential to encode contextual information more efficiently and effectively than transformer-based networks in natural image segmentation tasks. Inspired by the success of large-kernel depth-wise convolution networks and their variants, we propose a multi-scale lightweight depth-wise convolution network termed MS UX-Net, a U-shaped network mainly composed of convolution operations with different kernel sizes. Specifically, we design a multi-scale feature extraction module for feature encoding, which extracts features into four different scales and learns special features of different scales effectively. Furthermore, we adopt multi-scale depth-wise convolution rather than single-scale standard convolution during the decoding stages, which yields a notable reduction in both the number of model parameters and the computational complexity. The competitive results on two public FLARE2021 and Synapse datasets and a private Pancreatic tumor dataset demonstrate the effectiveness of our method.

Keywords: Multi-scale · Depth-wise convolution · Medical image segmentation

1 Introduction

Over the past few years, Vision Transformers (ViT)s [11] have achieved state-of-the-art(SOTA) performance in various medical image analysis tasks [8,21,31], especially for medical image segmentation benchmarks [3,5,13,14,18,37]. Different from traditional Convolutional Neural Networks (CNN)s, the self-attention

Q. Liu et al. (Eds.): PRCV 2023, LNCS 14429, pp. 357–368, 2024.
https://doi.org/10.1007/978-981-99-8469-5_28

mechanism in ViTs is able to compute pair-wise relations between patches over the whole image by treating an image as a collection of spatial patches, thus achieving a large receptive field and modeling long-range dependencies directly, which is considered as the key to effectively extracting image features [29]. However, Computational complexity that grows quadratically with input size limits the application of pure non-local self-attention models(e.g. ViT) in 3D data scenarios. Although some adoptations were proposed to reduce the operation scopes of self-attention, these methods still suffer from large model capacity and high complexity [13,14,37].

Recently, depth-wise convolution with large kernel (starting from 7×7) has been proven to be more powerful than Swin transformer with lower computation complexity and smaller model capacity [10,25]. These studies provide a new direction for designing 3D convolution on volumetric high-resolution tasks. Lee et al. have made the first attempt to adopt volumetric depth-wise convolutions with large kernel sizes to simulate the operation of large receptive fields for generating self-attention in Swin transformer, called 3D UX-Net [23]. The good performance of 3D UX-Net demonstrates the great potential of large kernel volumetric depth-wise convolution on extracting features with lower parameters. However, it still has the following darwbacks:1)Both depth-wise convolution and point-wise depth convolution scaling extract channel-wise features independently in 3D UX-Net encoder. The lack of cross-channel feature fusion limits model performance.2)Despite having fewer parameters, the up-sampling stages of 3D UX-Net produce a larger feature resolution (almost twice that of SwinUNETR), which results in a notable increase in FLOPs and correspondingly reduces efficiency during both training and inference.

To further explore the potential of large kernel volumetric depth-wise convolution and address the drawbacks of 3D UX-Net, we propose a Multi-scale lightweight depth-wise convolution network MS UX-Net. In our MS UX-Net, we introduce two types of blocks for encoder and decoder. First, the multi-scale depth-wise convolution block(MS-DWC) in the encoder employs four convolution layers simultaneously, each with distinct kernel sizes, to extract input image features at various scales. Subsequently, these features are combined through two point-wise convolution layers to facilitate cross-channel interaction and strengthen their collective representation. Second, the decoder block, which is connected to the corresponding encoder at different resolutions via skip connections, is generally composed of an attention gate and a MS-DWC block. Adopting the depth-wise convolution-based MS-DWC module for decoding, rather than a standard convolution-based residual block, significantly reduces the number of model parameters and computational complexity while preserving its high performance standards. Furthermore, inspired by [30], we utilize a linear layer to further assign suitable weights to the features from different levels, generating the final feature for segmentation.

The main contributions of this work are 1) We revisit the design of convolution network with large kernel size in a volumetric setting and present a multi-scale depth-wise convolutional architecture that effectively captures spatial information across diverse scales, which is of significance in image segmentation. 2) We design the MS-DWC block and adopt it in both encoder and decoder, which reduces both model parameters and complexity, especially in the decoding stages. To our best knowledge, this is the first multi-scale depth-wise convolution block design in decoder.3) we propose a linear weighting module to fuse these features. The ablation experiment demonstrates that adding this module incurs only minimal additional overhead, while significantly improving both convergence speed and overall performance.

2 Related Work

2.1 Depth-Wise Convolution Based Methods

The depth-wise separable convolutions were initially introduced in [32] and subsequently used in Inception models [19], Xception network [6] and MobileNet [15]. Following these, a number of methods based on depth-wise convolution were developed. For instance, 3D U^2Net utilizes deep-wise convolution to process datasets into multi-domain to extract common and specific features [17]. Alalwan et al. replace the standard convolution by the depth-wise convolution to decrease the memory consumption and computation cost [1]. Sharp U-Net introduces depthwise convolution to address the semantic gap issue between the encoder and decoder features in a UNet-like architecture [38]. However, limited studies have performed large-kernel in depth-wise convolution until ConvNeXt [26] was proposed. RepLKNet [9] and SLaK [25] further expand the kernel size to 31×31 and 51×51. LKAU-Net [24] adopts dilated depth-wise convolution in decoder to capture long-range relationships.3D UX-Net [23] leverage depth-wise convolution with $7\times7\times7$ kernel size in a volumetric setting for robust volumetric segmentation. Different from LKAU-Net [24] and UX-Net [23], our method adopt depth-wise convolution not only in encoder but also in decoder, these enable our model to achieve higher performance than the aforementioned methods.

2.2 Multi-scale Networks

It is commonplace to develop multi-scale networks in the field of computer vision and multi-scale interaction has been proven to be a key component for improving segmentation models' performance [12]. For medical image segmentation, multi-scale block appears in encoder, skip-connection and decoder. Md-Net [35] introduced multi-scale dilated convolution layer in encoder to facilitate CT image segmentation. DSM [36] proposed multi-scale connection block in the skip-connection to provide more spatial information to the decoder layers. MC-Unet [16] replaced the second convolution in standard U-Net with a multi-scale convolution block to process the image features at different scales. However, limited studies have been proposed to efficiently leverage depth-wise convolution with multi-scale in a volumetric setting, we believe that depth-wise

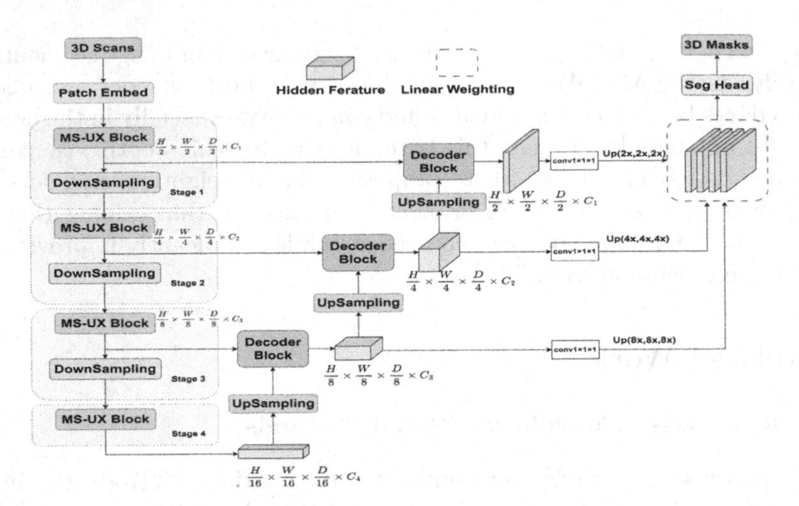

Fig. 1. Overall architecture of our MS UX-Net network. The input 3D scans are first projected into patch-wise embeddings through a patch embedding layer and then pass through four encoding stages. The encoded feature representations in each stage are fed to a decoder block via skip-connection at multiple resolutions. The final segmentation mask is obtained from the linear concatenation of features from different levels.

convolution has untapped potential that warrants further study. In this paper, we summarize the characteristics of those successful models designed for medical image segmentation and present a depth-wise convolution-based model, named MS UX-Net, a lightweight network for robust volumetric segmentation.

3 Method

3.1 The Overall Architecture

Our proposed MS UX-Net model is illustrated in Fig. 1, which is a hierarchical U-shaped architecture and mainly consists of two parts, i.e., the encoder and decoder. Specifically, the encoder involves one embedding layer, four MS-UX blocks (each block contains several successive layers) and three down-sampling layers. Symmetrically, the decoder branch includes three decoder blocks, three up-sampling layers, and the last segmentation head layer for making mask predictions. Inspired by [30], we develop an attention gate to focus more on fine-grained details from encoded features via skip-connections between corresponding feature pyramids of the encoder and decoder.

3.2 Patch Embedding

The input of MS UX-Net is a 3D patch $z \in R^{H \times W \times D}$ (usually randomly cropped from the original image), where H, W and D denote the height, width and depth of each input scan respectively. Similar to common medical image segmentation

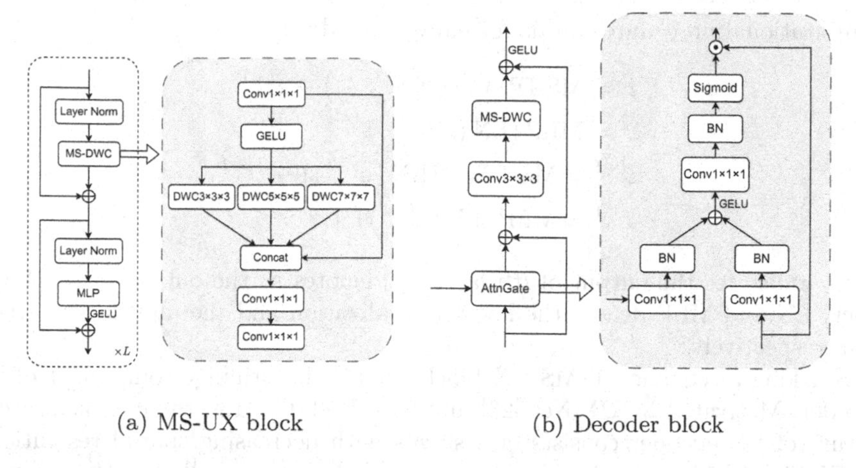

(a) MS-UX block (b) Decoder block

Fig. 2. The architecture of proposed MS-UX block and Decoder block. Here, *DWC* means depth-wise convolution. $\oplus$ means element-wise addition, $\odot$ means hadamard product.

networks in computer vision, the input 3D scan is first processed by a patch embedding layer. Different from 3D UX-Net [23] that leverages a large kernel convolution layer to extract features and ViT [11] that flattens image patches as a sequential input with linear layer, we adopt a $2 \times 2 \times 2$ standard convolution layer with $2 \times 2 \times 2$ stride to compute partitioned feature map with size $\frac{H}{2} \times \frac{W}{2} \times \frac{D}{2}$ that is projected into C channel dimensions (denoted by z_{em}). Note that the embedding feature z_{em} do not need to be reorganized into a sequence of size $(\frac{H}{2} \times \frac{W}{2} \times \frac{D}{2}) \times C$. Here we do this for two reasons, i.e., 1) Convolution layer encodes pixel-level spatial information more precisely than patch-wise positional encoding used in transformers. 2) compared to large-sized kernel, small kernel size helps reduce computational complexity.

3.3 MS-UX Block

After the embedding layer, z_{em} is fed into a MS-UX block. The main purpose behind is to fully capture long-term and short-term dependencies from features yielded by the initial embedding layer and down-sampling layers via performing convolution operations with various kernel sizes simultaneously. As shown in Fig. 2(a), a MS-UX block is composed of a LayerNorm layer [2], a MS-DWC block, another LayerNorm layer and a MLP layer with two residual connections to prevent gradient vanishing [33]. In MS-DWC block, there are four branches to capture multi-scale context, followed by two point-wise convolutions to model relationships between different channels and different scales. Here, the kernel size for each branch is set to 1, 3, 5 and 7, respectively. With these components, the computational procedure can be formally defined as:

$$\begin{aligned}
\hat{z}^l &= \text{MS-DWC}((\text{LN}\left(z^{l-1}\right)) + z^{l-1} \\
z^l &= \text{MLP}\left(\text{LN}\left(\hat{z}^l\right)\right) + \hat{z}^l \\
\hat{z}^{l+1} &= \text{MS-DWC}\left(\text{LN}\left(z^{l-1}\right)\right) + z^{l-1} \\
z^{l+1} &= \text{MLP}\left(\text{LN}\left(\hat{z}^{l+1}\right)\right) + \hat{z}^{l+1}
\end{aligned} \tag{1}$$

where z^ldenotes the output of lth layer, $\hat{z}^l$ denotes as the output of MS-DWC block. LN and MLP denote the layer normalization and the multilayer perceptron, respectively.

Stacking a sequence of MS-UX blocks yields the primary component of the encoder. Motivated by UX-Net [23] and SwinUNETR [13], the complete architecture of the encoder consists of 4 stages with decreasing spatial resolutions. On FLARE2021, the resolutions are set to $\frac{H}{2} \times \frac{W}{2} \times \frac{D}{2}$, $\frac{H}{4} \times \frac{W}{4} \times \frac{D}{4}$, $\frac{H}{8} \times \frac{W}{8} \times \frac{D}{8}$, $\frac{H}{16} \times \frac{W}{16} \times \frac{D}{16}$ respectively. Note that depending on different datasets, these may accordingly vary in practice. In each stage, we adopt a patch merging operation followed by a linear projection as a down-sample layer to downscale the feature resolution by a factor of 2. We opt for patch merging rather than a convolutional layer, as it is better suited for MLP and no need for feature rearrangement(see Fig. 1)

3.4 Decoder

The architecture of three decoder blocks in the decoder is highly symmetrical to those in the encoder. As shown in Fig. 2(b), in each decoder block, we adopt an Attention Gate to reweigh the up-sampling features with output features from the corresponding encoder, which helps to better capture both semantic and fine-grained information. Subsequently, the reweigh features are fed into a $3 \times 3 \times 3$ convolution layer followed by a MS-DWC block for further spatial information integration. There are two advantages of using multi-scale depth convolution block rather than residual block in decoder stage. First, It better captures both semantic and fine-grained information. Second, It helps reduce the number of model parameters by a large margin.

Between every two decoder blocks, a transpose convolution layer is adopted to recover image resolution. Finally, the output features of each decoder block are fed into a $1 \times 1 \times 1$ convolution layer and then up-sampled to the original resolution via the nearest neighbor interpolation layer. We also concatenate all these features together and input the features into a linear weighting layer followed by a $3 \times 3 \times 3$ convolution layer to predict the segmentation mask.(See Fig. 1). The reasons why we choose the nearest neighbor interpolation for up-sampling are two-fold. First, Nearest neighbor interpolation is able to preserve the information of a specific level more effectively. Second, compared to other up-sampling methods, it requires almost no additional computational overhead.

Table 1. Comparison of previous SOTA approaches on the FLARE 2021 dataset.

Methods	Spleen	Kidney	Liver	Pancreas	Mean
3D U-Net [7]	0.911	0.962	0.905	0.789	0.892
nn-UNet [20]	0.971	0.966	0.976	0.792	0.926
TransBTS [34]	0.964	0.959	0.974	0.711	0.902
UNETR [14]	0.927	0.947	0.960	0.710	0.886
nnFormer [37]	0.973	0.960	0.975	0.717	0.906
SwinUNETR [13]	0.979	0.965	0.980	0.788	0.929
3D UX-Net [23]	**0.981**	**0.969**	**0.982**	0.801	0.934
Ours	0.980	0.964	0.979	**0.821**	**0.936**

4 Experiments

4.1 Datasets

FLARE2021 Dataset. This dataset includes 361 cases of abdominal CT scans. Following the split used in [23], we perform five-fold cross-validations with 80% (train)/ 10% (validation)/ 10% (test) split. The average dice similarity coefficient (DSC) is used as the measure for evaluating the segmentation performances of the four target organs, including Spleen, Kidney, Liver, and Pancreas.

Synapse Dataset. This dataset includes 30 axial contrast-enhanced abdominal CT scans. Following the training-test split in [5], 18 of the 30 scans are used for training and the remaining ones are for testing. We report the model performance evaluated with DSC score on 8 abdominal organs, which are aorta(Aor), gallbladder(Gal), spleen(Spl), left kidney(LKid), right kidney(RKid), liver(Liv), pancreas(Pan) and stomach(Sto).

PDAC Dataset. This dataset contains 150 CT cases with the pancreatic tumor annotation. We use this dataset to test the ability of segment small size target of our method. And we use the split of 70% (train)/ 10% (validation)/ 20% (test) and report the DSC score on tumor segmentation for comparison.

Table 2. Comparison of previous SOTA approaches on the Synapse dataset.

Methods	Aor	Gal	LKid	RKid	Liv	Pan	Spl	Sto	Mean
ViTCUP [5]	0.702	0.451	0.747	0.674	0.913	0.420	0.818	0.704	0.679
TransUNet [5]	0.872	0.632	0.819	0.770	0.941	0.559	0.851	0.756	0.775
Swin-UNet [3]	0.855	0.665	0.833	0.796	0.943	0.566	0.907	0.766	0.791
UNETR [14]	0.900	0.606	0.857	0.848	0.945	0.593	0.878	0.740	0.796
MISSFormer [18]	0.870	0.687	0.852	0.820	0.944	0.657	0.919	0.808	0.820
SwinUNETR [13]	0.911	0.665	0.870	0.863	0.957	0.688	0.954	0.770	0.835
nnFormer [37]	0.920	0.702	0.866	0.863	**0.968**	**0.834**	0.905	**0.868**	0.866
Ours	**0.922**	**0.718**	**0.873**	**0.897**	0.965	0.811	**0.929**	0.854	**0.871**

Table 3. Comparison of previous SOTA approaches on the PDAC dataset.

Methods	DSC
UNETR [14]	0.444
3D UX-Net [23]	0.598
SwinUNETR [13]	0.617
nnFormer [37]	0.629
Ours	**0.679**

4.2 Implementation Details

To perform fair comparison, we conduct our MS UX-Net in two different environments. For the FLARE 2021 dataset, we implement it on pytorch 1.13.0, using the architecture of MONAI [4]. Following [23], we take AdamW [27] as the optimizer with an initial learning rate 0.0001 and train our model for 40000 iterations with batch size 2. For the Synapse dataset and PDAC dataset, we implement it on pytorch 1.8.1, using the architecture of nnUNet [20]. Following [37], we set the batch equal to 2 and train our model for 1000 epochs with 250 iterations for each epoch. Basic data augmentation techniques such as intensity shifting, rotation, and scaling are implemented in all experiments.

Table 4. Comparison of the numbers of parameters and FLOPs, the FLOPs are computed with the input image size of $96 \times 96 \times 96$.

Methods	#Params	FLOPs	Methods	#Params	FLOPs
3D U-Net	4.81M	135.9G	nnFormer	149.3M	240.2G
nn-UNet	31.2M	743.3G	SwinUNETR	62.2M	328.4G
TransBTS	31.6M	110.4G	3D UX-Net	53.0M	639.4G
UNETR	92.8M	82.6G	Ours	43.6M	153.0G

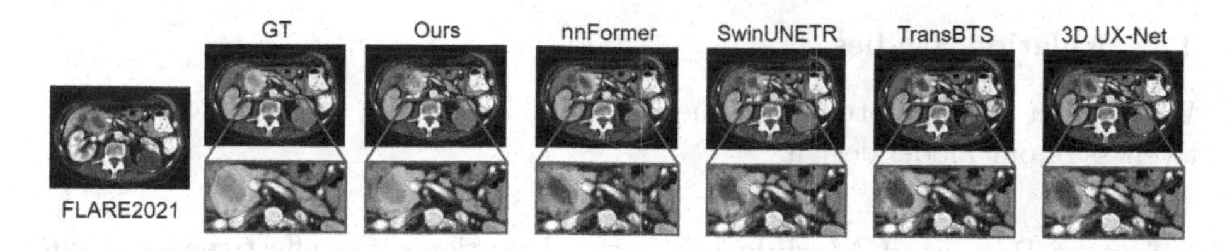

Fig. 3. Qualitative representations of multi-organ segmentation on FLARE2021 dataset. Our method shows the best segmentation quality compared to the ground-truth.

4.3 Quantitative Results

We compare our model with several previous convolution-based and transformer-based SOTA methods, including 3D U-Net [7], 3D UX-Net [23], nnFormer [37], SwinUNETR [13] and so on. As shown in Table 1 2 3, our method obtains excellent performance on each task and outperforms all the previous work. These results show that our method performs better in multi-scale segmentation. We visualize some segmentation results of mentioned method in Fig. 3, demonstrating the quality improvement in segmentation with MS UX-Net. More detail and visualizations of this dataset will be presented in the supplementary material.

In Table 4, we report the numbers of parameters and floating point operations (FLOPs) of our proposed method and several previous methods. It's clear that compared with SwinUNETR [13] and 3D UX-Net [23], our method has a lower computational cost(one-half of SwinUNETR and one-quarter of 3D UX-Net) and fewer model parameters while gains better performance (See Table 1)

Table 5. Ablation study of proposed modules.

Base	LinearWeight	AttenGate	Multi-scale	DSC
✓	✗	✗	✗	0.928
✓	✓	✗	✗	0.930
✓	✓	✓	✗	0.932
✓	✓	✓	✓	0.936

Table 6. Ablation study of different sizes of our method. C denotes the embedding dim and L denotes the network depth of each stage.

Methods	Setting	#Params	FLOPs	DSC
MS UX-Net-S	C=48,L=[3,3,12,3]	33.1M	104.2G	0.931
MS UX-Net-B	C=64,L=[3,3,5,3]	43.6M	153.0G	0.936
MS UX-Net-L	C=72,L=[3,3,8,3]	62.2M	203.9G	0.942

4.4 Ablation Studies

We conduct ablation studies on the FLARE2021 dataset to evaluate the effectiveness of our model design.

Effect of Proposed Module. To better investigate the effectiveness of our proposed module (e.g. linear weighting module, Attention Gate module, MS-DWC module), we first build a basic model without using the module above. Specifically, we remove the attention gate and linear weighting modules from the original architecture and replace the MS-DWC by using only one branch with kernel size $t \times 7 \times 7$. Then we add modules one by one to build the model for comparison. As we can see from Table 5, each module contributes to the final performance.

The Sizes of Different Architecture Variants. To further explore the performance with different encoder sizes, we develop three encoder models with different sizes named MS UX-Net-S, MS UX-Net-B and MS UX-Net-L. Detailed network settings are displayed in Table 6. By examining the table, we can observe that the performance improves as the network becomes deeper, but the number of parameters and FLOPs also increase accordingly.

5 Conclusions

In this paper, we revisit the design of convolution network with large kernel size in a volumetric setting and present the MS UX-Net for medical image segmentation. We propose a multi-scale features extraction block to model different scale information and adopt it in the decoder to reduce both the number of parameters and FLOPs. MS UX-Net outperforms previous SOTA methods with lower complexity and fewer parameters.

References

1. Alalwan, N., Abozeid, A., ElHabshy, A.A., Alzahrani, A.: Efficient 3D deep learning model for medical image semantic segmentation. Alexandria Eng. J. **60**(1), 1231–1239 (2021)
2. Ba, J.L., Kiros, J.R., Hinton, G.E.: Layer normalization (2016). arXiv preprint arXiv:1607.06450
3. Cao, H., et al.: Swin-Unet: Unet-like pure transformer for medical image segmentation. In: Karlinsky, L., Michaeli, T., Nishino, K. (eds.) Computer Vision – ECCV 2022 Workshops. ECCV 2022. LNCS, vol. 13803. Springer, Cham (2023). https://doi.org/10.1007/978-3-031-25066-8_9
4. Cardoso, M.J., et al.: MONAI: An open-source framework for deep learning in healthcare (2022). arXiv preprint arXiv:2211.02701
5. Chen, J., et al.: TransUNet: Transformers make strong encoders for medical image segmentation (2021). arXiv preprint arXiv:2102.04306

6. Chollet, F.: Xception: Deep learning with depthwise separable convolutions. In: Proceedings of the IEEE Conference on Computer Vision and Pattern Recognition, pp. 1251–1258 (2017)
7. Çiçek, Ö., Abdulkadir, A., Lienkamp, S.S., Brox, T., Ronneberger, O.: 3D U-Net: learning dense volumetric segmentation from sparse annotation. In: Ourselin, S., Joskowicz, L., Sabuncu, M.R., Unal, G., Wells, W. (eds.) MICCAI 2016. LNCS, vol. 9901, pp. 424–432. Springer, Cham (2016). https://doi.org/10.1007/978-3-319-46723-8_49
8. Dai, Y., Gao, Y., Liu, F.: TransMed: transformers advance multi-modal medical image classification. Diagnostics **11**(8), 1384 (2021)
9. Ding, X., Zhang, X., Han, J., Ding, G.: Scaling up your kernels to 31x31: revisiting large kernel design in CNNs. In: Proceedings of the IEEE/CVF Conference on Computer Vision and Pattern Recognition, pp. 11963–11975 (2022)a
10. Ding, X., Zhang, X., Han, J., Ding, G.: Scaling up your kernels to 31x31: revisiting large kernel design in CNNs. In: Proceedings of the IEEE/CVF Conference on Computer Vision and Pattern Recognition, pp. 11963–11975 (2022)b
11. Dosovitskiy, A., et al.: An image is worth 16x16 words: Transformers for image recognition at scale (2020). arXiv preprint arXiv:2010.11929
12. Guo, M.-H., Lu, C.-Z., Hou, Q., Liu, Z., Cheng, M.-M., Hu, S.-M.: SegNeXt: Rethinking convolutional attention design for semantic segmentation (2022). arXiv preprint arXiv:2209.08575
13. Hatamizadeh, A., Nath, V., Tang, Y., Yang, D., Roth, H.R., Xu, D.: Swin UNETR: swin transformers for semantic segmentation of brain tumors in MRI images. In: Crimi, A., Bakas, S. (eds.) Brainlesion: Glioma, Multiple Sclerosis, Stroke and Traumatic Brain Injuries. BrainLes 2021. LNCS, vol. 12962. Springer, Cham (2022)a. https://doi.org/10.1007/978-3-031-08999-2_22
14. Hatamizadeh, A., et al.: UNETR: transformers for 3D medical image segmentation. In: Proceedings of the IEEE/CVF Winter Conference on Applications of Computer Vision, pp. 574–584 (2022)b
15. Howard, A.G., et al.: MobileNets: Efficient convolutional neural networks for mobile vision applications (2017). arXiv preprint arXiv:1704.04861
16. Hu, H., Zheng, Y., Zhou, Q., Xiao, J., Chen, S., Guan, Q.: MC-Unet: multi-scale convolution Unet for bladder cancer cell segmentation in phase-contrast microscopy images. In: 2019 IEEE International Conference on Bioinformatics and Biomedicine (BIBM), pp. 1197–1199. IEEE (2019)
17. Huang, C., Han, H., Yao, Q., Zhu, S., Zhou, S.K.: 3D U^2-Net: a 3D universal U-Net for multi-domain medical image segmentation. In: Shen, D., et al. (eds.) MICCAI 2019. LNCS, vol. 11765, pp. 291–299. Springer, Cham (2019). https://doi.org/10.1007/978-3-030-32245-8_33
18. Huang, X., Deng, Z., Li, D., Yuan, X.: MISSFormer: An effective medical image segmentation transformer (2021). arXiv preprint arXiv:2109.07162
19. Ioffe, S., Szegedy, C.: Batch normalization: accelerating deep network training by reducing internal covariate shift. In: International Conference on Machine Learning, pp. 448–456. PMLR (2015)
20. Isensee, F., Jaeger, P.F., Kohl, S.A.A., Petersen, J., Maier-Hein, K.H.: nnU-Net: a self-configuring method for deep learning-based biomedical image segmentation. Nat. Methods **18**(2), 203–211 (2021)
21. Karimi, D., Vasylechko, S.D., Gholipour, A.: Convolution-free medical image segmentation using transformers. In: de Bruijne, M., et al. (eds.) MICCAI 2021. LNCS, vol. 12901, pp. 78–88. Springer, Cham (2021). https://doi.org/10.1007/978-3-030-87193-2_8

22. Landman, B., Xu, Z., Igelsias, J., Styner, M., Langerak, T., Klein, A.: MICCAI multi-atlas labeling beyond the cranial vault-workshop and challenge. In: Proceedings of MICCAI Multi-Atlas Labeling Beyond Cranial Vault-Workshop Challenge, vol. 5, pp. 12 (2015)
23. Lee, H.H., Bao, S., Huo, Y., Landman, B.A.: 3D UX-Net: A large kernel volumetric convnet modernizing hierarchical transformer for medical image segmentation (2022). arXiv preprint arXiv:2209.15076
24. Li, H., Nan, Y., Yang, G.: LKAU-Net: 3D Large-Kernel Attention-Based U-Net for Automatic MRI Brain Tumor Segmentation. In: Yang, G., Aviles-Rivero, A., Roberts, M., Schönlieb, CB. (eds.) Medical Image Understanding and Analysis. MIUA 2022. LNCS, vol. 13413. Springer, Cham (2022). https://doi.org/10.1007/978-3-031-12053-4_24
25. Liu, S., et al.: More convnets in the 2020s: Scaling up kernels beyond 51x51 using sparsity (2022)a. arXiv preprint arXiv:2207.03620
26. Liu, Z., Mao, H., Wu, C.-Y., Feichtenhofer, C., Darrell, T., Xie, S.: A convNet for the 2020s. In: Proceedings of the IEEE/CVF Conference on Computer Vision and Pattern Recognition, pp. 11976–11986 (2022)b
27. Loshchilov, I., Hutter, F.: Decoupled weight decay regularization (2017). arXiv preprint arXiv:1711.05101
28. Ma, J., et al.: AbdomenCT-1K: is abdominal organ segmentation a solved problem? IEEE Trans. Pattern Anal. Mach. Intell. **44**(10), 6695–6714 (2021)
29. Raghu, M., Unterthiner, T., Kornblith, S., Zhang, C., Dosovitskiy, A.: Do vision transformers see like convolutional neural networks? Adv. Neural. Inf. Process. Syst. **34**, 12116–12128 (2021)
30. Rahman, M.M., Marculescu, R.: Medical image segmentation via cascaded attention decoding. In: Proceedings of the IEEE/CVF Winter Conference on Applications of Computer Vision, pp. 6222–6231 (2023)
31. Shen, Z., Fu, R., Lin, C., Zheng, S.: COTR: convolution in transformer network for end to end polyp detection. In: 2021 7th International Conference on Computer and Communications (ICCC), pp. 1757–1761. IEEE (2021)
32. Sifre, L., Mallat, S.: Rigid-motion scattering for texture classification (2014). arXiv preprint arXiv:1403.1687
33. Vaswani, A., et al.: Attention is all you need. In: Advances in Neural Information Processing Systems, vol. 30 (2017)
34. Wang, W., Chen, C., Ding, M., Yu, H., Zha, S., Li, J.: TransBTS: multimodal brain tumor segmentation using transformer. In: de Bruijne, M., et al. (eds.) MICCAI 2021. LNCS, vol. 12901, pp. 109–119. Springer, Cham (2021). https://doi.org/10.1007/978-3-030-87193-2_11
35. Xia, H., Sun, W., Song, S., Mou, X.: MD-Net: multi-scale dilated convolution network for CT images segmentation. Neural Process. Lett. **51**, 2915–2927 (2020)
36. Zhang, G., et al.: DSM: a deep supervised multi-scale network learning for skin cancer segmentation. IEEE Access **7**, 140936–140945 (2019)
37. Zhou, H.-Y., Guo, J., Zhang, Y., Yu, L., Wang, L., Yu, Y.: nnFormer: Interleaved transformer for volumetric segmentation (2021). arXiv preprint arXiv:2109.03201
38. Zunair, H., Hamza, A.B.: Sharp U-Net: depthwise convolutional network for biomedical image segmentation. Comput. Biol. Med. **136**, 104699 (2021)

AnoCSR–A Convolutional Sparse Reconstructive Noise-Robust Framework for Industrial Anomaly Detection

Jie Zhong[1], Xiaotong Tu[1,2(✉)], Yue Huang[1,2], and Xinghao Ding[1,2]

[1] School of Informatics, Xiamen University, Xiamen, China
[2] Institute of Artificial Intelligence, Xiamen University, Xiamen, China
xttu@xmu.edu.cn

Abstract. Industrial anomaly detection involves the identification and localization of abnormal regions in images, of which the core challenge is modeling normal data in appropriate ways. Inspired by dictionary learning, we propose a convolutional sparse reconstructive noise-robust framework, named **AnoCSR**. The proposed convolutional sparse encoding block (**CSE-Block**) in AnoCSR treats the convolutional dictionary as learnable parameters, where the convolutional kernels serve as the atoms. By training on normal samples, we optimize the parameters of the CSE-Block to extract optimal sparse codes. The CSE-Block stacks to form a convolutional sparse reconstructive network (**CSR-Net**) to progressively extract the sparse code and reconstruct the input image inversely. This enables effective reconstruction of normal samples while inadequately reconstructing abnormal regions, thereby facilitating anomaly detection and localization. The CSR-Net is parallelly connected with the downstream **Localizer**, forming an end-to-end framework. Our experimental results demonstrate that AnoCSR achieves performance comparable to state-of-the-art image reconstruction-based methods on the MVTecAD dataset. Moreover, considering that noise may affect the modeling of normal data and the discrimination of anomalies, we conduct the simulated noise resistance experiment. The result demonstrates that AnoCSR significantly outperforms other similar methods, indicating its strong noise robustness in addition to its outstanding performance.

Keywords: Anomaly Detection · Sparse Code · Dictionary Learning

1 Introduction

Industrial anomaly detection involves the identification and localization of abnormal regions in images, which poses a significant challenge as anomalous regions often constitute only a small fraction of the total pixels in an image. Meanwhile,

Supplementary Information The online version contains supplementary material available at https://doi.org/10.1007/978-981-99-8469-5_29.

Q. Liu et al. (Eds.): PRCV 2023, LNCS 14429, pp. 369–380, 2024.
https://doi.org/10.1007/978-981-99-8469-5_29

as the anomalies are rare events, it is hard to directly train a classifier or model anomalous samples, as anomalous regions often exhibit unexpected textures or structures. As a result, how to model the normal samples becomes crucial.

AutoEncoder (AE) is a commonly used model structure in image reconstruction-based methods. Vanilla AE was initially used in brain MRI image segmentation [1]. However, the traditional AE-based methods [6,21] have not addressed a crucial problem: when the representation capability of AE becomes too strong, a "**short-cut**" issue may arise. This occurs because the AE overly generalizes and accurately reconstructs anomalous regions in the output image, as illustrated in Fig. 1. Therefore, effectively learning the primary features of normal samples remains the key to solving this problem. In addition to AE, dictionary learning methods have also been effective in the field of image restoration [24]. Recently, Xili Dai et al. [13] introduced differentiable optimization layers (CSC-Layer) defined from convolutional sparse coding, where the dictionary is treated as parameters of the neural network. This approach achieves faster convergence and the ability to reconstruct images compared to previous work [4]. However, their work has only been validated in image classification tasks, demonstrating performance and robustness surpassing traditional CNNs, without extension to other domains. Motivated by this and the goal of better capturing the features of normal samples, we propose the CSE-Block and extend it to the task of anomaly detection.

In recent years, most anomaly detection methods have used the MVTecAD dataset [3] as a benchmark and achieved excellent performance. However, we observed that the MVTecAD dataset has limited natural noise in both its training and test sets. In real industrial environments, noise is often present in captured images due to poor lighting conditions, suboptimal camera performance, or limitations of the imaging devices themselves. Therefore, we explored and compared the performance of recent state-of-the-art reconstruction-based methods under different simulated noise environments. The followings are the main contributions of this paper:

- We utilize the CSC-Layer and propose the CSE-Block and the CSR-Net that stacks CSE-Block to extract features from normal samples. It exhibits a simple architecture and fast inference speed.
- We propose the framework, AnoCSR, consisting of the upstream reconstruction network CSR-Net and the downstream anomaly localization network, Localizer. It enables end-to-end detection and localization of anomalies while being robust to noise and offering fast inference speed.
- We compared the impact of noise on the performance of recent state-of-the-art image reconstruction-based methods and ours under different simulated noise environments.

2 Related Works

Image reconstruction-based methods are the most fundamental approaches for anomaly detection and localization. While Vanilla AE was successfully used

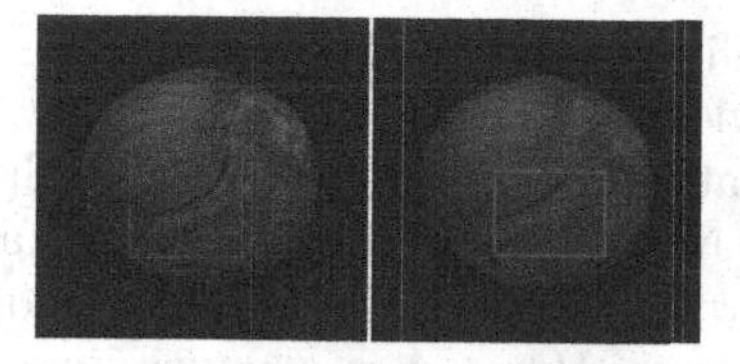

Fig. 1. Illustration of the "short-cut" issue

in brain MRI image segmentation [1], subsequent works [6,21] improved the reconstruction quality and representation power of AE. However, the "short-cut" problem of Vanilla AE was not addressed. Therefore, [10] proposed using memory banks to detect anomalies, where each element of the memory banks module encodes the feature of defect-free samples. During training, a limited number of elements are used for reconstruction, encouraging each element to represent each row. Thus, normal samples are indexed to the most similar element for good reconstruction, while the difference between anomalies and reconstruction is amplified as anomaly score. This method significantly alleviates the over-generalization problem of AE, i.e., the "short-cut" problem mentioned earlier. Subsequent works [17] improved and iterated on this basis and achieved promising performance. This opens up new possibilities for dictionary learning in anomaly detection and localization, but these methods cannot reconstruct complex textures and structures well. In subsequent work, some methods [22,23] based on self-supervised learning have incorporated the idea of image inpainting to enhance the representation capability of the models. In recent years, dictionary learning, as a method for learning the main components of signals, has been effectively applied in image restoration [24]. In the field of anomaly detection, some works [4] have proposed using dictionary learning, but their performance is inferior to deep neural network-based methods, and they do not reflect the robustness of dictionary learning to noise. Recent works [13] proposed a differentiable optimization layer defined from convolutional sparse coding layer, which can be inserted into a neural network as a layer for backpropagation and has similar performance to ResNet [11] in classification tasks with faster inference speed.

3 Method

3.1 Convolutional Sparse Reconstructive Net(CSR-Net)

Our proposed CSE-Block (in Fig. 2a) maps an input signal to a feature encoding, and can perform the inverse operation of reconstructing the input signal from the obtained feature encoding, such that the reconstructed signal is as close as possible to the original input signal. This process involves two steps: **forward encoding** and **dictionary parameters updating**.

Forward Encoding. Given an RGB input image $\boldsymbol{x} \in \mathbb{R}^{3\times H\times W}$, where H and W are the height and width of the image, respectively, and 3 is the number of channels. We assume that $\boldsymbol{x}$ can be reconstructed as the result of a convolution operation between a sparse encoding $\boldsymbol{z} \in \mathbb{R}^{C\times H\times W}$ and a multi-dimensional kernel $\boldsymbol{A} \in \mathbb{R}^{3\times C\times k\times k}$, where k is the size of the convolutional kernel and each kernel is treated as an atom. Here, $\boldsymbol{A}$ is a multi-dimensional tensor consisting of $3 \times C$ atoms and is referred to as a convolutional dictionary. We define the operation as follows:

$$\mathcal{A}(\boldsymbol{z}|\boldsymbol{A}) \doteq \sum_{c=1}^{C} (\boldsymbol{\alpha}_{1c} * \boldsymbol{z}_c, \boldsymbol{\alpha}_{2c} * \boldsymbol{z}_c, \boldsymbol{\alpha}_{3c} * \boldsymbol{z}_c) \quad \in \mathbb{R}^{3\times H\times W} \tag{1}$$

where $\boldsymbol{\alpha}_{ic} \in \mathbb{R}^{k\times k}$ is the convolution kernel in the i-th layer and c-th channel of $\boldsymbol{A}$, $\boldsymbol{z}_c \in \mathbb{R}^{H\times W}$ is the sparse encoding of the c-th channel in $\boldsymbol{z}$, and "$*$" denotes the convolution operation. Thus, under this definition, the optimization problem to be solved is given by:

$$\boldsymbol{z}_* = \arg\min_{\boldsymbol{z}} \frac{1}{2}\|\boldsymbol{x} - \mathcal{A}(\boldsymbol{z}|\boldsymbol{A})\|_2^2 + \lambda\|\boldsymbol{z}\|_1 + \frac{\mu}{2}\|\boldsymbol{z}\|_2^2 \quad \in \mathbb{R}^{C\times H\times W} \tag{2}$$

Here, λ and μ are regularization parameters, where μ is added for more stable convergence. Here, the input signal is assumed to have 3 channels for ease of definition and explanation. In general, for an input signal $\boldsymbol{x}^{[l-1]} \in \mathbb{R}^{C^{[l-1]}\times H^{[l-1]}\times W^{[l-1]}}$, given a convolutional dictionary $\boldsymbol{A}^{[l]} \in \mathbb{R}^{C^{[l-1]}\times C^{[l]}\times k^{[l]}\times k^{[l]}}$, the forward encoding process is given by:

$$\boldsymbol{z}_*^{[l]} = \arg\min_{\boldsymbol{z}^{[l]}} \frac{1}{2}\|\boldsymbol{z}^{[l-1]} - \mathcal{A}(\boldsymbol{z}^{[l]}|\boldsymbol{A}^{[l]})\|_2^2 + \lambda^{[l]}\|\boldsymbol{z}^{[l]}\|_1 + \frac{\mu^{[l]}}{2}\|\boldsymbol{z}^{[l]}\|_2^2 \quad \in \mathbb{R}^{C^{[l]}\times H^{[l]}\times W^{[l]}} \tag{3}$$

$$\boldsymbol{x}^{[l]} = ReLU(BN(\boldsymbol{z}_*^{[l]})) \quad \in \mathbb{R}^{C^{[l]}\times H^{[l]}\times W^{[l]}} \tag{4}$$

l denotes the l-th CSE-block, and $\boldsymbol{z}^{[l]}$ is the sparse coding of $\boldsymbol{z}^{[l-1]}$. $BN(\cdot)$ represents Batch Normalization operation, which is beneficial to accelerate the training process of neural networks and improve the generalization ability of the model. $ReLU(\cdot)$ represents the ReLU activation function, which introduces non-linearity and enhances the expressiveness of the model. For solving the optimization problem in Eq. 3, we use the fast iterative shrinkage-thresholding algorithm (FISTA) proposed in [2], because it can converge quickly and handle high-dimensional data. Its convergence speed is faster than other sparse representation algorithms and it has good stability and convergence performance. Additionally, its implementation is relatively simple, making it easy to apply to practical problems. We denote the entire forward encoding process as $G(\cdot)$ and the backward reconstruction process as $G^{-1}(\cdot)$:

$$\boldsymbol{x}^{[l]} = G(\boldsymbol{x}^{[l-1]}) \tag{5}$$

$$\hat{\boldsymbol{x}}^{[l-1]} = G^{-1}(\boldsymbol{x}^{[l]}) = \mathcal{A}(BN^{-1}(\boldsymbol{x}^{[l]})|\boldsymbol{A}^{[l]}) \tag{6}$$

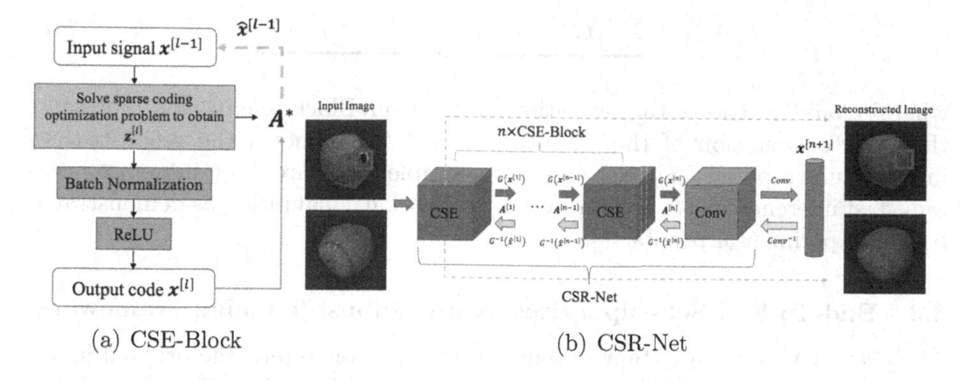

(a) CSE-Block (b) CSR-Net

Fig. 2. Illustration of the network architecture.

where, $BN^{-1}(\cdot)$ is the inverse operation of $BN(\cdot)$. It can be seen that the convolutional dictionary $\boldsymbol{A}^{[l]}$ is also used in the backward reconstruction operation, which indicates that once the parameters of the convolutional dictionary $\boldsymbol{A}^{[l]}$ are learned, it can be used in both the forward encoding and backward reconstruction processes. This makes the CSE-block have fewer parameters and relatively fast inference speed, which also gives our CSR-Net advantage of a lightweight model.

Dictionary Parameters Updating. During the backpropagation, the convolutional dictionary, which is composed of several learnable convolutional kernels, can be efficiently updated using automatic differentiation and GPU parallel computing. This results in much faster updates to the dictionary parameters compared to the traditional algorithm, making the CSE-Block an attractive option for faster inference speed and practical industrial applications.

3.2 Network Architecture

The proposed CSR-Net consists of $n+1$ modules, including the first n CSE-Blocks (in our experiments, we set $n=3$ or $n=4$) and the final convolutional block, as shown in Fig. 2b. The last block of CSR-Net is a convolutional block (ConvBlock) with a single convolutional layer. It compresses the feature encodings from the preceding n CSE-Blocks, facilitating faster convergence. Denoting the input image as $\boldsymbol{x}^{[1]}$, the forward encoding and reverse reconstruction process of CSR-Net:

$$\boldsymbol{x}^{[n]} = \underbrace{G(G(G(\cdots G(\boldsymbol{x}^{[1]})\cdots)))}_{n} \tag{7}$$

$$\boldsymbol{x}^{[n+1]} = Conv(\boldsymbol{x}^{[n]}) \tag{8}$$

$$\hat{\boldsymbol{x}}^{[n]} = Conv^{-1}(\boldsymbol{x}^{[n+1]}) \tag{9}$$

$$\hat{\boldsymbol{x}}^{[1]} = \underbrace{G^{-1}(G^{-1}(G^{-1}(\cdots G^{-1}(\hat{\boldsymbol{x}}^{[n]})\cdots)))}_{n} \tag{10}$$

where, $Conv(\cdot)$ denotes the operation of the ConvBlock, $Conv^{-1}(\cdot)$ represents the inverse operation of the ConvBlock, and $\hat{\boldsymbol{x}}^{[1]}$ denotes the reconstructed image. Our proposed CSR-Net exhibits a simple structure, high inference accuracy, fast inference speed, and strong generalization capabilities, as demonstrated by our experimental results.

3.3 End-To-End Self-supervised Noise-Robust Training Framework

The pretext task is important in our framework. Concretely, the original image is denoted as $\boldsymbol{x} \in \mathbb{R}^{3\times H\times W}$, random sampling of Perlin noise [15] is represented as $\boldsymbol{P} \in \mathbb{R}^{3\times H\times W}$, and by thresholding the noise, it can be binarized into an anomaly map $\boldsymbol{M}_a \in \mathbb{R}^{3\times H\times W}$, thus enabling the simulation of various shapes of anomalies. Therefore, the pretext task is to corrupt $\boldsymbol{x}$

$$\boldsymbol{x}_a = (1 - \boldsymbol{M}_a) \odot \boldsymbol{x} + (1-\beta)(\boldsymbol{M}_a \odot \boldsymbol{x}) + \beta(\boldsymbol{M}_a \odot \boldsymbol{A}) \tag{11}$$

where, $\odot$ denotes element-wise multiplication, $\beta \in [0,1]$ controls the transparency of the generated anomaly region, and $\boldsymbol{A} \in \mathbb{R}^{3\times H\times W}$ is a randomly sampled image from a dataset of object textures [7]. Then, $\boldsymbol{x}_a$ is fed into the reconstruction network to obtain the reconstructed image $\hat{\boldsymbol{x}}$, and a suitable loss function is defined to minimize the distance between $\hat{\boldsymbol{x}}$ and $\boldsymbol{x}$, enabling the learning of the feature encoding of $\boldsymbol{x}$.

Noise Perturbation. To simulate natural noise in industrial environments, such as low-light conditions or sensor limitations, we randomly select a proportion r of samples from the training set and add Gaussian noise $\boldsymbol{\epsilon}$ to them.

$$\tilde{\boldsymbol{x}} = \boldsymbol{x} + \boldsymbol{\epsilon} \tag{12}$$

The perturbed images, denoted as $\tilde{\boldsymbol{x}}$, undergo data augmentation and the pretext task to obtain the corrupted images, which serve as input to CSR-Net.

Loss Functions. For CSR-Net, the objective is to minimize the distance between the original image $\boldsymbol{x}$ and the reconstructed image $\hat{\boldsymbol{x}}$. In this study, we employ the L_2 loss function $L_2(\boldsymbol{x}, \hat{\boldsymbol{x}})$ and the Structural Similarity Index (SSIM) [18] loss function $L_{SSIM}(\boldsymbol{x}, \hat{\boldsymbol{x}})$. The combination of these two loss functions allows us to leverage the desirable convexity of L_2 loss and overcome the potential issue of excessive smoothness caused by L_2 loss alone. Inspired by R-Drop [19], to enhance the robustness of the CSE-Block in CSR-Net, we define a loss function $L_{CSE}(\boldsymbol{x}, \hat{\boldsymbol{x}})$

$$L_{CSE}(\boldsymbol{x}_1, \boldsymbol{x}_2) = \frac{1}{n}\sum_{l=1}^{n} L_2(\boldsymbol{z}_{1*}^{[l]}, \boldsymbol{z}_{2*}^{[l]}) \tag{13}$$

where, $\boldsymbol{x}_1$ and $\boldsymbol{x}_2$ represent the images obtained by applying data augmentation twice to $\boldsymbol{x}$. n denotes the number of CSE-Blocks in CSR-Net, and l corresponds to the l-th CSE-Block. $\boldsymbol{z}_*^{[l]}$ represents the sparse encoding obtained by the l-th CSE-Block. This approach aims to emphasize the features of normal samples in the training set and encourage more compact feature encoding by the CSE-Blocks. Therefore, the loss function of CSR-Net is defined as follows:

$$L_{CSR}(\boldsymbol{x}, \hat{\boldsymbol{x}}) = L_2(\boldsymbol{x}, \hat{\boldsymbol{x}}) + L_{SSIM}(\boldsymbol{x}, \hat{\boldsymbol{x}}) + \gamma L_{CSE}(\boldsymbol{x}, \hat{\boldsymbol{x}}) \tag{14}$$

where, $\gamma \in [0, 1]$ is a weight coefficient, and in our experiments, we set $\gamma = 0.8$.

In the downstream segmentation task, the presence of noise $\boldsymbol{\epsilon}$ in the input image and its impact on the quality of the reconstructed image make traditional threshold-based methods ineffective. Therefore, we employ the U-Net structure [16] as the Localizer $\boldsymbol{f}$, reducing the number of convolutional kernels in the downsampling and upsampling structures to mitigate overfitting and enhance robustness to noise. The anomaly score map $\hat{\boldsymbol{M}}$ and the anomaly score S of $\boldsymbol{x}$ are computed as follows:

$$\hat{\boldsymbol{M}} = \text{Softmax}(\boldsymbol{f}(\text{Concat}(\hat{\boldsymbol{x}}, \boldsymbol{x}))) \tag{15}$$

$$S = \max(\hat{\boldsymbol{M}}) \tag{16}$$

In segmentation tasks, the presence of a significant background class can result in the model focusing more on foreground classes and neglecting the background, leading to decreased accuracy and recall. To mitigate this issue, we utilize the FocalLoss [14] the loss function for the downstream segmentation task. Therefore, the loss function is defined as follows:

$$L_{seg}(\boldsymbol{M}, \hat{\boldsymbol{M}}) = FocalLoss(\hat{\boldsymbol{M}}, \boldsymbol{M}) \tag{17}$$

where, $\boldsymbol{M}$ is the ground truth mask. Therefore, the total loss is defined as:

$$L(\boldsymbol{x}, \hat{\boldsymbol{x}}, \boldsymbol{M}, \hat{\boldsymbol{M}}) = L_{CSR}(\boldsymbol{x}, \hat{\boldsymbol{x}}) + L_{seg}(\boldsymbol{M}, \hat{\boldsymbol{M}}) \tag{18}$$

The incorporation of backpropagation allows for the joint optimization of the parameters in CSR-Net and Localizer through the addition of the loss functions from the pretext task and downstream segmentation tasks. This end-to-end training procedure enables a seamless integration of anomaly detection and segmentation, leveraging the automatic differentiation capability. Furthermore, the dynamic update of the convolutional dictionary in CSR-Net during training facilitates the learning of more discriminative sparse encodings that better capture the characteristics of the training data. The introduction of noise perturbation during training enhances the model's ability to handle noisy inputs, thereby improving its noise robustness.

4 Experiments

Experimental Setup. All experiments of AnoCSR were conducted on an RTX 2080ti. In CSR-Net, the initial values of λ and μ in the CSE-Block were set to 0.1 and 0.05, respectively, and the number of CSE-Blocks was set to $n = 4$. During training, a random selection of 20% of the images was corrupted with additive Gaussian noise (AGN) of variance $\sigma^2 = 25$. In the Noise Resistance experiment, AGN with variances $\sigma_T^2 = 20$ and $\sigma_T^2 = 40$ were used for evaluation.

4.1 Comparison with Other Methods

AnoCSR was compared with recent image reconstruction-based methods DRAEM [22], RIAD [23], and the unsupervised method PaDiM [8] on the MVTecAD test set. The comparison was based on image AUROC and pixel AUROC, and the results are presented in Table 1, where, $\sigma_T^2 = 0, 20, 40$ represents the different AGN we add to the test set, and $\sigma_T^2 = 0$ means the noise perturbations is not performed. We denote this experiment as **Noise Resistance** experiment. In the following mentioned tables, we display the results in "**image AUROC/pixel AUROC**" format.

AnoCSR is more robust to noise compared to DRAEM, RIAD, and PaDiM, which are also image reconstruction and unsupervised methods. In the Noise Resistance experiment, DRAEM suffers the most significant performance decline, while it performs best on the original test set. This is because DRAEM's training focuses on learning embeddings of normal images, leading to a performance drop when the test set contains noise. At $\sigma_T^2 = 20$, DRAEM's reconstruction network mistakes noisy regions as normal, causing a substantial decrease in performance. At $\sigma_T^2 = 40$, almost the entire image is classified as an anomaly, as shown in Fig. 3. The experiments on CSR-Net demonstrate that each CSE-Block can learn convolutional dictionaries that capture features of normal images, and the Localizer can learn decision boundaries between normal and abnormal images. However, its performance is inferior to DRAEM. DRAEM has a larger parameter count (97.42M) compared to AnoCSR (20.84M), suggesting that CSR-Net may have lower expressive power than DRAEM. However, DRAEM is prone to overfitting on the training set and performs poorly on noisy test sets. AnoCSR introduces noise perturbations to increase the discrepancy between reconstructed and abnormal images, and utilizes ground truth masks to adjust decision boundaries, thereby improving its generalization to noise.

Based on the above, AnoCSR achieves comparable performance to state-of-the-art reconstruction-based methods on the MVTecAD test set. In the Noise Resistance experiment, AnoCSR significantly outperforms similar methods and exhibits superior noise robustness compared to the unsupervised method PaDiM, which incorporates a pre-trained backbone with resistance to noise. These experiments demonstrate that AnoCSR maintains high image AUROC and pixel AUROC while exhibiting strong noise robustness. In addition, AnoCSR exhibits

Table 1. Results on $\sigma_T^2 = 0, 20, 40$ for the task of detection and segmentation on MVTecAD(image AUROC/pixel AUROC)

(a) $\sigma_T^2 = 0$

class	DRAEM	RIAD	PaDiM	Ours
capsule	97.6/93.62	91.77/85.18	91.5/98.56	**100.0/98.67**
bottle	99.35/**98.82**	98.71/92.09	**99.84**/98.15	98.71/96.54
carpet	97.61/96.64	45.07/83.49	**99.92/99.0**	98.3/97.83
leather	**100.0**/99.24	99.57/95.07	**100.0**/98.92	**100.0/99.68**
pill	96.61/97.69	52.38/88.71	94.35/96.14	**98.64/97.81**
transistor	96.67/93.86	82.5/72.78	**97.75/97.48**	96.33/89.45
tile	**100.0/99.69**	79.88/72.95	97.37/93.85	**100.0**/99.64
cable	89.5/96.0	51.03/65.73	**92.19/96.76**	90.73/92.83
zipper	**99.89/98.77**	84.05/87.78	90.86/98.42	99.46/97.89
toothbrush	**100.0/97.72**	96.67/95.55	97.22/98.74	**100.0**/97.25
metal_nut	**100.0/98.75**	73.54/78.28	99.17/97.06	96.36/97.1
hazelnut	**100.0/99.57**	74.11/93.45	93.32/97.95	95.15/98.17
screw	**99.74/99.72**	86.38/94.84	84.38/98.32	94.83/94.88
grid	**100.0/99.48**	97.78/94.25	95.74/96.47	**100.0**/99.41
wood	**100.0**/97.49	95.78/75.62	98.77/94.09	**100.0/98.22**
mean	**98.47/97.8**	80.61/85.05	95.49/97.33	97.89/97.03

(b) $\sigma_T^2 = 20$

class	DRAEM	RIAD	PaDiM	Ours
capsule	60.38/89.92	91.94/75.73	69.52/69.13	**99.31/97.35**
bottle	87.74/85.22	99.03/85.58	63.19/69.41	**99.35/96.31**
carpet	84.17/89.03	52.38/83.57	68.61/89.19	**98.81/98.04**
leather	95.30/94.42	99.57/94.28	99.8/98.98	**100.0/99.69**
pill	76.58/83.75	50.11/63.54	**95.32/96.35**	95.25/93.57
transistor	82.0/81.43	78.83/66.82	95.43/**97.86**	**98.5**/86.21
tile	99.24/97.98	79.42/71.65	100.0/98.87	**100.0/99.67**
cable	61.46/67.71	60.54/65.7	51.71/94.47	**82.76/92.4**
zipper	88.03/72.94	81.47/83.5	94.78/95.67	**98.92/98.22**
toothbrush	76.67/71.0	88.89/76.55	84.63/**98.31**	**100.0**/98.27
metal_nut	77.37/69.93	87.07/51.88	**98.85**/94.17	94.34/**97.28**
hazelnut	44.12/46.62	77.06/77.52	**97.22/98.71**	91.91/97.85
screw	0.0/39.87	84.74/77.63	87.12/**97.00**	**90.34**/95.09
grid	57.04/37.05	99.26/92.2	97.46/95.15	**100.0/99.22**
wood	70.5/74.03	92.34/74.17	81.07/79.66	**100.0/96.85**
mean	70.71/73.39	81.51/76.02	85.65/91.53	**96.63/96.39**

(c) $\sigma_T^2 = 40$

class	DRAEM	RIAD	PaDiM	Ours
capsule	68.78/88.94	87.82/70.23	53.77/59.54	**95.88/96.85**
bottle	94.84/82.87	**99.03**/79.14	65.79/68.77	98.71/**96.05**
carpet	81.46/75.99	48.98/83.27	98.64/**98.92**	**99.32**/98.1
leather	93.04/85.25	99.15/93.10	**100.0**/98.92	**100.0/99.6**
pill	74.55/80.30	51.81/53.04	78.59/94.01	**92.86/91.25**
transistor	77.83/76.35	86.17/63.2	88.12/**96.74**	**93.0**/84.39
tile	95.73/88.75	84.45/71.03	97.55/94.34	**100.0/99.72**
cable	61.15/66.96	52.41/66.77	62.61/65.27	**81.38/91.96**
zipper	82.97/70.78	80.39/78.06	80.99/80.8	**98.17/97.88**
toothbrush	67.78/65.91	84.44/58.85	81.39/**98.63**	**96.67/98.63**
metal_nut	76.57/62.89	90.3/40.61	64.32/93.14	**93.54/96.07**
hazelnut	27.35/32.58	84.26/67.78	94.39/97.79	**96.32/98.06**
screw	0.0/35.70	89.48/63.71	**95.74**/87.82	87.5/**96.52**
grid	66.67/41.5	98.52/90.46	88.97/94.03	**100.0/99.2**
wood	77.01/66.71	96.93/71.44	86.49/92.2	**100.0/96.16**
mean	69.72/68.10	82.27/70.05	82.49/88.06	**95.56/96.03**

(d) Time and model complexity comparison

metric	DRAEM	RIAD	PaDiM	Ours
parameters (*M*)	97.42	28.8	68.88	**20.84**
speed (*s/image*)	0.016	0.119	1.417	**0.011**

(a) $\sigma_T^2 = 0$ (b) $\sigma_T^2 = 20$ (c) $\sigma_T^2 = 40$

Fig. 3. The sample results of $\sigma_T^2 = 0, 20, 40$ in (a), (b), (c) respectively. In each subplot, from left to right, we have the original image, reconstructed image, ground truth mask, and anomaly localization map. And, the first row is the result of DRAEM, the second row is RIAD and the last row is AnoCSR. The brighter pixel means higher anomaly score.

a smaller number of model parameters and faster inference speed with 256×256 RGB images (Table 1d) on a single RTX 2080ti. Note that inference speed is not only related to the number of parameters but also to the method itself.

4.2 Experimental Result on Other Datasets

AnoCSR conducted Noise Resistance experiments on NanoTWICE [5], FSSD-12 [9], and VisA datasets [25], and the results are shown in Table 2a, AnoCSR demonstrates strong noise robustness on both FSSD-12 and VisA datasets, and it exhibits high consistency in performance. AnoCSR also outperforms other unsupervised methods on VisA, the noise perturbation is not used in this comparison. In NanoTWICE experiment, we crop the raw 768×1024 image into 16 patches(192×256), and fed into AnoCSR, the total inferencing time is 0.163 s in average running on a single RTX 2080ti, which means AnoCSR has fast inference speed.

Table 2. Result of AnoCSR on other industrial anomaly detection datasets

(a) Noise Resistance results

Dataset	σ_T^2 =0	σ_T^2=20	σ_T^2=40
FSSD-12	99.72/92.16	99.45/92.75	99.35/91.36
NanoTWICE	99.55/98.19	87.64/96.37	53.42/93.17
VisA	91.60/96.32	89.64/95.69	88.1/95.12

(b) Comparison with other methods on VisA

method	image AUROC	pixel AUPRO
PaDiM	89.1	85.9
DRAEM	88.7	73.1
AnoDDPM [20]	78.2	60.5
CutPaste [12]	81.9	74.2
Ours	**91.6**	**87.1**

4.3 Ablation Study

We conducted ablation experiments on AnoCSR by varying the sample ratio r and the additive Gaussian noise σ^2. Firstly, we applied different random sampling ratios r to the training set images and added AGN with $\sigma^2 = 25$. The experimental results are shown in Table 3a. When $r = 0$, indicating no Noise Perturbation is applied, the results show a significant decrease in the network's robustness to noise. This is mainly due to overfitting of the Localizer. On the other hand, when $r = 100\%$, where all images in the training set are augmented with AGN, the network's performance decreases but stable metrics still demonstrate strong robustness. Moreover, we applied AGN with different σ_T^2 values to randomly selected $r = 20\%$ of the images. The results are shown in Table 3b, when $\sigma_T^2 = 0$, indicating no Noise Perturbation, the analysis results are similar to when $r = 0$. As σ_T^2 increases, the performance of the network decreases. This may be because excessively high noise intensity leads to significant differences between the reconstructed images and the original images, resulting in less accurate anomaly score maps.

Table 3. Results of ablation study

(a) Comparion on r

σ_T^2	r=0%	r=20%	r=100%
0	96.18/95.12	97.89/97.03	87.85/81.07
20	88.23/90.89	96.63/96.39	84.01/85.40
40	74.80/84.36	95.56/96.03	92.34/82.95

(b) Comparison on σ^2

σ_T^2	σ^2=0	σ^2=25	σ^2=50
0	96.18/95.12	97.89/97.03	96.45/96.79
20	88.23/90.89	96.63/96.39	94.66/95.31
40	74.80/84.36	95.56/96.03	93.56/92.02

5 Conclusion and Discussion

In this paper, we revisit the bottleneck of AE-based methods and address it from the perspective of convolutional dictionary learning. We propose AnoCSR and demonstrate through experiments that our method not only alleviates the "short-cut" issue but also exhibits strong noise robustness and fast inference speed. However, our experiments revealed that AnoCSR does not perform well on certain few-shot categories, such as the "transistor" category in MVTecAD. This issue could potentially be addressed by generating pseudo-normal samples using a diffusion model in future work.

Acknowledgement. The work was supported in part by the National Natural Science Foundation of China under Grant 82172033, U19B2031, 61971369, 52105126, 82272071, 62271430, and the Fundamental Research Funds for the Central Universities 20720230104.

References

1. Baur, C., Wiestler, B., Albarqouni, S., Navab, N.: Deep autoencoding models for unsupervised anomaly segmentation in brain MR images. In: Crimi, A., Bakas, S., Kuijf, H., Keyvan, F., Reyes, M., van Walsum, T. (eds.) BrainLes 2018. LNCS, vol. 11383, pp. 161–169. Springer, Cham (2019). https://doi.org/10.1007/978-3-030-11723-8_16
2. Beck, A., Teboulle, M.: A fast iterative shrinkage-thresholding algorithm for linear inverse problems. SIAM J. Imag. Sci. **2**(1), 183–202 (2009)
3. Bergmann, P., Fauser, M., Sattlegger, D., Steger, C.: MVTec AD-a comprehensive real-world dataset for unsupervised anomaly detection. In: Proceedings of the IEEE/CVF Conference on Computer Vision and Pattern Recognition, pp. 9592–9600 (2019)
4. Carrera, D., Boracchi, G., Foi, A., Wohlberg, B.: Detecting anomalous structures by convolutional sparse models. In: 2015 International Joint Conference on Neural Networks (IJCNN), pp. 1–8. IEEE (2015)
5. Carrera, D., Manganini, F., Boracchi, G., Lanzarone, E.: Defect detection in SEM images of nanofibrous materials. IEEE Trans. Industr. Inf. **13**(2), 551–561 (2016)
6. Chow, J.K., Su, Z., Wu, J., Tan, P.S., Mao, X., Wang, Y.H.: Anomaly detection of defects on concrete structures with the convolutional autoencoder. Adv. Eng. Inf. **45**, 101105 (2020)
7. Cimpoi, M., Maji, S., Kokkinos, I., Mohamed, S., Vedaldi, A.: Describing textures in the wild. In: Proceedings of the IEEE Conference on Computer Vision and Pattern Recognition, pp. 3606–3613 (2014)
8. Defard, T., Setkov, A., Loesch, A., Audigier, R.: PaDiM: a patch distribution modeling framework for anomaly detection and localization. In: Del Bimbo, A., et al. (eds.) ICPR 2021. LNCS, vol. 12664, pp. 475–489. Springer, Cham (2021). https://doi.org/10.1007/978-3-030-68799-1_35
9. Feng, H., Song, K., Cui, W., Zhang, Y., Yan, Y.: Cross position aggregation network for few-shot strip steel surface defect segmentation. IEEE Trans. Instrum. Meas. **72**, 1–10 (2023)
10. Gong, D., Liu, L., Le, V., Saha, B., Mansour, M.R., Venkatesh, S., Hengel, A.V.D.: Memorizing normality to detect anomaly: memory-augmented deep autoencoder for unsupervised anomaly detection. In: Proceedings of the IEEE/CVF International Conference on Computer Vision, pp. 1705–1714 (2019)

11. He, K., Zhang, X., Ren, S., Sun, J.: Deep residual learning for image recognition. In: Proceedings of the IEEE Conference on Computer Vision and Pattern Recognition, pp. 770–778 (2016)
12. Li, C.L., Sohn, K., Yoon, J., Pfister, T.: CutPaste: self-supervised learning for anomaly detection and localization. In: Proceedings of the IEEE/CVF Conference on Computer Vision and Pattern Recognition, pp. 9664–9674 (2021)
13. Li, M., et al.: Revisiting sparse convolutional model for visual recognition. Adv. Neural. Inf. Process. Syst. **35**, 10492–10504 (2022)
14. Lin, T.Y., Goyal, P., Girshick, R., He, K., Dollár, P.: Focal loss for dense object detection. In: Proceedings of the IEEE International Conference on Computer Vision, pp. 2980–2988 (2017)
15. Perlin, K.: An image synthesizer. ACM Siggraph Comput. Graph. **19**(3), 287–296 (1985)
16. Ronneberger, O., Fischer, P., Brox, T.: U-Net: convolutional networks for biomedical image segmentation. In: Navab, N., Hornegger, J., Wells, W.M., Frangi, A.F. (eds.) MICCAI 2015. LNCS, vol. 9351, pp. 234–241. Springer, Cham (2015). https://doi.org/10.1007/978-3-319-24574-4_28
17. Tan, D.S., Chen, Y.C., Chen, T.P.C., Chen, W.C.: TrustMAE: a noise-resilient defect classification framework using memory-augmented auto-encoders with trust regions. In: Proceedings of the IEEE/CVF Winter Conference on Applications of Computer Vision, pp. 276–285 (2021)
18. Wang, Z., Bovik, A.C., Sheikh, H.R., Simoncelli, E.P.: Image quality assessment: from error visibility to structural similarity. IEEE Trans. Image Process. **13**(4), 600–612 (2004)
19. Wu, L., et al.: R-Drop: regularized dropout for neural networks. Adv. Neural. Inf. Process. Syst. **34**, 10890–10905 (2021)
20. Wyatt, J., Leach, A., Schmon, S.M., Willcocks, C.G.: AnoDDPM: anomaly detection with denoising diffusion probabilistic models using simplex noise. In: Proceedings of the IEEE/CVF Conference on Computer Vision and Pattern Recognition, pp. 650–656 (2022)
21. Youkachen, S., Ruchanurucks, M., Phatrapomnant, T., Kaneko, H.: Defect segmentation of hot-rolled steel strip surface by using convolutional auto-encoder and conventional image processing. In: 2019 10th International Conference of Information and Communication Technology for Embedded Systems (IC-ICTES), pp. 1–5. IEEE (2019)
22. Zavrtanik, V., Kristan, M., Skočaj, D.: DRAEM-a discriminatively trained reconstruction embedding for surface anomaly detection. In: Proceedings of the IEEE/CVF International Conference on Computer Vision, pp. 8330–8339 (2021)
23. Zavrtanik, V., Kristan, M., Skočaj, D.: Reconstruction by inpainting for visual anomaly detection. Pattern Recogn. **112**, 107706 (2021)
24. Zha, Z., Yuan, X., Wen, B., Zhang, J., Zhou, J., Zhu, C.: Image restoration using joint patch-group-based sparse representation. IEEE Trans. Image Process. **29**, 7735–7750 (2020)
25. Zou, Y., Jeong, J., Pemula, L., Zhang, D., Dabeer, O.: Spot-the-difference self-supervised pre-training for anomaly detection and segmentation. In: Computer Vision-ECCV 2022: 17th European Conference, Tel Aviv, Israel, October 23–27, 2022, Proceedings, Part XXX, pp. 392–408. Springer (2022). https://doi.org/10.1007/978-3-031-20056-4_23

Brain Tumor Image Segmentation Based on Global-Local Dual-Branch Feature Fusion

Zhaonian Jia, Yi Hong, Tiantian Ma, Zihang Ren, Shuang Shi, and Alin Hou(✉)

Changchun University of Technology, Changchun 130102, China
alinhou@163.com

Abstract. Accurate segmentation of brain tumor medical images is important for confirming brain tumor diagnosis and formulating post-treatment plans. A brain tumor image segmentation method based on global-local dual-branch feature fusion is proposed to improve brain tumor segmentation accuracy. In target segmentation, multi-scale features play an important role in accurate target segmentation. Therefore, the global-local dual-branch structure is designed. The global branch and local branch are deep and shallow networks, respectively, to obtain the semantic information of brain tumor in the deep network and the detailed information in the shallow network. In order to fully utilize the obtained global and local feature information, an adaptive feature fusion module is designed to adaptively fuse the global and local feature maps to further improve the segmentation accuracy. Based on various experiments on the Brats2020 dataset, the effectiveness of the composition structure of the proposed method and the advancedness of the method are demonstrated.

Keywords: Brain tumor image segmentation · Transformer · Gated axial attention · Feature fusion

1 Introduction

Gliomas are the most common malignant tumors originating in the central nervous system, spanning all ages from children to the elderly, and they have the third highest mortality rate among systemic tumors within five years [1,2]. Currently, treatment of gliomas is based on surgical resection, but it is difficult for physicians to clearly identify glioma boundaries, leading to residual tumor and early recurrence [3]. With the use of Magnetic Resonance Imaging (MRI) technology in the medical field, MRI can show the internal structure of brain tumors

Supported by Jilin Provincial Education Department Science and Technology Research Project (Grant No.JJKH20210738KJ) and the Science and Technology Development Project (Grant No.20210201051GX) of Jilin Province.

Q. Liu et al. (Eds.): PRCV 2023, LNCS 14429, pp. 381–393, 2024.
https://doi.org/10.1007/978-981-99-8469-5_30

more clearly and reflect the nature of brain tumors more effectively [4], but MRI images are multimodal and complex, making expert physicians inefficient in manual labeling and may have large errors that affect subsequent medical diagnosis. Therefore, it is of great importance to design high-precision MRI brain tumor image segmentation algorithms.

Current brain tumor segmentation methods can be broadly classified into two categories: traditional brain tumor segmentation methods and segmentation methods based on deep learning. The brain tumor image segmentation methods based on traditional methods include region segmentation [5], threshold segmentation [6], and clustering-based segmentation etc [7]. Although great progress has been made, the traditional methods have high labor costs, are susceptible to noise and other factors, and have poor robustness, making it difficult to meet the standards for practical applications.

Segmentation methods based on deep learning have risen rapidly and been widely used in brain tumor segmentation in recent years. Sun et al. [8] proposed a multi-pathway feature extraction FCN, which together with cavity convolution extracted information under different receptive fields to obtain fine segmentation in brain MRI images. Liu et al. [9] designed Spatial Channel Fusion Block (SCFB) to aggregate brain tumor multimodal features and proposed a spatial loss to constrain the relationship between different tumor sub-regions. Bukhari et al. [10] coupled an encoder with three decoders, each of which accepts feature mappings directly from a generic encoder and segments different tumor subregions. Ding et al. [11] proposed a region-aware fusion network to effectively aggregate various available modalities for multimodal brain tumor segmentation.

The UNet [12] network employs a coding-decoding architecture with excellent performance and is widely used in medical image segmentation. However, the U-shaped architecture of convolutional neural networks follows the properties of visual information, which makes the network more inclined to focus on the connections between local features rather than the global context. In addition, existing feature fusion methods are commonly feature map channel stacking or feature map summing, which cannot refine the features. Therefore, a brain tumor image segmentation network model based on global-local dual-branch feature fusion is proposed.

In summary, our contributions can be summarized as follows:

1. Global-local branching structure is designed. The designed global branch acquires rich semantic information in the deep network. Local branches are designed to obtain fine-grained feature information in shallow networks.

2. Adaptive feature fusion module is proposed to refine the features and perform adaptive feature fusion to further improve the segmentation accuracy

3. Ablation experiments and comparison experiments are conducted in Brats2020 to demonstrate the effectiveness of the proposed method's composition structure and the advancement of the method.

2 Method

2.1 Overall Framework

The overall framework of the brain tumor segmentation network proposed in this paper is shown in Fig. 1. The model consists of global branches, local branches and adaptive feature fusion modules, in which the global branch structure is designed as a deep network with a larger sensory field to obtain more feature information of medium and large targets. The local branch structure is designed as a shallow network with a smaller perceptual field to obtain more texture and edge information in the feature map. The adaptive feature fusion module fuses the acquired global and local feature maps of brain tumors to obtain the target segmentation results.

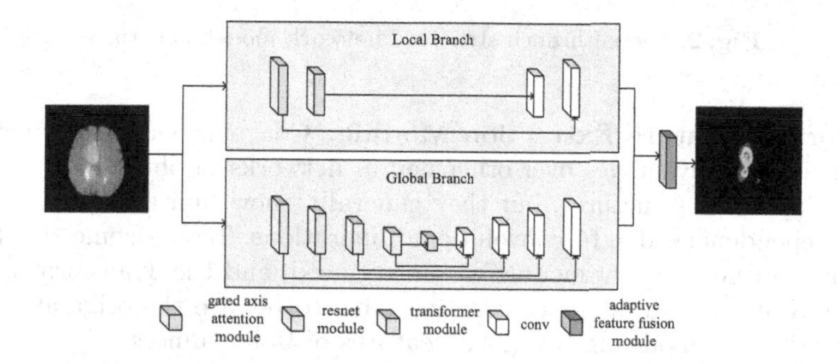

Fig. 1. Overall framework of brain tumor segmentation network

2.2 Global Branch

The global branching is based on UNet as the base framework, the contraction path part consists of deep residual network structure and Transformer module, and the extension path part consists of traditional convolutional layer and jump connection module. The deep network of the global branch can obtain rich semantic feature information, and the Transformer module is added to enhance the model to extract deep and shallow semantic features and obtain high-quality contextual semantic information, and the structure of the global feature extraction path network model is shown in Fig. 2.

Deep Residual Network. The jump connection of the deep residual network (ResNet) allows the network to preserve some shallow feature information in transmission, alleviating the problem of detailed feature loss due to deeper network layers. The global branching shrinkage path uses the ResNet50 network structure to increase the network depth, obtain rich global features, and capture more information of medium and large targets. As shown in Fig. 2. The contraction path is divided into 4 stages (ResNet-i,i $\in \{1, 2, 3, 4\}$). The bottom output dimension of the network is 1024.

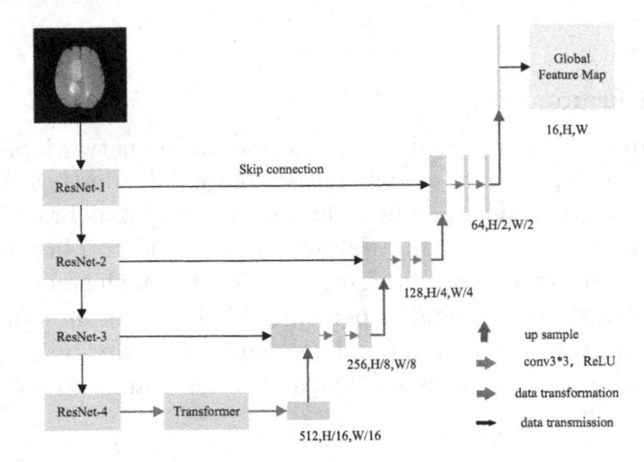

Fig. 2. Global branch structure network model structure

Transformer Feature Extraction Module. Convolutional neural networks have significant advantages over other neural networks in obtaining the underlying features and structures, but they generally show limitations in modeling remote dependencies due to convolutional operations. To overcome this limitation, the Transformer [13] module was introduced, and the Transformer model was placed at the end of the contraction path to replace the original convolutional module for modeling the global features of brain tumors.

The output of the deep residual network is a 3D matrix with 1024 channels and 10 widths and heights. First, the 3D feature information needs to be linearly transformed into a vector sequence that meets the input of the Transformer module to obtain linear features with a sequence length of 100 and a word vector length of 768, and the linear features are input into the Transformer feature extraction module for global attention calculation. The structure of Transformer module is shown in Fig. 3.
The Transformer module consists of Multi head Self Attention (MHSA), Multi layer perceptron (MLP) and Layer Normalization (LN) units through a residual structure. The data input Transformer module can be represented as:

$$\hat{M} = MHSA\left(LN\left(M\right)\right) + M \tag{1}$$

$$N = MLP\left(LN\left(\hat{M}\right)\right) + \hat{M} \tag{2}$$

M is the Transformer module input, N denotes the Transformer module output, and the structure of the multi-headed attention mechanism is shown in Fig. 3, whose specific process can be expressed as follows:

$$MHSA\left(Q, K, V\right) = soft\max\left(\frac{QK^T}{\sqrt{d}} + B\right)V \tag{3}$$

The multi-head attention mechanism is to match Q with K. The similarity between Q and K is obtained by the multi-head attention mechanism module,

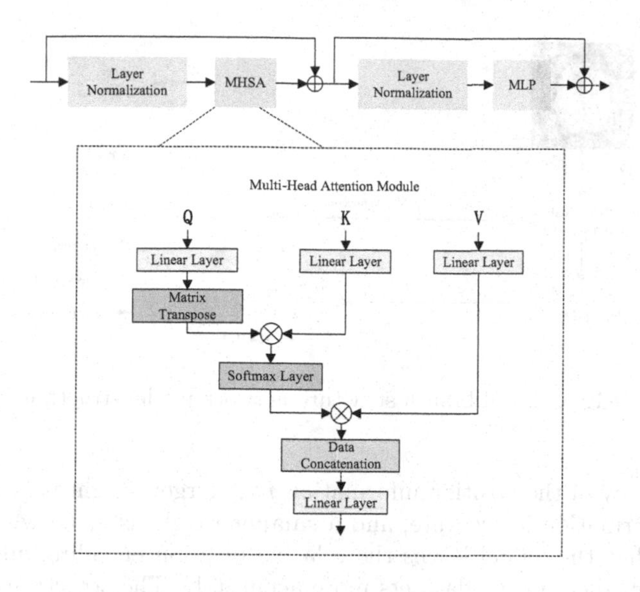

Fig. 3. Structural diagram of transformer module.

and then the attention weight matrix is obtained after the Softmax layer, and the output result is obtained by matrix multiplication of V with the weight matrix. Finally, the output feature sequence is reshaped to match the extended path, and the feature sequence is transformed into a feature map with a channel number of 512 and a width and height of 10.

2.3 Local Branch

The local branch is designed as a shallow U-shaped network, with the systolic path part consisting of a gated axial attention module and the extended path part consisting of a conventional convolutional layer and a jump connection module. The shallow network of local branches has rich information of detailed texture features, which can better capture the local detailed features of brain tumors. The introduction of gated axial attention module enhances the focus on the target region and further improves the ability to capture the detailed features. The structure of the local branching structure network model is shown in Fig. 4.

Gated Axial Attention Module. The Gated Axial-Attention [14] resolves the original two-dimensional self-attention into two one-dimensional self-attention along the height and width axes of the feature map, which improves the computational efficiency. When calculating the degree of association between elements, a position bias term is added to solve the problem that the self-attention is insensitive to the position information. The role of the gating unit in the gated axial attention is to regulate the weight of the position information according

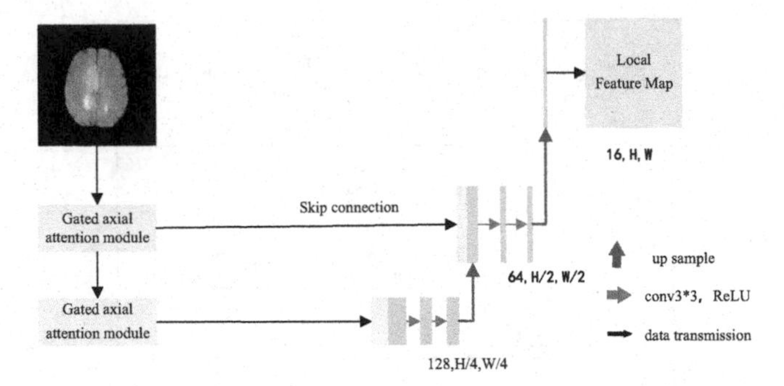

Fig. 4. Local branch structure network model structure.

to the accuracy of the position information r. A larger weight is given when the position information is accurate, and a smaller weight is given when it is inaccurate, so that the model learns the relative position encoding information on small-scale medical image datasets more accurately. The structure of the gated axial attention module is shown in Fig. 5.

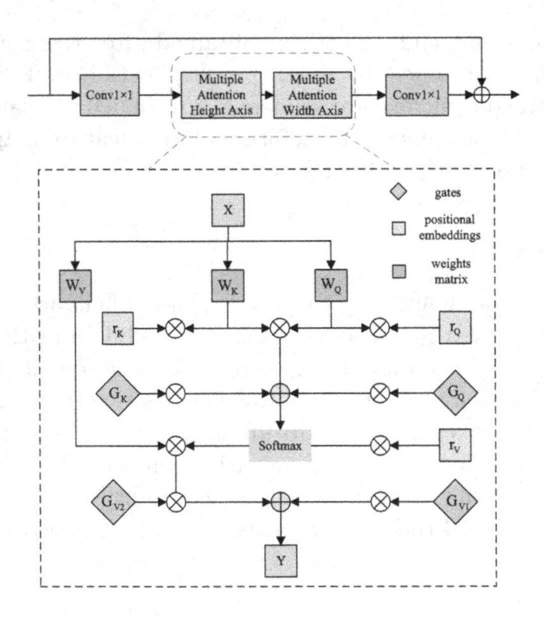

Fig. 5. Gated axial attention module structure.

The input features X generate q_p^T, q_n^T, and v_n after three transformation matrices of W_Q, W_K, and W_V. The relative position codes r_Q,r_K, and r_V can be updated during the model training process, and G_Q,G_K, $G_V \in R$ are gating units that control the impact of the learned relative position codes on the non-local context.

For the input feature map X, the multi-headed attention width axis module can be represented as follows:

$$y_m = \sum_{n \in w_{1\times C}(p)} soft\max\left(q_p^T k_n + G_Q q_p^T r_n^q + G_K k_n^T r_n^k\right)\left(G_{V1} v_n + G_{V2} r_n^v\right) \quad (4)$$

$w_{1\times C}(p)$ is the multi-headed attention width axis region where position $p = (i, j)$ is located, vectors r_n^q, r_n^k,r_n^v are relative position encodings, inner products $q_p^T r_n^q$, $k_n^T r_n^k$ are bias terms of Key and Query, and similarly the multi-headed attention height axis formula is similar.

2.4 Adaptive Feature Fusion Module

In order to make full use of the global feature information obtained in the global branch and the local detailed features obtained in the local branch, this paper, inspired by the literature [15], proposes an adaptive feature fusion module to refine the features and perform feature map fusion, the structure of which is shown in Fig. 6.

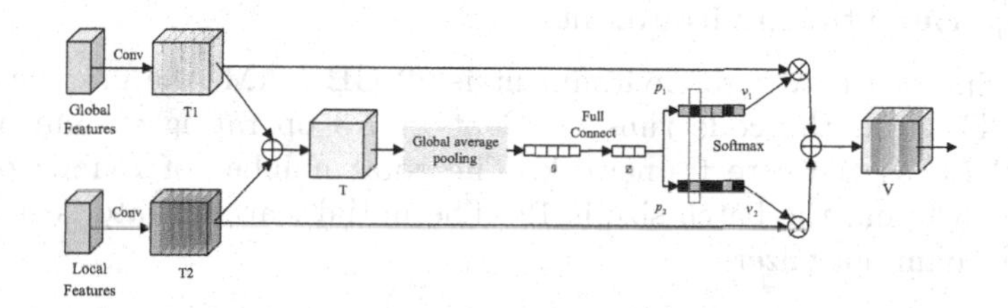

Fig. 6. Adaptive feature fusion module structure.

Firstly, the input global and local feature information is convolved separately to obtain its corresponding feature map, which is noted as T_1 and T_2, The two features are summed to obtain the feature map T. Then, the global average pooling (GAP) operation is performed on the feature map T to obtain the vector s indicating the importance of global and local features, which is calculated as follows:

$$s_c = \phi_{GAP}(T) = \frac{1}{H \times W} \sum_{i=1}^{H} \sum_{j=1}^{W} T_c(i, j) \quad (5)$$

s_c is for the c-th component of s. The dimensional vector z obtained by fusing the features of different layers of s using a fully connected layer with a scale compression of the pixel dimension to d dimensions is subsequently calculated as follows:

$$z = \sigma(B(Ws)) \quad (6)$$

$B()$ and $\sigma()$ denote batch normalization and ReLU operations, respectively, where $W \in R^{\mathrm{d} \times C}$ is fully connected, the value of d is taken as follows:

$$d = \max\left(\frac{C}{r}, L\right) \tag{7}$$

r is a reduced ratio, L is the minimum dimension of z, and r and L are used to control the output dimension. Then the vector z is re-varied back to length C by two fully connected layers to get $\{p_1,p_2\}$, and the components of the two vectors of the same latitude are subjected to Softmax operation to get the weight vectors of global and local information $\{v_1,v_2\}$. The extracted feature weights are dotted with T_1 and T_2 to obtain the optimized feature map and summed to obtain the global and local feature fused feature map V, which is calculated as follows:

$$V = \sum_{i=1}^{2} v_i T_i \tag{8}$$

3 Experimental Procedure

3.1 Experimental Environment

The experimental hardware environment is 32 GB RAM, the graphics card is NVIDIA Titan X, the code runs on CentOS 7.3 operating system and uses Pytorch 1.7.0 as the core framework. The total number of rounds of model training is 300 and the batch_size is 12. The initial learning rate is set to 1e-3 using the Adam optimizer.

3.2 Dataset

The dataset of the 2020 MICCAI Multimodal Brain Tumor Segmentation Challenge BraTs [16–18] is used in this paper. and the BraTs2020 dataset has a total of 369 cases. Only the training set of the dataset has real segmentation labels (Ground Truth, GT) manually labeled by experts, including background, Non-Enhancing Tumor(NET), Peritumoral Edema(ED), and Enhanced Tumor region (ET), respectively. In this experiment, 70% of the BraTs2020 training set was used for model training, and the remaining 30% was used for the test set.

3.3 Data Pre-processing

The brain tumor MRI images are 3D images, and the design network in this paper is a 2D network, so the 3D brain tumor MRI images need to be sliced and processed into 2D images, which are consistent with the input of the proposed 2D network model. The proportion of black background in the MRI images is large, resulting in small brain tumor marker regions, and the black background region is not the target marker region, so the area is cropped to remove the invalid region, and the image is 4×160×160 after cropped slices, the number of channels is 4, and the length and width are 160.

3.4 Evaluation Metrics

In order to analyze the segmentation effect of the model on brain tumors, two metrics, Dice coefficient and Hausdorff distance, are used as the evaluation metrics in this paper. The Dice coefficient is used to calculate the similarity between the predicted results and the real segmentation labels of brain tumors. Its calculation formula is as follows:

$$Dice = \frac{2TP}{FP + 2TP + FN} \tag{9}$$

TP is true positive, FP is true negative, and FN is false negative. The Hausdorff Distance is used to measure the distance between two point sets. Its calculation formula is as follows:

$$h(A, B) = \max_{a \in A} \min_{b \in B} \|a - b\| \tag{10}$$

$h(A, B)$ denotes the maximum mismatch distance between two point sets A and B. a and b are the points of sets A and B, respectively.

4 Analysis of Results

4.1 Results of Ablation Experiments

To verify the validity of each component structure of the model, ablation experiments were performed on the UNet network, global branch, global-local double branch (without adding the adaptive feature fusion module) and the complete model under the same experimental conditions, and the results were obtained as shown in Table 1, where WT, TC and ET are the tumor as a whole (composed of NET, ED, and ET), core tumor (composed of NET and ET), and enhancing tumor (composed of ET). By comparing the experimental results in the Table 1, the complete method is significantly better than the other compared methods, especially in the segmentation of the TC core tumor region is more obvious (Dice is improved by 1.3% ~ 4.92%, Hausdorff distance is decreased by 0.1681 ~ 0.2204) which fully demonstrates that the global-local dual-branching junction and adaptive feature fusion module have good comprehensive performance and help to improve the segmentation accuracy.

Table 1. Results of ablation experiment.

Model	Dice_WT	Dice_TC	Dice_ET	HD_WT	HD_TC	HD_ET
UNet	0.8440	0.8288	0.7775	2.6250	1.7279	2.7960
Global Branch	0.8450	0.8323	0.7804	2.6600	1.6917	2.7950
Global-Local branch	0.8540	0.8650	0.7964	2.5312	1.5598	2.6771
Complete method	**0.8589**	**0.8780**	**0.8055**	**2.5020**	**1.5075**	**2.6153**

4.2 Comparison Method Experimental Results

In order to show the effectiveness and advancement of the proposed method in this paper, the proposed algorithm is compared with other six algorithms on the Brats2020 dataset, including UNet [12], UNet++ [19], UNet3+ [20], SwinUNet [21], TA-Net [22], and TransUNet [23], and all experiments are derived under the same experimental environment for a fair comparison, and the code is the publicly released code of the paper. The data of segmentation result indexes are shown in Table 2, and the algorithm of this paper achieves the best results in all aspects, with Dice coefficients of 0.8589, 0.878, and 8055, and Hausdorff distances of 2.502, 1.5075, and 2.6153, respectively.

Table 2. Segmentation result indicator data of each model in the Brats 2020 dataset.

Model	Dice_WT	Dice_TC	Dice_ET	HD_WT	HD_TC	HD_ET
UNet [12]	0.8440	0.8288	0.7775	2.6250	1.7279	2.7960
UNet++ [19]	0.8370	0.8386	0.7813	2.5656	1.6550	2.7246
UNet3+ [20]	0.8407	0.8633	0.7888	2.5506	1.5579	2.6765
SwinUNet [21]	0.8317	0.8452	0.7698	2.6280	1.7022	2.7907
TA-Net [22]	0.8273	0.8468	0.7663	2.6335	1.6326	2.7885
TranUNet [23]	0.8577	0.8632	0.7985	2.5249	1.5823	2.6926
Ours	**0.8589**	**0.8780**	**0.8055**	**2.5020**	**1.5075**	**2.6153**

Dice coefficient and Hausdorff Distance comparison histograms of various algorithms on the Brats2020 dataset are shown in Fig. 7, and it can be visualized that the segmentation performance of this paper's method in the WT, TC, and ET regions is improved compared to other methods. In Dice coefficient, the segmentation performance of ET region is weakly improved, while the segmentation performance of WT and TC regions is relatively improved significantly. In Hausdorff Distance, this paper's algorithm achieves the lowest value, which is significantly better than other algorithms.

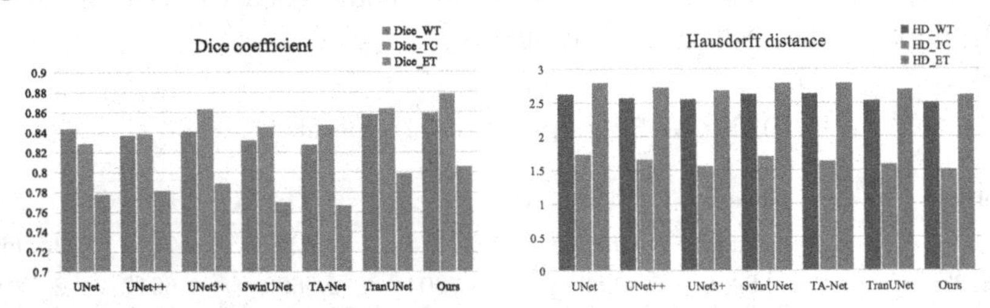

Fig. 7. Histogram of Dice and HD comparisons on the dataset BraTS2020.

The comparison graph of this paper's method with each model for brain tumor segmentation is shown in Fig. 8., where the ED region (green), NET region (red) and ET region (yellow), compared with other methods, this paper's method achieves better results in all aspects by, especially in the small and medium-sized target segmentation such as NET and ET, which achieves a greater improvement, predicts better shapes and edge details, and is more close to the real segmentation labels.

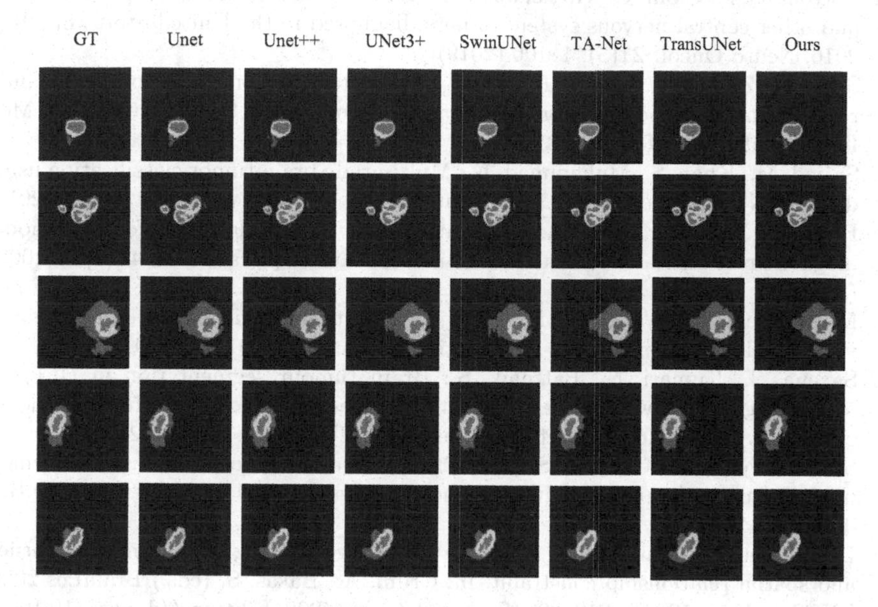

Fig. 8. Comparison of segmentation effects of various network models. (Color figure online)

5 Conclusion

A brain tumor image segmentation method based on global-local dual-branch feature fusion is proposed in this paper, which obtains deep semantic features and local detail information through dual branches respectively, and designs an adaptive feature fusion module for feature fusion of global-local features to obtain higher brain tumor MRI segmentation accuracy. To prove the effectiveness of the method, experiments are carried out on BraTs2020, a publicly available dataset for brain glioma segmentation, which shows that this method is significantly better than other segmentation methods in the segmentation task, and significantly improves the segmentation accuracy on the tumor as a whole, on the tumor core, and on the enhanced tumor region. In this paper, the network is a 2D network, which will lose the spatial information between the layers in the segmentation of 3D image data, and has some limitations, in the future work, we will continue to develop this method and apply this method to the 3D network, to further improve the network's segmentation ability for each lesion region of brain tumors.

References

1. Wen, J., Chen, W., Zhu, Y.: Clinical features associated with the efficacy of chemotherapy in patients with glioblastoma (GBM): a surveillance, epidemiology, and end results (SEER) analysis. BMC Cancer **21**(1), 81 (2021)
2. Ostrom, Q.T., Cioffi, G., Gittleman, H.: CBTRUS statistical report: primary brain and other central nervous system tumors diagnosed in the United States in 2012–2016. Neuro Oncol. **21**(5), 1–100 (2019)
3. Shen, B., Zhang, Z., Shi, X.: Real-time intraoperative glioma diagnosis using fluorescence imaging and deep convolutional neural networks. Eur. J. Nucl. Med. Mol. Imaging **48**(11), 3482–3492 (2021)
4. Sajjad, M., Khan, S., Muhammad, K.: Multi-grade brain tumor classification using deep CNN with extensive data augmentation. J. Comput. Sci. **30**, 174–182 (2019)
5. He, C.E., Xu, H.J., Wang, Z.: Automatic segmentation algorithm for multimodal magnetic resonance-based brain tumor images. Acta Optica Sinica. **40**(6), 0610001 (2020)
6. Mo, S., Deng, X., Wang, S.: Moving object detection algorithm based on improved visual background extractor. Acta Optica Sinica. **36**(6), 615001 (2016)
7. Saxena, S., Kumari, N., Pattnaik, S.: Brain tumour segmentation in DFLAIR MRI using sliding window texture feature extraction followed by fuzzy C-means clustering. Int. J. Healthc. Inf. Syst. Inf. (IJHISI) **16**(03), 1–20 (2021)
8. Sun, J., Peng, Y., Guo, Y.: Segmentation of the multimodal brain tumor image used the multi-pathway architecture method based on 3D FCN. Neurocomputing **423**, 34–45 (2021)
9. Liu, C., et al.: Brain tumor segmentation network using attention-based fusion and spatial relationship constraint. In: Crimi, A., Bakas, S. (eds.) BrainLes 2020. LNCS, vol. 12658, pp. 219–229. Springer, Cham (2021). https://doi.org/10.1007/978-3-030-72084-1_20
10. Bukhari, S.T., Mohy-ud-Din, H.: E1D3 U-Net for brain tumor segmentation: submission to the RSNA-ASNR-MICCAI BraTS 2021 challenge. In: Crimi, A., Bakas, S. (eds.) Brainlesion: Glioma, Multiple Sclerosis, Stroke and Traumatic Brain Injuries. BrainLes 2021. LNCS, vol. 12963. Springer, Cham (2022). https://doi.org/10.1007/978-3-031-09002-8_25
11. Ding Y., Yu X., Yang Y.: RFNet: region-aware fusion network for incomplete multimodal brain tumor segmentation. In: Proceedings of the IEEE/CVF International Conference on Computer Vision, pp. 3975–3984 (2021)
12. Ronneberger, O., Fischer, P., Brox, T.: U-Net: convolutional networks for biomedical image segmentation. In: Navab, N., Hornegger, J., Wells, W.M., Frangi, A.F. (eds.) MICCAI 2015. LNCS, vol. 9351, pp. 234–241. Springer, Cham (2015). https://doi.org/10.1007/978-3-319-24574-4_28
13. Vaswani A., Shazeer N., Parmar N., et al.: Attention is all you need. In: Advances in Neural Information Processing Systems, pp. 30 (2017)
14. Valanarasu, J.M.J., Oza, P., Hacihaliloglu, I., Patel, V.M.: Medical transformer: gated axial-attention for medical image segmentation. In: de Bruijne, M., et al. (eds.) MICCAI 2021. LNCS, vol. 12901, pp. 36–46. Springer, Cham (2021). https://doi.org/10.1007/978-3-030-87193-2_4
15. Li, X., Wang, W.H., Hu, X.I., et al.: Selective kernel network. In: 2019 IEEE CVF Conference on Computer Vision and Pattern Recognition, pp. 510–519. IEEE (2020)

16. Menze, B.H., Jakab, A., Bauer, S.: The multimodal brain tumor image segmentation benchmark (BRATS). IEEE Trans. Med. Imaging **34**(10), 1993–2024 (2014)
17. Bakas, S., Akbari, H., Sotiras, A.: Advancing the cancer genome atlas glioma MRI collections with expert segmentation labels and radiomic features. Sci. Data **4**(1), 1–13 (2017)
18. Bakas, S., Reyes, M., Jakab, A.: Identifying the best machine learning algorithms for brain tumor segmentation, progression assessment, and overall survival prediction in the BRATS challenge. Radiomics and Radiogenomics: Technical Basis and Clinical Application. New York: Chapman and Hall/ CRC, pp. 99–114. (2019)
19. Zhou, Z., Siddiquee, M., Tajbakhsh, N.: UNet++: a nested U-Net architecture for medical image segmentation. In: 4th Deep Learning in Medical Image Analysis (DLMIA) Workshop, pp. 3–11 (2018)
20. Huang H., Lin L., Tong R.: Unet 3+: A full-scale connected UNet for medical image segmentation. In: IEEE International Conference on Acoustics, Speech and Signal Processing (ICASSP), pp. 1055–1059. IEEE (2020)
21. Cao, H., et al.: Swin-Unet: Unet-like pure transformer for medical image segmentation. In: Karlinsky, L., Michaeli, T., Nishino, K. (eds.) Computer Vision – ECCV 2022 Workshops. ECCV 2022. LNCS, vol. 13803. Springer, Cham (2023). https://doi.org/10.1007/978-3-031-25066-8_9
22. Pang S., Du A., Orgun M. A.: Tumor attention networks: better feature selection, better tumor segmentation. Neural Netw. **140**(1), 203–222 (2021)
23. Chen J., Lu Y., Yu Q.: Transunet: Transformers make strong encoders for medical image segmentation. In: Computer Vision and Pattern Recognition, pp. 34–47 (2021)

PRFNet: Progressive Region Focusing Network for Polyp Segmentation

Jilong Chen[1], Junlong Cheng[1], Lei Jiang[1], Pengyu Yin[1], Guoan Wang[2], and Min Zhu[1](✉)

[1] College of Computer Science, Sichuan University, Chendu, China
zhumin@scu.edu.cn
[2] College of Computer Science, East China Normal University, Shanghai, China

Abstract. In clinical practice, colonoscopy serves as an efficacious approach to detect colonic polyps and aids in the early diagnosis of colon cancer. However, the precise segmentation of polyps poses a challenge due to variability in size and shape, indistinct boundaries, and similar feature representations with healthy tissue. To address these issues, we propose a concise yet very effective progressive region focusing network (PRFNet) that leverages progressive training to iteratively refine segmentation results. Specifically, PRFNet shares encoder parameters and partitions the feature learning process of decoder into various stages, enabling the aggregation of features at different granularities through cross-stage skip connections and progressively mining the detailed features of lesion regions at different granularities. In addition, we introduce a lightweight adaptive region focusing (ARF) module, empowering the network to mask the non-lesion region and focus on mining lesion region features. Extensive experiments have been conducted on several public polyp segmentation datasets, where PRFNet demonstrated competitive segmentation results compared to state-of-the-art polyp segmentation methods. Furthermore, we set up multiple cross-dataset training and testing experiments, substantiating the superior generalization performance of PRFNet.

Keywords: Polyp segmentation · Progressive feature refinement · Adaptive region focusing

1 Introduction

Colonoscopy allows for the investigation of polyps or pathological changes, particularly the detection of tumorous intestinal polyps, which bears significant impact on the early diagnosis of colorectal cancer (CRC) [14]. Once the location of tumorous polyps has been determined via colonoscopy, timely surgical removal can effectively avert the onset of colorectal cancer. Hence, accurate segmentation of polyp tissue in clinical scenarios is of paramount importance, with numerous

J. Chen and J. Cheng—Contributed equally and should be considered co-first authors.

Q. Liu et al. (Eds.): PRCV 2023, LNCS 14429, pp. 394–406, 2024.
https://doi.org/10.1007/978-981-99-8469-5_31

studies emerging to automate this process. However, due to issues such as indistinct polyp boundaries, variability in size and shape, and the similarity of feature representations between polyp tissue and healthy tissue [6,10], achieving precise segmentation of polyp still poses a challenge.

Numerous studies have utilized deep learning for medical image segmentation. Earlier works such as UNet [13], UNet++ [22], TransUNet [4], leveraged encoder-decoder network structures for relatively precise segmentation of medical images. For the more specialized field of polyp segmentation, PraNet [8] designed a parallel partial decoder (PD) for feature fusion and proposed a reverse attention (RA) module to introduce boundary constraints for performance enhancement. SSFormer [19] introduced modules such as local emphasis (LE), stepwise feature aggregation (SFA), and feature fusion units to smooth and emphasize local features. LDNet [21] proposed kernel generation (KG) and kernel update (KU) strategies, along with efficient self-attention (ESA) and lesion-aware cross-attention (LCA) modules to enhance polyp feature representations. Polyp-PVT [2] incorporated cascaded fusion module (CFM), camouflage identification module (CIM), and similarity aggregation module (SAM) into the transformer encoder to learn more robust representations from polyps. CaraNet [11] designed axial reverse attention (A-RA) and channel-wise feature pyramid (CFP) modules on top of the partial decoder to enhance the model's segmentation performance for smaller targets. To augment their model's learning capability, these studies have introduced numerous, complex functional modules based on the encoder-decoder structure, resulting in inferior generalization performance and a propensity towards overfitting, thereby leading to inaccuracies or even errors in segmentation. Therefore, proposing a method that can concisely and accurately automate the segmentation of polyp in the early stages of the disease, without the introduction of overly complex modules, carries significant importance.

In response to the above challenges, we propose a concise and effective progressive region focusing network (PRFNet). By leveraging a multiscale encoder-decoder structure coupled with a carefully designed training strategy, we mask background feature representations in the progressive training process, and based on learning different granularities of lesion features, we further extract fine-grained lesion features. Specifically, we first locate the lesion region in a progressive manner, integrating different levels of lesion features learned by the multiscale network, and refine the feature representations of the lesion stage-by-stage. To further enhance the model's learning of fine-grained features during the progressive training process, we propose an adaptive region focusing module. This module can adaptively mask the expression of background feature representations in feature maps during progressive training, enabling the network to focus more on the learning of local lesion region feature representations, thereby enhancing the model's generalization ability for more precise segmentation.

In summary, our contributions include the following aspects: (1) We propose a progressive learning method that accomplishes precise segmentation of polyp by mining and fusing feature representations at different levels. (2) We introduce a learnable adaptive region focusing module that, by adaptively masking a large number of non-lesion pixels during the progressive training process, enables the model to focus on the lesion region, thereby enhancing the model's generalization ability. (3) We perform comparative and ablation experiments on multiple public polyp segmentation datasets, demonstrating the learning and generalization abilities of our proposed PRFNet.

2 Method

In this chapter, we will introduce the concise and effective progressive region focusing network that we propose, which is primarily composed of progressive feature refinement and adaptive region focusing modules.

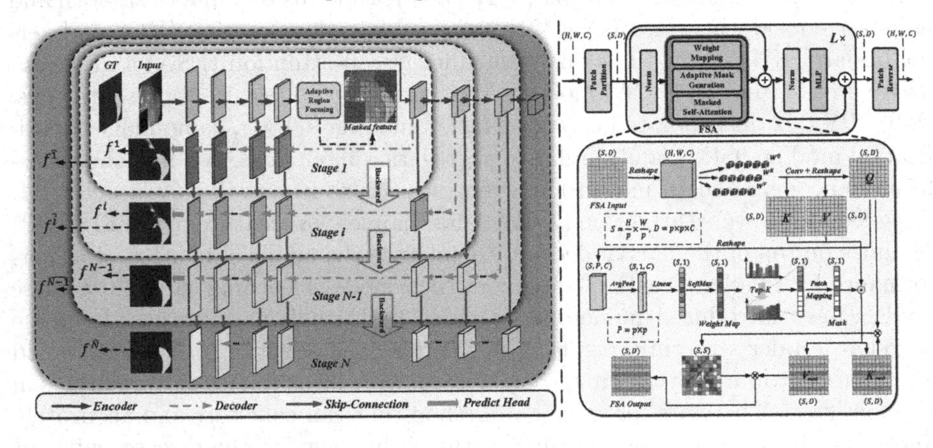

Fig. 1. Overview of the PRFNet. Left: progressive feature refinement which includes adaptive region focusing module. Right: implementation details of adaptive region focusing module.

2.1 Progressive Feature Refinement

There is a large number of work that has designed numerous functional modules to achieve medical image segmentation, not limited to polyps [5–7,20]. However, given the indistinct boundaries of polyps and its exceedingly similar feature representations to healthy tissue, it is necessary for the model to learn more discriminative features of polyp. Most semantic segmentation methods assume that the entire segmentation process can be performed through a single feed-forward process of the input image. The feature representations extracted in this manner is homogenous [12], and ineffective at extracting fine-grained feature representations.

Addressing these issues, as shown in Fig. 1, we propose a progressive feature refinement method that, under shared encoder-decoder parameters, trains multi-stage decoder networks in a progressive manner to mine and fuse features of different granularities. Specifically, we divide the progressive training process into N stages. The network of the latter stage will include the network of the previous stage after parameters updating. Therefore, each stage includes one or more encoder-decoder networks of different levels, with the maximum number of encoder or decoder layers in each stage being $L-N+1, L-N+2, ..., L$. All stage networks share a single lightweight encoder branch and use different decoder branches, with a skip connection between decoder branches of two adjacent stages for progressive information transfer. The progressive training strategy we propose is represented as follows:

$$\begin{cases} f^1 = \left(\prod_{l=1}^{L-N+1} d_l^1\right)\left(\prod_{l=1}^{L-N+1} e_l\right)(x), \\ f^2 = \left(\prod_{l=1}^{L-N+2} d_l^2\right)\left(e_{L-N+2}\left(\prod_{l=1}^{L-N+1} \widetilde{e_l}\right)\right)(x), \\ \dots, \\ f^N = \left(\prod_{l=1}^{L} d_l^N\right)\left(e_L\left(\prod_{l=1}^{L-1} \widetilde{e_l}\right)\right)(x) \end{cases} \tag{1}$$

where $x \in \mathbb{R}^{H\times W\times C}$ is the input image, and f^1, f^2, f^N are the prediction results of the $1st$, $2nd$, and Nth stage networks respectively. e_l represents the lth layer of the encoder, and d_l^1, d_l^2, d_l^N are the lth layer decoders of the $1st$, $2nd$, and Nth stage networks respectively (the sequence numbers l for the encoder and decoder are respectively numbered according to the direction of the arrows of the encoder and decoder in Fig. 1). The e_l, d_l^1 of the $1st$ stage network are obtained as $\widetilde{e_l}$ and $\widetilde{d_l^1}$, respectively, after calculating the loss via f^1 and updating parameters, where $l \in (1, L-N+1)$. Due to the shared encoder parameters and skip connections between decoders, the $2nd$ stage network inherits $\widetilde{e_l}$ and $\widetilde{d_l^1}$ from the $1st$ stage. x is encoded by $\widetilde{e_l}$ and the newly added e_{L-N+2} in the $2nd$ stage, and then enters both $\widetilde{d_l^1}$ and d_l^2 for decoding. There is a skip connection between the feature maps output by the two decoder branches, and the output of the d_l^2 branch is the $2nd$ stage prediction result, f^2. The subsequent stages are similar.

Through the above-mentioned progressive training strategy, we can obtain different granularity feature representations through the multi-scale network, and get the most discriminative fine-grained features of the lesion region. By sharing encoder parameters and skip connections between decoder branches, the network can initially locate the lesion region in the early stages, refine the fine-grained features inside the lesion region in the middle stages, and finally obtain accurate segmentation results in the last stage, thereby enhancing the learning and generalization ability of the entire network.

2.2 Adaptive Region Focusing

The progressive training method can gradually refine the feature representation of the lesion, but the features learned by the multi-scale network may focus on the non-lesion region. And for some polyp datasets, due to the collection reasons, some images may contain no-information areas. Therefore, we propose an adaptive region focusing module, which can adaptively generate masks for feature maps, masking the feature representation of non-lesion region, forcing the network to pay more attention to the lesion region during the training process. As shown in Fig. 1, ARF consists of L ViT blocks, in which we introduce a focusing self-attention (FSA) and adjust the module settings of the traditional ViT block to meet the needs of ARF.

Weight Mapping. For ViT block l, the feature map $f \in \mathbb{R}^{H \times W \times C}$ first goes through patch partition (not using patch embedding with a linear layer to preserve spatial information of feature maps) to divide the feature map into patches of a certain size, resulting in $f \in \mathbb{R}^{S \times D}$, where S is the number of patches and D is the product of patch size and number of channels. Then, f is passed through a normalization layer and rearranged into $f \in \mathbb{R}^{H \times W \times C}$, and then goes through 1×1 convolution layers and rearrange operations to generate $Q^l, K^l, V^l \in \mathbb{R}^{S \times D}$. And Q^l is unfolded along the dimension D to obtain $\widetilde{Q^l} \in \mathbb{R}^{S \times P \times C}$, where P is the patch size and C is the number of channels. Finally, based on $\widetilde{Q^l}$, the weight mapping $M^l \in \mathbb{R}^{S \times 1}$ is determined. M^l is defined as follows:

$$M^l = Softmax\left(Linear\left(AvgPool\left(\widetilde{Q^l}\right)\right)\right) \tag{2}$$

AvgPool refers to adaptive average pooling, which is used to aggregate the spatial information of each patch. The *Linear* layer is used to aggregate the channel information of each patch. By passing this through *Softmax*, the weights of the S patches M^l, can be obtained. The values in this mapping are all between 0 and 1, where the weights corresponding to the patches in the lesion region are higher, and vice versa.

Adaptive Mask Generation. Then we determine the adaptive mask based on M^l. We set different thresholds for the networks at each stage to control the number of non-lesion patches that are masked. The specific definition of the threshold is as follows:

$$threshold = Max\left(topkList\left(M^l\right)\right) \tag{3}$$

first, we sort M^l in ascending order and select the top-k list, which represents the smallest k patch weights, and the maximum value within this list is adopted as the threshold. The threshold size is controlled by adjusting the number of elements in the list with k. After determining the mask threshold used by the current stage network, we generate the adaptive mask $\widehat{M^l} \in \mathbb{R}^{S \times 1}$ for the current stage according to the following definition:

$$\widehat{M^l_{(i,j)}} = \begin{cases} 1, & if\ M^l_{(i,j)} > threshold \\ M^l_{(i,j)}, & otherwise \end{cases} \tag{4}$$

where (i, j) represents the row and column indices of an element in M^l, if the value at position (i, j) in M^l is greater than the threshold, then the value at the corresponding position in $\widehat{M^l}$ is set to 1. Otherwise, the value at that position remains the same as M^l. Furthermore, k will increase as the network stage progresses, which means that in later stages of the network with stronger learning ability, more very small patch weights will be preserved, gradually focusing on the lesion region during the progressive training process.

Masked Self-attention. Next, we apply the generated mask $\widehat{M^l}$ to K^l and V^l respectively, obtaining $K^l_{mask}, V^l_{mask} \in \mathbb{R}^{S\times D}$. We then calculate the attention map $A^l \in \mathbb{R}^{S\times S}$. The final output of the FSA is $O^l \in \mathbb{R}^{S\times D}$, defined as follows:

$$K^l_{mask} = \widehat{M^l} \circ K^l, V^l_{mask} = \widehat{M^l} \circ V^l \tag{5}$$

$$A^l = Softmax\left(\frac{Q^l k^{l^T}_{mask}}{\sqrt{D}}\right), O^l = A^l V^l_{mask} \tag{6}$$

where '$\circ$' represents element-wise multiplication with broadcasting mechanism.

2.3 Optimization and Inference

During the optimization process, we progressively update the parameters of the multi-stage network. For a network that includes N stages, we calculate the weighted IoU loss and binary cross entropy (BCE) loss at each stage to ensure fair comparison. Specifically, for networks in stages n from 1 to $N-1$, we adopt the following loss function:

$$\mathcal{L}^n = \mathcal{L}^w_{IoU}(f^n, G) + \mathcal{L}^w_{BCE}(f^n, G) \tag{7}$$

where f^n is the prediction result generated by the nth stage network, G is the corresponding ground truth for the input sample, and L^w_{IoU} and L^w_{BCE} represent the weighted IoU loss and BCE loss, respectively. In addition, the loss for the Nth stage network is defined as:

$$\mathcal{L}^N = \alpha\mathcal{L}^N_{Stage} + \beta\mathcal{L}^N_{Sum} \tag{8}$$

α and β are weight coefficients. For L^N_{Stage}, since the Nth stage inherits the encoder-decoder of all preceding stages, we supervise the outputs of the decoders corresponding to all 1 to N stage networks included in the Nth stage:

$$\mathcal{L}^N_{Stage} = \sum_{\hat{n}=1}^{N} \omega^{\hat{n}} \left(\mathcal{L}^w_{IoU}\left(f^{\hat{n}}, G\right) + \mathcal{L}^w_{BCE}\left(f^{\hat{n}}, G\right)\right) \tag{9}$$

where $f^{\hat{n}}$ represents the prediction result of the $\hat{n}th$ stage network included in the Nth stage network, and $\omega^{\hat{n}}$ is the weight coefficient. For L^N_{Sum}, we sum the outputs of each stage network included in the Nth stage to obtain f^{Sum}, and supervise f^{Sum}:

$$\mathcal{L}^N_{Sum} = \mathcal{L}^w_{IoU}\left(f^{Sum}, G\right) + \mathcal{L}^w_{BCE}\left(f^{Sum}, G\right) \tag{10}$$

During the inference process, we use the prediction result of the Nth stage network within the multi-stage network as the final output.

3 Experiments

3.1 Datasets and Evaluation Metrics

We conduct experiments on 4 polyp segmentation datasets, including Kvasir-SEG [9], CVC-ClinicDB [1] CVC-ColonDB [16], and EndoScene [18]. Kvasir-SEG and CVC-ColonDB contain 1000 and 380 polyp images respectively, while CVC-ClinicDB contains 612 polyp images extracted from 31 colonoscopy examination videos. The EndoScene is a combination of CVC-ClinicDB and CVC300. All datasets are split into training, validation, and test sets in a ratio of 80:10:10. To evaluate the learning ability of the model, we train and test it on the Kvasir-SEG and CVC-ClinicDB respectively to assess its performance.

To further evaluate the model's generalization ability, we first use the training set split from the Kvasir-SEG to train the model, and then testing it on the test set split from the CVC-ClinicDB. In addition, we follow the practices of PraNet [8] and LDNet [21], selecting 80% of the images from the Kvasir-SEG and CVC-ClinicDB to form our training set, and 10% of the images as the validation set. We then test the model on two unseen datasets, specifically the CVC-ColonDB and EndoScene. For CVC-ColonDB, we use all images as the test set. Notably, since the training set includes images from the CVC-ClinicDB, we only select images from the CVC300 portion of the EndoScene for testing to ensure that the model is tested on unseen datasets. For evaluation metrics, we use mean IoU (mIoU), mean Dice (mDice), precision (Prec), and recall.

3.2 Implementation Details

All models are trained using an NVIDIA GeForce RTX3090 GPU, with image sizes uniformly resized to 224×224 (except for a few networks that require a 256×256 size due to input limitations). We utilize simple data augmentation strategies, including random rotation, vertical flipping, and horizontal flipping. All models are trained with the same hyperparameters: a learning rate of 1e−4, decay rate of 1e−4, and a batch size of 16. We use the Adam optimizer for optimization, and the training lasts for 100 epochs.

For our proposed PRFNet, we use a progressive training strategy with $N = 3$ stages, and the encoder employs the lightweight backbone MobileViT, which has 5 layers, i.e., $L = 5$. The decoder is composed of Concatenation, Upsample, Conv2d, BatchNorm and ReLU operations, while the final layer of the decoder, termed the Predict Head layer, consists of Upsample and Conv2d operations. The ARF module is added after the $2nd$ layer of the encoder, and the 56×56 feature map input into the module. The patch size is set as 4×4, the number of blocks is 2, and this module masks 20% of patches when $N = 3$. α and β are set to 1.5 and 1.0, respectively, to make the distribution of prediction results at each stage consistent. Since the final output of the network is derived from the last decoder branch in the $3rd$ stage, $\omega^{\hat{1}}$, $\omega^{\hat{2}}$, and $\omega^{\hat{3}}$ are set to 1.0, 1.0, and 2.0 respectively, making the network focus more on the final output.

3.3 Results

Following the evaluation methods described in Sect. 3.1, we compare our PRFNet with 11 state-of-the-art methods used for polyp segmentation or general medical image segmentation. These include UNet [13], UNet++ [22], PraNet [8], TransUNet [4], HarDNet-DFUS [3], SSFormer-L [19], LDNet [21], GMSRFNet [15], TransResUNet [17], Polyp-PVT [2], and CaraNet [11].

Learning Ability. As shown in Table 1, our PRFNet achieves the best results on most metrics on the Kvasir-SEG and CVC-ClinicDB. Specifically, on the Kvasir-SEG, our PRFNet achieves the best results in mIoU, mDice, and Recall, with mIoU and mDice improving by 1.83% and 1.51% respectively compared to the second best model. For the CVC-ClinicDB, our PRFNet also achieves the best results on most metrics. By comparing with state-of-the-art methods on these datasets, we demonstrate the excellent learning ability of PRFNet in the field of polyp segmentation.

Generalization Ability. As shown in Table 1, when using Kavsir-SEG as the training set and CVC-ClinicDB as the test set for evaluation, PRFNet still achieves the best results on most metrics compared to other methods. As shown in Table 2, when mixing Kvasir-SEG and CVC-ClinicDB as the training set and testing on unseen datasets, PRFNet significantly outperforms other methods on the metrics of CVC-ColonDB and EndoScene, where the mIoU on the EndoScene dataset improved by 2.31% compared to the second best model. These experiments also demonstrate that our proposed ARF module combined with the progressive training strategy not only ensures the model's learning ability, but also possesses excellent generalization ability to unseen data.

Table 1. Quantitative results on Kvasir-SEG and CVC-ClinicDB datasets.

Architectures	**Train:** Kvasir-SEG				**Train:** CVC-ClinicDB				**Train:** Kvasir-SEG			
	Test: Kvasir-SEG				**Test:**CVC-ClinicDB				**Test:** CVC-ClinicDB			
	mIoU	mDice	Prec	Recall	mIoU	mDice	Prec	Recall	mIoU	mDice	Prec	Recall
UNet	75.17	83.3	88.06	85.45	80.72	87.53	88.44	89.63	55.24	63.71	77.84	70.47
UNet++	75.45	83.24	90.51	83.66	78.39	84.89	87.21	86.74	53.17	61.01	80.69	63.51
PraNet	81.28	87.95	89.99	90.45	84.81	89.82	88.74	91.88	69.16	77.02	82.82	82.33
TransUNet	81.39	88.00	91.24	91.24	84.18	89.72	87.12	96.05	71.23	78.59	84.43	80.32
HarDNet-DFUS	79.92	87.07	91.72	87.27	83.44	89.14	87.87	92.27	71.71	79.36	88.19	78.72
SSFormer-L	82.33	88.53	**92.09**	89.54	84.49	89.84	87.89	95.87	73.29	79.92	84.40	81.84
LDNet	82.44	88.98	89.53	92.15	84.12	89.42	87.98	92.18	71.93	79.79	84.88	82.18
GMSRFNet	78.50	86.12	87.74	89.30	84.21	90.42	85.80	**97.40**	63.18	71.21	83.54	75.26
TransResUNet	80.87	87.88	91.20	88.91	84.86	89.46	88.83	90.65	72.33	78.80	88.28	81.23
Polyp-PVT	82.23	88.84	91.80	89.68	85.01	90.49	88.78	95.05	73.39	80.01	86.96	81.55
CaraNet	80.78	87.81	90.50	89.44	83.48	89.11	88.39	90.94	71.40	78.50	**89.10**	79.05
PRFNet (Ours)	**84.27**	**90.49**	90.93	**92.57**	**85.63**	**90.92**	**89.96**	93.24	**73.89**	**80.88**	84.26	**83.21**

Table 2. Quantitative results on two unseen datasets.

Architectures	Train: Kvasir-SEG & CVC-ClinicDB							
	Test: CVC-ColonDB				Test: EndoScene			
	mIoU	mDice	Prec	Recall	mIoU	mDice	Prec	Recall
UNet	55.34	63.01	80.92	67.66	63.37	72.13	74.01	80.57
UNet++	55.79	64.69	73.43	73.19	66.32	75.18	71.38	90.15
PraNet	67.48	75.91	75.00	84.86	79.06	86.79	81.07	**97.27**
TransUNet	67.35	75.31	75.34	83.09	76.19	83.66	79.17	96.04
HarDNet-DFUS	66.59	74.16	79.05	79.06	77.98	84.93	81.36	95.32
SSFormer-L	67.91	76.07	77.60	82.09	76.67	84.37	80.30	94.87
LDNet	64.67	72.62	76.37	80.24	78.36	84.78	82.01	95.48
GMSRFNet	57.10	66.03	73.15	75.33	62.44	73.64	63.28	96.30
TransResUnet	65.49	73.74	78.54	78.33	76.91	84.97	79.15	96.90
Polyp-PVT	67.99	76.53	75.26	**85.52**	77.34	84.17	80.69	94.20
CaraNet	67.42	75.28	79.53	81.19	79.81	87.22	82.46	96.40
PRFNet (Ours)	**69.18**	**77.04**	**85.18**	78.28	**82.12**	**89.37**	**87.43**	93.65

Our qualitative analysis results, as shown in Fig. 2(a), reveal that other state-of-the-art methods only focus on the most prominent lesion parts in the polyp images, completely ignoring the lesion parts that are very similar to the healthy tissue. In contrast, our PRFNet pays attention to these regions, achieving outstanding polyp segmentation results.

3.4 Ablation Study

Effective of Progressive Feature Refinement. We set up different training strategies and adjust the structure of the network at each stage, as shown in Table 3. When only a single stage of the network is used independently, network performance gradually increases with the increase in layers, but it is inferior to the progressive training strategy. We also further modify the network structure at each stage. For instance, we remove the $1st$ stage's decoder included in the $2nd$ stage network (2 w/o 1). This lead to a loss of information of the corresponding granularity due to the removal of part of the network structure, resulting in a significant performance drop. Then, we adopt the progressive training strategy and only sequentially update the $1st$ and $2nd$ stage networks ($1 \rightarrow 2$), which also results in a significant drop in mIoU. In addition, we analyze the mIoU of the prediction results at each stage during the "$1 \rightarrow 2 \rightarrow 3$" training process, as shown in Fig. 3(a). It is noteworthy that the mIoU of the $2nd$ and $3rd$ stages are very similar in the later stage of training. However, from the results of "$1 \rightarrow 2$" and "$1 \rightarrow 2 \rightarrow 3$" in Table 3, it can be seen that the $3rd$ stage network should not be discarded in progressive training, as there is a significant difference in the generalization ability of the two processes.

Table 3. Ablation studies on progressive feature refinement.

Stage	mIoU	mDice	Prec	Recall
1	80.61	87.94	88.62	90.37
2	82.45	89.14	89.57	92.06
3	82.97	89.43	**91.01**	90.91
2 w/o 1	80.86	88.00	87.99	91.78
3 w/o 2	79.34	86.43	90.22	87.36
1 → 2	81.79	88.76	89.53	91.09
1 → 2 → 3	**84.27**	**90.49**	90.93	**92.57**

Table 4. Ablation studies on adaptive region focusing module.

Mask Rate	mIoU	mDice	Prec	Recall
w/o ARF	83.24	89.48	91.84	89.6
0%	82.84	89.25	90.75	91.17
10%	83.87	90.30	90.60	92.05
20%	**84.27**	**90.49**	**90.93**	**92.57**
30%	82.70	89.10	90.22	90.68
50%	81.42	88.18	88.80	90.83

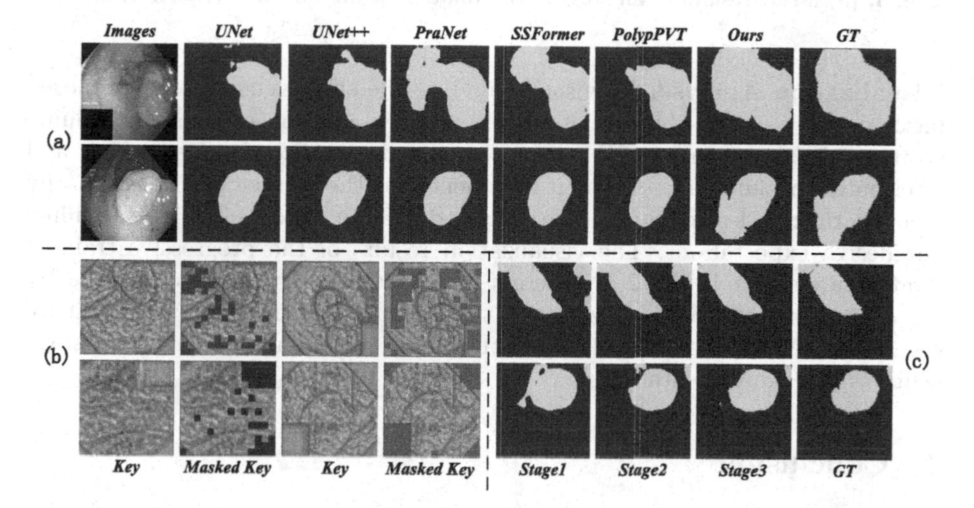

Fig. 2. (a) Qualitative results comparison on the Kvasir-SEG dataset. (b) Comparison of feature maps in ARF module before and after being masked. (c) Qualitative results of progressive feature refinement on Kvasir-SEG dataset.

Effective of Adaptive Region Focusing. We train and evaluate the network after removing ARF module and adjusting the patch masking ratio. The experimental results are shown in Table 4. After removing this module (w/o ARF), the mIoU is still 0.91% higher than the second best method in Sect. 3.3 for evaluating learning ability, indirectly proving the effectiveness of our proposed progressive training strategy. When the masking ratio is increased to more than 30%, it may mask some of the lesion regions, causing a significant drop in network performance. The network performance is optimal when the masking ratio is 20%. In addition, we also adjust the number of blocks. As shown in Fig. 3(b), it can be found that when the number of blocks is set to more than 2, the network performance significantly decreases. This might be because when the number of blocks is too many, the feature representation degrades, and the FSA cannot effectively mask the non-lesion patches.

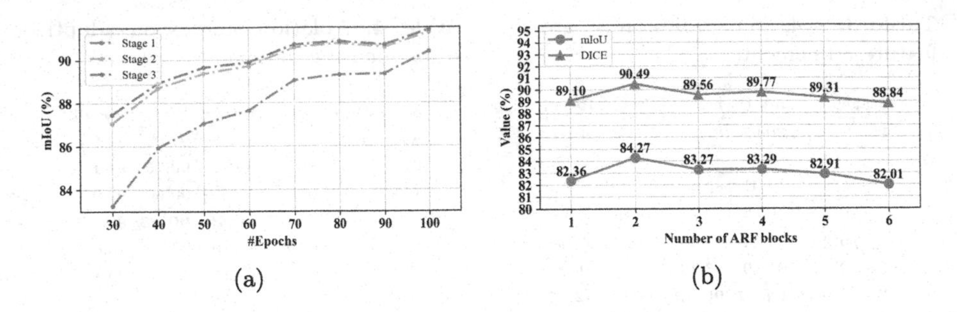

Fig. 3. (a) mIoU results at all stages. (b) Ablation studies on the number of blocks.

Visualization Analyse. To visually analyze the effectiveness of our proposed method, we visualize the output results of FSA and the progressive training strategy. As shown in Fig. 2(b), we reshape and visualize the Key before and after patch masking in FSA, and it is evident that the masked patches primarily focus on the non-lesion region, surrounding the lesion region, effectively fulfilling our design expectations. The visualization results of the progressive training strategy are shown in Fig. 2(c), where it can be seen that the network in the $1st$ stage can achieve initial localization of the lesion tissue, the predicted results of the $2nd$ stage further refine this based on the $1st$ stage, and finally the $3rd$ stage achieves precise segmentation.

4 Conclusion

In this paper, we propose the concise yet very effective PRFNet, which incorporates a progressive training strategy to gradually yield accurate segmentation results. We also introduce an adaptive region focusing module that masks patches corresponding to the non-lesion region, enabling the model to focus on the lesion region during the training process, thereby enhancing the network's generalization ability. We prove the learning and generalization ability of our proposed PRFNet on multiple polyp datasets and use systematic ablation studies to demonstrate the effectiveness of our proposed methods. For future work, we consider using a more lightweight backbone, increasing the number of stages, and further improving the network's learning ability. We also aim to experiment with and fine-tune PRFNet on other medical imaging tasks, with the intend to construct a generalized framework for medical image segmentation.

References

1. Bernal, J., Sánchez, F.J., Fernández-Esparrach, G., Gil, D., Rodríguez, C., Vilariño, F.: WM-DOVA maps for accurate polyp highlighting in colonoscopy: validation vs. saliency maps from physicians. Comput. Med. Imaging Graph. **43**, 99–111 (2015)
2. Bo, D., Wenhai, W., Deng-Ping, F., Jinpeng, L., Huazhu, F., Ling, S.: Polyp-PVT: polyp segmentation with pyramid vision transformers. In: CAAI AIR (2023)
3. Chao, P., Kao, C.Y., Ruan, Y.S., Huang, C.H., Lin, Y.L.: HarDNet: a low memory traffic network. In: Proceedings of the IEEE International Conference on Computer Vision, pp. 3552–3561 (2019)
4. Chen, J., et al.: TransUNet: transformers make strong encoders for medical image segmentation (2021)
5. Cheng, J., et al.: ResGANet: residual group attention network for medical image classification and segmentation. Med. Image Anal. **76**, 102313 (2022)
6. Cheng, J., et al.: DDU-Net: a dual dense U-structure network for medical image segmentation. Appl. Soft Comput. **126**, 109297 (2022)
7. Cheng, J., Tian, S., Yu, L., Lu, H., Lv, X.: Fully convolutional attention network for biomedical image segmentation. Artif. Intell. Med. **107**, 101899 (2020)
8. Fan, D.P., et al.: PraNet: parallel reverse attention network for polyp segmentation. In: Martel, A.L., et al. (eds.) MICCAI 2020. LNCS, vol. 12266, pp. 263–273. Springer, Cham (2020). https://doi.org/10.1007/978-3-030-59725-2_26
9. Jha, D., et al.: Kvasir-SEG: a segmented polyp dataset. arXiv arXiv:1911.07069 (2019)
10. Lai, H., Luo, Y., Zhang, G., Shen, X., Li, B., Lu, J.: Toward accurate polyp segmentation with cascade boundary-guided attention. Vis. Comput. **39**(4), 1453–1469 (2022)
11. Lou, A., Guan, S., Loew, M.: CaraNet: context axial reverse attention network for segmentation of small medical objects. J. Med. Imaging **10**(1), 014005 (2023)
12. Lou, M., Meng, J., Qi, Y., Li, X., Ma, Y.: MCRNet: multi-level context refinement network for semantic segmentation in breast ultrasound imaging. Neurocomputing **470**, 154–169 (2022)
13. Ronneberger, O., Fischer, P., Brox, T.: U-Net: convolutional networks for biomedical image segmentation. In: Navab, N., Hornegger, J., Wells, W.M., Frangi, A.F. (eds.) MICCAI 2015. LNCS, vol. 9351, pp. 234–241. Springer, Cham (2015). https://doi.org/10.1007/978-3-319-24574-4_28
14. Salmo, E., Haboubi, N.: Adenoma and malignant colorectal polyp: pathological considerations and clinical applications. EMJ Gastroenterol. **7**, 92–102 (2018)
15. Srivastava, A., Chanda, S., Jha, D., Pal, U., Ali, S.: GMSRF-Net: an improved generalizability with global multi-scale residual fusion network for polyp segmentation. arXiv preprint arXiv:2111.10614 (2021)
16. Tajbakhsh, N., Gurudu, S.R., Liang, J.: Automated polyp detection in colonoscopy videos using shape and context information. IEEE Trans. Med. Imaging **35**(2), 630–644 (2016)
17. Tomar, N.K., Shergill, A., Rieders, B., Bagci, U., Jha, D.: TransResU-Net: transformer based ResU-Net for real-time colonoscopy polyp segmentation. arXiv preprint arXiv:2206.08985 (2022)
18. Vázquez, D., et al.: A benchmark for endoluminal scene segmentation of colonoscopy images. J. Healthc. Eng. **2017**, 4037190 (2016)

19. Wang, J., Huang, Q., Tang, F., Meng, J., Su, J., Song, S.: Stepwise feature fusion: local guides global. In: Wang, L., Dou, Q., Fletcher, P.T., Speidel, S., Li, S. (eds.) Medical Image Computing and Computer Assisted Intervention, MICCAI 2022. LNCS, vol. 13433, pp. 110–120. Springer, Cham (2022). https://doi.org/10.1007/978-3-031-16437-8_11
20. Wang, W., et al.: Pyramid vision transformer: a versatile backbone for dense prediction without convolutions. In: Proceedings of the IEEE/CVF International Conference on Computer Vision, pp. 568–578 (2021)
21. Zhang, R., et al.: Lesion-aware dynamic kernel for polyp segmentation. In: Wang, L., Dou, Q., Fletcher, P.T., Speidel, S., Li, S. (eds.) Medical Image Computing and Computer Assisted Intervention, MICCAI 2022. LNCS, vol. 13433, pp. 99–109. Springer, Cham (2022). https://doi.org/10.1007/978-3-031-16437-8_10
22. Zhou, Z., Siddiquee, M.M.R., Tajbakhsh, N., Liang, J.: UNet++: redesigning skip connections to exploit multiscale features in image segmentation. IEEE Trans. Med. Imaging **39**, 1856–1867 (2019)

Pyramid Shape-Aware Semi-supervised Learning for Thyroid Nodules Segmentation in Ultrasound Images

Na Zhang[1], Juan Liu[1](✉), and Meng Wu[2]

[1] Institute of Artificial Intelligence, School of Computer Science, Wuhan University, Wuhan, China
{zhangna1999,liujuan}@whu.edu.cn
[2] Department of Ultrasound, Zhongnan Hospital, Wuhan University, Wuhan, China

Abstract. The accurate segmentation of thyroid nodules in ultrasound (US) images is critical for computer-aided diagnosis of thyroid cancer. While the fully supervised methods achieve high accuracy, they require a significant amount of annotated data for training, which is both costly and time-consuming. Semi-supervised learning can address this challenge by using a limited amount of labeled data in combination with a large amount of unlabeled data. However, the existing semi-supervised segmentation approaches often fail to account for both geometric shape constraints and scale differences of objects. To address this issue, in this paper we propose a novel Pyramid Shape-aware Semi-supervised Learning (PSSSL) framework for thyroid nodules segmentation in US images, which employs a dual-task pyramid prediction network to jointly predict the Segmentation Maps (SEG) and Signed Distance Maps (SDM) of objects at different scales. Pyramid feature prediction enables better adaptation to differences in nodule size, while the SDM provides a representation that encodes richer shape features of the target. PSSSL learns from the labeled data by minimizing the discrepancy between the prediction and the ground-truth and learns from unlabeled data by minimizing the discrepancy between the predictions at different scales and the average prediction. To achieve reliable and robust segmentation, two uncertainty estimation modules are designed to emphasize reliable predictions while ignoring unreliable predictions from unlabeled data. The proposed PSSSL framework achieves superior performance in both quantitative and qualitative evaluations on the DDTI and TN3k datasets to State-Of-The-Art semi-supervised approaches. The code is available at https://github.com/wuliZN2020/Thyroid-Segmentation-PSSSL.

Keywords: Semi-supervised learning · Thyroid nodules segmentation · Dual-task pyramid prediction network · Shape-aware · Uncertainty estimation

---

This work was partially supported by the Key R&D Project in Hubei Province (2023BCB024) and the Translational Medicine and Interdisciplinary Research Joint Fund of Zhongnan Hospital of Wuhan University (ZNJC202226).

Q. Liu et al. (Eds.): PRCV 2023, LNCS 14429, pp. 407–418, 2024.
https://doi.org/10.1007/978-981-99-8469-5_32

1 Introduction

Thyroid nodules are common disorders in the endocrine system and the incidence of the nodules has increased in recent years [1]. Ultrasound (US) is the primary imaging technique for diagnosing thyroid diseases [2]. Segmentation of thyroid nodules in US images is the initial step in computer-aided diagnosis systems, which helps physicians assess the nodules' size, morphology, boundary, local gland structure, and other characteristics, facilitating the diagnosis of benign and malignant thyroid nodules.

Though supervised deep learning methods have achieved high performance in thyroid nodules segmentation tasks, they require large amounts of high-quality annotated data, which is time-consuming and labor-intensive for specialists. Since the unlabeled data is easy and cheap to obtain, semi-supervised learning has emerged as a data-efficient strategy that utilizes limited labeled data and large amounts of unlabeled data to alleviate the label scarcity problem [3]. Semi-supervised segmentation methods are typically categorized into two types: pseudo labeling and consistency regularization. Pseudo labeling [4] generates proxy labels on the unlabeled data using a prediction model and combines them with the labeled data to provide additional information. Examples include meta pseudo labels [5], NoisyStudent [6], and TSSDL [7]. Consistency regularization enforces consistency constraints over unlabeled data, such as the Π-model and temporal ensembling strategy [8], mean-teacher framework [9,10], dual-task consistency network [11].

However, the scale of thyroid nodules in the images varies greatly, resulting that thyroid nodules exhibit significant variations in shape and size in the US images. The semi-supervised learning methods often face difficulties in adequately capturing these shape and size differences from the limited labeled data. This circumstance can potentially lead to issues of either under- or over-segmentation of nodules. Some methods, such as DAP [12], DPA-DenseBiasNet [13], and URPC [14], have attempted to address these challenges by incorporating probabilistic shape priors, learning anatomical features, and encouraging predictions at multiple scales to be consistent. However, they either only considered the geometric shape constraints or the scale differences. Besides, the existing shape constraint methods usually assume the input images are properly aligned, so that an anatomical prior can be incorporated on the object, which is infeasible in practice for nodules with large variations in size or shape.

In this work, we propose a novel Pyramid Shape-aware Semi-supervised Learning (PSSSL) framework for thyroid nodules segmentation. PSSSL utilizes multi-scale feature prediction and a flexible shape representation in the network architecture to enforce a shape constraint on the segmentation output and handle nodules with large variations in size or shape. PSSSL employs a dual-task pyramid prediction network to perform Segmentation Maps (SEG) and Signed Distance Maps (SDM) prediction at different scales simultaneously. Pyramid network prediction enables our model to capture multi-scale features of nodules, adapting to the scale variations of nodules in US images. The SDM assigns each pixel a value indicating its signed distance to the nearest object boundary, providing a shape-aware representation that encodes richer object shape features. PSSSL minimizes the discrepancy between the predicted SEG and SDM at different scales and the corresponding ground-truth SEG and SDM for labeled data. Multi-scale feature prediction enables better adaptation to differences in nodule size, while the SDM provides a representation that encodes richer shape features of the target. The combination of these two strategies reduces over- and under-segmentation issues caused by variations in nodule size and shape, thereby improving the accuracy and robustness of the segmentation results. For unlabeled data, PSSSL assumes consistent predictions across scales and minimizes the discrepancy between the predictions at different scales and the average prediction. We further designed separate uncertainty estimation modules for two tasks to emphasize reliable predictions from unlabeled data while ignoring unreliable predictions and ensuring reliable and robust segmentation. We evaluated our approach on two public thyroid nodules segmentation datasets. The quantitative results demonstrate that our PSSSL outperforms the State-Of-The-Art (SOTA) semi-supervised segmentation methods. The visualization results demonstrate that our approach produces more accurate boundaries and shapes, and effectively detects nodules of different sizes.

2 The Proposed Method

The PSSSL framework proposed for the semi-supervised segmentation of thyroid nodules in US images is shown in Fig. 1, which consists of a dual-task pyramid prediction network and two uncertainty estimation modules. The dual-task pyramid prediction network simultaneously performs SEG prediction and SDM prediction at different scales. For labeled data, PSSSL learns by minimizing the discrepancy between the predicted SEG and SDM at different scales and the corresponding ground-truth SEG and SDM. For unlabeled data, we assume that the predictions at different scales should be consistent. Therefore, PSSSL learns by minimizing the discrepancy between the predictions at different scales and the average prediction. Furthermore, we designed separate uncertainty estimation modules for each task to evaluate the confidence of the predicted results from unlabeled data at different scales.

2.1 Dual-Task Pyramid Prediction Network

Signed Distance Map (SDM). SDM can describe the distance of each pixel to the boundary in an image. A positive value means the pixel is inside the segmented region, a negative value means outside, and 0 means on the boundary [15]. The calculation formula of SDM is as follows:

$$\mathrm{Z}(x) = \begin{cases} 0, \quad x \in \Delta S \\ +inf_{y \in \Delta S} \|x - y\|_2, x \in S_{in} \\ -inf_{y \in \Delta S} \|x - y\|_2, x \in S_{out}, \end{cases} \tag{1}$$

where $\|x - y\|_2$ is the Euclidean distance between pixel x and y. ΔS, S_{in}, S_{out} denote the boundary, inside, and outside of the object, respectively.

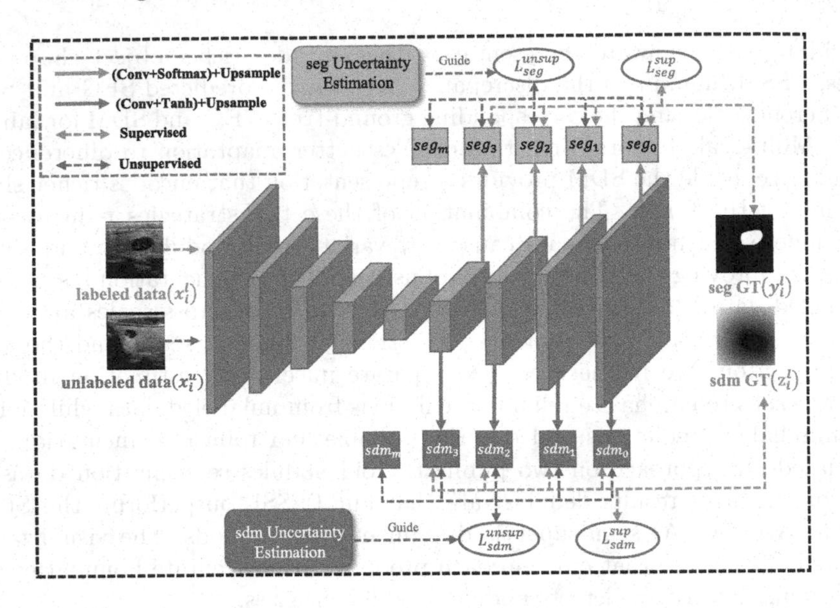

Fig. 1. Overview of the proposed PSSSL framework.

Network Architecture. Our network has two characteristics: a pyramid structure for multi-scale prediction outputs in the backbone network and two prediction tasks for each scale, SEG and SDM. Concretely, we modified the U-Net [16] backbone network by adding two prediction heads after each upsampling block in the decoder: the SEG and SDM prediction heads. The SEG prediction head consists of a convolutional block and a softmax activation function, while the SDM prediction head consists of a convolutional block and a tanh activation function. Finally, both SEG and SDM multi-scale prediction outputs are upsampled to the same size as the input image.

For our semi-supervised learning task, we have M labeled examples in a set $\mathcal{D}^l = \{(x_i^l, y_i^l, z_i^l)\}_{i=1}^{M}$ and N unlabeled examples in a set $\mathcal{D}^u = \{(x_i^u)\}_{i=N+1}^{N+M}$. z_i^l is the SDM ground-truth derived from SEG ground-truth y_i^l by Eq. 1. Given an input image x_i, PSSSL produces two sets of multi-scale predictions $\{seg_0, seg_1, ...seg_s, ..., seg_{S-1}\}$ and $\{sdm_0, sdm_1, ...sdm_s, ..., sdm_{S-1}\}$. Where $x_i \in \mathbb{R}^{H*W}$, $y_i \in \{0,1\}^{H*W}$, $z_i \in [-1,1]^{H*W}$, $seg_s \in [0,1]^{H*W}$, $sdm_s \in [-1,1]^{H*W}$, seg_s and sdm_s denote the SEG and SDM prediction at scale s, respectively. S is the number of scales.

2.2 Supervised Pyramid Shape-Aware Loss

For the labeled data, we combine dice loss and cross-entropy loss for multi-scale SEG outputs and a separate mean square loss for multi-scale SDM outputs.

$$L_{seg}^{sup} = \frac{1}{S}\sum_{s=0}^{S-1}\frac{L_{\text{dice}}\left(seg_s, y_i^l\right) + L_{ce}\left(seg_s, y_i^l\right)}{2}, \tag{2}$$

$$L_{sdm}^{sup} = \frac{1}{S}\sum_{s=0}^{S-1} L_{\text{mse}}\left(sdm_s, z_i^l\right), \tag{3}$$

$$L_{sup} = L_{seg}^{sup} + \alpha L_{sdm}^{sup}, \tag{4}$$

where y_i^l and z_i^l denote the SEG and SDM ground-truth of input x_i^l. L_{dice}, L_{ce}, and L_{mse} denote the dice loss, cross-entropy loss, and mean square loss, respectively. L_{seg}^{sup}, L_{sdm}^{sup}, and L_{sup} denote the supervised SEG loss, supervised SDM loss, and total supervised loss. α is a weighting coefficient balancing two supervised loss terms. In our experiment, α is set to 0.3 [17].

2.3 Unsupervised Shape-Aware Pyramid Consistency Loss

Uncertainty Estimation Module. For the unlabeled data, due to the different spatial resolutions of the prediction results at different scales in the PSSSL framework, upsampling them to the same size as the input image may result in varying degrees of loss of fine details. Enforcing consistency among prediction results at different scales directly may introduce noise and unreliable guidance to the model. Therefore, we have developed separate modules to evaluate the uncertainty of the consistency predictions for both SEG and SDM. Specifically, we estimate uncertainty for prediction at scale s by calculating the Jensen-Shannon (JS) divergence [18] between the prediction at scale s and the average prediction. JS divergence incorporates an average distribution in its calculation, making it smoother and more robust to outliers and noise compared to other methods for uncertainty estimation.

$$seg_m = \frac{1}{s}\sum_{s=0}^{S-1} seg_s, U_s^{seg} = \frac{1}{2}\sum_{i=1}^{C}\left(seg_s^i \log\frac{seg_s^i}{M(seg)_s^i} + seg_m^i \log\frac{seg_m^i}{M(seg)_s^i}\right); \tag{5}$$

$$sdm_m = \frac{1}{s}\sum_{s=0}^{S-1} sdm_s, U_s^{sdm} = \frac{1}{2}\sum_{i=1}^{C}\left(sdm_s^i \log\frac{sdm_s^i}{M(sdm)_s^i} + sdm_m^i \log\frac{sdm_m^i}{M(sdm)_s^i}\right); \tag{6}$$

where seg_m and sdm_m are the average SEG and SDM predictions across these scales, respectively. seg_s^i and sdm_s^i denote the ith channel of seg_s and sdm_s, respectively. C is the class (i.e., channel) number. U_s^{seg} and U_s^{sdm} are confidence maps of SEG and SDM consistency predictions at scale s, respectively. $M(seg)_s^i = \frac{1}{2}\left(seg_s^i + seg_m^i\right)$, and $M(seg)_s$ denote the average distribution of seg_s and seg_m. Similarly, $M(sdm)_s$ denote the average distribution of sdm_s and sdm_m. A high value in U_s indicates high uncertainty in pixel prediction at scale s due to significant deviation from other scale predictions.

Unsupervised Shape-Aware Pyramid Consistency Loss. With the guidance of the confidence map U, we emphasize the relatively reliable predictions and ignore unreliable predictions. Uncertainty-aware unsupervised SEG consistency loss L_{seg}^{unsup} consists of two terms. The first term represents unsupervised SEG consistency loss with uncertainty estimation. The second term represents SEG uncertainty loss, which aims to minimize the prediction uncertainty of each scale. L_{sdm}^{unsup} has a similar structure to L_{seg}^{unsup}.

$$L_{seg}^{unsup} = \frac{1}{S}\frac{\sum_{s=0}^{S-1}(seg_s - seg_m)^2 * w_s^{seg}}{\sum_{s=0}^{S-1} w_s^{seg}} + \frac{1}{S}\sum_{s=0}^{S-1}\|U_s^{seg}\|_2, \tag{7}$$

$$L_{sdm}^{unsup} = \frac{1}{S}\frac{\sum_{s=0}^{S-1}(sdm_s - sdm_m)^2 * w_s^{sdm}}{\sum_{s=0}^{S-1} w_s^{sdm}} + \frac{1}{S}\sum_{s=0}^{S-1}\left\|U_s^{sdm}\right\|_2, \tag{8}$$

$$L_{unsup} = L_{seg}^{unsup} + \alpha L_{sdm}^{unsup}, \tag{9}$$

where L_{unsup} is unsupervised shape-aware pyramid consistency loss. α is the same weighting coefficient as α in L_{sup}. $w_s^{seg} = 1/(1+U_s^{seg})$ and $w_s^{sdm} = 1/(1+U_s^{sdm})$ denote the weight of SEG and SDM prediction at scale s, respectively. For a given pixel at scale s, a higher uncertainty leads to a lower weight.

2.4 Overall Semi-Supervised Loss

The proposed PSSSL framework learns from both labeled and unlabeled data by minimizing the following combined objective function:

$$L = L_{sup} + \lambda L_{unsup}. \tag{10}$$

Following, we use the time-dependent Gaussian warming up function $\lambda(t) = 0.1 * e^{\left(-5(1-(t/t_{max}))^2\right)}$ [17] to control the balance between the supervised loss and unsupervised consistency loss, where t denotes the current training step and t_{max} is the maximum training step.

3 Experiments and Results

3.1 Datasets and Implementations

We evaluated our approach on two public thyroid nodule segmentation datasets: DDTI [19] and TN3k [20]. The DDTI dataset consists of 637 thyroid US images with a resolution size of 256 * 256, while the TN3k dataset contains 3493 US images with varying resolutions. We further performed online data augmentations, including random reshaping, flipping, and cropping, which resulted in a resolution size of 256 * 256 for all of the US images that we finally used as input to the model. The datasets are randomly split into training, validation, and test sets in a 6:2:2 ratio.

Our PSSSL framework is implemented in Pytorch and trained on a computer with 8 GeForce RTX 3090Ti GPUs. U-Net is adopted as the backbone network for easy implementation and fair comparison with other methods. All models are trained using the SGD optimizer with a momentum of 0.9. The poly learning rate strategy was used for learning rate decay $lr = 0.01 * (1 - (t/t_{max}))^{0.9}$. The batch size is set to 4, and the maximum training epoch is 1000. During testing, we take the SEG output at scale 0 (i.e., seg_0) for evaluation. We choose four evaluation metrics: Dice Score (%), Jaccard Score (%), 95% Hausdorff Distance (95HD), and Average Surface Distance (ASD). Given two object regions, Dice and Jaccard mainly compute the percentage of overlap between them, 95HD measures the closest point distance between them, and ASD computes the average distance between their boundaries [21].

Table 1. Quantitative segmentation results on the DDTI and TN3k dataset.

Labeled	Method	DDTI				TN3k			
		Dice↑	Jaccard↑	95HD↓	ASD↓	Dice↑	Jaccard↑	95HD↓	ASD↓
100%	U-Net (FS)	78.32	67.06	31.21	11.88	80.03	70.26	17.73	4.61
20%	U-Net (LS)	66.19	53.52	45.61	19.84	67.52	55.24	48.45	18.23
	UA-MT	71.96	61.04	33.42	12.49	72.65	61.10	33.82	12.05
	SASSNet	72.84	61.90	33.82	12.50	74.64	64.10	25.01	7.33
	DTC	73.76	62.54	36.82	15.33	74.26	63.61	26.11	7.73
	URPC	72.00	61.19	33.38	12.41	73.94	63.15	26.24	8.10
	SS-Net	75.05	64.39	30.76	11.19	76.51	66.03	23.93	7.56
	MC-Net	71.09	59.76	35.02	11.89	71.95	61.77	26.11	7.32
	PSSSL(ours)	**76.77**	**66.41**	**28.82**	**10.13**	**79.43**	**69.77**	**17.59**	**4.48**
10%	U-Net(LS)	56.01	45.28	48.53	18.46	64.64	53.80	30.81	8.95
	UA-MT	66.98	55.92	44.44	17.50	70.47	59.49	29.39	9.05
	SASSNet	69.01	57.36	40.71	15.61	71.56	61.28	26.05	**7.10**
	DTC	69.34	57.48	40.88	15.45	71.87	60.45	41.65	15.76
	URPC	67.65	56.39	45.69	17.21	70.94	59.36	32.95	11.98
	SS-Net	71.36	58.86	45.68	17.50	72.59	61.85	28.14	9.44
	MC-Net	66.77	55.29	39.25	14.81	69.27	58.47	31.09	10.01
	PSSSL(ours)	**72.36**	**60.85**	**32.81**	**12.60**	**75.60**	**65.37**	**25.18**	7.76

3.2 Quantitative Evaluation and Comparison

Comparison with Supervised Methods. Firstly, we compared our method with the limited supervised learning method that uses 20% and 10% of labeled data, which is denoted as U-Net (LS). Similarly, U-Net (FS) is denoted as utilizing all labeled data for fully supervised learning, which sets the performance upper bound. Table 1 reports quantitative results on the DDTI dataset and TN3k dataset, respectively, using only 20% and 10% labeled data for training. It can be seen that compared to U-Net (LS), our proposed method effectively utilizes

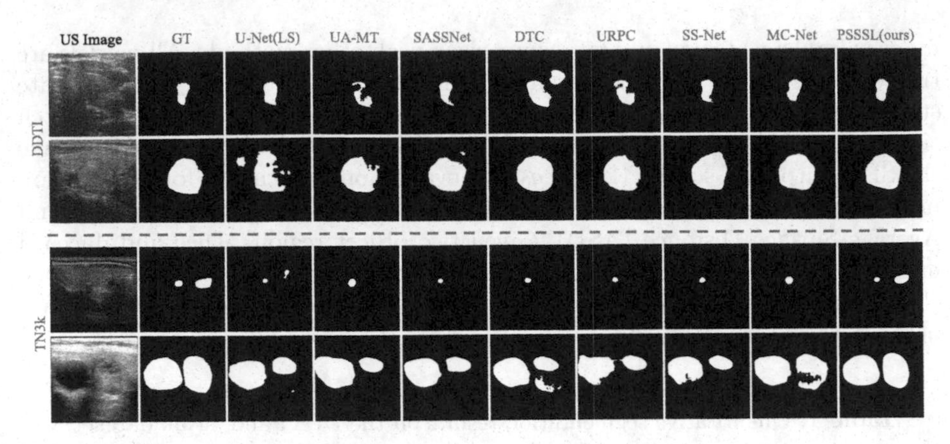

Fig. 2. Visual comparisons with other methods.

unlabeled data, resulting in a significant improvement in its performance for segmenting thyroid nodules in US images (Dice: +16.35%, Jaccard: +15.57%, 95HD: −15.72, ASD: −5.86 on DDTI; Dice: +10.96%, Jaccard: +11.57%, 95HD: −5.63, ASD: −1.19 on TN3k when using only 10% of labeled data).

Comparison with Other Semi-supervised Methods. Furthermore, we compared our method with several SOTA approaches for semi-supervised segmentation, including UA-MT [10], SASSNet [17], DTC [11], URPC [14], SS-Net [22], MC-Net [23]. The quantitative results in Table 1 show that compared to U-Net (LS), all semi-supervised segmentation methods achieved better segmentation performance. In particular, our method achieved the best segmentation performance on both datasets, which demonstrates that our method not only efficiently utilizes unlabeled data but also has stronger robustness. In Fig. 2, we visualized the thyroid nodules segmentation results of limited supervised and semi-supervised methods using 20% labeled data on the two datasets. It can be observed that our approach exhibits improved accuracy in delineating boundaries and shapes, enabling effective detection of nodules across various sizes. This can be attributed to the utilization of multi-scale prediction and shape constraints.

3.3 Ablation Study

Investigating the Impacts of SDM. To validate the effectiveness of SDM in thyroid nodules segmentation, we compared the performance of U-Net and U-Net with SDM heads (Row 1 vs. Row 2 in Table 2) in limited supervised segmentation. The experimental results show that the SDM head improved the model's performance (Dice: +2.83%, Jaccard: +4.13%, 95HD: −4.84, ASD: −3.6 on DDTI; Dice: +2.22%, Jaccard: +3.24%, 95HD: −17.69, ASD: −7.73 on TN3k). It should be noted that because there is no multi-scale prediction, the model with S = 1 (i.e., U-Net) can only learn from labeled data. Furthermore, we

Table 2. Ablation study of the proposed PSSSL framework on the DDTI and TN3k datasets when using 20% of labeled data. UE denotes the Uncertainty Estimation module.

Row	Method			DDTI				TN3k			
	S	SDM	UE	Dice↑	Jaccard↑	95HD↓	ASD↓	Dice↑	Jaccard↑	95HD↓	ASD↓
1	S = 1			66.19	53.52	45.61	19.84	67.52	55.24	48.45	18.23
2	S = 1	✓		**69.02**	**57.65**	**40.77**	**16.24**	**69.74**	**58.48**	**30.76**	**10.50**
3	S = 2	✓	✓	72.18	60.31	34.34	12.55	73.08	62.68	25.77	7.87
4	S = 3	✓	✓	75.41	64.47	30.97	10.58	75.91	65.46	23.56	7.07
5	S = 4	✓	✓	**76.77**	**66.41**	**28.82**	**10.13**	**79.43**	**69.77**	**17.59**	**4.48**
6	S = 4			69.66	57.89	40.50	15.32	71.80	60.23	31.52	11.08
7	S = 4	✓		73.40	61.45	37.57	14.54	76.82	66.47	23.97	7.75
8	S = 4		✓	72.11	60.53	33.14	12.51	73.84	63.21	25.73	7.56
9	S = 4	✓	✓	**76.77**	**66.41**	**28.82**	**10.13**	**79.43**	**69.77**	**17.59**	**4.48**

compared the performance of the model without and with the SDM head at S = 4 (Row 6 vs. Row 7; Row 8 vs. Row 9 in Table 2). The SDM head also improved the performance of semi-supervised segmentation, where the model learns from both labeled and unlabeled data at S = 4. This shows that introducing SDM can impose shape constraints on the segmentation nodules, leading to better segmentation performance.

Investigating the Impacts of Different Numbers of Scales. To investigate the impact of the number of scales on the performance of the dual-task pyramid prediction in PSSSL, we set S to 2, 3, and 4 (Row 3, 4, and 5 in Table 2), respectively. As S increased from 2 to 4, the performance of PSSSL continued to improve. This can be attributed to the fact that the predictions at different scales allowed the model to better perceive the differences in nodule sizes in US images, thereby enhancing the segmentation performance of the model.

Investigating the Impacts of Uncertainty Estimation. To investigate the effectiveness of the uncertainty estimation module in PSSSL, we removed the uncertainty estimation module from PSSSL with and without the SDM head at S = 4. The experimental results demonstrate that the uncertainty module boosts the semi-supervised segmentation performance of thyroid nodules (Row 6 vs. Row 8; Row 7 vs. Row 9 in Table 2). It indicates that the uncertainty estimation module emphasizes reliable predictions from unlabeled data while ignoring unreliable predictions.

Investigating the Robustness of PSSSL on Other Modalities. Considering that all the experiments were conducted on US modality datasets, to verify the effectiveness of AmmH on datasets of other modalities, we conducted comparative experiments with limited supervised learning on the Left Atrium (LA) dataset [24]. The LA dataset contains 100 3D gadolinium-enhanced MR imaging scans (GE-MRIs) and LA segmentation masks. The experimental results are presented in Table 3. It should be noted that, due to the LA dataset being a 3D dataset, we chose the V-Net model as our backbone in this experiment. Compared to V-Net (LS), our method still achieved significant improvements in the segmentation of MRI modality images, especially when using only 10% of the labels (Dice: +12.9%, Jaccard: +15.8%, 95HD: −11.98, ASD: −3.23). This fully demonstrates the robustness of the PSSSL model in the semi-supervised segmentation of medical images.

Table 3. Quantitative segmentation result on the LA datasets.

Labeled	Method	Metrics			
		Dice↑	Jaccard↑	95HD↓	ASD↓
100%	V-Net (FS)	90.74	83.12	8.09	1.48
20%	V-Net (LS)	86.81	77.21	10.59	2.59
	PSSSL (ours)	**89.69**	**81.38**	**12.31**	**1.69**
10%	V-Net (LS)	74.59	62.29	23.94	5.52
	PSSSL (ours)	**87.49**	**78.09**	**11.96**	**2.29**

4 Conclusion

In this paper, we have proposed a novel efficient semi-supervised learning framework, PSSSL, for thyroid nodules segmentation in US images. We have developed a dual-task pyramid prediction network that jointly performs SEG and SDM predictions of the object at multiple scales. Pyramid network prediction enables to capture multi-scale features of nodules, adapting to the scale variations of nodules in US images. The SDM prediction provides a flexible shape-aware representation that encodes richer object shape features. Separate uncertainty estimation modules for two tasks are designed to emphasize reliable predictions from unlabeled data. The proposed framework achieves superior performance in both quantitative and qualitative evaluations on the DDTI and TN3k datasets compared to SOTA semi-supervised approaches. The visual results show that PSSSL achieves superior performance with more accurate borders and shapes, indicating our method has alleviated the problems of over- and under-segmentation in the semi-supervised segmentation of thyroid nodules in US images.

References

1. Sun, J., et al.: TNSNet: thyroid nodule segmentation in ultrasound imaging using soft shape supervision. Comput. Meth. Program. Biomed. **215**, 106600 (2022)
2. Chen, J., You, H., Li, K.: A review of thyroid gland segmentation and thyroid nodule segmentation methods for medical ultrasound images. Comput. Meth. Program. Biomed. **185**, 105329 (2020)
3. Van Engelen, J.E., Hoos, H.H.: A survey on semi-supervised learning. Mach. Learn. **109**(2), 373–440 (2020)
4. Lee, D.-H., et al.: Pseudo-label: the simple and efficient semi-supervised learning method for deep neural networks. In: Workshop on Challenges in Representation Learning, ICML, vol. 3, p. 896 (2013)
5. Pham, H., Dai, Z., Xie, Q., Le, Q.V.: Meta pseudo labels. In: Proceedings of the IEEE/CVF Conference on Computer Vision and Pattern Recognition, pp. 11557–11568 (2021)
6. Xie, Q., Luong, M.-T., Hovy, E., Le, Q.V.: Self-training with noisy student improves ImageNet classification. In: Proceedings of the IEEE/CVF Conference on Computer Vision and Pattern Recognition, pp. 10687–10698 (2020)
7. Shi, W., Gong, Y., Ding, C., Ma, Z., Tao, X., Zheng, N.: Transductive semi-supervised deep learning using min-max features. In: Ferrari, V., Hebert, M., Sminchisescu, C., Weiss, Y. (eds.) ECCV 2018. LNCS, vol. 11209, pp. 311–327. Springer, Cham (2018). https://doi.org/10.1007/978-3-030-01228-1_19
8. Laine, S., Aila, T.: Temporal ensembling for semi-supervised learning. arXiv preprint arXiv:1610.02242 (2016)
9. Tarvainen, A., Valpola, H.: Mean teachers are better role models: weight-averaged consistency targets improve semi-supervised deep learning results. In: Advances in Neural Information Processing Systems, vol. 30 (2017)
10. Yu, L., Wang, S., Li, X., Fu, C.-W., Heng, P.-A.: Uncertainty-aware self-ensembling model for semi-supervised 3D left atrium segmentation. In: Shen, D., et al. (eds.) MICCAI 2019. LNCS, vol. 11765, pp. 605–613. Springer, Cham (2019). https://doi.org/10.1007/978-3-030-32245-8_67
11. Luo, X., Chen, J., Song, T., Wang, G.: Semi-supervised medical image segmentation through dual-task consistency. Proc. AAAI Conf. Artif. Intell. **35**, 8801–8809 (2021)
12. Zheng, H., et al.: Semi-supervised segmentation of liver using adversarial learning with deep atlas prior. In: Shen, D., et al. (eds.) MICCAI 2019. LNCS, vol. 11769, pp. 148–156. Springer, Cham (2019). https://doi.org/10.1007/978-3-030-32226-7_17
13. He, Y., et al.: DPA-DenseBiasNet: semi-supervised 3D fine renal artery segmentation with dense biased network and deep priori anatomy. In: Shen, D., et al. (eds.) MICCAI 2019. LNCS, vol. 11769, pp. 139–147. Springer, Cham (2019). https://doi.org/10.1007/978-3-030-32226-7_16
14. Luo, X., et al.: Efficient semi-supervised gross target volume of nasopharyngeal carcinoma segmentation via uncertainty rectified pyramid consistency. In: de Bruijne, M., et al. (eds.) MICCAI 2021. LNCS, vol. 12902, pp. 318–329. Springer, Cham (2021). https://doi.org/10.1007/978-3-030-87196-3_30
15. Xue, Y., et al.: Shape-aware organ segmentation by predicting signed distance maps. Proc. AAAI Conf. Artif. Intell. **34**, 12565–12572 (2020)

16. Ronneberger, O., Fischer, P., Brox, T.: U-Net: convolutional networks for biomedical image segmentation. In: Navab, N., Hornegger, J., Wells, W.M., Frangi, A.F. (eds.) MICCAI 2015, Part III. LNCS, vol. 9351, pp. 234–241. Springer, Cham (2015). https://doi.org/10.1007/978-3-319-24574-4_28
17. Li, S., Zhang, C., He, X.: Shape-aware semi-supervised 3D semantic segmentation for medical images. In: Martel, A.L., et al. (eds.) MICCAI 2020, Part I. LNCS, vol. 12261, pp. 552–561. Springer, Cham (2020). https://doi.org/10.1007/978-3-030-59710-8_54
18. Fuglede, B., Topsoe, F.: Jensen-Shannon divergence and Hilbert space embedding. In: Proceedings of the International Symposium on Information Theory, ISIT 2004, p. 31. IEEE (2004)
19. Pedraza, L., Vargas, C., Narváez, F., Durán, O., Muñoz, E., Romero, E.: An open access thyroid ultrasound image database. In: 10th International Symposium on Medical Information Processing and Analysis, vol. 9287, pp. 188–193. SPIE (2015)
20. Gong, H., et al.: Multi-task learning for thyroid nodule segmentation with thyroid region prior. In: 2021 IEEE 18th International Symposium on Biomedical Imaging (ISBI), pp. 257–261. IEEE (2021)
21. Wang, J., Lukasiewicz, T.: Rethinking Bayesian deep learning methods for semi-supervised volumetric medical image segmentation. In: Proceedings of the IEEE/CVF Conference on Computer Vision and Pattern Recognition, pp. 182–190 (2022)
22. Wu, Y., Wu, Z., Wu, Q., Ge, Z., Cai, J.: Exploring smoothness and class-separation for semi-supervised medical image segmentation. In: Wang, L., Dou, Q., Fletcher, P.T., Speidel, S., Li, S. (eds.) Medical Image Computing and Computer Assisted Intervention, MICCAI 2022, Part V. LNCS, vol. 13435, pp. 34–43. Springer, Cham (2022). https://doi.org/10.1007/978-3-031-16443-9_4
23. Wu, Y., Xu, M., Ge, Z., Cai, J., Zhang, L.: Semi-supervised left atrium segmentation with mutual consistency training. In: de Bruijne, M., et al. (eds.) MICCAI 2021, Part II. LNCS, vol. 12902, pp. 297–306. Springer, Cham (2021). https://doi.org/10.1007/978-3-030-87196-3_28
24. Xiong, Z., et al.: A global benchmark of algorithms for segmenting the left atrium from late gadolinium-enhanced cardiac magnetic resonance imaging. Med. Image Anal. **67**, 101832 (2021)

AgileNet: A Rapid and Efficient Breast Lesion Segmentation Method for Medical Image Analysis

Jiaming Liang[1], Teng Huang[1](✉), Dan Li[1], Ziyu Ding[1], Yunhao Li[1], Lin Huang[2], Qiong Wang[3], and Xi Zhang[4,5](✉)

[1] Institute of Artificial Intelligence and Blockchain, Guangzhou University, Guangzhou, China
huangteng1220@gzhu.edu.cn
[2] Department of Engineering and Engineering Technology, Metropolitan State University of Denver, Denver, USA
[3] Guangdong Provincial Key Laboratory of Computer Vision and Virtual Reality Technology Shenzhen Institute of Advanced Technology, Chinese Academy of Sciences, Beijing, China
[4] School of Arts, Sun Yat-sen University, Guangzhou, China
xizhangpiano@gmail.com
[5] College of Music, University of Colorado Boulder, Boulder, USA

Abstract. Current medical image segmentation approaches have shown promising results in the field of medical image analysis. However, their high computational demands pose significant challenges for resource constrained medical applications. We propose AgileNet, an efficient breast lesion segmentation that achieves a balance between accuracy and efficiency by leveraging the strengths of both convolutional neural networks and transformers. The proposed Agile block facilitates efficient information exchange by aggregating representations in a cost-effective manner, incorporating both global and local contexts. Through extensive experiments, we demonstrate that AgileNet outperforms state-of-the-art models in terms of accuracy, model size, and throughput when deployed on resource-constrained devices. Our framework offers a promising solution for achieving accurate and efficient medical image segmentation in resource-constrained settings.

Keywords: Medical Image Analysis · Segmentation · Resource-constrained Medical Application

---

Supported by the National Natural Science Foundation of China under Grant 62002074 and 62072452; Supported by the Shenzhen Science and Technology Program JCYJ20200109115627045, in part by the Regional Joint Fund of Guangdong under Grant 2021B1515120011.

Q. Liu et al. (Eds.): PRCV 2023, LNCS 14429, pp. 419–430, 2024.
https://doi.org/10.1007/978-981-99-8469-5_33

1 Introduction

Motivated by the potential of highlighting important regions in medical image analysis, semantic segmentation has garnered significant attention in recent years [1]. Advances in medical image segmentation methods have demonstrated their scalability in handling extensive training on high-resolution medical images across divergent organs. However, the complexity and resource requirements of these approaches present significant challenges [2]. Considering the limitations imposed by the medical application environment, there is a pressing need to prioritize the development of architectures that strike a balance between computational efficiency and high performance. Thus, the improvement of an efficient model for accurate segmentation tasks on resource-constrained medical devices becomes both essential and imperative.

With the rapid advancement of medical imaging technology, the quest for achieving accurate segmentation tasks using high-quality medical image data has become paramount. In the realm of medical image segmentation, two primary approaches have emerged: transformer-based and CNN-based methods. Transformer-based approaches have garnered significant attention from researchers due to their capability to extract global information from long-range sequences using self-attentive mechanisms, thereby enhancing feature extraction. Notably, UNETR [2] has pioneered the integration of a transformer as the backbone of the encoder, leading to substantial improvements in segmentation accuracy. However, the computational complexity associated with transformer-based approaches poses challenges in terms of execution efficiency, making them less suitable for resource-constrained scenarios.

On the other hand, CNN-based approaches have gained considerable traction due to their lightweight nature, making them highly computationally efficient. Frameworks such as [3,4] and [5] are notable examples in this category. These models prioritize efficient computations and are specifically designed to handle resource-limited scenarios encountered in medical imaging applications. However, a drawback of CNN-based approaches lies in their limited ability to capture long-range dependencies compared to transformer-based methods [6]. Nonetheless, ongoing research in CNN architecture design aims to address this limitation by exploring innovative techniques that enable improved learning of contextual information and long-range dependencies, thus enhancing the segmentation performance of CNN-based models [7]. The development of CNN-based approaches is driven by the pursuit of striking a balance between computational efficiency and the ability to capture essential information necessary for accurate segmentation in the domain of medical imaging.

Our goal is to contribute to the advancement of efficient medical segmentation frameworks suitable for resource-constrained medical units. We aim to achieve this by harnessing the strengths of both transformer-based and CNN-based approaches. We recognize that such frameworks need to meet two critical requirements: high accuracy and high inference efficiency, especially in the context of resource-constrained medical tasks [8]. However, achieving high accuracy in segmentation models often comes at the cost of increased computational complexity, which poses a significant challenge when striving for both accuracy and efficiency in applications with limited healthcare resources [9]. Therefore, it is crucial to address this challenge and develop medical segmentation frameworks that strike a balance between accuracy and efficiency, ensuring optimal performance within the constraints of resource-limited environments [10,11].

This paper presents AgileNet, an efficient and rapid approach for breast lesion segmentation. AgileNet leverages 3D convolution operations to effectively analyze medical images by coordinating hierarchical representation. The key component of AgileNet is the Agile block, which plays a crucial role in aggregating representations from each layer in an economical and efficient manner during global processing. The Agile block incorporates a contextual integration process that enables efficient learning of representations through two key components: the global sparse calculator (GSC) and the reversed diffuser (RD). The GSC selectively identifies globally significant features of authorized markers, while the RD mechanism efficiently propagates the updated representations from higher to lower layers. This well-designed framework enables AgileNet to strike a better balance between high precision and high inference efficiency in breast lesion segmentation tasks.

The main contributions in this paper are summarized as follows:

- AgileNet is a novel framework designed for efficient and rapid breast lesion segmentation in 3D medical image analysis, aiming to achieve a balance between segmentation precision and high inference speed.
- The key component of AgileNet is the AgileNet block, which incorporates a contextual integration process. This process enables fast information exchange by leveraging the global sparse calculator and the reversed diffuser.
- Experimental evaluations on two public datasets, BraTS 2021 and MSD, showcase that AgileNet achieves state-of-the-art performance, demonstrating its effectiveness in breast lesion segmentation.

2 Related Work

Semantic segmentation is employed in the analysis of medical images to examine anatomical structures [12]. The U-Net architecture, comprising an encoder and decoder, has demonstrated notable performance in segmentation tasks [13–16]. Expanding upon this foundation, nn-UNet further enhances segmentation precision by incorporating skip connections to capture features at different levels of stages [1]. Nevertheless, CNN-based approaches face constraints in capturing long-range dependencies due to their limited receptive fields [2,17]. To capture global representations, UNETR integrates transformers as encoders, however, this introduces quadratic complexity as a trade-off [17]. Thus, the development of an efficient framework that effectively learns contextual information while maintaining computational efficiency is necessary.

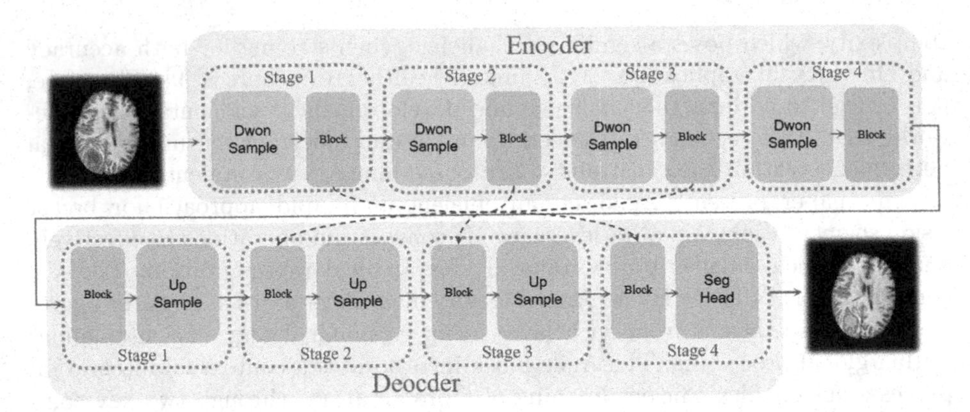

Fig. 1. Our approach leverages a U-shaped encoder-decoder architecture, with the incorporation of skip connections to establish hierarchical feature maps. As a fundamental element within each stage, an AgileNet block (illustrated in detail in Fig. 2) undertakes the vital function of facilitating efficient information exchange.

3 Methods

3.1 Framework

The AgileNet framework is built upon a deep learning architecture that comprises an encoder network and a decoder network, as illustrated in Fig. 1. These networks are connected through skip connections, facilitating the seamless flow of information across different levels of the model hierarchy. A key component of the framework is the Agile block, which plays a crucial role in efficiently extracting and aggregating representations from multimodal feature maps. For more detailed insights into the Agile block, please refer to Sect. 3.2. The loss function, depicted in Sect. 3.3, employed in the framework aims to ensure the predicted mask closely aligns with the ground truth, thus contributing to the achievement of high accuracy in segmentation tasks.

3.2 AgileNet Block

In scenarios where device processing capabilities are constrained, there is a need for lightweight segmentation models that can deliver satisfactory performance while minimizing resource requirements. The proposed AgileNet block, designed to address this need, depicted in Fig. 2, includes three basic components, Patch Partition, global sparse calculator (GSC), and reversed diffuser (RD).

Patch Partition. The AgileNet Block is a specialized module designed for processing 3-dimensional images denoted as X, with dimensions representing width W, height H, depth D, and channel C. Through a patch partitioning layer, the input image X is divided into smaller patches, resulting in N tokens. Unlike other

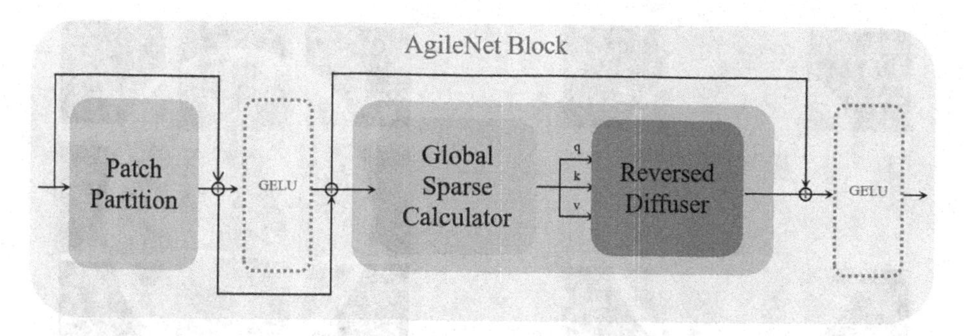

Fig. 2. Structure of AgileNet block. This block consists of three basic components, Patch Partition, global sparse calculator, and reversed diffuser.

transformer-based methods that convert these patches into a linear sequence of embeddings, our approach utilizes GSC to explore and capture global features from the entire X. By leveraging GSC, the AgileNet Block gains a comprehensive understanding of the image and effectively utilizes global information. This innovative methodology enhances the block's ability to capture relevant features, leading to improved performance and superior feature representation.

Global Sparse Calculator. The Global Sparse Calculator (GSC) serves the purpose of implementing attentional computations for input features through the utilization of sparse computation techniques. By employing the GSC, global representations are structured using windows, facilitating an efficient and effective approach. This module comprises two key designs that contribute to the sparsification of the computational transformer. The first design incorporates hierarchical downsampling operations across multiple stages, resulting in the generation of a sparse set of uniformly distributed delegate tokens. This is achieved through a 3D pointwise convolution with a stride rate denoted as $r = (1, 2, 2, 4)$. On the other hand, the second design involves locally-grouped multi-head self-attention mechanisms (L-MSA) that manage token length and accommodate shared parameters. In this approach, the delegate tokens play a crucial role in computing global sparse attention among the central tokens. Figure 3(a) provides a visual illustration of this process. Through the synergy of these designs, the GSC enables efficient attention computations and facilitates the generation of structured global representations for improved performance in various tasks. The computation of global sparse attention coefficients is performed in accordance with Eq. 1 [18].

$$Attention(q, k, v) = softmax(\frac{qk^T}{\sqrt{d_k}})v \tag{1}$$

Reversed Diffuser. Incorporating global contextual representations into the central tokens of the GSC requires propagating this information to their neighboring tokens. This task is accomplished by the reversed diffuser (RD) module. The RD is responsible for distributing the updated feature knowledge from the central tokens to the entire window embedding, as shown in Fig. 3(b).

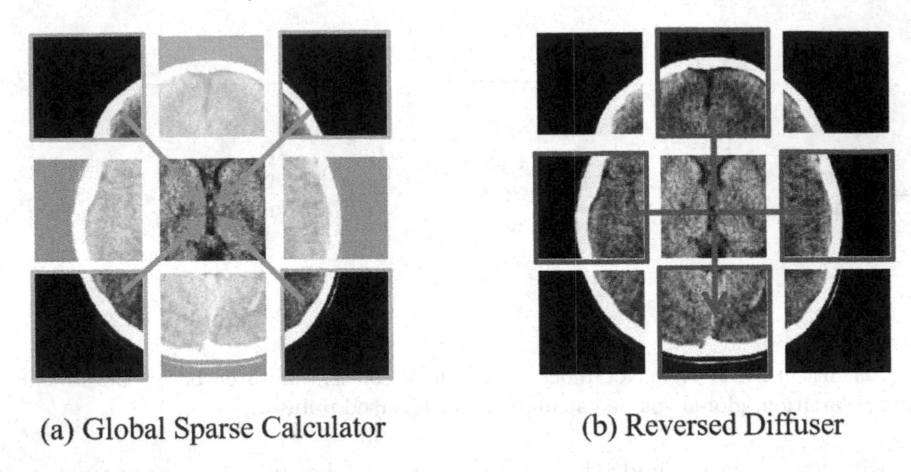

Fig. 3. Contextual integration process.

We achieve the sharing of global representations across windows by utilizing 3D transposed convolution and pointwise convolution. These operations employ a different rate of $r = (1, 2, 2, 4)$ across different stages. This ensures that each local central token effectively shares global representations with its corresponding window. The complete process of the RD module, including information propagation and knowledge distribution, is illustrated in Fig. 2.

By utilizing this mechanism, the GSC comprehensively integrates global contextual information, thereby enhancing the overall understanding and representation of the input data.

3.3 Loss Function

Soft dice loss is a commonly used loss function in semantic segmentation [19]. It compares the predicted segmentation mask with the ground truth by evaluating the overlap between their pixel-wise binary masks. This loss encourages accurate and precise segmentation results by penalizing the differences between the predicted and ground truth masks. The calculation can be represented by the Eq. 2.

$$\mathcal{L}_{dice}(Y, Y') = 1 - \frac{2}{J} \sum_{j=1}^{M} \frac{\sum_{i=1}^{N} Y_{i,j} Y'_{i,j}}{\sum_{i=1}^{N} Y_{i,j} + \sum_{i=1}^{N} Y'_{i,j}} \tag{2}$$

In the given context, I and J indicate the account of voxels and classes.

4 Experiments and Discussion

4.1 Datasets

BraTS 2021. The BraTS 2021 training dataset comprises 1251 data samples of 3D MRI scans, consisting of four modalities: native (T1), post-contrast T1-weighted (T1Gd), T2-weighted (T2), and T2 Fluid-attenuated Inversion

Recovery (T2-FLAIR). These modalities provide comprehensive imaging information for analysis. The dataset includes three distinct tumor sub-regions: the enhancing tumor, peritumoral edema, and the necrotic and non-enhancing tumor core. Each sub-region is accompanied by explanatory notes that provide additional context and insights. The dataset follows a standardized image size of $240 \times 240 \times 155$ pixels to ensure consistency across the samples. Furthermore, the explanatory notes are organized into three hierarchical sub-regions: Whole Tumor (WT), Tumor Core (TC), and Enhancing Tumor (ET), allowing for a deeper understanding of the different aspects and characteristics of the tumors within the dataset.

MSD. The Hepatic Vessel dataset, specifically the MSD No.8 task dataset, was chosen to evaluate the generalizability of our approach. Unlike brain data samples, these images exhibit noticeable differences in terms of anatomical structure and image features due to the focus on the liver region. The images were reconstructed using standard convolution kernels, with slice thickness ranging from 2.5–5 mm and a reconstruction diameter between 360–500 mm. To ensure accurate annotations, contour adjustments were manually performed by radiology experts specializing in abdominal imaging. This dataset serves as a robust testbed for evaluating the efficacy and adaptability of our approach in the context of hepatic vessel analysis.

4.2 Evaluation Metrics

Hardware Metrics. Three important metrics are commonly used to evaluate the performance and efficiency of models: 1. FPS, measured in images/second, provides insights into the practical processing efficiency of a model and is essential for assessing its lightweight efficacy; 2. FLOPs (floating point operations per second) represent the model's inference efficiency by measuring its ability to perform a large number of floating point operations within a specific time frame; 3. Model parameters, typically measured in millions (MParams), quantify the size of the model, which is a critical factor to consider when deploying models on devices with limited resources. These metrics collectively contribute to assessing the overall performance and resource requirements of models in various contexts.

Segmentations Dice. The Dice similarity coefficient (Dice) is a widely used metric for evaluating the performance of segmentation tasks. It quantifies the agreement or overlap between the predicted segmentation mask and the ground truth mask. The Dice coefficient is calculated by taking twice the intersection of the predicted and ground truth masks and dividing it by the sum of their sizes. The metric can be formulated as shown in Eq. 3.

$$Dice = \frac{\sum_{i=1}^{I} Y_i Y_i'}{\sum_{i=1}^{I} Y_i + \sum_{i=1}^{I} Y_i'} \tag{3}$$

where Y and Y_i' denote the ground truth and prediction of voxel values.

4.3 Implementation Details

The training of AgileNet was conducted on parallel NVIDIA V100 GPUs, while the testing phase utilized an Intel(R) Xeon(R) Gold 6240R CPU. Data augmentation techniques were applied during training, with probabilities set as (0.5, 0.2, and 0.2) for random flips at each axis, intensity scaling, and shifting. Additionally, random flips, rotations, and intensity scaling were applied with probabilities of 0.5, 0.25, and 0.5, respectively. For the BraTS dataset, 3D patches were randomly cropped to sizes of $128 \times 128 \times 128$, whereas for the Hepatic Vessel dataset, the cropped patch sizes were $96 \times 96 \times 96$. The initial learning rate was set to 0.001, and training utilized a linear warmup, cosine annealing learning rate scheduler, and AdamW optimizer with a weight decay of 0.4.

4.4 Results and Discussion

Segmentation Results on BraTS 2021. This section presents an evaluation of AgileNet's efficiency and effectiveness on the BraTS 2021 dataset in a comparable computing environment, as shown in Table 1. AgileNet demonstrates a significant advantage over state-of-the-art algorithms in terms of hardware metrics. It achieves a processing speed of 4.663 images per second, outperforming the fastest VitAutoEnc model by 0.179 images per second in segmentation tasks. Furthermore, AgileNet exhibits the smallest MParams and FLOPs computations compared to similar models. In terms of segmentation dice analysis, AgileNet achieves a top dice accuracy of 91.02% on the BraTS 2021 dataset, surpassing the accuracy of the second-best model, UNETR, by 0.71%. Overall, AgileNet effectively balances high accuracy, and high inference efficiency, making it a notable improvement over other methods.

Table 1. Quantitative performance comparison of different networks in BraTS 2021. The performance metrics of AgileNet are highlighted in gray , while the best performance in terms of hardware metrics and segmentation is recorded in **blue** and **purple**.

Methods	Hardware Metrics			Brain tumor Segmentations Dice			
	FPS↑	FLOPs↓	MPara.↓	Avg.(%)↑	WT(%)↑	ET(%)↑	TC(%)↑
VNet [19]	0.828	78.0	4.56	89.09	91.38	86.90	89.01
AttUNet [20]	1.042	16.65	0.27	90.40	92.02	88.28	90.94
UNet 3D [13]	0.32	238.21	1.92	87.93	92.69	84.10	87.10
nn-UNet [1]	1.4198	44.6	1.38	90.84	92.71	88.34	91.39
TransVW [21]	0.359	238.13	1.91	88.21	92.32	82.09	90.21
SegResNet [22]	1.298	33.02	1.06	90.78	**92.73**	88.31	91.31
UNETR [2]	1.531	20.3	10.2	90.31	92.53	87.59	90.78
TransBTS [23]	0.891	26.38	3.06	89.18	91.05	86.75	98.76
TransUNet [24]	0.891	24.64	18.7	84.59	87.68	83.34	82.75
VitAutoEnc [25]	3.217	12.56	18.34	76.14	81.41	68.35	78.66
AgileNet	**4.663**	**1.24**	**0.18**	**91.02**	91.06	**88.58**	**93.42**

Table 2. Quantitative performance comparison of different networks in MSD(Hepatic Vessel). The indicator color design is the same as in Table 1.

Methods	Hardware Metrics			Segmentations Dice		
	FPS↑	FLOPs↓	MPara.↓	Avg.(%)↑	Tumors(%)↑	Vessel(%)↑
VNet [19]	1.859	32.37	4.56	66.38	63.55	69.21
AttUNet [20]	2.253	6.98	0.27	61.81	56.94	66.68
UNet 3D [13]	0.675	100.47	1.92	60.62	57.29	63.95
nn-UNet [1]	4.505	18.58	1.38	69.12	**66.46**	71.78
TransVW [21]	0.74	100.23	1.91	68.62	65.80	71.44
SegResNet [22]	3.942	13.76	1.06	67.52	64.08	71.01
UNETR [2]	2.192	8.24	9.26	66.96	62.84	69.08
TransUNet [24]	2.092	10.07	17.76	65.89	60.95	70.83
VitAutoEnc [25]	3.723	5.01	17.39	52.28	48.26	56.30
AgileNet	**6.609**	**0.82**	**0.14**	**69.14**	65.43	**72.85**

Segmentation Results on MSD (Hepatic Vessel). This section presents the evaluation of AgileNet on the MSD dataset, as shown in Table 2. Despite the differences in modalities between CT images for MSD and MRI images for BraTS, AgileNet demonstrates excellent performance. It achieves the best inference efficiency and model size, outperforming the second-best model in terms of throughput with 2.104 images per second. Furthermore, AgileNet achieves

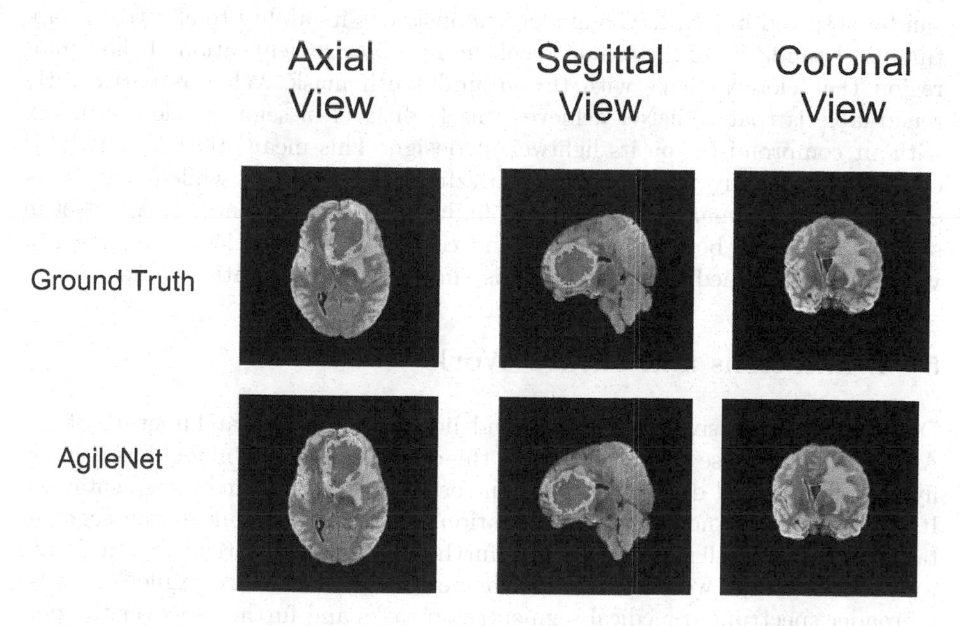

Fig. 4. Qualitative visualizations of the proposed AgileNet and ground truth.

SOTA performance in terms of accuracy, model size, and inference efficiency on this dataset. This experiment highlights the versatility of AgileNet across different datasets and modalities.

Ablation Study Results. Table 3 outlines the contributions of AgileNet blocks to medical segmentation performance. Although the employment of AgileNet Block incurs some level of throughput consumption, the apparent enhancement in segmentation performance is noteworthy. Specifically, the segmentation performance witnessed a substantial upsurge from 73.87% to 91.02%, marking a remarkable surge of 17.18%.

Table 3. Quantitative performance comparison with/without AgileNet Block in the BraTS 2021 dataset. In this table, we have indicated superiority in purple and disadvantage in blue.

Modules	Thr.↑	Flops↓	Mparams↓	Dice↑
w/o	315.87+112.09	1.07-0.8	0.07-0.07	73.84-17.18
w	**203.78**	**1.87**	**0.14**	**91.02**

Qualitative Visualization with Ground Truth. The segmentation performance of the AgileNet model in accurately identifying the primary tumor region can be observed in Fig. 4. The model demonstrates its ability to effectively capture the boundaries of the tumor, resulting in a precise delineation of the tumor region that closely aligns with the ground truth mask. What is particularly remarkable is that AgileNet achieves this high level of segmentation accuracy without compromising on its lightweight design. This means that the model is capable of efficiently processing and analyzing medical images while maintaining exceptional performance. These results highlight the effectiveness of AgileNet in striking a balance between accuracy and computational efficiency, making it a valuable tool for medical image analysis and tumor segmentation tasks.

5 Limitations and Future Work

Despite the impressive performance and lightweight model architecture of the AgileNet model in segmentation tasks, the evaluation of this model has primarily centered around dense segmentation tasks, specifically tumor segmentation. Its performance in non-dense segmentation tasks, such as multi-organ segmentation, remains a direction requiring further in-depth investigation. In future research endeavors, we aim to explore the performance bounds of AgileNet across a broader spectrum of medical segmentation tasks and further uncover the performance potential of AgileNet Block.

6 Conclusion

This study introduces AgileNet, an efficient and rapid breast lesion segmentation approach that combines the advantages of transformer-based and CNN-based methods. The key component of AgileNet is the AgileNet block, which facilitates contextual integration and streamlined information exchange. This makes AgileNet a practical solution for medical imaging analysis, particularly on devices with limited hardware resources. Experimental results demonstrate that AgileNet achieves cutting-edge performance on both the BraTS 2021 and MSD (Hepatic Vessel) datasets. These findings highlight the potential of AgileNet in medical scenarios where computational complexity needs to be reduced and inference efficiency needs to be improved. The research opens up promising avenues for further advancements in medical imaging analysis under resource constraints.

References

1. Isensee, F., Jäger, P.F., Full, P.M., Vollmuth, P., Maier-Hein, K.H.: nnU-Net for brain tumor segmentation. In: Crimi, A., Bakas, S. (eds.) BrainLes 2020. LNCS, vol. 12659, pp. 118–132. Springer, Cham (2021). https://doi.org/10.1007/978-3-030-72087-2_11
2. Hatamizadeh, A., et al.: UNETR: transformers for 3D medical image segmentation. In: Proceedings of the IEEE/CVF Winter Conference on Applications of Computer Vision, pp. 574–584 (2022)
3. Sandler, M., Howard, A., Zhu, M., Zhmoginov, A., Chen, L.-C.: MobileNetV2: inverted residuals and linear bottlenecks. In: Proceedings of the IEEE Conference on Computer Vision and Pattern Recognition, pp. 4510–4520 (2018)
4. Pang, Y., et al.: Graph decipher: a transparent dual-attention graph neural network to understand the message-passing mechanism for the node classification. Int. J. Intell. Syst. **37**(11), 8747–8769 (2022)
5. Zhang, X., Zhou, X., Lin, M., Sun, J.: ShuffleNet: an extremely efficient convolutional neural network for mobile devices. In: Proceedings of the IEEE Conference on Computer Vision and Pattern Recognition, pp. 6848–6856 (2018)
6. Tan, M., Le, Q.: EfficientNet: rethinking model scaling for convolutional neural networks. In: International Conference on Machine Learning, pp. 6105–6114. PMLR (2019)
7. Koonce, B.: MobileNetV3. In: Convolutional Neural Networks with Swift for Tensorflow, pp. 125–144. Apress, Berkeley, CA (2021). https://doi.org/10.1007/978-1-4842-6168-2_11
8. Li, L., Zhou, T., Wang, W., Li, J., Yang, Y.: Deep hierarchical semantic segmentation. In: Proceedings of the IEEE/CVF Conference on Computer Vision and Pattern Recognition, pp. 1246–1257 (2022)
9. Ma, N., Zhang, X., Zheng, H.-T., Sun, J.: ShuffleNet V2: practical guidelines for efficient CNN architecture design. In: Ferrari, V., Hebert, M., Sminchisescu, C., Weiss, Y. (eds.) Computer Vision – ECCV 2018. LNCS, vol. 11218, pp. 122–138. Springer, Cham (2018). https://doi.org/10.1007/978-3-030-01264-9_8
10. Xu, L., Ouyang, W., Bennamoun, M., Boussaid, F., Xu, D.: Multi-class token transformer for weakly supervised semantic segmentation. In: Proceedings of the IEEE/CVF Conference on Computer Vision and Pattern Recognition, pp. 4310–4319 (2022)

11. Ziegler, A., Asano, Y.M.: Self-supervised learning of object parts for semantic segmentation. In: Proceedings of the IEEE/CVF Conference on Computer Vision and Pattern Recognition, pp. 14 502–14 511 (2022)
12. Monteiro, M., et al.: Multiclass semantic segmentation and quantification of traumatic brain injury lesions on head CT using deep learning: an algorithm development and multicentre validation study. Lancet Digit. Health **2**(6), e314–e322 (2020)
13. Ronneberger, O., Fischer, P., Brox, T.: U-Net: convolutional networks for biomedical image segmentation. In: Navab, N., Hornegger, J., Wells, W.M., Frangi, A.F. (eds.) MICCAI 2015. LNCS, vol. 9351, pp. 234–241. Springer, Cham (2015). https://doi.org/10.1007/978-3-319-24574-4_28
14. Yu, L., Yang, X., Chen, H., Qin, J., Heng, P.A.: Volumetric convnets with mixed residual connections for automated prostate segmentation from 3D MR images. In: Thirty-First AAAI Conference on Artificial Intelligence (2017)
15. Li, X., Chen, H., Qi, X., Dou, Q., Fu, C.-W., Heng, P.-A.: H-DenseUNet: hybrid densely connected UNet for liver and tumor segmentation from CT volumes. IEEE Trans. Med. Imaging **37**(12), 2663–2674 (2018)
16. Dou, Q., Chen, H., Jin, Y., Yu, L., Qin, J., Heng, P.-A.: 3D deeply supervised network for automatic liver segmentation from CT volumes. In: Ourselin, S., Joskowicz, L., Sabuncu, M.R., Unal, G., Wells, W. (eds.) MICCAI 2016. LNCS, vol. 9901, pp. 149–157. Springer, Cham (2016). https://doi.org/10.1007/978-3-319-46723-8_18
17. Hatamizadeh, A., Nath, V., Tang, Y., Yang, D., Roth, H.R., Xu, D.: Swin UNETR: Swin transformers for semantic segmentation of brain tumors in MRI images. In: Crimi, A., Bakas, S. (eds.) Brainlesion: Glioma, Multiple Sclerosis, Stroke and Traumatic Brain Injuries, BrainLes 2021. LNCS, vol. 12962, pp 272-284. Springer, Cham (2022). https://doi.org/10.1007/978-3-031-08999-2_22
18. Khan, S., Naseer, M., Hayat, M., Zamir, S.W., Khan, F.S., Shah, M.: Transformers in vision: a survey. ACM Comput. Surv. (CSUR) **54**(10s), 1–41 (2022)
19. Milletari, F., Navab, N., Ahmadi, S.-A.: V-Net: fully convolutional neural networks for volumetric medical image segmentation. In: Fourth International Conference on 3D Vision (3DV), vol. 2016, pp. 565–571 . IEEE (2016)
20. Oktay, O., et al.: Attention U-Net: learning where to look for the pancreas. arXiv preprint arXiv:1804.03999 (2018)
21. Haghighi, F., Taher, M.R.H., Zhou, Z., Gotway, M.B., Liang, J.: Transferable visual words: exploiting the semantics of anatomical patterns for self-supervised learning. IEEE Trans. Med. Imaging **40**(10), 2857–2868 (2021)
22. Myronenko, A.: 3D MRI brain tumor segmentation using autoencoder regularization. In: Crimi, A., Bakas, S., Kuijf, H., Keyvan, F., Reyes, M., van Walsum, T. (eds.) BrainLes 2018. LNCS, vol. 11384, pp. 311–320. Springer, Cham (2019). https://doi.org/10.1007/978-3-030-11726-9_28
23. Wang, W., Chen, C., Ding, M., Yu, H., Zha, S., Li, J.: TransBTS: multimodal brain tumor segmentation using transformer. In: de Bruijne, M., et al. (eds.) MICCAI 2021. LNCS, vol. 12901, pp. 109–119. Springer, Cham (2021). https://doi.org/10.1007/978-3-030-87193-2_11
24. Chen, J., et al.: TransUNet: transformers make strong encoders for medical image segmentation. arXiv preprint arXiv:2102.04306 (2021)
25. Cardoso, M.J., et al.: MONAI: an open-source framework for deep learning in healthcare. arXiv preprint arXiv:2211.02701 (2022)

CFNet: A Coarse-to-Fine Framework for Coronary Artery Segmentation

Shiting He[1], Yuzhu Ji[1(✉)], Yiqun Zhang[1], An Zeng[1], Dan Pan[2], Jing Lin[1], and Xiaobo Zhang[1]

[1] Guangdong University of Technology, Guangzhou, China
2112105092@mail2.gdut.edu.cn, andrewchiyz@gmail.com, {yqzhang,zengan,lj,zxb_leng}@gdut.edu.cn

[2] Guangdong Polytechnic Normal University, Guangzhou, China
pandan@gpnu.edu.cn

Abstract. Coronary Artery (CA) segmentation has become an important task to facilitate coronary artery disease diagnosis. However, existing methods have not effectively addressed the challenges posed by the thin and complex structure of CA, leading to unsatisfactory performance in grouping local detailed vessel structures. Therefore, we proposed a novel coarse-to-fine segmentation framework, namely CFNet, to refine the CA segmentation results progressively. The global structure targeting module aims to capture the spatial structure of the CA by introducing dilated pseudo labels as supervision. In addition, a lightweight transformer-based module is designed to refine the coarse results and produce more accurate segmentation results by capturing the long-range dependencies. Our model exploits both local and global contextual features by integrating a convolutional neural network and visual Transformer. These two modules are cascaded using a center-line patching strategy, which filters out unnecessary features and mitigates the sparsity of CA annotation. Experimental results demonstrate that our model performs well in CA segmentation, particularly in handling challenging cases for fine vessel structures, and achieves competitive results on a large-scale dataset, i.e., ImageCAS, in comparison to state-of-the-art methods.

Keywords: coronary artery segmentation · coarse-to-fine framework · multi-stage training · medical image analysis

1 Introduction

Automated coronary artery (CA) segmentation plays a crucial role in facilitating the diagnosis of coronary artery disease. It can accurately identify CA and clarify the location and extent of vascular disease, which is useful to make treatment plans for patients. In addition, automated CA segmentation reduces the need for manual segmentation by doctors, thus improving disease diagnosis efficiency. For example, coronary heart disease caused by CA stenosis can clearly show the location and the severity of the narrowing through automated segmentation. It helps physicians in the diagnosis and treatment. However, the CA predicted by

Q. Liu et al. (Eds.): PRCV 2023, LNCS 14429, pp. 431–442, 2024.
https://doi.org/10.1007/978-981-99-8469-5_34

current models will rupture and miss some branches. Therefore, it has become increasingly important to propose automated segmentation methods for CA.

In recent years, convolutional neural network (CNN) based models have gained significant attention due to their excellent performance in medical image segmentation, for example, Res-Unet [26] for Retina Vessel segmentation, Multi-Unet [4] for CA segmentation. Generally, CNN-based models use kernels with fixed sizes of receptive fields to learn discriminative features within local volumes, thus its limitations in learning long-range dependencies. Moreover, Visual Transformers (ViTs) have made remarkable progress in both high and low-level computer vision tasks for natural scene images [6,14]. Transformer-based models break the limitation of the inductive bias of CNN, they are able to capture long-range contextual dependency. Thanks to this ability, the transformer has also been widely applied for medical image segmentation, such as TransUnet [3] and SwinUnetr [7] for multi-organ segmentation. Albeit long-range contextual relations can be learned for capturing the overall CA structure, it is difficult to recover the complex and fine-detailed structure of the CA vessels. Overall, in practice, for CA segmentation, 3D structures of CA are intricate, and tissues surrounding the vessels are also cluttered and noisy. Therefore, it remains challenging and non-trivial to achieve accurate CA segmentation in CT volumes by directly adopting a CNN or ViT-based model for CA segmentation.

In this paper, we propose a novel coarse-to-fine coronary artery segmentation framework with better targeting of the global structure and recovering local structures of the CA. Our framework takes advantage of both CNN and ViTs for capturing local and global contextual features. Specifically, our network consists of two main modules: (1) a global structure targeting module with a U-shape CNN-based encoder-decoder model tailored to produce the coarse structure of the CA; (2) a local structure recovering module with a lightweight visual transformer for refining the first-stage coarse segmentation results. In practice, the global structure targeting module is trained with dilated pseudo labels. These two modules are cascaded and trained by using a center-line patch cropping strategy, which filters out irrelevant and noisy voxels and achieves coarse-to-fine vessel segmentation. Experimental results on a large-scale dataset, namely ImageCAS [28], show that our proposed framework can get benefits from both CNN and ViT models for coarse vessel targeting and fine-grained local structure segmentation, respectively. Challenging cases, including interlaced local vessel structures, crushed thin CA, etc., can also be accurately segmented in comparison with state-of-the-art methods. In summary, our contributions are three-fold:

- We propose a new coarse-to-fine coronary artery segmentation framework by combining the CNN- and Transformer-based encoder-decoder models. It progressively refines rough CA and achieves detailed vessel structure segmentation for CA.
- We employ a multi-stage training strategy by incorporating the dilated pseudo label generating strategy and center-line patch cropping strategy. It effectively addresses the issues of fragmentation or broken branches in segmentation

results caused by the complexity of anatomical structures and surrounding irrelevant tissues.
- Experimental results demonstrate that our model can achieve competitive performance and outperform state-of-the-art methods for 3D CA segmentation on the ImageCAS dataset [28].

2 Related Work

2.1 CNN-Based Models for Medical Image Segmentation

Recent advances in CNN-based encoder-decoder models have achieved significant performance in medical image segmentation tasks [15,18]. To effectively capture and fuse contextual information on different scales for better recovering the details of the segmentation mask, Ronneberger *et al.* [18] proposed Unet. It consists of a symmetric encoder-decoder architecture with skip connections. The success of the U-Net in medical image segmentation has inspired the development of its variants, including Res-UNet [26], Unet++ [30], and Unet 3+ [9].

The attention mechanism [16,22] was also introduced to automatically learn and align the features that are relevant to the target in segmentation tasks. It helps improve the performance and expressive power of the model. Specifically, Ozan *et al.* [17] proposed Attention U-Net, introduced an attention module between the encoder and decoder, which dynamically adjusts the weights of features. In order to encourage the network to learn discriminative features for targets such as tumours [13], organs [12], etc. However, medical images often have a 3D structure, and the additional dimension in 3D images contains rich spatial information compared to 2D images. Therefore, 3D-Unet [5] was proposed to better utilize spatial information and capture contextual information. In addition, many researchers have proposed 3D medical image segmentation methods, such as V-net [15], among others.

Influenced by the attention mechanism and 3D volume segmentation, Shen *et al.* [20] proposed a 3D Fully Convolutional Network (FCN) integrating an Attention Module to focus on the CA in the noisy volumes. Apart from that, CA has a distinctive branching structure, and many methods have been designed to address this characteristic. Bin *et al.* [10] designed a tree-structured ConvRNN network introducing the center-line of CA for CA segmentation. For the special structure of blood vessels, Zhao *et al.* [29] proposed a cascaded network composed of a graph convolutional network (GCN) and a CNN. It can fuse the features across the network to improve performance. However, these methods lack the ability to capture the detailed structure of the CA.

2.2 Vision Transformers for Medical Image Segmentation

The emergence of ViT has garnered significant attention in the field of computer vision and has proven to be an effective alternative, breaking the limitations of traditional CNN-based models [1,2,6]. While ViT is good at capturing long-range context dependencies, it may struggle with capturing local information due to its tokenization process. Therefore, Liu *et al.* [14] proposed a hierarchical Transformer, namely Swin Transformer, to address the limitations of the standard ViT model in capturing fine-grained details and local context. Due to the outstanding performance of ViTs [21,24,25,27], many researchers have started to apply it to the field of medical image segmentation [1,3,7,8,11,23]. Specifically, Chen *et al.* [3] advanced Trans-UNet, a novel architecture that combines the strengths of Transformer and U-Net for medical image segmentation. TransBTS [23] is the first 3D ViTs segmentation network which divides the image into a sequence of tokens by using 3D CNN and utilizes the Transformer to capture long-range dependencies among 3D tokens. A novel architecture called UNETR [8], proposed by Hatamizadeh *et al.*, captures long-range dependencies by pure Transformer encoders and obtains global multi-scale information through residual connections.

Moreover, Cao *et al.* [1] proposed Swin-Unet which is the first U-Net-like segmentation network based on pure Transformers. Swin Transformer with shifted windows is able to capture local and global contextual features. In practice, since the parameters of ViTs are in a quadratic relationship with the image size. The medical images are generally 3D having a large resolution, so the calculation cost of the ViTs-based model is very high. Lee *et al.* [11] introduced a lightweight network designed for 3D medical image segmentation, called 3D UX-Net. It improves the transformer block by using ConvNet modules, which improves segmentation performance and reduces computational costs.

In summary, CNN-based models are good at obtaining the local information and the Transformer-based model performs better in capturing the long-range dependencies. However, since the structure of CA is long and interlaced, both the local features and the long-range dependencies are important to achieve high accuracy of segmentation. Therefore we designed a coarse to fine framework getting benefits from both CNNs and ViTs. Considering the tissues surrounding the vessels are cluttered and noisy, we introduced the dilated pseudo label generating strategy and center-line patch cropping strategy to target the main structure of CA and recover the local fine-grained structure progressively.

3 Method

3.1 Overview

Given an input 3D CT volume with resolution $D \times H \times W$, where D, H and W are the spatial depth, height and width, respectively. Our model aims to use the Global Structure Targeting Module (GSTM) to predict the coarse result and refine it into the final prediction by the Local Structure Recovering Module (LSRM). The overall architecture of our proposed CFNet is illustrated in Fig. 1. Firstly, a dilated pseudo label generating strategy is applied to produce the pseudo label by dilating the ground-truth (GT) label. Next, the pseudo label is used as supervision to encourage the GSTM to predict the complete structure of CA. Then, a center-line patch cropping strategy is utilized to crop the input volume into patches by locating regions with high confidence according to the

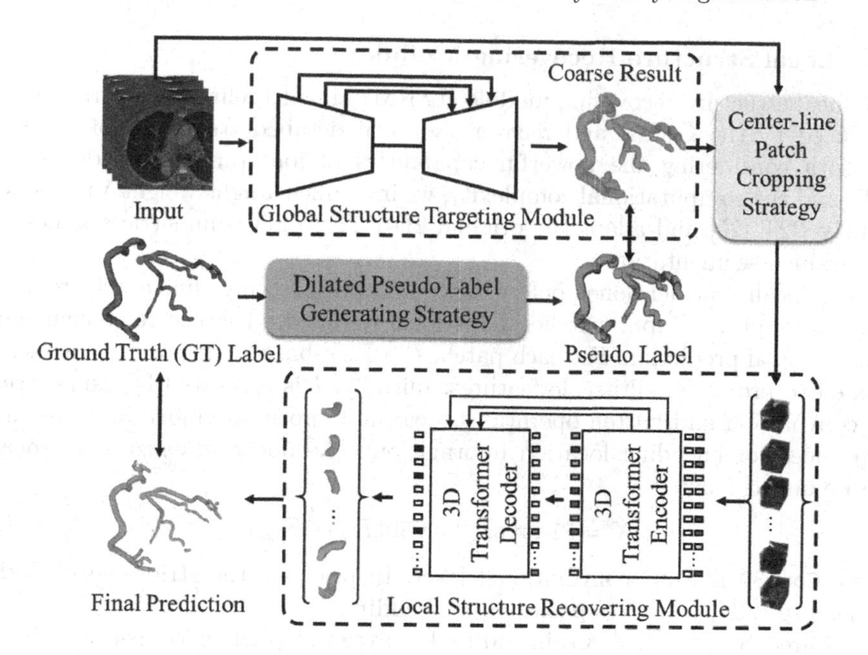

Fig. 1. Overview of our proposed CFNet.

center-line extracted from the coarse prediction. After that, these local patches were fed into the LSRM to generate patch-wise predictions. The final prediction of the CA was reconstructed by merging patch-wise predictions.

3.2 Global Structure Targeting Module

Global Structure Targeting Module (GSTM) aims to obtain the global structure of the CA by capturing spatial feature information. This module is a CNN-based encoder-decoder structure. Thanks to CNN's ability to capture local information, GSTM can get a complete structure including small branches in the whole volumetric image by using the dilated pseudo label as supervision.

Concretely, the building block of the encoder part consists of two 3D convolutional layers followed by using ReLU as the activation function and a max-pooling layer to downsample the feature maps. For each 3D convolutional layer, we used $3 \times 3 \times 3$ kernels to perform convolution. For the decoder part, the building block is constructed by using an upsampling layer with a stride of 2 followed by a feature concatenation layer and two 3D convolutional layers with $3 \times 3 \times 3$ kernels. The ReLU activation function is also applied after each 3D convolutional layer. Similar to the 3D Unet, we stack five encoder building blocks to extract multi-level features and symmetrically five decoder building blocks to fuse multi-level features, respectively. Note that the pseudo labels are used as supervision during the training of GSTM, which guides GSTM to roughly target the whole structure and shape of the CA with more boundary information.

3.3 Local Structure Recovering Module

The local structure recovering module (LSRM) aims to refine the coarse results produced by the GSTM and recover the local detailed structures of the CA. By both considering the powerful capabilities of long-range dependencies of ViTs and the computational complexity, we introduce a light-weight ViT model, namely P2T [25] and adapt the original 2D-P2T to 3D volumetric patches for fine-grained segmentation.

Specifically, as mentioned before, we first applied a center-line patch cropping strategy to obtain input patches I for local detailed structure refinement and accurate final prediction. For each patch, I, a P2T-based transformer encoder is applied to produce multi-scale features. Initially, I is tokenized by using stride 3D convolution and flatten operations. Learnable position embeddings are also introduced for encoding location information. The above tokenization process can be denoted by:

$$X = \texttt{Flatten}(\texttt{Conv3D}(I)) + E_{pos}, \tag{1}$$

where `Conv3D` is a 3D convolutional layer. In practice, the stride is set as the kernel size s. E_{pos} denotes positional embedding.

Inspired by P2T [25], we introduced a pyramid pooling transformer block to obtain rich long-range contextual dependencies within the local patches for fine-grained CA structure recovery. The core idea of the P2T block is to produce multi-scale contextual features by applying multiple average pooling w.r.t different pooling ratios. The attention matrix is calculated by introducing the pyramid pooling layer into the original self-attention block. Therefore, the original multi-head self-attention (MHSA) block is converted to a pooling-based MHSA (P-MHSA). The self-attention block in P2T can be denoted by:

$$\text{Attention}(Q, K, V) = \text{softmax}(\frac{QK^T}{\sqrt{d_k}}V) = \text{softmax}(\frac{XW_q \times PW_k}{\sqrt{d_k}}PW_v), \tag{2}$$

where W_q, W_k and W_v are projection parameters for Q, K and V, respectively. P represents the multi-sale feature maps by concatenating different scales of pooling results. Concretely, $P = \texttt{Concat}(P_1^{'}, P_2^{'}, ..., P_n^{'})$, where $P_i^{'}$ denotes the i-th layer of pyramid feature maps. $P_i^{'}$ is obtained by using depth-wise separable convolution and a residual connection, *i.e.* $P_i^{'} = \texttt{ConvD}(P_i) + P_i$, where P_i is obtained from the i-th average pooling layer. Finally, the output of Attention(Q, K, V) was fed into a feed-forward network to generate the feature map Z.

Similar to the Unet, we stack four stages of P2T encoding blocks to generate multi-level features, *i.e.* $[Z^{(1)}, Z^{(2)}, Z^{(3)}, Z^{(4)}]$. In the decoder part, we adopted the same upsampling block from GSTM to recover the spatial resolution of the feature map and predict the patch segmentation result.

3.4 Dilated Pseudo Label Generating Strategy

Considering the complex anatomical structure of CA and the surrounding tissues, the boundaries of CA are often not enough detail. However, to encourage the GSTM to target the overall structure of the CA, we introduce the

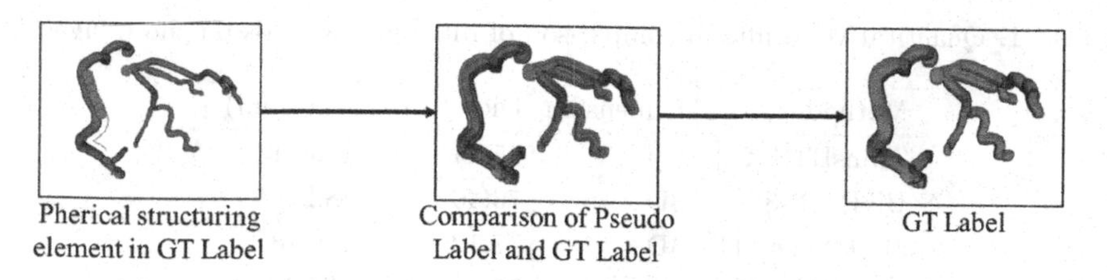

Fig. 2. Dilated Pseudo Label Generating Process.

Dilated Label Pseudo Generating Strategy (DLPGS) to generate pseudo labels for GSTM. Specifically, the DLPGS applies a morphological dilation operator to generate the dilated pseudo label based on the GT label. Figure 2 shows an example of applying the dilation operation on a GT label for generating a dilated pseudo label.

3.5 Center-Line Patch Cropping Strategy

Considering the vessels of the heart are crisscrossed, a simple split strategy will lead the network to predict the other arteries as CA. The center-line of the CA, as a strong prior knowledge, can be used to locate the CA. Therefore, we introduce the Center-line Patch Cropping Strategy (CPCS) to filter out volumetric data without CA. In our implementation, the coarse result produced by GSTM was used to extract the center-line of CA by using an erosion operation. The CT volume and GT Label were cropped into patches along with the center-line. After that, only sub-volumes containing CA were extracted and utilized to train the LSRM.

4 Experiments

4.1 Dataset and Evaluation Metrics

The ImageCAS dataset was collected from a hospital, and all the Computed Tomography Angiography (CTA) images were captured by a Siemens 128-slice dual-source scanner. The dataset collected a total of 1000 cases, including 414 female and 586 male patients. Among these cases, 800 cases are selected for training and the reset 200 cases are for testing. The input image size is $512 \times 512 \times$ (206–275), with a planar resolution ranging from $0.29\,\text{mm}^2$ to $0.43\,\text{mm}^2$, and a spacing ranging from 0.25 mm to 0.45 mm. The left and right CA in each image were independently annotated by two experienced radiologists, and their results were cross-checked. If the annotations were inconsistent, a third radiologist got involved and provided annotations until the final annotations reached a consensus. In our experiment, we use the Dice coefficient (Dice) to evaluate the accuracy of the segmentation and the Hausdorff Distance (HD) to measure the accuracy of the CA boundaries.

Table 1. Quantitative results in comparison of different methods on the ImageCAS.

Method	Dimension	Dice (%) ↑	HD (mm) ↓
TransBTS [23]	3D	75.24	129.61
UNETR [8]	3D	76.46	88.24
3D UXNet [11]	3D	77.12	179.78
TransUnet [3]	2D	77.18	17.98
SwinUnet [1]	2D	67.78	41.37
P2T [25]	2D	75.87	19.88
FCN [20]	3D	79.80	18.09
FCN_AG [20]	3D	80.39	17.20
3D U-Net [5]	3D	77.00	**16.80**
U-Net [18]	2D	74.99	20.01
CFNet(Ours)	3D	**82.67**	18.83

4.2 Implementation Details and Hyper-parameter Settings

The proposed CFNet is implemented by using PyTorch. The Input and GT Label were resized to $512 \times 512 \times 256$ by linear interpolation. Specifically, the GSTM is trained for 30 epochs with a batch size of 1. In addition, the local structure recovering module is trained for another 30 epochs using a batch size of 8. During the training phase, we adopted the Adam optimizer to update the parameters. The learning rate is set to 1e−4 and 1e−5 for the coarse and fine training stage, respectively. A mixture of the Dice loss and cross-entropy loss is employed to train the network. Moreover, the training policy and hyper-parameter settings of the models, including FCN [20], FCN_AG [20], 3D U-Net [5] are consistent with GSTM. For TransBTS [23], UNETR [8] and 3D UXNet [11], we follow the same data augmentation strategy and hyper-parameter settings reported in their papers. For U-Net [18], TransUnet [3], SwinUnet [1] and P2T [25], 2D slices were extracted from 3D volumes and used as training data. Other hyper-parameter settings, including optimizer, loss function, learning rate, etc., are the same as our coarse stage model.

4.3 Comparison with State-of-the-Arts

We conduct experiments on the ImageCAS dataset by comparing our CFNet with ten state-of-the-art models for medical image segmentation, including FCN [19], FCN_AG [20], 3D U-Net [5], TransBTS [23], UNETR [8], 3D UXNet [11], U-Net [18], TransUnet [3], SwinUnet [1] and P2T [25]. Table 1 shows the quantitative comparison results of the proposed CFNet and state-of-the-art methods.

Experimental results demonstrate that our CFNet achieves the best performance with a Dice of 0.8267, but a slightly higher HD of 18.83. We believe it is caused by the error in segmenting the dense vessels of the heart area. The

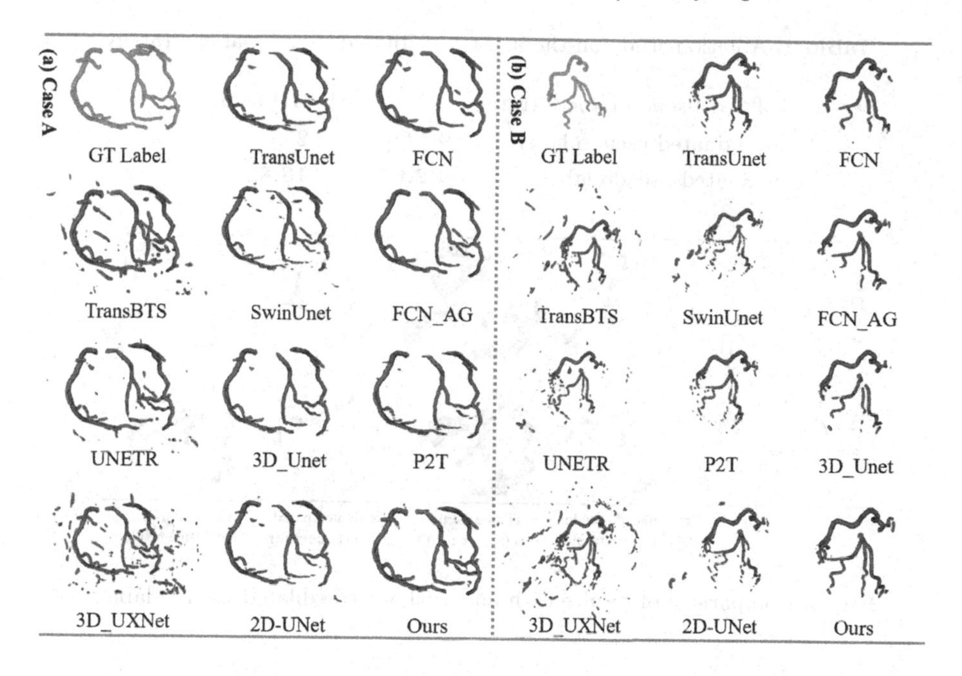

Fig. 3. Qualitative results in comparison of state-of-the-art methods.

patch-wise segmentation strategy may fail to accurately segment regions without considering the relative position of the target in volume. Albeit the HD is not the lowest, our proposed model performs well in comparison with various state-of-the-art methods. To better visualize the CA segmentation results, we provide Fig. 3 to show some quantitative results in comparison to state-of-the-art methods. It can be observed that our model can produce more accurate segmentation results with fewer broken vessels.

4.4 Ablation Study

To further validate the effectiveness of our proposed modules, we ablated our model by incrementally introducing the proposed global structure targeting and local structure recovering modules into our framework. In addition, hyper-parameter settings, including dilation rate for generating pseudo labels, and cropping size for patch-wise prediction, are also analyzed.

Ablation on Dilated Pseudo Label Generation Strategy. To investigate the effectiveness of the dilated label strategy, we conducted experiments by directly using GT labels as the supervision to train GSTM. The patch cropping size of all experiments is 32, and the dilation rate is 7. The experimental results in Table 2 show that the proposed model with the first stage of dilated pseudo label generation strategy can obtain better performance in terms of Dice and

Table 2. Ablation study on the impact of dilated pseudo label strategy.

Dilated Pseudo Label Strategy	Dice (%) ↑	HD (mm) ↓
w/o dilated pseudo label	81.06	25.74
w dilated pseudo label	**82.67**	**18.83**

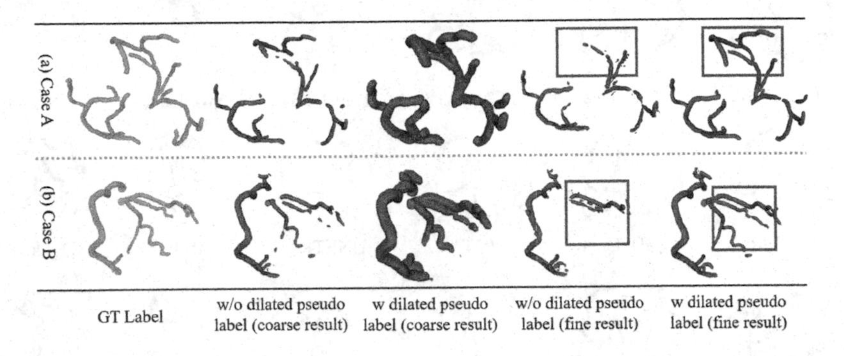

Fig. 4. Comparison of results with and without the dilated pseudo label.

HD. Thanks to the capability of coarsely targeting the global structure using the dilated pseudo label as supervision, Fig. 4 illustrates that local fine-grained structures of cases with severely broken blood vessels can be better recovered using our CFNet.

Ablation on Dilation Rate. To explore the impact of different dilation rates, we ablate dilation rates w.r.t 5, 7, and 9 for generating pseudo labels. Figure 5(a) shows that the HD increase following the dilation rate. The best performance in terms of Dice can be obtained by setting the dilation rate to 7. The quantitative results show that the performance can be improved by increasing the dilation rate for generating pseudo labels. However, a larger dilation rate may corrupt the final predictions. We believe the reason is that a larger dilation rate (*i.e.* 9) may introduce more noisy samples in the 3D data.

Ablation on Patch Cropping Size. To investigate the impact of patch cropping sizes on the segmentation performance, we performed ablation on patch cropping sizes 24, 32, 40, 48, and 56. As shown in Fig. 5(b), similarly, the Dice coefficient increases along with the cropping size, but drops when setting the cropping size larger than 40. In practice, since the center-line cropping strategy filters out parts of the volume without CA, a tradeoff should be made for accurately including the local structures without introducing more irrelevant voxels. Therefore, in our implementation, we set the cropping size to 32 to achieve the best Dice performance.

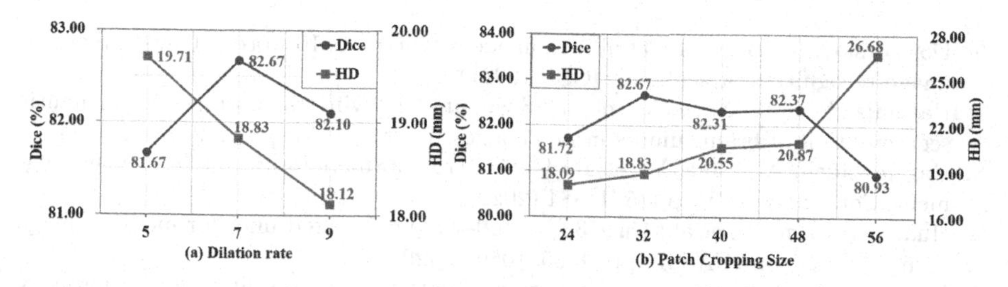

Fig. 5. Ablation study on (a) Dilation rate and (b) Patch cropping size.

5 Conclusion and Discussion

In this paper, we proposed a new coarse-to-fine framework for CA segmentation, namely CFNet, which includes two main modules and two strategies. Specifically, a global structure targeting module and a dilated pseudo label generating strategy are proposed to roughly target the overall structure of the CA. Moreover, a local structure recovering module and a center-line patch cropping strategy were introduced to encourage the model to focus and recover the local detailed structures of the CA. The whole model can be trained using pseudo-dilated labels and ground-truth labels in multiple training stages. It achieves accurate CA segmentation by capturing the global and local 3D structure of CA for CT volumes in a coarse-to-fine manner. Experimental results on large-scale dataset, ImageCAS, show our framework can consistently outperform other state-of-the-art models. However, the hyperparameter choices of the DPLGS and CLPCS are the bottleneck of the runtime efficiency in our method. In practice, it needs to trade off the efficiency and accuracy. Moreover, to demonstrate the robustness and applicable scenarios of our proposed model, the performance of different model variations on different datasets should be further investigated in our future work.

Acknowledgement. This work was supported in part by the National Natural Science Foundation of China under Grants 61976058 and 62302104, the Science and Technology Planning Project of Guangdong under Grants 2019A050510041, 2021A1515012300, 2021B0101220006, 2022A1515011592, and 2023A1515012884, and the Science and Technology Planning Project of Guangzhou under Grant 202103000034.

References

1. Cao, H., Wang, Y., et al.: Swin-unet: Unet-like pure transformer for medical image segmentation. In: ECCV. pp. 205–218 (2023)
2. Carion, N., Massa, F., et al.: End-to-end object detection with transformers. In: ECCV. pp. 213–229 (2020)
3. Chen, J., Lu, Y., et al.: Transunet: Transformers make strong encoders for medical image segmentation. CoRR (2021), https://arxiv.org/abs/2102.04306
4. Chen, Y.C., Lin, Y.C., et al.: Coronary artery segmentation in cardiac ct angiography using 3d multi-channel u-net. arXiv preprint arXiv:1907.12246 (2019)
5. Çiçek, Ö., Abdulkadir, A., et al.: 3d u-net: learning dense volumetric segmentation from sparse annotation. In: MICCAI. pp. 424–432 (2016)

6. Dosovitskiy, A., Beyer, L., et al.: An image is worth 16x16 words: Transformers for image recognition at scale. In: ICLR (2021)
7. Hatamizadeh, A., Nath, V., et al.: Swin unetr: Swin transformers for semantic segmentation of brain tumors in mri images. In: MICCAI. pp. 272–284 (2021)
8. Hatamizadeh, A., Tang, Y., et al.: Unetr: Transformers for 3d medical image segmentation. In: WACV. pp. 574–584 (2022)
9. Huang, H., Lin, L., et al.: Unet 3+: A full-scale connected unet for medical image segmentation. In: ICASSP. pp. 1055–1059 (2020)
10. Kong, B., Wang, X., et al.: Learning tree-structured representation for 3d coronary artery segmentation. Comput Med Imaging Graph p. 101688 (2020)
11. Lee, H.H., Bao, S., et al.: 3d ux-net: A large kernel volumetric convnet modernizing hierarchical transformer for medical image segmentation (2022)
12. Li, C., Tan, Y., et al.: Attention unet++: A nested attention-aware u-net for liver ct image segmentation. In: ICIP. pp. 345–349 (2020)
13. Li, S., Dong, M., et al.: Attention dense-u-net for automatic breast mass segmentation in digital mammogram. IEEE Access pp. 59037–59047 (2019)
14. Liu, Z., Lin, Y., et al.: Swin transformer: Hierarchical vision transformer using shifted windows. In: ICCV. pp. 10012–10022 (2021)
15. Milletari, F., Navab, N., et al.: V-net: Fully convolutional neural networks for volumetric medical image segmentation. In: 3DV. pp. 565–571 (2016)
16. Mnih, V., Heess, N., et al.: Recurrent models of visual attention. In: NeurIPS. pp. 2204–2212 (2014)
17. Oktay, O., Schlemper, J., et al.: Attention u-net: Learning where to look for the pancreas. arXiv preprint arXiv:1804.03999 (2018)
18. Ronneberger, O., Fischer, P., et al.: U-net: Convolutional networks for biomedical image segmentation. In: MICCAI. pp. 234–241 (2015)
19. Shelhamer, E., Long, J., Darrell, T.: Fully convolutional networks for semantic segmentation. IEEE TPAMI pp. 640–651 (2017)
20. Shen, Y., Fang, Z., et al.: Coronary arteries segmentation based on 3d fcn with attention gate and level set function. IEEE Access **7**, 42826–42835 (2019)
21. Touvron, H., Cord, M., et al.: Training data-efficient image transformers & distillation through attention. In: ICML. pp. 10347–10357 (2021)
22. Vaswani, A., Shazeer, N., et al.: Attention is all you need. In: NeurIPS. pp. 5998–6008 (2017)
23. Wang, W., Chen, C., et al.: Transbts: Multimodal brain tumor segmentation using transformer. In: MICCAI. pp. 109–119 (2021)
24. Wang, Z., Cun, X., et al.: Uformer: A general u-shaped transformer for image restoration. In: CVPR. pp. 17683–17693 (2022)
25. Wu, Y.H., Liu, Y., et al.: P2t: Pyramid pooling transformer for scene understanding. IEEE TPAMI (2022)
26. Xiao, X., Lian, S., et al.: Weighted res-unet for high-quality retina vessel segmentation. In: ITME. pp. 327–331 (2018)
27. Xie, E., Wang, W., et al.: Segformer: Simple and efficient design for semantic segmentation with transformers. In: NeurIPS. pp. 12077–12090 (2021)
28. Zeng, A., Wu, C., et al.: Imagecas: A large-scale dataset and benchmark for coronary artery segmentation based on computed tomography angiography images. CoRR (2022)
29. Zhao, G., Liang, K., et al.: Graph convolution based cross-network multi-scale feature fusion for deep vessel segmentation. IEEE TMI (2022)
30. Zhou, Z., Siddiquee, M.M.R., et al.: Unet++: A nested u-net architecture for medical image segmentation. In: MICCAI. pp. 3–11 (2018)

Edge-Prior Contrastive Transformer for Optic Cup and Optic Disc Segmentation

Yaowei Feng[1], Shijie Zhou[1], Yaoxing Wang[1], Zhendong Li[1,2,3](✉), and Hao Liu[1,2,3]

[1] School of Information Engineering, Ningxia University, Yinchuan 750021, China
{yaoweifeng,zhoushijie1,wangyaoxing}@stu.nxu.edu.cn,
{lizhendong,liuhao}@nxu.edu.cn

[2] Collaborative Innovation Center for Ningxia Big Data and Artificial Intelligence Co-founded by Ningxia Municipality and Ministry of Education, Yinchuan 750021, China

[3] Key Laboratory of the Internet of Water and Digital Water Governance of the Yellow River in Ningxia, Yinchuan 750021, China

Abstract. Optic Cup and Optic Disc segmentation plays a vital role in retinal image analysis, with significant implications for automated diagnosis. In fundus images, due to the difference between intra-class features and the complexity of inter-class features, existing methods often fail to explicitly consider the correlation and discrimination of target edge features. To overcome this limitation, our method aims to capture interdependency and consistency by involving differences in pixels on the edge. To accomplish this, we propose an Edge-Prior Contrastive Transformer (EPCT) architecture to augment the focus on the indistinct edge information. Our method incorporates pixel-to-pixel and pixel-to-region contrastive learning to achieve higher-level semantic information and global contextual feature representations. Furthermore, we incorporate prior information on edges with the Transformer model, which aims to capture the prior knowledge of the location and structure of the target edges. In addition, we propose an anchor sampling strategy tailored to the edge regions to achieve efficient edge features. Experimental results on three publicly available datasets demonstrate the effectiveness of our proposed method, as it achieves excellent segmentation performance.

Keywords: Medical segmentation · Transformer · Contrastive learning

1 Introduction

Glaucoma, a chronic and progressive ocular disease, ranks among the foremost causes of global blindness, substantially influencing patients' visual function and overall quality of life. In pursuing early glaucoma detection and treatment, segmenting the Optic Cup (OC) and Optic Disc (OD) from color fundus images is

Y. Feng and S. Zhou—Equal contribution.

Q. Liu et al. (Eds.): PRCV 2023, LNCS 14429, pp. 443–455, 2024.
https://doi.org/10.1007/978-981-99-8469-5_35

paramount as a pivotal diagnostic procedure [1]. However, fundus images often exhibit low contrast and significant inter-individual variability, manifesting as variations in the sizes and shapes of the OC and OD. A discernible level of visual resemblance is observed in OC and OD, particularly at the object's boundaries. Presently, clinical practice predominantly relies on manual segmentation, necessitating the involvement of ophthalmologists for subsequent analysis. However, this laborious and time-consuming process impairs efficiency and accuracy, rendering it impractical for large-scale screening initiatives. As a result, several segmentation algorithms are under investigation, aiming to assist ophthalmologists and artificial intelligence systems in the diagnostic process.

These methodologies can be categorized into traditional approaches and deep learning approaches. Traditional approaches predominantly rely on color, texture, edge characteristics and other manually crafted features to segment the OC and OD [2,15,19,23]. In recent years, with the development of deep learning, significant progress has been made in the task [8,10,14,16]. Deep learning approaches can automatically learn intricate and comprehensive feature representations from the original fundus images, leading to more precise OC and OD segmentation. However, the complexity and variability of the OC and OD led to insufficient segmentation or excessive segmentation of the target's edge regions. Specifically, the complex texture, size and shape of the OC and OD exhibit considerable inter-individual variability (see Fig. 1(a–d)). The edge regions between the OC and OD may exhibit blurriness or indistinctness (see Fig. 1(e–f)).

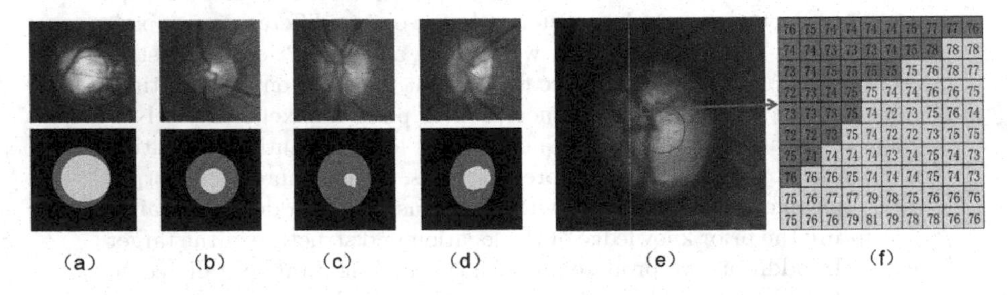

Fig. 1. (a)–(d) show the examples of fundus images, where the top row represents the original images and the bottom row represents the corresponding ground truth images. (e) Gray-scale fundus images. The blue circle indicates the edge of the OC. (f) Magnified view of the boundary region showing the gray-scale values. (Color figure online)

In this paper, we propose a Transformer-based network for OC and OD Segmentation. Motivated by the fact that the feature correlation and differentiation related to pixels along the edge is helpful regarding the variations of the sizes, shapes and indistinctness over individual variability. We observed over-segmentation or under-segmentation in the initial segmentation result images. Therefore, our model aims to exploit the semantic feature interdependencies among edge pixels and the intra-class and inter-class feature relationships. To achieve this, we carefully designed the Edge-Prior Contrastive Transformer (EPCT), which consists of the Edge-Prior Transformer Module (EPTM) and the Edge-wise Supervised Contrastive Learning method. The Edge-wise Supervised Contrastive Learning method focuses on complex edge samples, learning the consistency and differences among pixels belonging to different classes. The Edge-Prior Transformer Module learns to model the relationships between key positional features as the structural relationships of the target by incorporating the positional information of the edge prior points. Experimental results demonstrate the remarkable improvement of our EPCT on RIM-ONE v3 [9], REFUGE [25] and DRISHTI-GS [28] datasets.

The main contributions are summarized as follows:

- We propose a supervised pixel-region-wise and pixel-wise contrastive learning method for OC and OD segmentation, which explores the semantic embedding space of pixels and regions in the indistinct edge regions. As a result, our model explicitly leverages the global semantic similarity between labeled edge pixels to learn a well-structured pixel semantic embedding space.
- We carefully develop an Edge-Prior Transformer Module, which implicitly models the semantic structure relationships between areas and edge points, thereby addressing the issue of varying target sizes and diverse shapes caused by individual differences.
- We conducted comprehensive experiments on widely recognized benchmarks. Extensive experiments on these datasets demonstrate that the proposed EPCT advances state-of-the-art performance. Specifically, our quantitative experimental analysis reveals that EPCT enhances the discriminability of edge regions among different classes.

2 Related Works

2.1 OC and OD Segmentation

The purpose of the task is to separate the OC and OD in color fundus images and assign labels to each pixel. Some work [8,10,14,20,22] have been carried out based on Fully Convolutional Network (FCN) [18] and U-Net [27]. M-Net [8] converting the fundus image into a polar coordinate system, fully using space constraints and balancing the ratio of OC and OD. However, it does not explicitly consider the correlation between the circular shape priors. To better use prior information, BGA-Net [20] integrated boundary prior and adversarial learning, reducing the disorder entropy along the boundary region. Nevertheless, it may still face challenges in effectively capturing fine details and handling complex boundary cases. To address this, our proposed method is designed to incorporate prior information on the edge to capture the prior knowledge of the location and structure of the target edges.

2.2 Transformer

The success of the Transformer in natural language processing has led to its application in medical segmentation tasks [3,4,21,32]. Trans-UNet [4] was the first paper to study the application of Transformer in medical image segmentation. However, it faces challenges in effectively handling small-scale objects and intricate boundaries in capturing fine details. To address this limitation, a new architecture called Eformer [21] was proposed using the Sobel-Feldman operator for edge enhancement. However, it is restricted to boosting edges horizontally and vertically, possibly neglecting diagonal or other directions. Inspired by this, we consider the edge point as the position of the OC and OD and model the relationships between key positional features as the structural relationships of the target by incorporating the positional information on the edge prior points.

2.3 Contrastive Learning

Contrastive learning has attracted wide attention in computer vision processing in recent years [5,11,29–31]. Its core idea is to learn representations by comparing the similarity relationship between positive (similar) pairs and negative (dissimilar) pairs. A pixel-by-pixel contrastive learning paradigm for semantic segmentation is proposed to capture the global semantic relationship and guide pixels embedded into the category discriminant representation across images [29]. Inspired by the above works, due to the unique characteristics of fundus images, we design pixel-to-pixel and pixel-to-region contrastive learning focusing on complex edge samples, learning the consistency and differences among pixels belonging to different classes.

3 Method

The proposed method consists of a CNN backbone, an Edge-Prior Transformer Module (EPTM) and an Edge-wise Supervised Contrastive Learning method with a Hard Example Selection (HES). As in Fig. 2, we employed a ResNet50 model pre-trained on ImageNet as a CNN backbone to obtain input features. EPTM is a fusion of traditional Transformer-based with edge-prior in Transformer encoder. The Edge-wise Supervised Contrastive Learning method incorporates hard example selection, pixel-to-pixel and pixel-to-region.

3.1 Edge-Prior Transformer Module

We introduce the structural boundary of the target through the edge points, allowing the Transformer to effectively capture edge-wise correlated features in the image through the self-attention mechanism. Given an input image $I \in \mathbb{R}^{H \times W \times 3}$, with width W and height H, output an image feature map $I_f \in \mathbb{R}^{\frac{H}{16} \times \frac{W}{16} \times C}$, the input of the Transformer layer expects a sequence. Therefore, we converted the feature mapping plane into 1-dimensional tokens, which is the feature map F, introduced a position embedding to encode the position information. Then we fused it with the feature mapping I_f through direct addition to obtaining the following feature embedding:

$$Z_0 = F + E = W \times I_f + E, \tag{1}$$

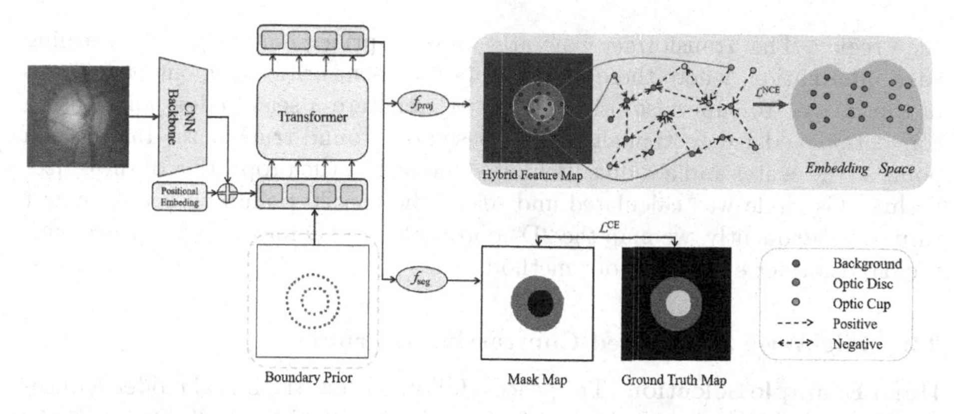

Fig. 2. EPCT architecture. It first uses a CNN backbone to extract visual features, combines them with positional encodings and flattens them into a sequence of local feature vectors. In the EPTM, the local feature is contextualized by a few Edge-Prior Transformer layers to derive global edge-point relations. Then, a Project Head f_{proj} maps each high-dimensional pixel embedding into a feature vector to compute the contrastive loss L_{NCE}. Segmentation Head f_{seg} up-sampling feature map into a mask map. In the HES, the negative/positive pixels will be projected and contrasted to further differentiate the embedding in the edges of OC and OD.

where W is the linear projection matrix, E denotes the position embeddings and $Z_0 \in \mathbb{R}^{L \times C}$, $L = \frac{HW}{256}$ refers to the feature embeddings. We employ a standard Transformer encoder to capture the long-range context in fundus images.

Edge-Prior Module. We propose an Edge-Prior Module (EPM) that integrates prior knowledge of edges into the Transformer encoder. The Transformer encoder comprises L Transformer layers, each Transformer layer consists of a Self Attention (MSA) and a Feed Forward Network (FFN). MSA operates in parallel to capture the semantic and structural features within the input sequence. It contains a query Q, a key K and a value V as input and outputs a refined feature as follows:

$$SA(z_i) = Softmax(\frac{q_i k^T}{\sqrt{d_h}})v, \tag{2}$$

where $[q, k, v] = zW_(qkv)$, $W_(qkv) \in \mathbb{R}^{D_0 \times 3D_h}$ is the projection matrix and vector $z_i \in \mathbb{R}^{1 \times D_0}$, $q_i \in \mathbb{R}^{1 \times D_h}$ are the ith row of z and q, respectively.

Therefore, the contextual feature of the Transformer encoder output is expressed as:

$$\begin{aligned} X^{j-1} &= MSA(Z^{j-1}) \oplus FFN(MSA(Z^{j-1})), \\ Z^j &= X^{j-1} \oplus (X^{j-1} \otimes \hat{P}^{j-1}). \end{aligned} \tag{3}$$

Our EPM is introduced after the FFN in the Transformer encoder. Specifically, the values of 0 and 1 are used to indicate the patches of the Transformer, where a value of 1 represents that the corresponding patch is located in the fuzzy edge region. The Transformer generates an edge prior point map E by learning edge information, where these prior points represent critical regions or features in the image. To supervise prior points, we generate a set of edge points using conventional edge detection algorithms based on ground-truth data. Taking each point as the center and a radius (default value of 10), the proportion of the object within this circle was calculated and select the top 20 points with the highest values. Subsequently, we map the 2D coordinates to 1D for correspondence with the Transformer encoder in our method.

3.2 Edg-Wise Supervised Contrastive Learning

Hard Example Selection. The process follows: First, the initial model deduces the entire training set and obtains the initial segmentation result. Based on the initial segmentation result, most pixels with segmentation errors are at the edge. Therefore, we choose to sample the edge parts of the target for comparative learning. Our anchor sampling strategy focuses on two specific categories of pixels: those with incorrect predictions and those at the edge. By selecting pairs of pixels within a certain distance (within 30 pixels of the ground truth point), the selected positive sample pairs contain boundary information and can provide useful feature contrast signals. We extract a mask map M and a feature map $F \in \mathbb{R}^{\frac{H}{8} \times \frac{W}{8} \times C}$, from the last two layers in Transformer. We use trilinear interpolation to upsample the feature map to the size of $W \times H$ and concatenate it with the mask map M into a hybrid feature map. The feature vectors of representative points in the hybrid feature map can be extracted according to their point index as anchor embeddings.

Pixel-to-Pixel Contrastive Learning. According to the ground truth, pixels are divided into four categories, including OC (represented by C), OD (represented by D), background pixel (represented by G) and edge pixel set (represented by E). The edge pixel set E is obtained from the pixels D near C and the background pixels B near D. Specifically, the boundary of the OC and OD is obtained according to ground truth and the edge pixel set is obtained by expanding 30 pixels outward and inward. These three circles form two annular regions for selecting anchors.

According to our anchor sampling strategy, pixel-to-pixel contrastive learning is performed on hard examples and feature sets, making the edge region's features more cohesive and discriminative. For a given hard anchor embedding i with its corresponding ground truth semantic label c, the positive samples consist of other pixels that also pertain to the same class c, while the negative samples comprise pixels from the remaining classes C. Our pixel-wise contrastive loss is defined as:

$$L_{NCE}^{i} = \frac{1}{|P_i|} \sum_{i^+ \in P_i} -\log \frac{\exp\left(i \cdot i^+/\tau\right)}{\exp\left(i \cdot i^+/\tau\right) + \sum_{N_i} \exp\left(i \cdot i^-/\tau\right)}, \tag{4}$$

where P_i and N_i denote pixel embedding collections of the positive and negative samples.

Pixel-to-Region Contrastive Learning. In order to carry out regional contrast learning, it is necessary to extract regional features. Firstly, we divide the image into different regions according to the ground truth of pixels and each region contains pixels with the same category label. Then, we carry out average pooling operations for each region and average the features of all embedded pixels in the region. In this way, we get a regional D-dimensional feature vector, which represents the semantic information of the region. For that, when calculating the anchored pixels, we compare the category label of the pixel with the regional features. If the anchored pixels belong to class c, the regional features with the same class c are considered positive samples, while the regional features with other class c are considered negative samples. It is computing in Eq. (4).

3.3 Training Objective

We optimize our framework by the standard Cross-Entropy loss L_{CE} reducing the predicted prior map from Transformer and its ground truth map; the Dice loss L_{DICE} minimizes the difference between the ground truth map and the predicted mask map and contrastive loss L_{NCE}. Formally, the segmentation loss function L_{Total} is defined as:

$$L_{Total} = L_{NCE} + L_{DICE} + L_{Prior}. \tag{5}$$

Specifically, the L_{Prior} is defined as:

$$L_{Prior} = (1 - K) \cdot L_{PriorD} + K \cdot L_{PriorC}, \tag{6}$$

$$\begin{aligned} L_{PriorD} &= \phi_{CE}(M_{GT}, \hat{M}_{PredD}), \\ L_{PriorC} &= \phi_{CE}(M_{GT}, \hat{M}_{PredC}), \end{aligned} \tag{7}$$

where L_{PriorC} denotes the prior point of OC, L_{PriorD} denotes the prior point of OD. $\hat{M}_{PredC}$ denotes the predicted OC prior map, $\hat{M}_{PredD}$ denotes the predicted OD prior map, K denotes the category of the prior map. ϕ_{CE} denotes the Cross-Entropy loss function.

4 Experiments

4.1 Datasets and Metrics

Datasets. To evaluate the effectiveness of the proposed method, we conducted experiments on the following public datasets: REFUGE Challenge [25], Drishti-GS [28] and RIM-ONE v3 [9].

REFUGE [25]: It consists of 1200 fundus images and corresponding expert annotations. It is divided into a training set (400 images), a validation set (400 images) and a testing set (400 images). We added the original validation set for the training.

Drishti-GS [28]: It contains 101 retinal fundus images of identities, each of which includes manual labels by four ophthalmologists with different clinical experiences. In our experiments, fundus images were randomly selected as the test dataset, leaving the other 50 images for training.

RIM-ONE v3 [9]: There are 159 images in the dataset, which are not divided into the training set and test set. In this paper, 99 images are randomly selected as the training set and 60 images as the testing set.

Evaluation Metrics. Our test sets were evaluated by the standard Dice coefficient as the evaluation metric. The overlap of the algorithm segmentation results and the ground truth labels was measured using the Dice score. For each image, we calculated the prediction result of the OD and OC Dice scores, respectively.

$$Dice = \frac{2|X \sqcap Y|}{|X| + |Y|}, \tag{8}$$

where X is the ground truth, Y is the prediction result.

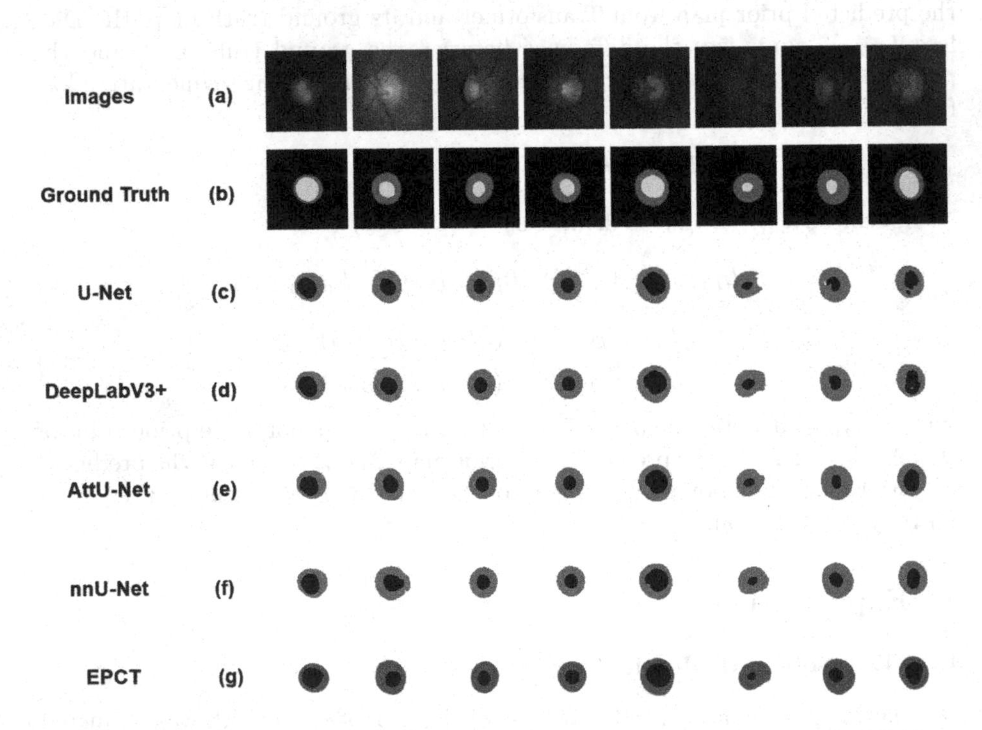

Fig. 3. Visualization of OD and OC segmentation results on RIM-ONE v3 [9]. From top to bottom: (a) RIM-ONE v3 images. (b) The ground truth. (c) The segmentation results of U-Net. (d) The segmentation results of Deeplabv3+. (e) The segmentation results of AttU-Net. (f) The segmentation results of nnU-Net. (g) The segmentation results of our EPCT.

4.2 Implementation Details

Each model was trained using PyTorch 1.10.0 on the NVIDIA GeForce RTX 3090 GPU. All images were resized to 512 × 512. Data augmentation included vertical flip, horizontal flip and random scale change. For all networks, the batch size was 4. The encoder of each network was initially pre-trained on ImageNet and 500 epochs of parameter fine-tuning were performed. The number of Transformer encoder layers was set to 4. In the comparative experiment, the learning rate was set to 0. 0001. The iterations of all models were 200 epochs.

Table 1. Comparisons of our approach compared with different state-of-the-art methods on the DRISHTI-GS dataset, RIM-ONE v3 dataset and REFUGE dataset.

Method	RIM-ONE v3		DRISHTI-GS		REFUGE	
	$Dice_{cup}$	$Dice_{disc}$	$Dice_{cup}$	$Dice_{disc}$	$Dice_{cup}$	$Dice_{disc}$
U-Net [27]	0.837	0.948	0.830	0.945	0.835	0.951
U-Net3+ [12]	0.843	0.955	0.833	0.952	0.837	0.959
DeepLabV3+ [7]	0.857	0.961	0.842	0.951	0.855	0.943
AttU-Net [24]	0.852	0.965	0.845	0.950	0.857	0.964
M-Net [8]	0.862	0.952	0.859	0.948	0.864	0.952
PreNet [6]	0.856	0.961	0.841	0.953	0.857	0.966
nnU-Net [13]	0.865	0.966	0.862	0.960	0.876	0.965
SERT [32]	0.877	0.965	0.880	0.954	0.878	0.955
TransU-Net [4]	0.874	0.954	0.883	0.944	0.877	0.964
BGA-Net [20]	0.872	0.967	0.898	0.956	✗	✗
NENet [26]	✗	✗	0.840	0.963	✗	✗
Eff-S Net [17]	✗	✗	0.912	0.980	0.887	0.959
Ours	**0.898**	**0.973**	**0.914**	**0.982**	**0.892**	**0.978**

4.3 Comparison with State-of-the-Arts

We conducted a comparative analysis to compare several other methods, as shown in Table 1. RIM-ONE v3 [9], REFUGE [25] and DRISHTI-GS [28] datasets were used for comparison experiments. The proposed method achieves state-of-the-art performance from the experiments and performs better in challenging cases compared with the considered methods. Figure 3 illustrates the qualitative comparison of segmentation results between the baseline and our method on the RIM-ONE v3 [9]. The black parts represent the OC and the gray parts represent the OD. By comparing the ground truth and our visualization results in the figures, the segmentation results show different models have differences, particularly in the edge of the OD and OC. It can be seen that the edge part of the OC and OD partition is more complete.

4.4 Ablation Study

We performed an ablation analysis to demonstrate the contribution of each module (EPTM and Edge Supervised Contrastive Learning). Table 2 shows that each module of the proposed contributes to increased performance, with the addition of our Edge Supervised Contrastive Learning (WSCL), including the anchor sampling strategy, pixel-to-pixel (PTP) and pixel-to-region (PTR). Figure 4 and Fig. 5 show the visual examples of the segmentation results, where the first row is the fundus images, the middle row is the ground truth and the last row is our results. We visualized on REFUGE [25] and DRISHTI-GS [28] datasets.

Table 2. Ablation study of our proposed method on the DRISHTI-GS dataset [28] and RIM-ONE v3 [9].

Method	RIM-ONE v3		DRISHTI-GS	
	Dice_{cup}	$\text{Dice}_{\text{disc}}$	Dice_{cup}	$\text{Dice}_{\text{disc}}$
Baseline	0.870	0.963	0.878	0.965
Baseline+EPTM	0.882	0.967	0.890	0.972
Baseline+PTP	0.878	0.965	0.888	0.968
Baseline+PTR	0.870	0.968	0.894	0.970
Baseline+WSCL (PTP+PTR)	0.888	0.971	0.903	0.976
Ours	**0.898**	**0.974**	**0.914**	**0.982**

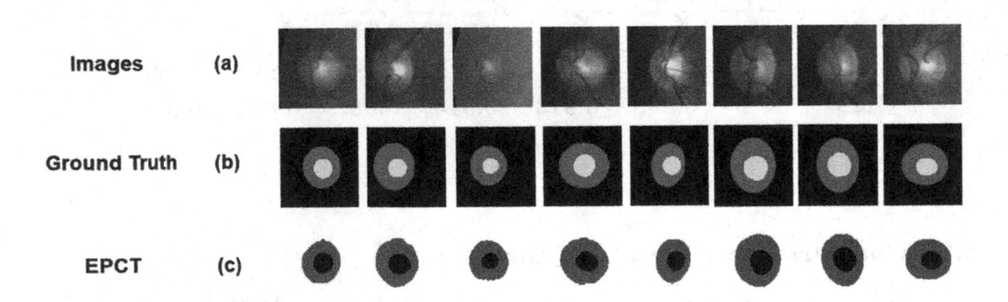

Fig. 4. Visualization of OD and OC segmentation results on REFUGE [25].

Our method has accurately partitioned the OC and OD from the low-contrast area and the processing of the edge part has been smoother and more accurate. From the analysis of the results, the performance of PTR and PTP is the best when they work together and PTR further explores the relations on edge and complements PTP. Moreover, the introduction of the EPTM also enhances the baseline performance with a better understanding of the semantic relationships and contextual information between edges and surrounding areas.

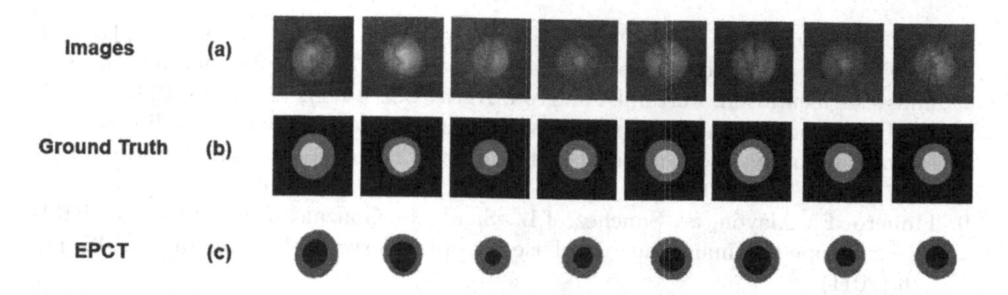

Fig. 5. Visualization of OD and OC segmentation results on DRISHTI-GS [28].

5 Conclusion

In this paper, we have proposed a new framework for OC and OD segmentation. Our EPCT captures target structural relationships and improves image comprehension and perception of edge features, effectively addressing issues of excessive or insufficient segmentation resulting from indistinct edges. The experimental evidence from a publicly available dataset shows that the introduced method is effective, working better than existing methods. In future works, we will focus on self-supervised architecture by including fewer labels.

Acknowledgment. This work was supported in part by the National Science Foundation of China under Grants 62076142 and 62241603, in part by the National Key Research and Development Program of Ningxia under Grant 2023AAC05009, 2022BEG03158 and 2021BEB0406.

References

1. Almazroa, A., Burman, R., Raahemifar, K., Lakshminarayanan, V.: Optic disc and optic cup segmentation methodologies for glaucoma image detection: a survey. J. Ophthalmol. **2015**, 180972 (2015)
2. Aquino, A., Gegúndez-Arias, M.E., Marín, D.: Detecting the optic disc boundary in digital fundus images using morphological, edge detection, and feature extraction techniques. TMI **29**(11), 1860–1869 (2010)
3. Cao, H., et al.: Swin-Unet: unet-like pure transformer for medical image segmentation. In: Karlinsky, L., Michaeli, T., Nishino, K. (eds.) Computer Vision, ECCV 2022 Workshops. LNCS, vol. 13803, pp. 205–218. Springer, Cham (2023). https://doi.org/10.1007/978-3-031-25066-8_9
4. Chen, J., et al.: TransUNet: transformers make strong encoders for medical image segmentation. arXiv preprint arXiv:2102.04306 (2021)
5. Chen, T., Kornblith, S., Norouzi, M., Hinton, G.: A simple framework for contrastive learning of visual representations. In: ICML, pp. 1597–1607 (2020)
6. Fan, D.-P., Ji, G.-P., Zhou, T., Chen, G., Fu, H., Shen, J., Shao, L.: PraNet: parallel reverse attention network for polyp segmentation. In: Martel, A.L., et al. (eds.) MICCAI 2020. LNCS, vol. 12266, pp. 263–273. Springer, Cham (2020). https://doi.org/10.1007/978-3-030-59725-2_26

7. Firdaus-Nawi, M., Noraini, O., Sabri, M., Siti-Zahrah, A., Zamri-Saad, M., Latifah, H.: DeepLabv3+ _encoder-decoder with atrous separable convolution for semantic image segmentation. Pertanika J. Trop. Agric. Sci. **34**(1), 137–143 (2011)
8. Fu, H., Cheng, J., Xu, Y., Wong, D.W.K., Liu, J., Cao, X.: Joint optic disc and cup segmentation based on multi-label deep network and polar transformation. TMI **37**(7), 1597–1605 (2018)
9. Fumero, F., Alayón, S., Sanchez, J.L., Sigut, J., Gonzalez-Hernandez, M.: RIM-ONE: an open retinal image database for optic nerve evaluation. In: CBMS, pp. 1–6 (2011)
10. Gu, Z.: CE-Net: context encoder network for 2D medical image segmentation. TMI **38**(10), 2281–2292 (2019)
11. He, K., Fan, H., Wu, Y., Xie, S., Girshick, R.: Momentum contrast for unsupervised visual representation learning. In: CVPR, pp. 9729–9738 (2020)
12. Huang, H., et al.: UNet 3+: a full-scale connected Unet for medical image segmentation. In: ICASSP, pp. 1055–1059 (2020)
13. Isensee, F., Jaeger, P.F., Kohl, S.A., Petersen, J., Maier-Hein, K.H.: nnU-Net: a self-configuring method for deep learning-based biomedical image segmentation. Nat. Meth. **18**(2), 203–211 (2021)
14. Jiang, Y., et al.: JointRCNN: a region-based convolutional neural network for optic disc and cup segmentation. TBE **67**(2), 335–343 (2019)
15. Joshi, G.D., Sivaswamy, J., Krishnadas, S.: Optic disk and cup segmentation from monocular color retinal images for glaucoma assessment. TMI **30**(6), 1192–1205 (2011)
16. Li, S., Sui, X., Luo, X., Xu, X., Liu, Y., Goh, R.: Medical image segmentation using squeeze-and-expansion transformers. arXiv preprint arXiv:2105.09511 (2021)
17. Liu, B., Pan, D., Shuai, Z., Song, H.: ECSD-Net: a joint optic disc and cup segmentation and glaucoma classification network based on unsupervised domain adaptation. Comput. Meth. Prog. Bio. **213**, 106530 (2022)
18. Long, J., Shelhamer, E., Darrell, T.: Fully convolutional networks for semantic segmentation. In: CVPR, pp. 3431–3440 (2015)
19. Lu, S.: Accurate and efficient optic disc detection and segmentation by a circular transformation. TMI **30**(12), 2126–2133 (2011)
20. Luo, L., Xue, D., Pan, F., Feng, X.: Joint optic disc and optic cup segmentation based on boundary prior and adversarial learning. IJARS **16**(6), 905–914 (2021)
21. Luthra, A., Sulakhe, H., Mittal, T., Iyer, A., Yadav, S.: Eformer: edge enhancement based transformer for medical image denoising. arXiv arXiv:2109.08044 (2021)
22. Misra, I., Maaten, L.: Self-supervised learning of pretext-invariant representations. In: CVPR, pp. 6707–6717 (2020)
23. Mittapalli, P.S., Kande, G.B.: Segmentation of optic disk and optic cup from digital fundus images for the assessment of glaucoma. Biomed. Sig. Process. **24**, 34–46 (2016)
24. Oktay, O., et al.: Attention U-Net: learning where to look for the pancreas. arXiv preprint arXiv:1804.03999 (2018)
25. Orlando, J.I., Fu, H., Breda, J.B., Van Keer, K., Bathula, D.R., et al.: Refuge challenge: a unified framework for evaluating automated methods for glaucoma assessment from fundus photographs. Media **59**, 101570 (2020)
26. Pachade, S., Porwal, P., Kokare, M., Giancardo, L., Mériaudeau, F.: NENet: Nested EfficientNe and adversarial learning for joint optic disc and cup segmentation. Media **74**, 102253 (2021)

27. Ronneberger, O., Fischer, P., Brox, T.: U-Net: convolutional networks for biomedical image segmentation. In: Navab, N., Hornegger, J., Wells, W.M., Frangi, A.F. (eds.) MICCAI 2015. LNCS, vol. 9351, pp. 234–241. Springer, Cham (2015). https://doi.org/10.1007/978-3-319-24574-4_28
28. Sivaswamy, J., Krishnadas, S., Joshi, G.D., Jain, M., Tabish, A.U.S.: Drishti-GS: retinal image dataset for optic nerve head (ONH) segmentation. In: ISBI, pp. 53–56 (2014)
29. Wang, W., Zhou, T., Yu, F., Dai, J., Konukoglu, E., Van Gool, L.: Exploring cross-image pixel contrast for semantic segmentation. In: ICCV, pp. 7303–7313 (2021)
30. Wang, X., Zhang, R., Shen, C., Kong, T., Li, L.: Dense contrastive learning for self-supervised visual pre-training. In: CVPR, pp. 3024–3033 (2021)
31. Zhao, X., et al.: Contrastive learning for label efficient semantic segmentation. In: ICCV, pp. 10623–10633 (2021)
32. Zheng, S., et al.: Rethinking semantic segmentation from a sequence-to-sequence perspective with transformers. In: CVPR, pp. 6881–6890 (2021)

BGBF-Net: Boundary-Guided Buffer Feedback Network for Liver Tumor Segmentation

Ying Wang[1], Kanqi Wang[1], Xiaowei Lu[1], Yang Zhao[2](✉), and Gang Liu[3](✉)

[1] Institute of Artificial Intelligence, Xiamen University, Xiamen 361102, China
[2] Pen-Tung Sah Institute of Micro-Nano Science and Technology, Xiamen University, Xiamen 361102, China
zhaoy@xmu.edu.cn
[3] State Key Laboratory of Molecular Vaccinology and Molecular Diagnostics Center for Molecular Imaging and Translational Medicine, School of Public Health, Xiamen University, Xiamen 361102, China
gangliu.cmitm@xmu.edu.cn

Abstract. Medical images such as CT can provide important reference value for doctors to diagnose diseases. Identifying and segmenting lesions from medical images is crucial for its diagnosis and treatment. However, unlike other segmentation tasks, medical image has the characteristics of blurred boundaries and variable lesions sizes, which poses challenges to medical image segmentation. In this paper, we propose a Boundary-Guided Buffer Feedback Network(BGBF-Net), using the boundary guidance module to combine the low-level feature map rich in boundary information and the high-level semantic segmentation feature map generated by the encoder module, and output the features that focus on the boundary, which is used to enhance the attention of the decoder to boundary features. The buffer feedback module is used to strengthen the network's supervision of the decoder while speeding up convergence of the model. We apply the proposed BGBF-Net on the LiTS dataset. Comprehensive results show that the BGBF-Net improves by 2.36% compared to other methods in terms of Dice.

Keywords: Liver tumor segmentation · Boundary guidance · Cross attention

1 Introduction

In recent years, the incidence of liver cancer has gradually increased. According to the World Health Organization, liver cancer ranks second in cancer-related deaths after lung cancer. Medical imaging technology is usually used to diagnose liver cancer. Among them, CT is used more commonly in medical imaging. Correctly interpreting medical images and accurately segmenting the liver and liver lesions are critical to the diagnosis of liver diseases. In the diagnosis of

Financial Supported by Fujian Science and Technology Project (No. 2022I0003).

Q. Liu et al. (Eds.): PRCV 2023, LNCS 14429, pp. 456–467, 2024.
https://doi.org/10.1007/978-981-99-8469-5_36

liver tumors, professional doctors usually need to manually mark the location of organs and lesions in CT and other medical image to assist diagnosis. Manual labeling not only relies on the professional level and experience of doctors, but also wastes precious time and energy of doctors. Therefore, it is very necessary to use computer technology to assist medical image segmentation, which will effectively improve the efficiency and accuracy of image segmentation, and help the treatment and diagnosis of diseases.

Medical image segmentation algorithms mainly include traditional machine learning algorithms and deep learning algorithms. Traditional segmentation algorithms [1,12] usually perform segmentation based on the characteristics of medical image. Although these algorithms can achieve the segmentation of target objects to a certain extent, they are all semi-automatic methods, which require researchers to manually adjust parameters and have poor generalization. Because the method based on deep learning has the advantages of full automation and strong generalization, it is widely used in image segmentation tasks, such as fully convolutional network FCN [14], U-Net [10,17,21], DeepLabv3+ [4], TransUNet [3] etc. [5,7,8,11,13,18]

The above algorithms have achieved good results in the field of segmentation, but medical images have their own particularity. First, the tumors are multiple and their locations vary widely. Secondly, the medical image background is relatively complex, and the border of the tumor is blurred, and its gray value is similar to that of the surrounding tissue. Third, unlike organ segmentation with a larger area, the tumor area is much smaller than the image size, which tends to cause models to ignore the tumor. Therefore, liver tumor segmentation is considered to be a more challenging task (Fig. 1).

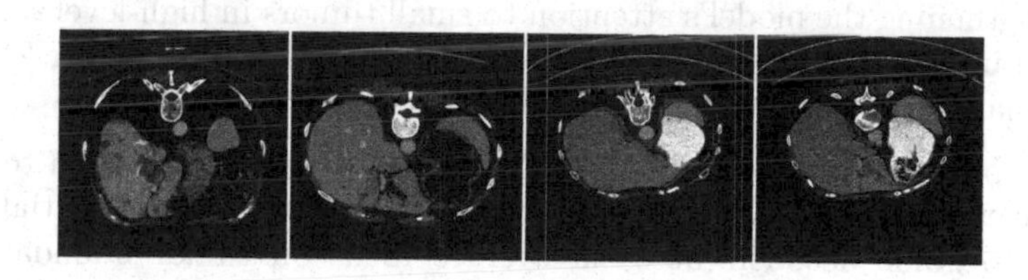

Fig. 1. The figure above shows multiple, blurred, and small, irregular tumors.

In order to solve the problem of multiple tumors and large differences in location, cascaded networks [6,20] are widely used. The cascaded network is a multi-stage segmentation method from coarse to fine, which locates liver tumors step by step and improves segmentation accuracy. However, such methods are often complex. Although the location of the tumor varies widely, it is always within the area of the liver. Therefore, we propose a Full Scale Feature Extraction Module(FSFEM). This module consists of a dilated convolution block and a transformer block, which are used to extract multi-level features and global correlation information of images respectively. By capturing the relative positional relationship between the tumor and the liver, FSFEM can promote the model to better locate the tumor, thereby improving the segmentation accuracy. And compared to the multi-stage method, our method is more streamlined.

Usually, deep learning-based segmentation methods [9,17] use downsampling to extract deep semantic information. However, the downsampling operation results in the loss of spatial and positional details of the image. Although skip connections can pass the encoder feature map to the decoder network at the same layer, enriching the spatial information of the segmented output map, the edge details will be inevitably lost. To solve this problem, a large number of methods have been proposed. Dense skip connections [21] and full scale skip connections [10] pass high-level semantics and low-level semantics of different scales to the decoder, but this will lead to excessive network parameters and a reduction in computational efficiency. Some networks [15,19] only choose to transfer the low-level feature map of the encoder to each scale of the decoder, and the high-level semantic feature map generated by the encoder is not effectively used. Therefore, we propose a Boundary Guidance Module(BGM), which fuses the boundary features of the bottom layer into the feature map rich in high-level semantic information at the top level through the cross attention mechanism, so as to promote the model to pay better attention to the boundary feature and improve the segmentation accuracy.

Furthermore, the size of tumors varies greatly from patient to patient. Since small tumors account for a small proportion of the entire image, small tumors are likely to be ignored by the model during upsampling. Some methods use different local attention mechanisms [16] to solve this problem, and it will lead to an increase in computational complexity. Therefore, we propose a Buffer Feedback Module(BUF), by outputting image features at different scales and calculating the loss, thereby shortening the loss feedback path length of each layer of decoder and strengthening the model's attention to small tumors in high-level semantics. And then improves the model's ability to segment small tumors (Fig. 2).

Our main contributions can be highlighted as follows:

1. We propose a Boundary Guidance Module(BGM) that uses cross attention to effectively fuse high-level semantic feature maps and low-level spatial feature maps to guide decoders at each layer to focus on tumor boundary while focusing on central part of the tumor.
2. We propose a Full Scale Feature Extractor Module(FSFEM) that combines global and local information of image. The former part of FSFEM uses the 4th ResNet-50 block to extract the local relationship features of the image. The latter part uses a Dilated Block to increase the receptive field of the network to obtain multi-scale information, and uses a Transformer Block to obtain the global features of the image, so as to obtain the full-scale features of the image efficiently.
3. We propose a Buffer Feedback Module(BUF), which shortens the loss feedback path length of each layer decoder, and provides it with an additional gradual and gentle boundary information feedback path, which is used to improve the segmentation accuracy of the network to the boundary and speed up the convergence of the network.

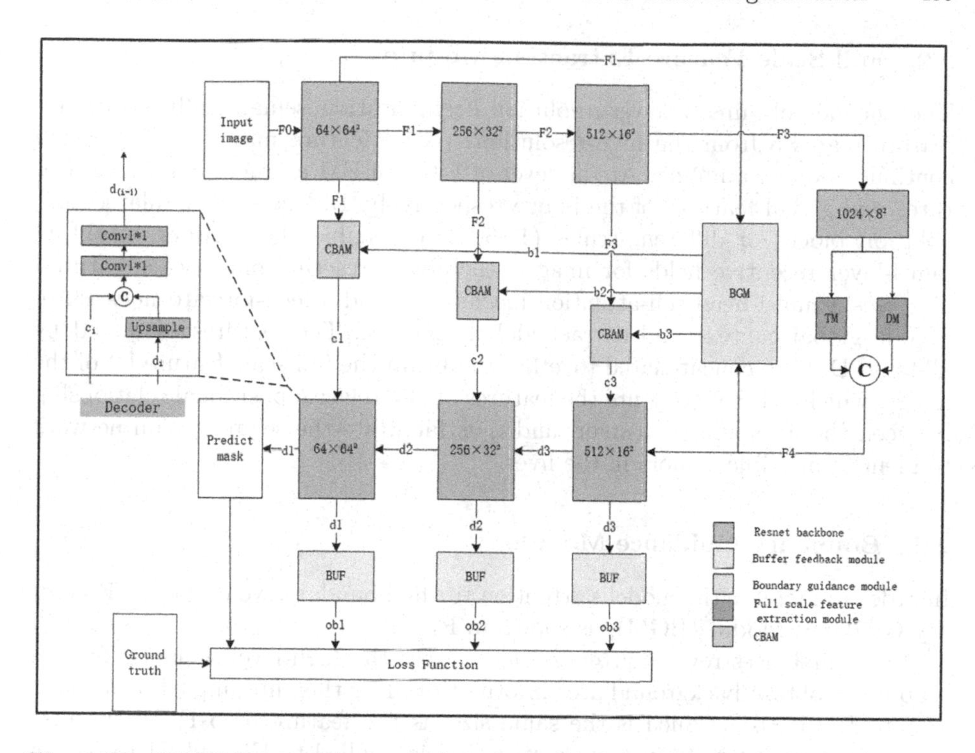

Fig. 2. The BGBF-Net model proposed in this paper adopts the encoder-decoder structure. FEFSM contains the 4th Resnet-50 block and Transformer Module(TM) and Dilated Module(DM) to obtain full scale feature from the high-level semantic feature map F4. BGM extracts boundary information from F1 and F4 and uses it to guide the decoder to segment tumor boundary. BUF takes the output of the decoder and feeds it back the direction to optimize the network.

2 Methodology

2.1 Overall Architecture

The encoder uses the first three feature extraction blocks of ResNet-50 [9] as the backbone, and then uses FSFEM to further extract the full scale features of the image. The decoder upsamples the high-level semantic feature maps extracted by the encoder to high-resolution segmentation maps. BGM takes the low-level feature map(F1) and the high-level feature map(F4) as input, and outputs three boundary feature maps of different sizes(b1, b2, b3). Then c1, c2, c3 obtained by multiplying b1, b2, b3 with the encoder feature maps F1, F2, F3 are passed to the decoder through a skip connection using CBAM [18]. It is used to guide the decoder to pay attention to the edge area of the tumor. BUF accepts the output of the decoder and outputs prediction of the tumor. The feedback information from the loss function will pass through a buffer layer for smooth correction and optimization of the BGBF-Net.

2.2 Full Scale Feature Extraction Module

The encoder obtains a lower-resolution segmentation semantically significant feature map F3 from the high-resolution spatial feature map F1, F2 through continuous downsampling. At the level of F3, FSFEM extracts multi-level features and global features of the image respectively. DM uses three dilated convolution blocks of different scales (1, 3, 5) to enable the network to obtain multi-layer receptive fields for images, thereby extracting multi-scale features. TM uses a multi-head self-attention mechanism and a feed-forward network to encode global context and extract global features. The features extracted by TM and DM are concatenated together to obtain the full scale feature F4 of the image, which will help capture the features of the relative positional relationship between the liver and the tumor, and thus facilitate the segmentation network to identify multiple tumors in the liver.

2.3 Boundary Guidance Module

In order to increase the model's attention to the boundary, we propose a Boundary Guidance Module(BGM), as shown in Fig. 3.

BGM first uses reverse attention to process the high-level semantic feature map F4 to obtain background area information. It is then upsampled by a factor of 8 to U5(64×64), which is the same size as the feature map F1 in the first layer of the encoder. Then a cross attention is applied to U5 and F1 to obtain boundary features. U5 with background high-level semantic information is used as query, F1 with spatial information is used as key and value, and cross attention is used through Formula 4 to increase the attention on the boundary features in F1. Afterwards, the obtained features are passed to the decoders through downsampling($\frac{1}{2}$, $\frac{1}{4}$), and multiplied with the feature maps(F1,F2,F3), so that the feature maps of this layer pay more attention to the boundary. In order to reduce computing costs and speed up network training, BGM discards the feature map of the middle layer(F2,F3) of the encoder that is similar to the high-level and low-level features, and only the low-level feature map F1 containing rich boundary information and the feature map F4 containing rich segmentation information are adopted as the input of the boundary guidance module.

$$U5 = upsample_8(Conv((1 - Conv_{ra}(F4))) \tag{1}$$

$$Q = Conv_q(Conv(U5)) \tag{2}$$

$$K, V = Conv_{k,v}(Conv(F1)) \tag{3}$$

$$b_{(i=1,2,3)} = Conv_{(i=1,2,3)}(V \odot Sigmoid(Q \odot K)) \tag{4}$$

In Formula 1,2,3, $Conv_{ra,q,k,v}$ represents the convolution operation at different module positions respectively. In Formula 4, i represents three stages of both the decoder and BUF.

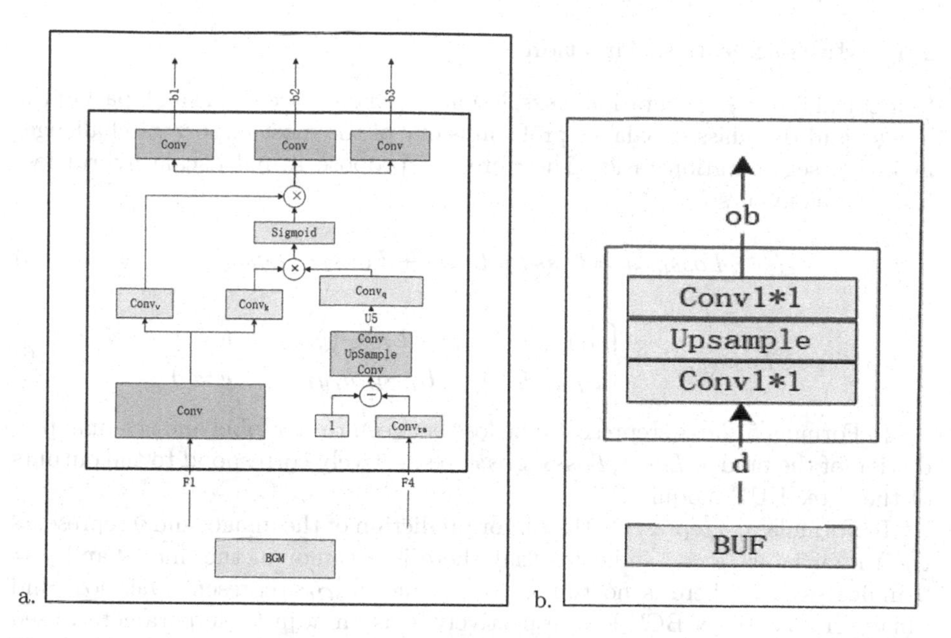

Fig. 3. (a): The BGM first uses reverse attention to obtain attention to the boundary from the high-level semantic feature F4, and then uses cross attention to extracted boundary information from F1 and U5. (b): The BUF contains a buffer convolution layer and a feedback block after upsampling.

2.4 Buffer Feedback Module

BUF contains a buffer layer and a feedback layer. The buffer layer consists of a 1×1 convolutional layer to maintain the stability of the decoder parameters. Compared with directly upsampling for multi-scale supervision, the buffer layer smoothes the influence of the feedback layer on the decoder, which does not make the network fluctuate, and at the same time facilitates the smooth convergence of the network while increasing the number of feedback paths. The feedback layer consists of an upsampling layer and a 1×1 convolutional layer used to adjust the number of channels to output the final segmentation map, using tumor label for supervision and obtaining loss feedback.

BUF provides a new loss feedback path for each layer of decoders, enriches the learning path of the decoder, and reduces the influence of incomplete and biased feedback information from the single supervision path of the last layer of decoder. In addition, BUF also shortens the length of the loss feedback path of the underlying decoder, which enables the decoders of each layer to directly receive information about small tumors from the ground truth, thereby improving the network's ability to segment small tumors.

2.5 Multi-scale Loss Function

In medical image segmentation tasks, lesions often only occupy a small part of the image, and this class imbalance problem is one of the most important challenges in image segmentation tasks. Therefore, we propose a multi-scale hybrid loss function as follows:

$$Loss_{total} = Loss_0 + Loss_1 + Loss_2 + Loss_3 \tag{5}$$

$$Loss(\hat{y}, y) = \begin{cases} \alpha(L_{Dice}(\hat{y}, y)) + L_{BCE}(\hat{y}, y)), & y = 1 \\ L_{Dice}(\hat{y}, y)) + L_{BCE}(\hat{y}, y), & y = 0 \end{cases} \tag{6}$$

In Formula 5, $Loss_0$ represents the loss between tumor label and the final prediction of the model. $Loss_1, Loss_2, Loss_3$ respectively correspond to the outputs of the three BUF modules.

In Formula 6, $\hat{y}$ represents the tumor prediction of the image, and y represents the tumor label. $y = 1$ indicates that there is a tumor in the image, and $y = 0$ indicates that there is no tumor. L_{Dice} and L_{BCE} represent Dice loss and binary cross-entropy BCE loss respectively. α is an adjustable parameter, used to strengthen the model's attention to images with tumors, α is set to 1.1 in this paper.

L_{BCE} and L_{Dice} are as follows: N is the total number of pixels in the image, y and $\hat{y}$ represent the tumor label and the model's prediction of the tumor respectively. ε is a small constant that prevents the denominator from being 0.

$$L_{BCE} = -\frac{1}{N}\sum_{i=1}^{N}(y_i \cdot log(\hat{y}_i) + (1 - y_i) \cdot log(1 - \hat{y}_i)) \tag{7}$$

$$L_{Dice} = 1 - \frac{2 * (\sum_{i=1}^{N} y_i \cap \sum_{i=1}^{N} \hat{y}_i)}{\sum_{i=1}^{N} y_i + \sum_{i=1}^{N} \hat{y}_i + \varepsilon} \tag{8}$$

3 Experiments

3.1 Data Preprocess

The dataset used in this paper is the liver tumor segmentation challenge dataset LiTS, which comes from the 2017 IEEE International Symposium on Biomedical Imaging (ISBI) and the 2017 International Conference on Medical Image Computing and Computer-Aided Intervention (MICCAI). The dataset comes from multiple medical centers and contains a total of 201 clinical samples with various liver diseases, 131 of which contain manual annotations of the liver and tumor regions by radiologists. And the original CT data are saved as files in NIFTI format. The 131 cases of data containing manual annotations were divided into a training set of 111 cases and a test set of 20 cases. Each case has hundreds of slices, but the slices containing the liver and tumor are relatively few, so the slices containing the liver in each case are saved in a two-dimensional image format, and at the same time, for the cases with less than 48 liver slices, the data were evenly collected before and after the liver slices to supplement the number of 48 slices. In the end, a total of 16940 training sets and 2218 test sets were collected. The LiTS dataset contains two types of labels, liver and tumor. Because the goal of the task is to directly segment the tumor, the liver label is set as the background, and the tumor is the only label.

3.2 Implementation Details

The experimental model building environment is Python3.9.12, Pytorch1.12.1 framework, using NVIDIA 3060 12G GPU. The initial learning rate is 0.0001, the batch size is 16, and the image size is (128, 128).

Table 1. Results of different methods on LiTS

Methods	Dice	Jaccard	Precision	Recall
UNet++	0.7428	0.7181	0.9123	0.7928
Attention U-Net	0.6958	0.6873	0.9879	0.6959
TransUNet	0.7495	0.7265	0.9580	0.7576
DeepLabv3+	0.7248	0.7045	0.9451	0.7444
Swin-Unet	0.7145	0.6995	0.9688	0.7231
Proposed	0.7731	0.7489	0.9521	0.7876

3.3 Experiments and Results

Segmentation Experiments. In the experiment, we compared the proposed BGBF-Net method with five different methods, including Attention U-Net [16], TransUNet [3], DeepLabv3+ [4], UNet++ [21] and Swin-UNet [2]. We analyzed and compared the 4 indicators on the LiTS dataset, For fair comparison, these methods use the same parameters for training and testing. It can be seen from Table 1 that the BGBF-Net proposed by us has achieved the best results in Dice, Jaccard, and Precision. Among them, the accuracy of Dice and Jaccard are 2.36% and 2.24% higher than that of TransUNet, which is the algorithm with the highest accuracy among the comparison algorithms.

It can be seen from case 1 in Fig. 4 that our proposed method BGBF-Net can segment most of the tumors more completely in the segmentation of multiple tumors, false positives and false negatives have occurred in tumor segmentation by other methods. In case 2, BGBF-Net correctly identified the blurred border region of the tumor which shows the effectiveness of our boundary processing modules BGM and BUF. For the small tumor slices in case 3, the Swin-Unet and the Attention U-Net lost a large number of edge parts of small tumors, and our

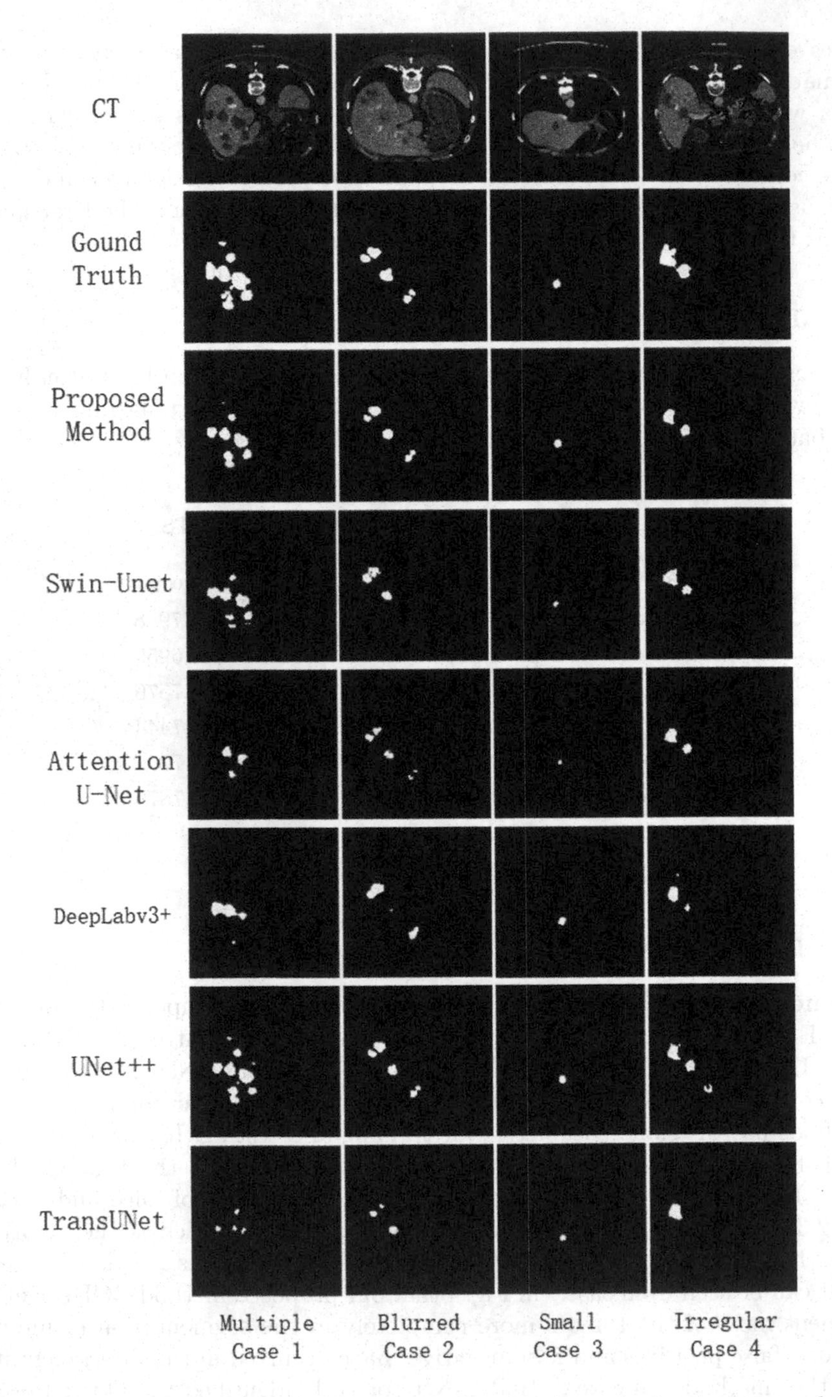

Fig. 4. The first row is the input CT image, the second row is the liver tumor label, The rows from 3 to 8 represent for different methods. Each column is a slice sample. respectively shows the multiple tumors, tumor with blurred borders, small tumors, and tumors with irregular edges.

BGBF-Net also made more accurate segmentation for small tumors, which shows that BGBF-Net can better solve the problem that the gray level of the tumor and the surrounding tissue are similar. BGBF-Net, Swin-Unet, and Attention U-Net have similar segmentation accuracy for the tumor with irregular edges in case 4, but the latter two methods have obvious shortcomings in the segmentation of the first three cases. Generally speaking, our method better solves the three problems of multiple tumors, blurred boundary, and small size, and it can be said to be a relatively powerful segmentation algorithm.

Ablation Experiments. Table 2 shows the results of the ablation experiments, in which we compare the segmentation results of BGBF-Net and BGBF-Net without the BGM, FSFEM, BUF, CBAM on LiTS, respectively. The experimental results show that BGBF-Net is 1.93%, 0.99%, 2%, and 0.14% higher than the above four networks in Dice's evaluation index. In the ablation experiment, BUF and FSFEM were the two modules with the greatest improvement, which shows that extracting full scale features and shortening the decoder feedback path help the model to segment tumors. BGM has been improved by 1%, which is not as much as BUF and FSFEM. This is because the boundary accounts for fewer pixels in the overall tumor, and the evaluation indicators such as Dice are positively correlated with the number of pixels, so compared with the two modules of BUF and FSFEM, BGM is a little lower in the evaluation index of quantitative analysis. We can see from the ablation experiment without CBAM that when with BGM, BUF, and FSFEM, the Dice value is very close to the method we proposed, and the gap is only 0.14%, so it can be concluded that the above three modules have a great contribution to the improvement of experimental segmentation accuracy.

Table 2. Ablation study of the proposed BGBF-Net on LiTS

Methods	Dice	Jaccard	Precision	Recall
Proposed wo BUF	0.7538	0.7315	0.9623	0.7606
Proposed wo BGM	0.7632	0.7398	0.9611	0.7655
Proposed wo FSFEM	0.7531	0.7296	0.9325	0.7789
Proposed wo CBAM	0.7717	0.7480	0.9599	0.7772
Proposed	0.7731	0.7489	0.9521	0.7876

4 Conclusions

In this paper, a new liver tumor segmentation framework BGBF-Net is proposed to solve the three challenges of blurred tumor borders, multiple tumors and small tumors in the field of liver tumor segmentation. To address the above challenges, we propose a Full Scale Feature Extraction Module(FSFEM), a Boundary Guidance Module(BGM), and a Buffer Feedback Module(BUF). FSFEM combines the advantages of CNN and Transformer to extract local features, multi-level features and global features in the image. Using the full scale features, the model can effectively capture the relative positional relationship between the tumor and the liver, thereby improving the ability of the model to localize the tumor. BGM uses cross attention to enhance the decoder's attention to the boundary contour information, which helps to improve the accuracy of boundary segmentation. BUF is a simple and effective way to improve the accuracy of tumor segmentation, by adding a path to smoothly adjust the decoder network, it can speed up the network convergence while improving the segmentation accuracy. Ablation experiments confirm the effectiveness of the above three modules. Experiments on the LiTS dataset show that the accuracy of the model is 2.36% higher than other methods. We can conclude that the BGBF-Net can accurately segment liver tumors and is a relatively powerful liver tumor segmentation method. Afterwards, we can try 3D segmentation methods to capture the relationship features between CT slices to guide clinical liver tumor segmentation.

References

1. Adams, R., Bischof, L.: Seeded region growing. IEEE Trans. Pattern Anal. Mach. Intell. **16**(6), 641–647 (1994)
2. Cao, H., et al.: Swin-Unet: Unet-like pure transformer for medical image segmentation. In: Karlinsky, L., Michaeli, T., Nishino, K. (eds.) Computer Vision – ECCV 2022 Workshops. ECCV 2022. LNCS, vol. 13803. Springer, Cham (2022). https://doi.org/10.1007/978-3-031-25066-8_9
3. Chen, J., et al.: TransUNet: transformers make strong encoders for medical image segmentation. arXiv preprint arXiv:2102.04306 (2021)
4. Chen, L.-C., Zhu, Y., Papandreou, G., Schroff, F., Adam, H.: Encoder-decoder with atrous separable convolution for semantic image segmentation. In: Ferrari, V., Hebert, M., Sminchisescu, C., Weiss, Y. (eds.) ECCV 2018. LNCS, vol. 11211, pp. 833–851. Springer, Cham (2018). https://doi.org/10.1007/978-3-030-01234-2_49
5. Chlebus, G., Meine, H., Moltz, J.H., Schenk, A.: Neural network-based automatic liver tumor segmentation with random forest-based candidate filtering. arXiv preprint arXiv:1706.00842 (2017)
6. Gruber, N., Antholzer, S., Jaschke, W., Kremser, C., Haltmeier, M.: A joint deep learning approach for automated liver and tumor segmentation. In: 2019 13th International conference on Sampling Theory and Applications (SampTA), pp. 1–5. IEEE (2019)
7. Gu, Z., et al.: CE-Net: context encoder network for 2D medical image segmentation. IEEE Trans. Med. Imaging **38**(10), 2281–2292 (2019)
8. Han, X.: MR-based synthetic CT generation using a deep convolutional neural network method. Med. Phys. **44**(4), 1408–1419 (2017)
9. He, K., Zhang, X., Ren, S., Sun, J.: Deep residual learning for image recognition. In: Proceedings of the IEEE Conference on Computer Vision and Pattern Recognition, pp. 770–778 (2016)
10. Huang, H., et al.: UNet 3+: a full-scale connected UNet for medical image segmentation. In: ICASSP 2020-2020 IEEE International Conference on Acoustics, Speech and Signal Processing (ICASSP), pp. 1055–1059. IEEE (2020)

11. Isensee, F., Jaeger, P.F., Kohl, S.A., Petersen, J., Maier-Hein, K.H.: nnU-Net: a self-configuring method for deep learning-based biomedical image segmentation. Nat. Methods **18**(2), 203–211 (2021)
12. Kass, M., Witkin, A., Terzopoulos, D.: Snakes: active contour models. Int. J. Comput. Vision **1**(4), 321–331 (1988)
13. Li, X., Chen, H., Qi, X., Dou, Q., Fu, C.W., Heng, P.A.: H-Denseunet: hybrid densely connected UNet for liver and tumor segmentation from CT volumes. IEEE Trans. Med. Imaging **37**(12), 2663–2674 (2018)
14. Long, J., Shelhamer, E., Darrell, T.: Fully convolutional networks for semantic segmentation. In: Proceedings of the IEEE Conference on Computer Vision and Pattern Recognition, pp. 3431–3440 (2015)
15. Ma, H., Xu, C., Nie, C., Han, J., Li, Y., Liu, C.: DBE-Net: dual boundary-guided attention exploration network for polyp segmentation. Diagnostics **13**(5), 896 (2023)
16. Oktay, O., et al.: Attention U-Net: learning where to look for the pancreas. arXiv preprint arXiv:1804.03999 (2018)
17. Ronneberger, O., Fischer, P., Brox, T.: U-Net: convolutional networks for biomedical image segmentation. In: Navab, N., Hornegger, J., Wells, W.M., Frangi, A.F. (eds.) MICCAI 2015. LNCS, vol. 9351, pp. 234–241. Springer, Cham (2015). https://doi.org/10.1007/978-3-319-24574-4_28
18. Woo, S., Park, J., Lee, J.-Y., Kweon, I.S.: CBAM: convolutional block attention module. In: Ferrari, V., Hebert, M., Sminchisescu, C., Weiss, Y. (eds.) ECCV 2018. LNCS, vol. 11211, pp. 3–19. Springer, Cham (2018). https://doi.org/10.1007/978-3-030-01234-2_1
19. Yang, H., et al.: Is-Net: automatic ischemic stroke lesion segmentation on CT images. IEEE Trans. Radiat. Plasma Med. Sci. **7**(5), 483–493 (2023)
20. Zhang, J., Saha, A., Zhu, Z., Mazurowski, M.A.: Hierarchical convolutional neural networks for segmentation of breast tumors in MRI with application to Radiogenomics. IEEE Trans. Med. Imaging **38**(2), 435–447 (2018)
21. Zhou, Z., Rahman Siddiquee, M.M., Tajbakhsh, N., Liang, J.: UNet++: a nested U-net architecture for medical image segmentation. In: Stoyanov, D., et al. (eds.) DLMIA/ML-CDS -2018. LNCS, vol. 11045, pp. 3–11. Springer, Cham (2018). https://doi.org/10.1007/978-3-030-00889-5_1

MixU-Net: Hybrid CNN-MLP Networks for Urinary Collecting System Segmentation

Zhiyuan Liu[1], Mingxian Yang[1], Hao Qi[1], Ming Wu[1], Kaiyun Zhang[1], Song Zheng[2], Jianhui Chen[2], Yinran Chen[1(✉)], and Xiongbiao Luo[1]

[1] Department of Computer Science and Technology, Xiamen University, Xiamen, Fujian, China
{liuzy,yangmingxian,qihao,wuming,zhangkaiyun}@stu.xmu.edu.cn, yinran_chen@xmu.edu.cn

[2] Fujian Medical University Union Hospital, Fuzhou, Fujian, China
zhengwu_99@aliyun.com

Abstract. Segmenting the urinary collecting system based on preoperative contrast-enhanced computed tomography urography volumes is necessary for assisting flexible ureterorenoscopy. The urinary collecting system consists of complex elongated tubular structures and irregular tree-like structures, making it challenging for precise segmentations using current deep-learning-based methods. Existing deep learning-driven methods face challenges in accurately segmenting the urinary collecting system from contrast-enhanced computed tomography urography volumes. In this work, we propose a novel MixU-Net by embedding global feature mix blocks. Particularly, the global feature mix blocks allow wider receptive fields based on fused multi-layer-perception and 3D convolutions across different dimensions. The experimental validations on the clinical computed tomography urography volumes demonstrate that our method achieves state-of-the-art in terms of dice similarity coefficients, intersection over union, and Hausdorff distance when compared with other methods that use pure convolutional neural networks or hybrid convolutional neural networks and Transformers. In addition, preliminary experiments conducted on the navigation system demonstrate the improved accuracy of the virtual depth maps when adopting the segmented urinary collecting system obtained by our MixU-Net.

Keywords: Urinary Collecting System Segmentation · Contrast-enhanced Computed Tomography Urography · Flexible Ureteroscopic Navigation

1 Introduction

With the steady increase of worldwide incidence, renal calculi has become one of the main causes of chronic kidney damage (CKD) [24]. Shock wave lithotripsy,

Q. Liu et al. (Eds.): PRCV 2023, LNCS 14429, pp. 468–479, 2024.
https://doi.org/10.1007/978-981-99-8469-5_37

ureteroscopy, and percutaneous nephrolithotomy are commonly adopted as the basic treatment options for renal calculi [20]. Particularly, ureteroscopy is a minimally-invasive treatment option with the advantages of less damage and higher flexibility when compared with percutaneous nephrolithotomy. Clinically, ureteroscopy-based retrograde intrarenal surgery (RIRS) is suitable for treating non-multiple renal calculi and helps alleviate patients' suffering [1,20].

Before RIRS, contrast-enhanced CT urography (CTU) is usually performed to image the urinary collecting system includes the ureter, renal pelvis, and renal calyces. During RIRS, a ureteroscope is inserted through the urinary collecting system to locate the lesions. However, successful RIRS requires a high level of proficiency for the physicians [5,6]. Particularly, the complex structure of the urinary collecting system and poor visual conditions can interfere a physician's judgment of the ureteroscope's pose. Therefore, precisely segmenting the renal collecting system based on the preoperative CTU volumes can effectively assist the physicians to perceive the operation scene and perform accurate ureteroscope navigation and operations.

Currently, the automatic segmentation of kidneys, renal tumors, and blood vessels based on CT or MRI volumes have shown promising results. Taha et al. [25] proposed a 3D fully convolutional neural network (CNN), KidNet, for end-to-end segmentation of the renal arteries, veins, and partial ureters on CTU volumes. Xia et al. [29] designed a twin neural network based on ResNet [13] and used dense SIFT [18] matching to automatically segment the kidney and kidney lesions. Heller et al. [14] summarized the results of the 2019 MICCAI Kidney and Kidney Tumor Segmentation Challenge, in which Fabian et al. [16] achieves the first place using the pre-activation nnU-Net. To address the challenges of insufficient annotating CT image datasets, Kim et al. [17] proposed a 3D U-Net based incremental learning strategy to assist the annotation of CT datasets. El-Melegy et al. [10] combined fuzzy C-means clustering and Markov random fields to automatically segment the kidneys in DCE-MRI images.

According to our literature review, reports on the segmentation of the urinary collecting system are substantially missing. In general, the urinary collecting system presents a continuous tree-like structure with complex morphology and a large spatial span. Moreover, the collecting system structure varies significantly among patients. Therefore, precisely segmenting the urinary collecting system requires methods with large receptive fields and the capabilities of generalization.

Recently, self-attention-based neural networks have become one of the hot topics in the research fields. For example, Vision Transformer (ViT) [9] and DETR [30] have demonstrated the capabilities of self-attention mechanisms. As a result, researchers tried to incorporate the self-attention mechanisms into the processing of medical images. TransUNet [4] combined self-attention mechanisms with CNNs for multiple segmentation using a U-shaped structure. Swin-UNet [2] discarded CNNs and adopted an encoder-decoder structure with pure self-attention mechanisms in multiple segmentation tasks.

Although self-attention mechanisms have proven their advantages over CNNs in some specific tasks, they generally require a large amount of data in the training phase, resulting in heavier computational burden. Consequently, the requirement of a large amount of annotative data limits the further application of self-attention mechanisms in medical imaging tasks. More recently, Multi-layer perceptron (MLP) models are regaining more attention in the research fields. Particularly, Tolstikhin et al. [26] designed a channel-position mixer MLP model MLP-Mixer and demonstrated its effectiveness in terms of a faster inference speed and fewer parameters than the self-attention-based models. In this study, we collected a dataset consisting of 40 CTU volumes and annotated them under the guidance of radiologists and urologists. We propose a U-shaped network with a hybrid CNN-MLP architecture, namely MixU-Net, for the precise segmentation of the urinary collecting system. The MixU-Net leverages the wide receptive fields of MLPs to compensate for the inductive bias of CNNs. To validate the performance of our method on segmenting the structurally complex and spatially extensive urinary collecting system, we collected an in-house dataset consisting of 40 CTU volumes and annotated them under the guidance of experienced radiologists and urologists.

The technical contributions of this paper are clarified as follows:

1. We propose a novel network MixU-Net. Specifically, we design a U-shaped backbone composed of local feature extractor (LFE) modules and introduce an MLP-based global feature mix block (GFMB) to assist the backbone in mixing global information and expanding the receptive field.
2. We establish a well-annotated CTU dataset under the authoritative guidances. All the experiments are conducted on this dataset. We will release the dataset on our laboratory website soon.
3. We combine the proposed method with the RIRS navigation system to assist in the generation of accurate virtual depth maps and the proposed method improves the accuracy of navigation.

The rest of this paper is organized as follows. Section 2 introduces the proposed method. Sections 3 and 4 present the experiment configurations and the results, respectively. Finally, the results are discussed and concluded in Sects. 5 and 6, respectively.

2 Method

2.1 Overview

Figure 1 shows the overview of our MixU-Net architecture. In this section, we firstly introduce the proposed GFMB embedded in the U-shaped network. Next, we introduce the U-shaped hybrid CNN-MLP networks containing CNN-based LFE modules and the GFMBs.

2.2 GFMB

Since the urinary collecting system consists of complex continuous tree-like structures, to better capture the features during the segmentation of the urinary collecting system, we design the GFMB. As illustrated in Fig. 2, the GFMB mainly

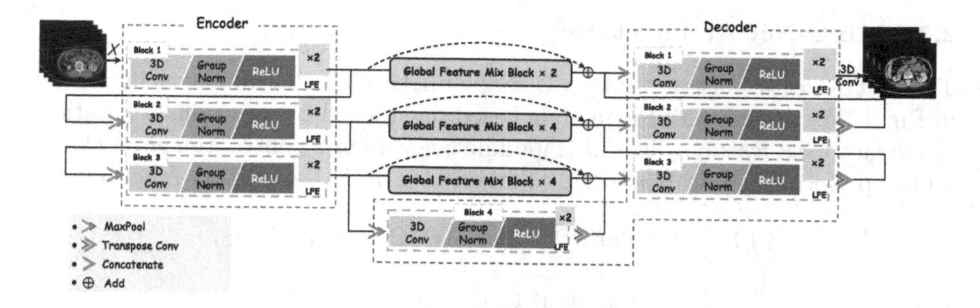

Fig. 1. The architecture of MixU-Net.

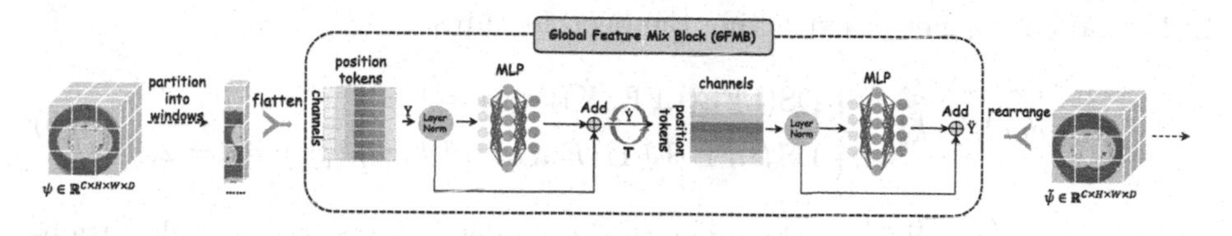

Fig. 2. The architecture of GFML. It contains several GFMBs.

consists of two MLPs. Each MLP consists of an input layer, an output layer, and a hidden layer with the size twice of the input and output layers. The GELU [15] is used between the input layer and the hidden layer. The first MLP is the token-mixing MLP, which operates on the column vectors of the input feature map $Y \in \mathbb{R}^{\widetilde{C} \times S}$ and extracts the relevance in the spatial dimension of Y.

$$\dot{Y} = Y \oplus \text{MLP}_{spatial}(\text{LayerNorm}(Y)). \tag{1}$$

The second MLP is the channel-mixing MLP which operates on the row vectors of the input feature map $\ddot{Y} \in \mathbb{R}^{\widetilde{C} \times S}$. The role of this MLP is to calculate the relevance in the channel dimension of $\dot{Y}$.

$$\ddot{Y} = \dot{Y} \oplus \text{MLP}_{channel}(\text{LayerNorm}(\dot{Y})), \tag{2}$$

where $\oplus$ represents the addition operation.

Several GFMBs are concatenated to form a global feature mix layer (GFML). At the input of the GFML, the feature map $\psi \in \mathbb{R}^{C \times H \times W \times D}$ needs to be divided into patches of size $P \times P \times P$. Each feature map patch is flatten and concatenated to $Y \in \mathbb{R}^{\widetilde{C} \times S}$.

$$\psi \in \mathbb{R}^{C \times H \times W \times D} \xrightarrow{Patch, Flatten, Concat} Y \in \mathbb{R}^{\widetilde{C} \times S}. \tag{3}$$

Similarly, the size of output $\ddot{Y} \in \mathbb{R}^{\widetilde{C} \times S}$ of GFML also needs to be rearranged to the original input size $\widetilde{\psi} \in \mathbb{R}^{C \times H \times W \times D}$.

$$\ddot{Y} \in \mathbb{R}^{\widetilde{C} \times S} \xrightarrow{Rearrange} \widetilde{\psi} \in \mathbb{R}^{C \times H \times W \times D}, \tag{4}$$

where $S = P^3$ and P is the size of the patches.

2.3 MixU-Net Architecture

The MixU-Net is constructed by a U-shaped network as backbone, as illustrated in Fig. 1. We design the LFE module based on stacked convolutional operations to extract local features. The LFE module consists of a convolutional block and a Group Normalization block [28], followed by a ReLU activation function:

$$\mathrm{LFE}\,(X) = \mathrm{ReLU}(\mathrm{GroupNorm}(\mathrm{Conv}_{3\times3\times3}(X))), \tag{5}$$

where $X \in \mathbb{R}^{C\times H\times W\times D}$ is the input feature map.

The U-shaped structure contains 4 layers, each of which consists of two consecutive LFE modules. In the encoder, downsampling is used to concatenate the LFE modules for better extracting the deep features:

$$EncoderBlock_i = \begin{cases} \mathrm{DS}(\mathrm{LFE}(\mathrm{LFE}(X))) \text{ if } i = 1 \\ \mathrm{DS}\,(\mathrm{LFE}\,(\mathrm{LFE}\,(EncoderBlock_{i-1}))) \text{ if } i = 2, 3 \end{cases} \tag{6}$$

where $X \in \mathbb{R}^{C\times H\times W\times D}$ is the input to the model and DS indicates downsampling.

In the decoder, upsampling is used to concatenate the LFE modules and restores the sizes of feature maps. Each decoder block also receives global feature from the GFMLs:

$$DecoderBlock_i = \begin{cases} \mathrm{LFE}\,(\mathrm{LFE}\,(EncoderBlock_3)) \text{ if } j = 4 \\ \mathrm{LFE}\,(\mathrm{LFE}\,(\mathrm{US}\,(DecoderBlock_{j+1}) \otimes \mathrm{GFML}_i)) \text{ if } j = 1, 2, 3 \end{cases} \tag{7}$$

where $\otimes$ represents concatenation along the channel dimension, and US represents upsampling.

2.4 Loss Function

We incorporate the mean structure similarity (MSSIM) index [27], cross-entropy, and dice in the loss function to train the MixU-Net. While the cross-entropy and dice are commonly used in segmentation tasks, the MSSIM is more sensitive to differences in image brightness and contrast:

$$\mathcal{L}_{MSSIM} = 1 - \frac{1}{M}\sum_{i=1}^{M} \frac{(2\mu_x^i\mu_y^i + C_1)(2\sigma_{xy}^i + C_2)}{({\mu_x^i}^2 + {\mu_y^i}^2 + C_1)({\sigma_x^i}^2 + {\sigma_y^i}^2 + C_2)}, \tag{8}$$

$$\mu_x = \sum_{i=1}^{N} \omega_i x_i, \tag{9a}$$

$$\sigma_x = \sqrt{\sum_{i=1}^{N} \omega_i (x_i - \mu_x)}, \tag{9b}$$

$$\sigma_{xy} = \sum_{i=1}^{N} \omega_i (x_i - \mu_x)(y_i - \mu_y), \tag{9c}$$

where x, y are the network predictions and ground truths, respectively. ω represents the weights of a Gaussian kernel, N is the total number of voxels, and M is the total number of windows, for which the window size is $11 \times 11 \times 11$. $C_1 = 0.0001$, $C_2 = 0.0009$.

Therefore, the loss function employed in the MixU-Net is forlumated as:

$$\mathcal{L} = \mathcal{L}_{dice} + \mathcal{L}_{CE} + \mathcal{L}_{MSSIM}, \tag{10}$$

where $\mathcal{L}_{CE}$ and $\mathcal{L}_{dice}$ represent the cross-entropy loss and the dice loss.

3 Experimental Configurations

3.1 Datasets and Evaluation Metrics

In this paper, we collected a dataset of clinical CTU volumes, and video images during the RIRS. Under the guidance of experienced radiologists and urologists, we selected 40 CTU volumes in the excretory phase and annotated the urinary collecting systems. The annotated dataset underwent careful examination by radiologists and urologists. The slice thickness of each volume is either 0.625mm or 1.25mm, and the spatial dimensions of each slice are 512512 pixels. The total number of slices in the 40 CTU volumes was 21,360.

We evaluated segmentation quality by dice similarity coefficient (DSC), Hausdorff distance (HD), surface distance (SD), and intersection over union (IoU).

3.2 Implementation Details

We implemented the MixU-Net on PyTorch v2.0.0, MONAI [3], and nnU-Net libraries [16]. To ensure fairness of the comparisons, we standardized the data pre-processing: First, we resampled the spacing of all volumes to the average value of the dataset. Next, we normalized the intensity of the volumes. Specifically, we designed an intensity window for each volume that ignored the top 0.5% of intensity in each volume and normalized the intensity to [0, 1]. Finally, we performed region cropping on each volume to exclude regions without annotations.

We adopted the default image augmentation methods from nnU-Net [16], which includes random 3-D elastic deformation, random scaling, random mirroring, and random gamma adjustment. The augmented volumes were randomly cropped into patches of size $128 \times 128 \times 128$ as inputs to the network. Due to the hardware limitations, the batch size was set to 1. We used SGD [23] as the optimizer and trained the MixU-Net for 400 epochs. Each epoch consisted of 250 training iterations, with a validation performed every 50 iterations. The initial learning rate was set to 0.00005, and the learning rate decay was specified as follows:

$$Lr = Lr_0 \times \left(1 - \frac{epoch}{\max epoch}\right)^{0.9}, \tag{11}$$

where Lr_0 represents the initial learning rate.

We trained our MixU-Net on an NVIDIA RTX 3090Ti GPU with 24GB of memory. During inference, we utilized the sliding window approach with an overlap ratio of 0.5. We saved the model that performed best on the validation set and tested on the test set to obtain the final segmentations.

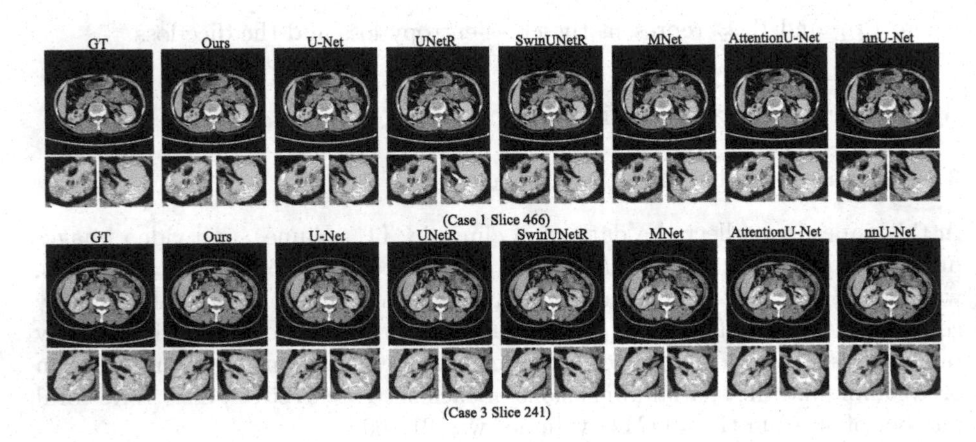

Fig. 3. Two examples for visual comparison of the proposed method with the other six medical segmentation methods.

4 Results

4.1 Comparisons

Figure 3 shows the typical frames of segmentations obtained by U-Net [22], UNetR [12], SwinUNetR [11], MNet [8], AttentionU-Net [21], and nnU-Net [16]. Figure 4 presents the corresponding 3D volume renderings of the segmented urinary collecting system. In general, our method performs more accurate in segmentations of the urinary collecting system. Compared to other methods, the MixU-Net can extract small, textureless, and weak boundaries. For the boundary between the kidneys and the urinary collecting systems, MixU-Net can effectively distinguish and establish clear edges. Furthermore, our method also exhibits better continuity in segmenting the elongated ureter structures.

Table 1 summarizes the quantitative results in accordance with the four metrics of the seven methods. Evidently, the MixU-Net outperforms the compared

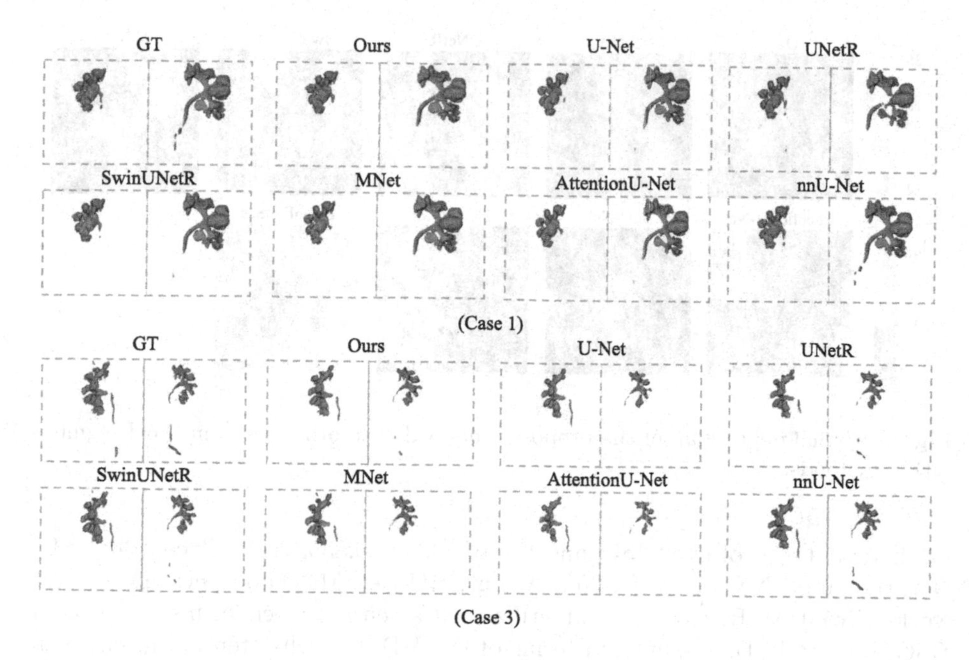

Fig. 4. Visual 3D segmentation results of the proposed method with other six medical segmentation methods.

Table 1. Quantitative comparison of using the seven segmentation methods for urinary collecting system segmentation. The bold results in the table indicate the best performance, while the underlined results indicate the second-best performance.

Approaches	Collecting System			
	DSC↑	IoU↑	HD↓	SD↓
U-Net	0.873 ± 0.099	0.783 ± 0.148	6.40 ± 6.17	0.36 ± 0.05
Attention U-Net	0.797 ± 0.170	0.683 ± 0.220	17.90 ± 23.56	0.53 ± 0.11
MNet	0.898 ± 0.075	0.820 ± 0.119	18.80 ± 27.52	**0.29 ± 0.04**
UNetR	0.850 ± 0.075	0.744 ± 0.111	4.31 ± 1.29	1.32 ± 0.67
SwinUNetR	0.838 ± 0.050	0.725 ± 0.084	4.59 ± 1.96	0.93 ± 0.44
nnU-Net	0.904 ± 0.070	0.831 ± 0.113	2.28 ± 1.30	0.69 ± 0.25
Ours	**0.908 ± 0.059**	**0.835 ± 0.096**	**2.11 ± 1.34**	0.77 ± 0.65

Table 2. Ablation experiments for MixU-Net. The bold results in the table indicate the best performance.

Approaches	DSC↑	IoU↑	HD↓	SD↓
Without GFMBs and MSSIM	0.8079	0.6848	4.6386	1.4997
Without GFMBs	0.8221	0.7040	59.400	7.5612
Without MSSIM	0.9063	0.8326	2.4140	**0.5747**
MixU-Net	**0.9078**	**0.8345**	**2.1125**	0.7733

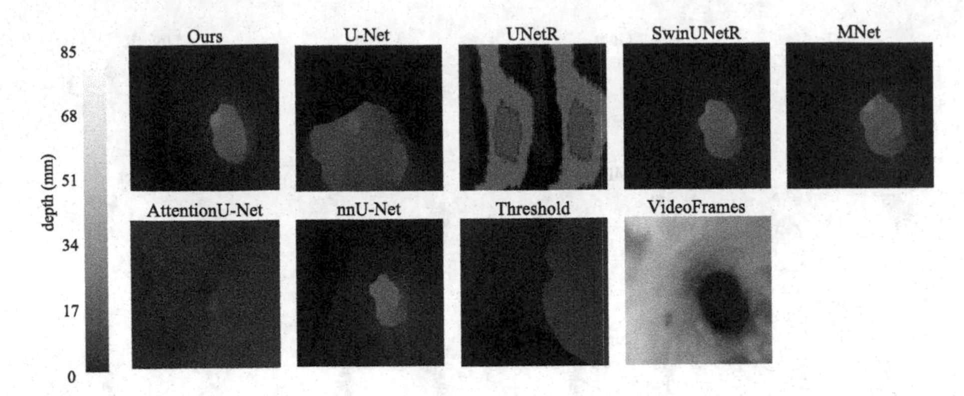

Fig. 5. Virtual depth map of the proposed method with other seven medical segmentation methods.

methods in terms of DSC, IoU, and HD of (0.908, 0.835, 2.11). Except for nnU-Net, the pure CNN models (U-Net, AttentionU-Net, MNet) outperform the networks (SwinUNetR, UNetR) that utilize self-attention mechanisms in terms of DSC, IoU, and SD. However, in terms of the HD, the self-attention mechanism performs better than the pure CNNs.

In addition, we combined our method with a home-made vision-based navigation system used for RIRS. The navigation system consists several steps: (1) ureteroscopic structure extraction, (2) virtual depth map generation, and (3) structural point similarity and optimization. We replaced the primitive threshold-based raw segmentations with segmentations from different networks in (1) and compared the performance of the depth maps generating in (2).

Figure 5 shows the generation virtual depth maps from segmentations of different networks and threshold. Threshold, Attention U-Net, and UNetR get depth maps with significant cumulative errors due to their low segmentation accuracy. For the results of the remaining networks, it can be observed that our method generates virtual depth maps that show most smooth in terms of depth transitions and exhibit consistent matching with target cavities.

4.2 Ablation Study

To evaluate the respective effectiveness of the building blocks in the MixU-Net, we conducted the following ablation study. Specifically, we compared the MixU-Net without the GFMBs and MSSIM loss, the MixU-Net without the GFMBs, the MixU-Net without the MSSIM loss, and the full version of MixU-Net. The datasets and the training settings were the same as mentioned in Sect. 3.1.

Table 2 lists the results of the ablation study. It can be observed that after removing the GFMBs and the MSSIM loss, the MixU-Net suffers from significant performance reduction in terms of DSC, IoU, and HD. These experimental results demonstrate the necessity of our proposed method in improving performance.

5 Discussion

In this paper, we propose a GFMBs-embedded MixU-Net for the segmentation of urinary collecting system based on CTU volumes. The experimental results demonstrated the out performance of our method when compared with other state-of-the-art networks. Particularly, the MixU-Net achieves the best DSC, IoU, and HD in the comparisons. The ablation study validates the respective effectiveness of the GFMB and the proposed term in the loss function. Since we strictly unified the pre- and post-processing of the dataflow across different methods, the major cause of the performance differences can be attributed to the receptive fields of different methods. Since the self-attention mechanism significantly increases the receptive field of a network, SwinUNetR and UNetR achieve better HD at the expense of worse SD than the methods that use pure CNNs (U-Net, Attention U-Net, and MNet). On the other hand, the MixU-Net designs a soft attention mechanism for a better balance between the local and global feature perceptions, which is particularly useful to segment the complex urinary collecting system.

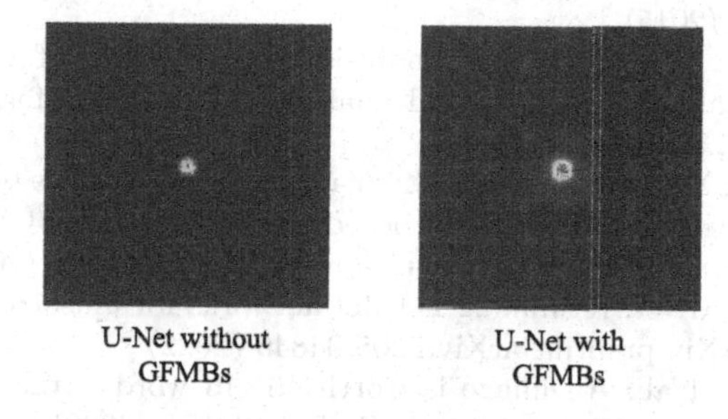

Fig. 6. Visual receptive field comparison of the U-Net with and without the GFMBs.

Figure 6 compares the visualizations of the receptive fields without or with the GFMBs obtained by a visualization method reported in [7,19]. We can see that the GFMBs effectively expands the receptive field of the convolution-based U-shaped network.

6 Conclusion

In this paper, we propose a novel MixU-Net embedding GFMBs to perform efficient local and global feature fusion for precise segmentation of the complex urinary collecting system based on preoperative CTU volumes. Our method is validated to outperform other methods that employ pure CNNs or hybrid CNNs and Transformers, showing better DSC, IoU, and HD when segmenting the urinary collecting system. The ablation study also demonstrates the respective effectiveness of the GFMBs and the proposed MSSIM term in the loss function. Integrating our method to the navigation system for high-efficient RIRS is one of the directions of our future work.

Acknowledgements. This work was supported in part by the National Natural Science Foundation of China (Grant No. 62001403 and 61971367), Natural Science Foundation of Fujian Province of China (No. 2020J05003 and 2020J01004), and the Fujian Provincial Technology Innovation Joint Funds under Grant 2019Y9091

References

1. Breda, A., Ogunyemi, O., Leppert, J.T., Schulam, P.G.: Flexible ureteroscopy and laser lithotripsy for multiple unilateral intrarenal stones. Eur. Urol. **55**(5), 1190–1197 (2009)
2. Cao, H., et al.: Swin-Unet: Unet-like pure transformer for medical image segmentation. In: Karlinsky, L., Michaeli, T., Nishino, K. (eds.) Computer Vision – ECCV 2022 Workshops. ECCV 2022. Lecture Notes in Computer Science, vol. 13803, pp. 205–218. Springer, Cham (2023). https://doi.org/10.1007/978-3-031-25066-8_9
3. Cardoso, M.J., et al.: MONAI: an open-source framework for deep learning in healthcare. arXiv preprint arXiv:2211.02701 (2022)
4. Chen, J., et al.: TransUNet: transformers make strong encoders for medical image segmentation. arXiv preprint arXiv:2102.04306 (2021)
5. Cho, S.Y.: Current status of flexible ureteroscopy in urology. Korean J. Urol. **56**(10), 680–688 (2015)
6. Cho, S.Y., et al.: Cumulative sum analysis for experiences of a single-session retrograde intrarenal stone surgery and analysis of predictors for stone-free status. PLoS ONE **9**(1), e84878 (2014)
7. Ding, X., Zhang, X., Han, J., Ding, G.: Scaling up your kernels to 31×31: revisiting large kernel design in CNNs. In: Proceedings of the IEEE/CVF Conference on Computer Vision and Pattern Recognition, pp. 11963–11975 (2022)
8. Dong, Z., et al.: MNet: rethinking 2D/3D networks for anisotropic medical image segmentation. arXiv preprint arXiv:2205.04846 (2022)
9. Dosovitskiy, A., et al.: An image is worth 16×16 words: transformers for image recognition at scale. arXiv preprint arXiv:2010.11929 (2020)
10. El-Melegy, M., Kamel, R., El-Ghar, M.A., Shehata, M., Khalifa, F., El-Baz, A.: Kidney segmentation from DCE-MRI converging level set methods, fuzzy clustering and Markov random field modeling. Sci. Rep. **12**(1), 18816 (2022)
11. Hatamizadeh, A., Nath, V., Tang, Y., Yang, D., Roth, H.R., Xu, D.: Swin UNETR: swin transformers for semantic segmentation of brain tumors in MRI images. In: Crimi, A., Bakas, S. (eds.) Brainlesion: Glioma, Multiple Sclerosis, Stroke and Traumatic Brain Injuries. BrainLes 2021. Lecture Notes in Computer Science, vol. 12962, pp. 272–284. Springer (2022). https://doi.org/10.1007/978-3-031-08999-2_22
12. Hatamizadeh, A., et al.: UNETR: transformers for 3D medical image segmentation. In: Proceedings of the IEEE/CVF Winter Conference on Applications of Computer Vision, pp. 574–584 (2022)
13. He, K., Zhang, X., Ren, S., Sun, J.: Deep residual learning for image recognition. In: Proceedings of the IEEE Conference on Computer Vision and Pattern Recognition, pp. 770–778 (2016)
14. Heller, N., et al.: The state of the art in kidney and kidney tumor segmentation in contrast-enhanced CT imaging: results of the kits19 challenge. Med. Image Anal. **67**, 101821 (2021)
15. Hendrycks, D., Gimpel, K.: Gaussian error linear units (GELUs). arXiv preprint arXiv:1606.08415 (2016)

16. Isensee, F., Jaeger, P.F., Kohl, S.A., Petersen, J., Maier-Hein, K.H.: nnU-Net: a self-configuring method for deep learning-based biomedical image segmentation. Nat. Methods **18**(2), 203–211 (2021)
17. Kim, T., et al.: Active learning for accuracy enhancement of semantic segmentation with CNN-corrected label curations: evaluation on kidney segmentation in abdominal CT. Sci. Rep. **10**(1), 366 (2020)
18. Lowe, D.G.: Object recognition from local scale-invariant features. In: Proceedings of the Seventh IEEE International Conference on Computer Vision, vol. 2, pp. 1150–1157. IEEE (1999)
19. Luo, W., Li, Y., Urtasun, R., Zemel, R.: Understanding the effective receptive field in deep convolutional neural networks. In: Advances in Neural Information Processing Systems, vol. 29 (2016)
20. Miller, N.L., Lingeman, J.E.: Management of kidney stones. Bmj **334**(7591), 468–472 (2007)
21. Oktay, O., et al.: Attention U-Net: learning where to look for the pancreas. arXiv preprint arXiv:1804.03999 (2018)
22. Ronneberger, O., Fischer, P., Brox, T.: U-Net: convolutional networks for biomedical image segmentation. In: Navab, N., Hornegger, J., Wells, W.M., Frangi, A.F. (eds.) MICCAI 2015. LNCS, vol. 9351, pp. 234–241. Springer, Cham (2015). https://doi.org/10.1007/978-3-319-24574-4_28
23. Ruder, S.: An overview of gradient descent optimization algorithms. arXiv preprint arXiv:1609.04747 (2016)
24. Rule, A.D., Bergstralh, E.J., Melton, L.J., Li, X., Weaver, A.L., Lieske, J.C.: Kidney stones and the risk for chronic kidney disease. Clin. J. Am. Soc. Nephrol. **4**(4), 804–811 (2009)
25. Taha, A., Lo, P., Li, J., Zhao, T.: Kid-Net: convolution networks for kidney vessels segmentation from CT-volumes. In: Frangi, A.F., Schnabel, J.A., Davatzikos, C., Alberola-López, C., Fichtinger, G. (eds.) MICCAI 2018. LNCS, vol. 11073, pp. 463–471. Springer, Cham (2018). https://doi.org/10.1007/978-3-030-00937-3_53
26. Tolstikhin, I.O., et al.: MLP-Mixer: An all-MLP architecture for vision. Adv. Neural. Inf. Process. Syst. **34**, 24261–24272 (2021)
27. Wang, Z., Bovik, A.C., Sheikh, H.R., Simoncelli, E.P.: Image quality assessment: from error visibility to structural similarity. IEEE Trans. Image Process. **13**(4), 600–612 (2004)
28. Wu, Y., He, K.: Group normalization. In: Proceedings of the European Conference on Computer Vision (ECCV), pp. 3–19 (2018)
29. Xia, K.j., Yin, H.s., Zhang, Y.d.: Deep semantic segmentation of kidney and space-occupying lesion area based on SCNN and ResNet models combined with SIFT-flow algorithm. J. Med. Syst. **43**, 1–12 (2019)
30. Zhu, X., Su, W., Lu, L., Li, B., Wang, X., Dai, J.: Deformable DETR: deformable transformers for end-to-end object detection. arXiv preprint arXiv:2010.04159 (2020)

Self-supervised Cascade Training for Monocular Endoscopic Dense Depth Recovery

Wenjing Jiang[1(✉)], Wenkang Fan[1(✉)], Jianhua Chen[3], Hong Shi[3(✉)], and Xiongbiao Luo[1,2(✉)]

[1] Department of Computer Science and Technology, Xiamen University, Xiamen, China
{jiangwj,23020210156921}@stu.xmu.edu.cn, xiongbiao.luo@gmail.com
[2] National Institute for Data Science in Health and Medicine, Xiamen University, Xiamen, China
[3] Fujian Cancer Hospital, Fujian Medical University, Fuzhou, China
endoshihong@hotmail.com

Abstract. Dense depth prediction for 3-D reconstruction of monocular endoscopic images is an essential way to expand the surgical field and augment the perception of surgeons in robotic endoscopic surgery. However, it is generally challenging to precisely estimate the monocular dense depth and reconstruct such a field due to complex surgical fields with a limited field of view, illumination variations, and weak texture information. This work proposes a new framework of self-supervised learning with a two-stage cascade training strategy for dense depth recovery of monocular endoscopic images. While the first stage is to train an initial deep-learning model through sparse depth consistency supervision, the second stage introduces photometric consistency supervision to further train and refine the initial model for improving its capability. Our framework was evaluated on patient data of monocular endoscopic images acquired from colonoscopic procedures, with the experimental results demonstrating that our self-supervised learning model with cascade training provides a promising strategy outperforming other models. On the one hand, both visual quality and quantitative assessment of our method are better than current monocular dense depth estimation approaches. On the other hand, our method relies less on sparse depth data for supervision than other self-supervised methods.

Keywords: Monocular depth estimation · 3-D reconstruction · self-supervised learning · robotic-assisted endoscopy · structure from motion · simultaneous localization and mapping

1 Introduction

Robotic-assisted endoscopic procedures are increasingly performed to diagnose and treat intestinal diseases [17]. Unfortunately, the endoscope itself cannot provide surgeons with any depth information of the surgical scene or field, resulting

W. Jiang and W. Fan—Equally contributed.

X. Luo and H. Shi—Corresponding authors.

Q. Liu et al. (Eds.): PRCV 2023, LNCS 14429, pp. 480–491, 2024.
https://doi.org/10.1007/978-981-99-8469-5_38

in additional surgical risks such as accidental injuries and inaccurate manipulation of surgical tools. To address these limitations, endoscopic field three-dimensional (3-D) reconstruction is widely discussed as an effective way to expand the endoscopic viewing and augment the perception of the surgeon [2,5].

Precise endoscopic 3-D reconstruction unavoidably requires accurate dense depth estimation. Because of the limited space in the human body, monocular depth estimation is appropriate using only one camera to acquire a sequence of 2-D images to estimate the depth and camera pose. Conventional monocular depth recovery methods, such as feature-based and direct methods, generally rely on multi-view camera geometric principles. For example, direct sparse odometry [3] basically uses the photometric consistency loss to predict the camera pose and sparse 3-D reconstruction of images, while they suffer from the photometric inconsistency with poor effectiveness. Although feature-based approaches(e.g., structure from motion (SfM) [15,21] and simultaneous localization and mapping (SLAM) [9,13]) are less affected by illumination variations than the direct methods, they limit themselves to insufficient and incorrect feature detection and matching due to textureless endoscopic images.

More recently, deep learning methods are increasingly used for dense depth estimation. Supervised methods usually require many annotated or ground-truth data, which are particularly unrealistic for monocular surgical endoscopic videos. To this end, unsupervised and self-supervised learning methods are commonly introduced to predict depth maps from monocular cameras. In particular, the distinction between self-supervised and unsupervised learning is whether or not the sparse data generated by the conventional 3-D reconstruction method is used to supervise the model training in the absence of ground truth. Unsupervised learning methods usually employ the photometric loss for training to predict dense depth and camera pose simultaneously [18–20]. These methods hardly work on monocular endoscopic video images because of limited and nonuniform illumination conditions. To address this limitation, Ozyoruk et al. [14] proposed an Endo-SfMLearner method that uses global illumination affine transformation and introduces structural similarity index (SSIM) to calculate the photometric loss. Self-supervised learning methods for monocular depth estimation are a most active research area in the literature [8,11,12]. These methods obtain sparse depth data by feature-based detection and matching with epipolar geometry constraints (e.g., SfM) and use them to supervise the network training [1,10]. Unfortunately, these self-supervised learning based depth recovery approaches depend critically on the quality of sparse depth data.

Although most unsupervised and self-supervised learning methods work well on monocular natural indoor or outdoor image depth estimation for various computer vision tasks, they remain challenging for monocular surgical endoscopic images due to weak textures and illumination variations. The objective of this work is to precisely expand and reconstruct 3-D surgical field or viewing, empowering surgeons to intuitively perceive the depth distribution in robotic endoscopic procedures. We focus on self-supervised learning for monocular depth estimation. The highlights of this work are clarified as follows.

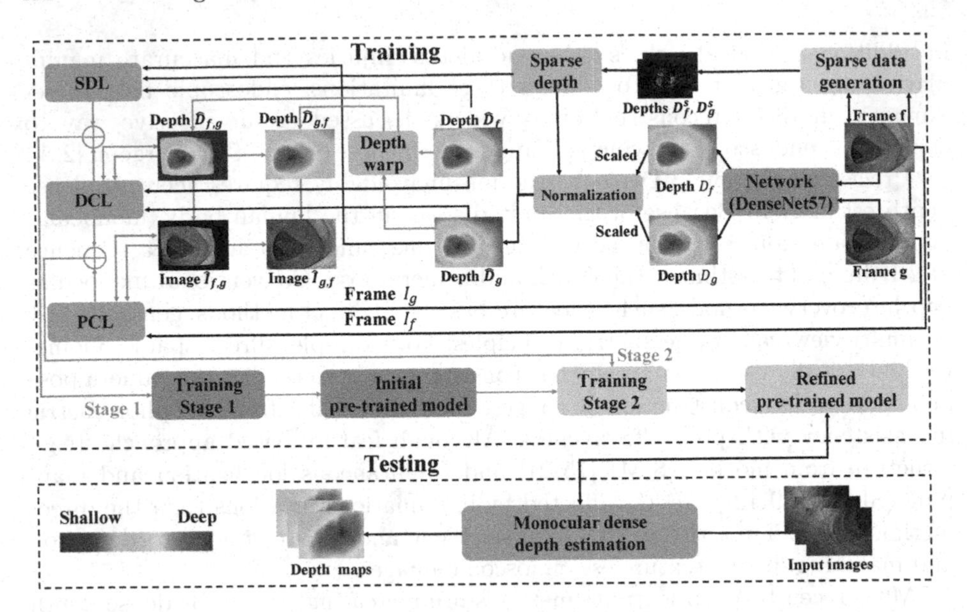

Fig. 1. Pipeline of our proposed self-supervised learning with cascade training based monocular depth estimation method.

- This work proposes a self-supervised learning method with cascade training for monocular endoscopic dense depth prediction. This framework implements a two-stage training strategy to take advantages of both sparse depth consistency supervision and photometric consistency supervision.
- Furthermore, we successfully address the problems of weak textures and nonuniform illumination on endoscopic images. Compared to sole sparse depth consistency supervision, our method can employ more information on tissue structure on the basis of photometric consistency supervision to reduce the quality requirements of sparse depth data for supervision to solve the problem of weak texture. Moreover, our method uses sparse depth consistency supervision to provide a good initialization for sole photometric consistency supervision, which can deal with nonuniform illumination.

2 Approaches

This section details our self-supervised and cascade-trained network model for monocular endoscopic dense depth prediction. The first training (Stage 1) uses SfM-driven sparse depth data to train the network model through sparse depth loss and depth consistency loss and obtain an initial pre-trained model, while the second training (Stage 2) performs photometric consistency-loss and depth consistency-loss supervisions to further refine the initial model. Figure 1 illustrates the pipeline of training and testing our proposed method.

2.1 Sparse Depth Generation

Accurate generation of sparse depth data is important for dense depth recovery. This step outputs sparse depth maps and camera poses to supervise the training procedure. This work uses an SfM-based method to generate sparse depth data. Such an SfM method consists of two main steps (1) image feature point detection and matching and (2) epipolar geometric analysis.

The quality and quantity of detected and matched feature points are important to camera pose estimation and sparse reconstruction. However, it is hard to find accurate feature point correspondences between the continuous frames because of poor texture or structural information on endoscopic images. This work employs a local region expansion (LRE) based feature detection and matching method [4] which can find more accurate feature correspondences.

After obtaining feature-matching pairs, epipolar geometric analysis is implemented to calculate the fundamental and essential matrices, $\mathbf{F}$ and $\mathbf{E}$ [15]:

$$p_2^T \mathbf{F} p_1 = 0, \mathbf{E} = (K')^T \mathbf{F} K, \tag{1}$$

where p_1 and p_2 are a pair of feature matching points, K and K' are the camera intrinsic matrices of the two images involved. After that, we implement singular value decomposition on $\mathbf{E}$ to obtain the camera pose $\mathbf{T}_w^f$ at frame f.

Next, we can use the camera pose $\mathbf{T}_w^f$ (w represents the world coordinate system) to reconstruct sparse point cloud $\mathbf{P}_w$ by triangulation. Eventually, we project the sparse 3-D point cloud $\mathbf{P}_w$ onto image f to obtain sparse depth map $D_f^s = K\mathbf{T}_w^f \mathbf{P}_w$. The sparse data is then used to train the model with sparse depth consistency supervision.

2.2 Network Architecture

This work uses FC-DenseNet-57 [7] as the CNN architecture to predict the dense depth map, which is the same as the previous work [10] for a fair comparison in our experiments. FC-DenseNet-57 performs better than other network architectures in feature extraction and parameter reduction.

2.3 Self-supervised Cascade Training

This section discusses the details of the two-stage training procedure with sparse depth consistency and photometric consistency optimizations.

Sparse Depth Consistency Supervision. Stage 1 uses a sparse depth loss (SDL) function and a depth consistency loss (DCL) function to train the network model in a self-supervised manner to obtain an initial pre-trained model (Fig. 1).

The training requires two input images at frames or times f and g, respectively. Since there exists inconsistency between the numerical distributions of the sparse depth map D_f^s and dense depth map D_f predicted by DenseNet57, we first normalize the dense depth map D_f to obtain $\hat{D}_f$

$$\hat{D}_f = F_{normalization}(D_f, D_f^s). \quad (2)$$

SDL is to compare the sparse depth map D_f^s to the normalized or scaled dense depth map $\hat{D}_f$ and can be calculated by

$$L_{SDL}(D_f^s, \hat{D}_f) = \frac{\sum(D_f^s - \hat{D}_f)^2}{\sum(D_f^s)^2 + (\hat{D}_f)^2}. \quad (3)$$

Note that SDL only takes the feature points or positions with depth information on the sparse depth map D_f^s into the calculation, resulting in a fraction of point pixels in practice. As a result, many pixels on the sparse depth map are not properly guided. Therefore, DCL is introduced to maintain depth consistency.

DCL actually uses geometric constraints to optimize the depth consistency between the two frames. We first normalize the two depth maps D_f and D_g by Eq. (2) and obtain the scaled depth maps $\hat{D}_f$ and $\hat{D}_g$. Next, we warp $\hat{D}_f$ and $\hat{D}_g$ from one to another using the geometry transformation matrices (or camera poses) $\mathbf{T}_w^f$ and $\mathbf{T}_w^g$ and achieve the warped depth maps $\hat{D}_{f,g}$ and $\hat{D}_{g,f}$. Then, DCL can be computed by

$$L_{DCL}(f, g) = \frac{\sum(\hat{D}_{f,g} - \hat{D}_g)^2}{\sum(\hat{D}_{f,g}^2 + \hat{D}_g^2)} + \frac{\sum(\hat{D}_{g,f} - \hat{D}_f)^2}{\sum(\hat{D}_{g,f}^2 + \hat{D}_f^2)}. \quad (4)$$

Eventually, the total loss $L_{S1}(f, g)$ for input images f and g at the first training (Stage 1) can be written as:

$$L_{S1}(f, g) = \alpha_1(L_{SDL}(f) + L_{SDL}(g)) + \alpha_2 L_{DCL}(f, g), \quad (5)$$

where constants α_1 and α_2 represent the weights to balance between SDL and DCL, and $\alpha_1 + \alpha_2 = 1$.

The initial pre-trained model is simultaneously supervised by DCL and SDL at Stage 1. While DCL can keep the depth consistency and structure smoothness, SDL depends critically on the number and accuracy of the sparse depth points. This implies that the performance of the initial model will be degraded if it was supervised by incorrect sparse depth maps.

Photometric Consistency Supervision. Stage 2 combines DCL with PCL to further train the initial model, refining it to be able to predict the depth at the positions with incorrect depth supervision and without depth supervision.

We first use DenseNet57 and normalization to compute the scaled depth map $\hat{D}_f$ and $\hat{D}_g$ for two endoscopic images I_f and I_g. Similar to DCL, we also employ the geometry transformation (camera pose) T_f^g or T_g^f to warp I_f and I_g mutually to obtain the warped images $I_{f,g}$ and $I_{g,f}$.

Table 1. Pre-trained models with different weights to combine various loss functions of SDL, PCL, and DCL

Methods	Epochs	SDL	PCL	DCL
Model-1	20	–	0.7	0.3
Model-2	20	0.4	0.4	0.2
Model-3	1:10 at Stage 1	0.7	–	0.3
	11:20 at Stage 2	–	0.7	0.3

Basically, PCL aims to measure the photometric inconsistency between the endoscopic image I_g and warped image $I_{f,g}$. To address illumination variations, we introduce a brightness transformation [14] to adjust the brightness of the input images for better supervision during training:

$$\hat{I}_{f,g} = (I_{f,g} - Ave(I_{f,g}))\frac{Var(I_{f,g})}{Var(I_g)}) + Ave(I_g), \tag{6}$$

where $Ave(I)$ and $Var(I)$ represent the average and variance of image I, respectively. Additionally, we also introduce the structural similarity index (SSIM) to calculate the structural similarity between images $\hat{I}_{f,g}$ and I_g because the brightness on the two images is still inconsistent even after the brightness transformation. Finally, we can calculate PCL by

$$L_{PCL}(f,g) = 1 - \frac{SSIM(\hat{I}_{f,g}, I_g) + SSIM(\hat{I}_{g,f}, I_f)}{2}. \tag{7}$$

We still combine PCL with DCL in the second training procedure because only using photometric consistency loss leads to large depth bias at adjacent positions on endoscopic images where tissue deformation and occlusion are observed during robotic surgery. Eventually, the total loss $L_{S2}(f,g)$ involved photometric consistency supervision at the second training (Stage 2) can be defined as

$$L_{S2}(f,g) = \beta_1 L_{PCL}(f,g) + \beta_2 L_{DCL}(f,g), \tag{8}$$

where constants β_1 and β_2 is to balance PCL and DCL.

While only using $L_{S1}(f,g)$ for training suffers from incorrect sparse-depth supervision or lack of sparse-depth supervision at many pixels on endoscopic images, only using $L_{S2}(f,g)$ to supervise and train DenseNet57 for monocular depth prediction easily get trapped into local optimum because the intensity distribution of endoscopic images is strongly non-convex. This motivates our new framework of self-supervised learning with a cascade training strategy for monocular endoscopic depth recovery.

2.4 Pre-trained Models

This work will generate three pre-trained models to compare different loss functions used in DenseNet57. Without loss of generality, Model-1 is a one-stage model trained by the self-supervised PCL and DCL function $L_{M1}(f, g)$, while Mode-2 is also a one-stage trained by the self-supervised SDL, PCL, and DCL function $L_{M2}(f, g)$ with respect to Eqs.(3), (4), and (7):

$$L_{M1}(f, g) = \gamma_1 L_{PCL} + \gamma_2 L_{DCL}, \tag{9}$$

$$L_{M2}(f, g) = \lambda_1 L_{SDL} + \lambda_2 L_{PCL} + \lambda_3 L_{DCL}, \tag{10}$$

where constants γ_1, γ_2, λ_1, λ_2, and λ_3 are the weights of loss functions.

Model-3 is generated by our proposed self-supervised learning methods with a cascade training strategy. Note that Liu et al. [10] proposed method used SDL with DCL for one-stage training. Table 1 summarizes all the pre-trained models that employ different weights during training in this work.

Our experiments skip data augmentation for the generalization of these models. We downsampled all the images to a size of 320×256 and simply remove highlight pixels on endoscopic images by an in-painting operation, avoiding the degradation of the photometric consistency supervision. We set the learning rate from 10^{-5} to 10^{-4} and use the stochastic gradient descent algorithm as an optimizer with the momentum of 0.9 during training [16]. The batch size, epoch, and iterations were set to 4, 20, and 1000, respectively.

3 Validation

We recorded monocular endoscopic video data from surgical procedures. All the endoscopic video images in a size of 1920×1080. We manually selected a large number of images that contain much structural information as our experimental data. For these selected images, we generated sparse depth maps using the local region expansion based SfM algorithm [4,15]. Finally, we obtained 96 video sequences with relatively good sparse results from inconsistent scenarios and each sequence contains from 15 to 30 frames. Each sparse depth map usually contains numerous projected feature key points ranging from 1000 to 3500.

Since we cannot generate ground truth for the patient data, we use two metrics for quantitative evaluation: SSIM and peak signal to noise ratio (PSNR) [6]. Specifically, we use the estimated camera pose and monocular dense depth map to warp one frame into the perspective of another and calculate SSIM and PSNR between the original and warped images. Additionally, we calculate back-projection error (BPE) to measure the average distance between reconstructed points and their corresponding matched points on endoscopic images.

To evaluate the robustness of the models in Table 1, we compare them to two monocular dense depth recovery methods: (1) Liu et al. [10], a self-supervised learning method with SDL and DCL training, and (2) Ozyoruk et al. [14], an unsupervised learning method that simultaneously predicts camera pose and dense depth on the basis of photometric consistency loss. Note that we used the same sparse depth maps and camera poses to normalize dense depth maps and calculate the metrics for a fair comparison.

We also implement ablation study to evaluate our cascade training strategy using different sparse reconstruction results. For each video sequence, we

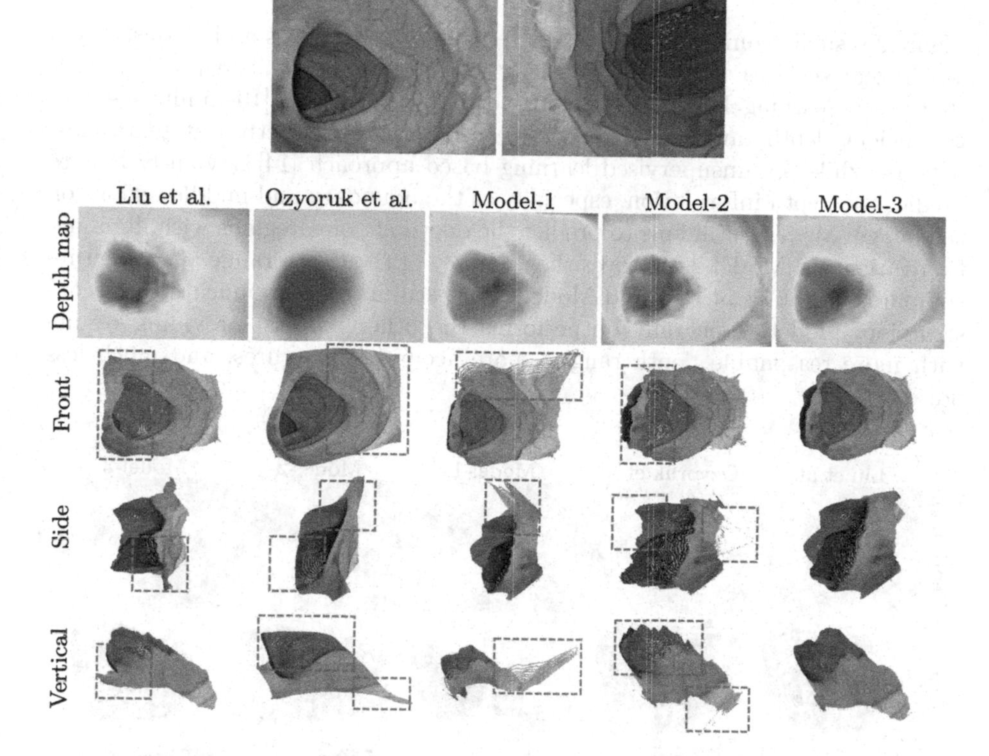

Fig. 2. Image A's estimated monocular dense depth maps and reconstructed 3-D surfaces using five models, respectively. *Red* boxes indicates inaccurate reconstructed structures. The results of image B are compared in Fig. 3 (Color figure online)

Table 2. Quantitative results of SSIM, PSNR, and back-projection errors (BPE) of using the five methods. Note that *<1* represents the proportion of the feature points whose back-projection errors are less than 1 pixel.

Methods	SSIM	PSNR	<1	<2	<3	BPE
Liu et al.	0.601± 0.078	21.362 ± 2.773	0.655	0.835	0.898	1.292 ± 1.463
Ozyoruk et al.	0.522± 0.072	20.352 ± 2.581	0.438	0.665	0.784	2.036 ± 2.174
Model-1	0.605± 0.076	21.821 ± 2.797	0.639	0.819	0.877	1.357 ± 1.504
Model-2	0.616± 0.080	22.235 ± 2.879	0.699	0.851	0.923	1.185 ± 1.256
Model-3	**0.629± 0.071**	**22.448 ± 2.761**	**0.713**	**0.887**	**0.946**	**1.032 ± 1.105**

generate multiple feature groups with different matches at a certain increasing number. Based on different numbers of feature matches, we perform the SfM algorithm to obtain various sparse depth maps and camera poses for training.

4 Results

Figure 2 visually compares their estimated dense depth maps and reconstructed 3-D image surfaces using the five deep learning methods. The previous self-supervised learning based method proposed by Liu et al. [10] limits itself to insufficient depth range and inaccurate reconstructed structure at textureless regions, while the unsupervised learning based approach [14] obviously fails to predict the depth information, especially at the boundary and middle regions on the image. Model-1 still fails to predict the depth at some regions with illumination variations. Model-2 improves the accuracy of the depth range and structure compared to Liu et al. [10] and Model-1, but still introduces much noise reconstruction. Model-3 generally outperforms the other models, providing results with more reasonable depth ranges, more accurate structures, and much less noise.

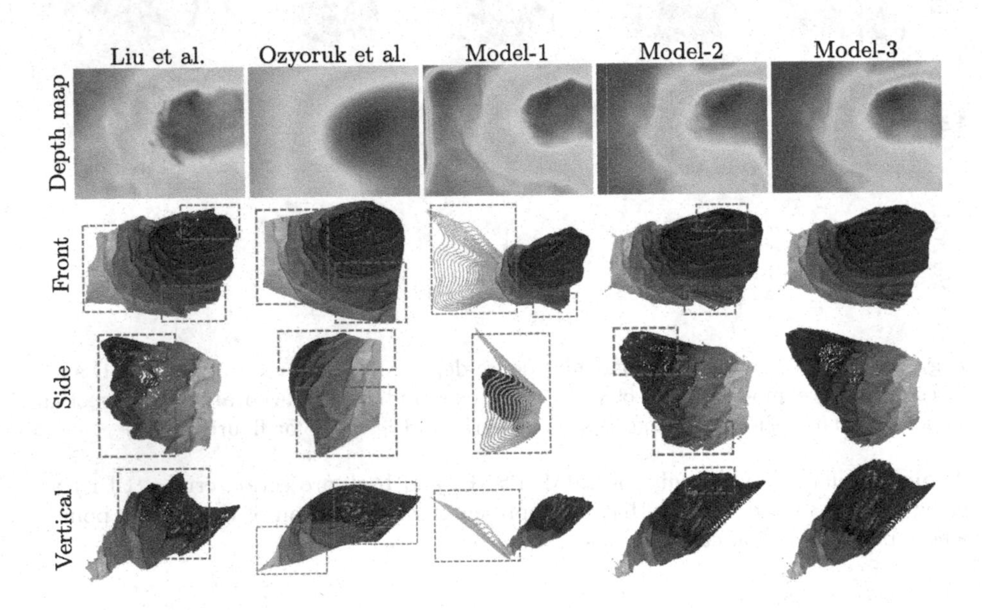

Fig. 3. Compared SSIM and PSNR of the five methods tested on 29 pairs of monocular endoscopic video sequences.

Figure 3 plots the SSIM and PSNR results of the five methods evaluated on 29 monocular endoscopic video sequences. Table 2 further summarizes quantitative assessment results of using the five deep learning methods. SSIM, PSNR, and average back-projection errors of Model-3 were slightly better than Model-2 and much better than Liu et al. [10], Ozyoruk et al. [14], and Model-1.

Figure 4 displays the dense depth maps with different sparse reconstruction results. The dense depth accuracy of using Liu et al. [10] and Model-2 clearly depends on the quality of the sparse depth estimation. The more numerous feature points for spare depth supervision, the more accurate monocular depth distribution prediction. More interestingly, our cascade training method (Model-3) can effectively improve the dense depth estimation map only supervised by the low-quality sparse depth supervision (20%).

5 Discussion

The objective of this work is to predict reasonable dense depth distribution and reconstruct 3-D surfaces of monocular endoscopic video sequences. The effectiveness and limitations of different models are discussed as follows.

Self-supervised learning methods generally work better than unsupervised learning such as Ozyoruk et al. [14]. The unsupervised method [14] limits itself to getting trapped into local minima or optimum because its used photometric consistency training loss is sensitive to illumination variation. By using sparse depth consistency loss for self-supervision, Liu et al. [10] still suffers from insufficient depth range and incomplete structure due to an insufficient number and limited quality of sparse feature points. Model-1 remains challenging to illumination variations. Model-2 and Model-3 certainly outperform the other three methods since they use both sparse depth and photometric consistency supervision for training. Unfortunately, Model-2 introduces much noise on predicted depth maps, resulting from a profound depth difference among image pixel locations supervised only by sparse depth loss but their surrounding areas only supervised by photometric consistency loss. Model-3 can achieve better results

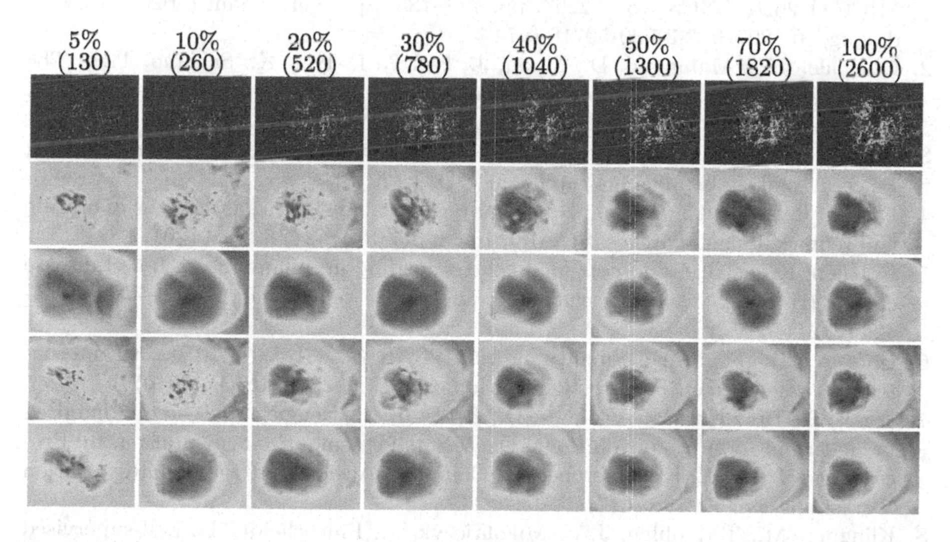

Fig. 4. $n\%$ is the percentage and (m) represents the average number of projected points of all training data on the sparse depth maps, while the third row shows the sparse depth maps. Rows 4∼7 illustrate all the dense depth maps of Fig. 2 (Image A) predicted by Liu et al. [10], Model-1, Model-2, and Model-3 under different sparse depth supervisions from the results at Row 3, respectively.

because it uses a cascade training strategy to build an initial-to-refine learning model. Additionally, Model-3 can generally obtain precise depth structures even if the quality of sparse data was inaccurate since the initial depth map of our proposed model can be optimized by the photometric consistency supervision.

Our proposed methods still suffer from some limitations. Our methods still cannot estimate the depth and reconstruct 3-D surfaces for monocular endoscopic video data with large blur and specular highlight regions. Additionally, the computational speed of our method was about 10 frames per second, which can be improved up to real-time processing for clinical applications.

In summary, this work presents a new self-supervised learning with a two-stage cascade training strategy for dense depth prediction of monocular endoscopic images. The experimental results show our cascade-trained model outperforms other unsupervised and self-supervised methods.

Acknowledgements. This work was supported in part by the National Natural Science Foundation of China under Grants 61971367, 82272133, and 62001403, in part by the Natural Science Foundation of Fujian Province of China under Grants 2020J01004 and 2020J05003, and in part by the Fujian Provincial Technology Innovation Joint Funds under Grant 2019Y9091.

References

1. Bae, G., Budvytis, I., Yeung, C.-K., Cipolla, R.: Deep multi-view stereo for dense 3D reconstruction from monocular endoscopic video. In: Martel, A.L., et al. (eds.) MICCAI 2020. LNCS, vol. 12263, pp. 774–783. Springer, Cham (2020). https://doi.org/10.1007/978-3-030-59716-0_74
2. Crismale, J.F., Mahmoud, D., Moon, J., Fiel, M.I., Iyer, K., Schiano, T.D.: The role of endoscopy in the small intestinal transplant recipient: a review. Am. J. Transplant. **21**(5), 1705–1712 (2021)
3. Engel, J., Koltun, V., Cremers, D.: Direct sparse odometry. IEEE Transactions on Pattern Analysis & Machine Intelligence, p. 1 (2016)
4. Farhan, E., Meir, E., Hagege, R.: Local region expansion: a method for analyzing and refining image matches. SIAM J. Imag. Sci. **8**(4), 2771–2813 (2017)
5. Gottlieb, K., et al.: Endoscopy and central reading in inflammatory bowel disease clinical trials: achievements, challenges and future developments. Gut **70**(2), 418–426 (2021)
6. Hore, A., Ziou, D.: Image quality metrics: PSNR vs. SSIM. In: 2010 20th International Conference on Pattern Recognition, pp. 2366–2369. IEEE (2010)
7. Jégou, S., Drozdzal, M., Vazquez, D., Romero, A., Bengio, Y.: The one hundred layers tiramisu: Fully convolutional DenseNets for semantic segmentation. In: Proceedings of the IEEE Conference on Computer Vision and Pattern Recognition Workshops, pp. 11–19 (2017)
8. Klingner, M., Termöhlen, J.A., Mikolajczyk, J., Fingscheidt, T.: Self-supervised monocular depth estimation: solving the dynamic object problem by semantic guidance. ArXiv abs/2007.06936 (2020)
9. Leonard, S., et al.: Evaluation and stability analysis of video-based navigation system for functional endoscopic sinus surgery on in vivo clinical data. IEEE Trans. Med. Imaging **37**(10), 2185–2195 (2018)

10. Liu, X., et al.: Dense depth estimation in monocular endoscopy with self-supervised learning methods. IEEE Trans. Med. Imag. **PP**(99), 1 (2019)
11. Liu, X., et al.: Reconstructing sinus anatomy from endoscopic video – towards a radiation-free approach for quantitative longitudinal assessment. In: Martel, A.L., et al. (eds.) MICCAI 2020. LNCS, vol. 12263, pp. 3–13. Springer, Cham (2020). https://doi.org/10.1007/978-3-030-59716-0_1
12. Ma, R., et al.: RNNSLAM: reconstructing the 3d colon to visualize missing regions during a colonoscopy. Med. Image Anal. **72**, 102100 (2021)
13. Mur-Artal, R., Tardós, J.D.: ORB-SLAM2: an open-source slam system for monocular, stereo, and RGB-D cameras. IEEE Trans. Rob. **33**, 1255–1262 (2017)
14. Ozyoruk, K.B., et al.: EndoSLAM dataset and an unsupervised monocular visual odometry and depth estimation approach for endoscopic videos. Med. Image Anal. **71**, 102058 (2021)
15. Schonberger, J.L., Frahm, J.M.: Structure-from-motion revisited. In: Proceedings of the IEEE Conference on Computer Vision and Pattern Recognition, pp. 4104–4113 (2016)
16. Smith, L.N.: Cyclical learning rates for training neural networks. In: 2017 IEEE Winter Conference on Applications of Computer Vision (WACV), pp. 464–472. IEEE (2017)
17. Sung, H., et al.: Global cancer statistics 2020: Globocan estimates of incidence and mortality worldwide for 36 cancers in 185 countries. CA: Cancer J. Clin. **71**(3), 209–249 (2021)
18. Wang, R., Pizer, S.M., Frahm, J.M.: Recurrent neural network for (un-)supervised learning of monocular videovisual odometry and depth (2019)
19. Yin, Z., Shi, J.: Geonet: Unsupervised learning of dense depth, optical flow and camera pose. In: 2018 IEEE/CVF Conference on Computer Vision and Pattern Recognition (CVPR) (2018)
20. Zhou, T., Brown, M., Snavely, N., Lowe, D.G.: Unsupervised learning of depth and ego-motion from video. Arxiv (2017)
21. Zhu, S., et al.: Very large-scale global SFM by distributed motion averaging. In: 2018 IEEE/CVF Conference on Computer Vision and Pattern Recognition, pp. 4568–4577 (2018)

Pseudo-Label Clustering-Driven Dual-Level Contrast Learning Based Source-Free Domain Adaptation for Fundus Image Segmentation

Wei Zhou[1], Jianhang Ji[1], Wei Cui[2], and Yugen Yi[3](✉)

[1] Shenyang Aerospace University, Shenyang 110136, China
[2] The Agency for Science, Technology and Research (A*STAR), Singapore, Singapore
[3] Jiangxi Normal University, Nanchang 330022, China
yiyg510@jxnu.edu.cn

Abstract. Source-Free Domain Adaptation (SFDA) has gained attention as a promising solution to address the domain shift issue, eliminating the requirement for labeled data from the source domain. However, current SFDA methods heavily rely on self-training, which are confronted with two main challenges: inevitable occurrence of noisy pseudo-labels and insufficient adaptation across a single scale or level. To overcome these limitations, a novel SFDA method is developed for fundus image segmentation across different datasets. Our method encompasses two essential phases: the generation phase and the adaptation phase. In the generation phase, we introduce clustering to SFDA segmentation and propose a feature-enhanced clustering method to generate robust pseudo-labels. This process improves adaptation quality particularly when the source model's feature learning capability is limited in the target domain. In the adaptation phase, we develop a dual-level contrast learning method aimed at mitigating domain shift through self-supervision. First, we present a full-scale feature-level contrast loss that utilizes low-level and high-level features from both the target domain data and its augmented version. This enables the model to acquire discriminative characteristics while minimizing disparities between the original and augmented data. Second, we design a clinical prior-guided label-level contrast loss to filter out low-quality pseudo-labels, providing favorable guidance for the segmentation model. Extensive experiments on cross-domain datasets of fundus images demonstrate its superiority over mainstream SFDA methods. In the challenging Drishti-GS target domain, our method surpasses SOTA models by 3.14% and 2.18% in optic disc and optic cup Dice scores, respectively. Codes are available at https://github.com/M4cheal/PCDCL-SFDA.

Keywords: Unsupervised domain adaptation · Source-free · Fundus image segmentation · Clustering · Dual-level contrast learning

Q. Liu et al. (Eds.): PRCV 2023, LNCS 14429, pp. 492–503, 2024.
https://doi.org/10.1007/978-981-99-8469-5_39

1 Introduction

In recent studies, deep learning models have shown impressive results in medical image segmentation [10]. However, the deployment of deep learning models in real-world scenarios does indeed come with challenges, and one prominent challenge is the domain shift issue [23]. The domain shift issue refers to the scenario where a machine learning model, which has been trained on a specific source domain such as data from a particular hospital, struggles to effectively apply its learned knowledge to a different target domain, such as data from another hospital. In this situation, the model's effectiveness may degrade significantly due to the differences in image properties and characteristics between the source and target domains. To address this issue, researchers have extensively investigated methods of Unsupervised Domain Adaptation (UDA) [4,9,17,19,22]. These methods typically rely on accessing to labeled data from the source domain and unlabeled data from the target domain. However, the availability of source domain data is restricted due to concerns regarding privacy and security [1]. As a solution with limited exploration, Source-Free Domain Adaptation (SFDA) has emerged [12,15], focusing on adapting the target domain using pre-trained source models and unlabeled target domain data.

Self-training is a popular technique employed in SFDA, where a source model is utilized to produce pseudo-labels for the target domain. These pseudo-labels are then utilized to train the target domain in a supervised manner. Achieving accuracy and reliability in the generated pseudo-labels is vital for the effectiveness of self-training in SFDA. To address this, several techniques have been developed to rectify the pseudo-labels generated by the source model in a denoising manner. Adaptive thresholding and uncertainty correction techniques are widely employed to enhance the quality of pseudo-labels [5,24]. Entropy minimization is another commonly used method in self-training. It combines entropy loss and prior knowledge predictors to transfer valuable knowledge between domains [1,2], which helps reduce uncertainty and improves the reliability of pseudo-labels. Recently, drawing inspiration from the successful application of contrast learning, researchers have incorporated this technique into the adaptation process for target domains. In existing methods, contrast learning is commonly employed to facilitate feature-level adaptation and label-level adaptation by utilizing target domain images alongside their enhanced counterparts. For instance, the researchers employ Batch Normalisation (BN) statistics to regenerate source-like images and compare the target domain image with the corresponding source-like images for contrast learning [25]. Similar techniques are also proposed in [14].

Although these techniques are beneficial in enabling the model to adapt to target domain, there remain certain challenges when dealing with substantial domain shift in real-world applications. One of these limitations is the generation of overconfident probability maps by the source model, leading to noisy pseudo-labels that can weaken the effectiveness of the adaptation phase. Additionally, medical imaging data can vary greatly in terms of acquisition protocols, imaging modalities, patient populations, and disease manifestations. As a result, relying solely on single-scale or single-level adaptation is insufficient [13]. Moreover, it is crucial to acknowledge the significance of low-level features that capture fundamental visual information and local patterns in tackling domain shift. Disregarding the influence of these features, extracted from shallow networks, would overlook their substantial impact on the model's ability to handle domain shift effectively [3,6].

To tackle the aforementioned challenges, this study presents a novel SFDA method, named Pseudo-label Clustering-driven Dual-level Contrast Learning, for fundus image segmentation. Our method comprises a generation phase and an adaptation phase. In the generation phase, a feature-enhanced clustering technique is proposed to generate reliable pseudo-labels by addressing the challenge of overconfident predictions in target domains. Although clustering-based methods have been developed for SFDA in classification tasks [7,11,26], no SFDA segmentation algorithm utilizing clustering currently exists. The introduction of clustering into the SFDA segmentation task promotes similar predictions among local nearest neighbors in the feature space, enhancing adaptability and robustness. However, insufficient adaptation of the pre-trained source model to the target domain can hinder its feature learning capability, leading to inaccurate estimation of pseudo-labels. To tackle this limitation, entropy and prediction probabilities are incorporated into the feature enhancement process to refine the features, ensuring that the generated pseudo-labels are more accurate. Furthermore, feature-level clustering is performed to enhance the quality of the pseudo-labels, thereby improving their reliability. In the adaptation phase, a dual-level contrast learning method is developed to address the limitations of single-scale or single-level adaptation. To ensure better feature alignment, a full-scale feature-level contrast loss is designed to diminish the differences between the source and target domains across all scales. Additionally, a clinical prior-guided label-level contrast loss is introduced to filter out low-quality pseudo-labels and offer dependable assistance for the segmentation model.

Notable contributions from our work include: (1) We introduce clustering into the SFDA segmentation task for the first time, along with a feature enhancement method, to generate robust pseudo-labels and greatly enhance the quality of the subsequent adaptation process. (2) We devise a dual-level contrast learning method, consisting of a full-scale feature-level contrast loss and a clinical prior-guided label-level contrast loss, enabling more effective adaptation in situations with significant domain shift. (3) Comprehensive experiments showcase the remarkable adaptability of our method towards various target domains.

2 Method

2.1 Overview

This study introduces a novel SFDA method for fundus image segmentation, comprising a generation phase and an adaptation phase, as illustrated in Fig. 1. The source model $M_s : x_s \rightarrow y_s$ is assumed to be trained using labeled images x_s from the source domain along with their corresponding labels y_s. The generation phase focuses on producing pseudo-labels using enhanced feature maps and clustering. In the adaptation phase, we incorporate a dual-level contrast loss to emphasize domain-invariant features and facilitate adaptation of the unfrozen

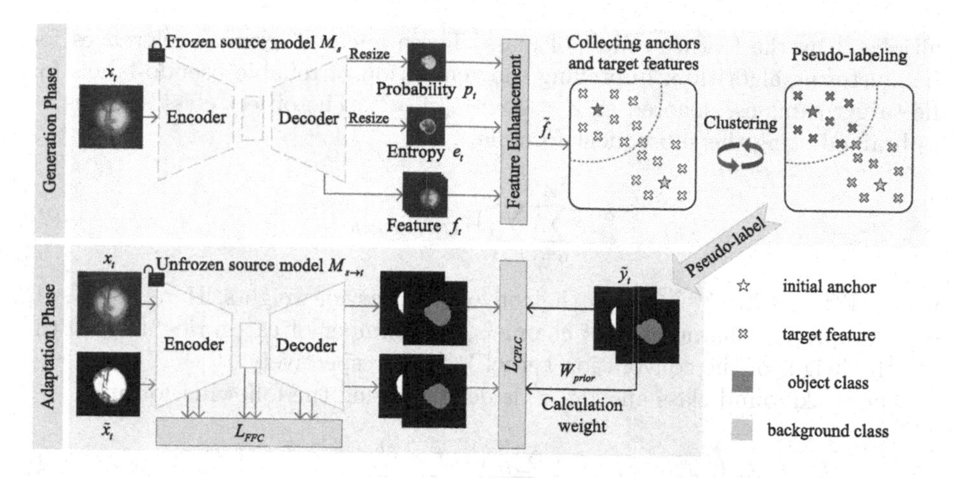

Fig. 1. Overview of the proposed SFDA method.

source model $M_{s\rightarrow t}$ to unlabelled target domain images x_t. Specifically, the dual-level contrast loss consists of a full-scale feature-level contrast loss L_{FFC} and a clinical prior-guided label-level contrast loss L_{CPLC}, contributing to improved performance in cross-domain OD and OC segmentation tasks.

2.2 Generation Phase

In the generation phase, ensuring the reliability of pseudo-labels is crucial. This paper designs a feature-enhanced clustering method to generate reliable pseudo-labels. We introduce clustering into the SFDA segmentation task for the first time, which encourages local nearest neighbors in the feature space to exhibit similar predictions, enhancing the model's adaptability and robustness in handling domain shift. However, a potential issue arises when directly utilizing the output feature map of the target domain for clustering. This is due to the limited target domain feature learning capability of the unadapted pre-trained source model, which may lead to inaccurate estimation of pseudo-labels for target data. To overcome this limitation, we advocate to minimize the false-negative regions (high entropy regions) and therefore propose a feature enhancement method. Our method enhances the output feature map f by utilizing additional information from the model's predictions, including the output probability map p and the entropy map e. The enhanced feature map $\tilde{f}$ is generated using the following specific equation:

$$\tilde{f} = f \times Normalise(e) + f \times p \tag{1}$$

where $Normalise(\cdot)$ denotes the normalization process.

Based on the obtained enhanced features, we apply feature-level k-means clustering to encourage dissimilar predictions for samples belonging to different clusters, thereby generating more accurate and reliable pseudo-labels. During the clustering process, the object class anchors z^o are initialized by the convolutional kernel weights of the output layer [7] and the background class anchors z^b are initialized by the feature center of mass. These anchors serve as references for the clustering algorithm, improving the generation of reliable pseudo-labels for the target samples, denoted as $Z = concat(z^o, z^b)$. The object class anchors z^o is obtained using the subsequent formula:

$$z^o = \sum_{a=1}^{A} \sum_{b=1}^{B} W^{conv}_{C \times K \times a \times b} \tag{2}$$

where $W^{conv} \in \mathrm{IR}^{C \times K \times A \times B}$ is the convolution kernel weights. Here, C, K, A, and B represent the number of channels, the number of categories, the width, and the height of the convolution kernel W^{conv}, respectively.

The background class anchors z^b is derived using the following formula:

$$z^b = \frac{\sum_{i=1}^{H \times W} \tilde{f} \times \hat{y}_i^b}{\sum_{i=1}^{H \times W} \hat{y}_i^b} \tag{3}$$

where the background class prediction mask, denoted as $\hat{y}^b$, is generated by performing a logical AND operation on the complement of the target class binary mask and the binary high entropy map. The target class binary mask is generated by applying a thresholding operation to the probability map p, while the binary high entropy map is obtained by applying a threshold to the entropy map e.

Finally, an iterative k-means clustering procedure is performed on the enhanced features $\tilde{f}$ and anchors Z to obtain the final pseudo-labels $\tilde{y}$. Notably, only the OC region is significantly affected by domain shift within the fundus image segmentation task. Therefore, the clustering process is applied only to the OC region for efficiency reasons, while the OD region utilizes the prediction mask of the source model as the pseudo-label.

2.3 Adaptation Phase

To address the insufficient adaptation across a single scale or level in target domains with significant domain shift, this study proposes a dual-level adaptation method, which contains a full-scale feature-level contrast loss and a clinical prior-guided label-level contrast loss. In contrast to the existing single-level or single-scale adaptation methods, our full-scale feature-level contrast loss aids in aligning features across multiple layers by measuring the L1 norm of the difference between target domain images and their enhanced versions. Notably, our method considers the influence of both low-level and high-level features, ensuring effective mitigation of domain shift. The formulation of the full-scale feature-level contrast loss L_{FFC} is defined as follows:

$$L_{FFC} = \sum_i \lambda_i \cdot \left| f_i^t - \tilde{f}_i^t \right|_1 \tag{4}$$

where f_i^t refers to the target feature of the i-th layer of the network, and $\tilde{f}_i^t$ is the enhanced feature. The term λ_i signifies the weight assigned to each layer. This weight adjustment enables effective feature alignment across multiple layers and enhances the overall adaptation process, contributing to improved performance in handling domain shift. Empirically, the encoder, global, and decoder features in the feature-level contrast loss are assigned weights of 0.4, 0.5, and 0.1, respectively, based on their relative importance in capturing domain-invariant features.

Despite the inherent stylistic differences between the target images x_t and their enhanced versions $\tilde{x}_t$, it is crucial for them to produce consistent segmentation results. Although existing methods commonly employ label-level contrast learning to ensure consistency, the effectiveness of pseudo-labeling can be hindered by domain shifts, thereby reducing the effectiveness of these approaches. To mitigate this issue, we propose a clinical prior-guided label-level contrast loss. This loss function leverages clinical knowledge and domain-specific information to guide the model's learning process and effectively filters out low-quality pseudo-labels. Therefore, by incorporating the proposed clinical prior-guided label-level contrast loss, the segmentation results can be enhanced. In this study, we introduce the concept of compactness, represented as C^{prior}, to impose a constraint on the pseudo-labels. Additionally, to account for the circular shape prior of the target region, we design a circular compactness metric V^{prior} as an additional constraint. The weight W^{prior} is defined as the ratio of the circular compactness metric C^{prior} to V^{prior}:

$$W^{prior} = \frac{C^{prior}}{V^{prior}} = \frac{4\pi a/p^2}{\sqrt{\sum_{i=1}^{n}\left(d_i - \bar{d}\right)/n}} \tag{5}$$

where p represents the perimeter of the segmented target, while a denotes its area. d_i signifies the distance from the boundary pixel point to the center, and $\bar{d}$ corresponds to the mean value of d_i. The variable n reflects the count of boundary pixel points. Lower-quality pseudo-labels yield smaller values of C^{prior}, resulting in reduced corresponding weights. Similarly, smaller values of V^{prior} are associated with lower weights. The clinical prior-guided label-level contrast loss L_{CPLC} is formulated as follows:

$$L_{CPLC} = W^{prior}\left(L^{t}seg + L^{\tilde{t}}seg\right) \tag{6}$$

where L^{t}_{seg} and $L^{\tilde{t}}_{seg}$ represent the discrepancies between the output labels and pseudo-labels of the target domain image and its enhanced image, respectively. Following [21], our L_{seg} involves a Dice loss and a smoothing loss. The Dice loss measures the overlap between the predicted binary masks (y and $\tilde{y}$), while the smoothing loss encourages uniform predictions in the neighborhood by considering the interaction between predicted and true values. The weight W^{prior} dynamically adjusts the significance of the label-level contrast loss within the overall optimization process.

Finally, in the adaptation phase, the total loss is computed as the sum of the individual components described above, which is formulated as:

$$L_{adapt} = L_{FFC} + \gamma \times L_{CPLC} \tag{7}$$

where γ is a weighting factor and its value is specifically set to 0.001.

3 Experiments

3.1 Datasets and Evaluation Metric

Experiments are carried out on the RIGA+ dataset [9], a multi-domain dataset designed for OD and OC segmentation. The dataset comprises five domains: BinRushed, Magrabia, BASE1, BASE2, and BASE3. To ensure fairness, we use annotations from the first physician for training and evaluation. BinRushed and Magrabia serve as the source domains, while BASE1, BASE2, and BASE3 are designated as the target domains. Among the target domains, BASE1, BASE2, and BASE3 consist of 173, 148, and 133 labeled samples, respectively, along with 227, 238, and 252 unlabeled samples. The unlabeled data is utilized for training the model, while the labeled data is reserved for testing purposes. To enhance realism, we introduce two additional fundus image datasets as target domains: the Drishti-GS [18] and the REFUGE training [16] datasets. The Drishti-GS dataset is split into 50 training images and 51 test images. The REFUGE training dataset consists of 400 labeled images, which we further divide into training and test subsets using 4-fold cross-validation due to the absence of official divisions. We evaluate the segmentation performance using the Dice score ($D, \%$). A higher D value indicates better segmentation performance.

3.2 Implementation Setup

Our method is implemented using the PyTorch 1.8.1 framework and trained on an NVIDIA TITAN RTX GPU. We resize the RIGA+ dataset from 800×800 to 512×512. For the Drishti-GS and REFUGE datasets, we utilize a pre-trained model to identify the centers of the OD regions. Subsequently, we crop the images to a size of 512×512 for further processing. The segmentation network employs a UNet framework with a ResNet34 backbone. The adaptation phase utilizes the SGD optimizer with a momentum of 0.99 and an initial learning rate of $lr_0 = 0.001$. The learning rate follows a polynomial decay rule, defined as $lr = lr_0 \times (1 - t/T)^{0.9}$, where $T = 100$ denotes the maximum epoch and t represents the current epoch. The batch size is set to 16.

3.3 Experimental Results

In our study, our SFDA method is compared with five SOTA Domain Adaptation (DA) methods in various setups. These setups include the "Intra-Domain" setup, where both training and testing utilize target domain markers, the "w/o DA" setup, using the source model to test directly on the target domain, and the "Baseline" setup, which relies solely on self-training for DA. We evaluate three SFDA methods (AdaMI [1], DPL [5], and FSM [25]) and two UDA methods (BEAL [20] and CLR [8]). To ensure a fair comparison, we report their best performance and re-implement all SFDA methods using the same source model.

Table 1. The performance of our SFDA method and other compared methods in joint OD and OC segmentation. "SF" represents whether the method is in source-free setting. The best results for SFDA methods are marked in bold.

Methods	SF	BASE1		BASE2		BASE3		Drishti-GS		REFUGE	
		D_{OD}	D_{OC}	D_{OD}	D_{OC}	D_{OD}	D_{OC}	D_{OD}	D_{OC}	D_{OD}	D_{OC}
w/o DA	✓	94.70	82.04	89.73	70.84	92.85	78.91	92.17	66.65	86.49	74.70
Intra-Domain	×	95.65	86.00	95.65	87.61	95.35	88.59	96.09	84.20	95.47	86.60
Baseline	✓	95.33	84.22	92.18	76.58	94.76	83.10	94.89	74.13	88.71	78.27
BEAL [20]	×	95.98	83.73	96.18	85.13	96.38	85.31	90.81	78.73	91.24	80.39
CLR [8]	×	93.45	86.28	96.16	85.09	95.91	87.17	93.99	75.74	94.03	82.44
AdaMI [1]	✓	95.28	83.76	93.93	83.12	94.46	85.85	91.39	62.39	84.95	75.67
DPL [5]	✓	92.31	81.96	91.77	80.81	92.41	84.30	92.49	81.91	85.70	76.75
FSM [25]	✓	93.45	83.62	85.15	83.30	92.70	82.51	93.28	79.23	89.52	81.11
Ours	✓	**95.42**	**85.19**	**94.07**	**84.34**	**95.07**	**86.30**	**95.63**	**84.09**	**90.35**	**82.61**
w/o GP	✓	95.32	83.17	94.05	82.31	95.50	85.90	95.13	77.92	90.45	81.08
w/o AP*	✓	95.51	82.73	91.75	82.42	94.55	85.24	94.44	80.98	89.35	81.74
w/o AP	✓	95.44	84.24	91.09	82.93	94.22	86.12	94.52	79.68	88.45	81.69

Table 1 presents the achieved Dice similarity coefficients by our method and comparative methods. By observing the performance gap between "w/o DA" and "Intra-Domain", we gain a comprehensive understanding of the domain shift. Although AdaMI shows overall improvement, it performs worse on the Drishti-GS target domain compared to the "w/o DA" approach, possibly due to significant domain shift resulting in different category proportions. Despite the significant domain shift, DPL achieves competitive results. However, crudely discarding suspect noise can compromise the representativeness of the learned features and alter the class distribution [24]. The performance of DPL on the BASE1 and REFUGE target domains is lower than that of the "Baseline". In contrast, FSM obtains better overall performance. However, the FSM method solely relies on single-scale contrast learning methods, limiting its adaptability. Consequently, the performance of FSM is sub-optimal. Overall, our method exhibits superior performance, particularly in scenarios with substantial domain shift between the source and target domains, as evident from the results obtained on the Drishti-GS dataset. Our method achieves outstanding results, with a Dice score of 95.63% for OD (D_{OD}) and 84.09% for OC (D_{OC}). These results highlight the effectiveness of our proposed method in mitigating domain shift, comparable even to UDA methods.

Figure 2 visually presents segmentation maps produced by our method, as well as other SFDA methods, in comparison to the Ground Truth ("GT") and "w/o DA" results. This visual comparison serves as compelling evidence, showcasing the ability of our method to generate segmentation results that closely match GT and provides further validation and support for the effectiveness of our approach.

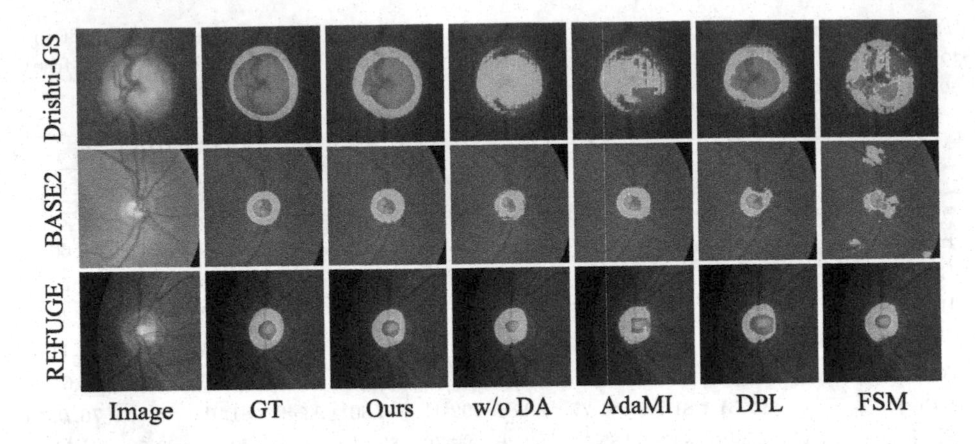

Fig. 2. Visualization of segmentation results by our SFDA method and competing methods, along with GT.

3.4 Ablation Experiments

Our method consists of two distinct phases designed to reduce domain shift: the Generation Phase (GP) and the Adaptation Phase (AP). We perform ablation experiments on three datasets (RIGA+, Drishti-GS, and REFUGE) to evaluate the effectiveness of GP and AP in our method. Table 1 demonstrates the performance of our SFDA method and its variants. The results show that the performance of the "w/o GP" variant decreases when the GP-generated pseudo-label is not used as supervision. Additionally, when the AP is completely removed and only GP-generated pseudo-labels are used, the "w/o AP*" variant performs better than the "Baseline" setup. These findings suggest that our GP produces effective pseudo-labels. Furthermore, when the AP is removed and replaced with the Baseline setup, the performance of the "w/o AP" variant significantly decreases. This observation suggests that incorporating full-scale feature alignment is advantageous for segmentation tasks, and the utilization of prior constraints proves effective in mitigating the inclusion of low-quality pseudo-labels. The "Ours" (combined GP and AP) yields the best segmentation performance, confirming their effectiveness. In summary, our ablation experiments demonstrate the effectiveness of the GP and AP phases in reducing domain shift and improving segmentation performance.

We also compare the variants "w/o Entropy" (without entropy) and "w/o Output" (without output probability) within the feature-enhanced clustering method. Figure 3(A) visually represents the results, highlighting the importance of incorporating both entropy and output probability enhancement techniques for reliable pseudo-label generation. Furthermore, we analyze the impact of dual-level contrast learning in the adaptation phase by removing specific feature contrasts. The removal of any feature contrast (global "W/o G", decoder output "W/o D", or encoder "W/o E") results in performance degradation, as shown in Fig. 3(B). Removing prior weighting in label-level contrasts "W/o W" leads

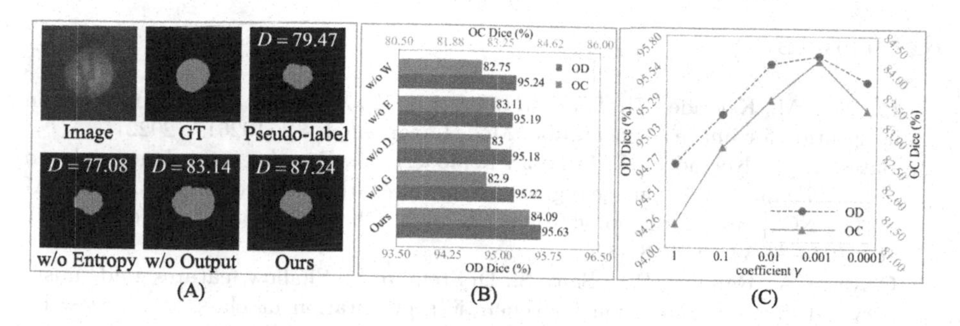

Fig. 3. Visual analysis of ablation experiments. (A) Comparison of GP variants without entropy and output probability. (B) Impact of feature contrast removal in the AP. (C) AP performance for different values of gamma.

to a significant decrease in performance, emphasizing the crucial role of prior weighting in achieving accurate segmentation results. Figure 3(C) displays the performance of the adaptation phase for different choices of γ, indicating the optimal choice of 0.001 for maximizing the score.

4 Conclusion

This study develops a novel SFDA method specifically designed for fundus image segmentation. Our method tackles the domain shift problem in both the generation phase and adaptation phase. In the generation phase, robust pseudo-labels are generated through the utilization of feature-enhanced clustering. These pseudo-labels serve as valuable annotations for the subsequent adaptation phase. In the adaptation phase, a dual-level contrast learning method is employed in a self-supervised manner to enhance the segmentation model's adaptability, resulting in improved performance. We conduct a comprehensive evaluation of the effectiveness of our method on three publicly available fundus image datasets. The experimental results indicate that our method outperforms SOTA SFDA methods, particularly on datasets characterized by significant domain shift. This highlights the superiority of our method in addressing the challenges posed by domain shift in fundus image segmentation task.

In future research, it is recommended to conduct studies on larger and more diverse datasets, including multi-center datasets, to validate the performance and generalizability of our method in practical clinical scenarios.

Acknowledgments. This work is supported in part by grants from the National Natural Science Foundation of China (Nos. 62062040, 62102270, 62041702), the project of Natural Science Foundation of Liaoning province (No. 2023-MS-246), the Outstanding Youth Project of Jiangxi Natural Science Foundation (No. 20212ACB212003), and the Jiangxi Province Key Subject Academic and Technical Leader Funding Project (No. 20212BCJ23017).

References

1. Bateson, M., Kervadec, H., Dolz, J., Lombaert, H., Ayed, I.B.: Source-free domain adaptation for image segmentation. Med. Image Anal. **82**, 102617 (2022)
2. Bateson, M., Kervadec, H., Dolz, J., Lombaert, H., Ben Ayed, I.: Source-relaxed domain adaptation for image segmentation. In: Martel, A.L., et al. (eds.) MICCAI 2020. LNCS, vol. 12261, pp. 490–499. Springer, Cham (2020). https://doi.org/10.1007/978-3-030-59710-8_48
3. Cardace, A., Ramirez, P.Z., Salti, S., Di Stefano, L.: Shallow features guide unsupervised domain adaptation for semantic segmentation at class boundaries. In: Proceedings of the IEEE/CVF Winter Conference on Applications of Computer Vision, pp. 1160–1170 (2022)
4. Chen, C., Dou, Q., Chen, H., Qin, J., Heng, P.A.: Unsupervised bidirectional cross-modality adaptation via deeply synergistic image and feature alignment for medical image segmentation. IEEE Trans. Med. Imaging **39**(7), 2494–2505 (2020)
5. Chen, C., Liu, Q., Jin, Y., Dou, Q., Heng, P.-A.: Source-free domain adaptive fundus image segmentation with denoised pseudo-labeling. In: de Bruijne, M., et al. (eds.) MICCAI 2021. LNCS, vol. 12905, pp. 225–235. Springer, Cham (2021). https://doi.org/10.1007/978-3-030-87240-3_22
6. Chen, Z., Pan, Y., Xia, Y.: Reconstruction-driven dynamic refinement based unsupervised domain adaptation for joint optic disc and cup segmentation. IEEE J. Biomed. Health Inf. **27**, 3537–3548 (2023)
7. Ding, N., Xu, Y., Tang, Y., Xu, C., Wang, Y., Tao, D.: Source-free domain adaptation via distribution estimation. In: Proceedings of the IEEE/CVF Conference on Computer Vision and Pattern Recognition, pp. 7212–7222 (2022)
8. Feng, W., et al.: Unsupervised domain adaptive fundus image segmentation with category-level regularization. In: Wang, L., Dou, Q., Fletcher, P.T., Speidel, S., Li, S. (eds.) Medical Image Computing and Computer Assisted Intervention – MICCAI 2022. MICCAI 2022. Lecture Notes in Computer Science, vol. 13432, pp. 497–506. Springer, Cham (2022). https://doi.org/10.1007/978-3-031-16434-7_48
9. Hu, S., Liao, Z., Xia, Y.: Domain specific convolution and high frequency reconstruction based unsupervised domain adaptation for medical image segmentation. In: Wang, L., Dou, Q., Fletcher, P.T., Speidel, S., Li, S. (eds.) Medical Image Computing and Computer Assisted Intervention – MICCAI 2022. MICCAI 2022. Lecture Notes in Computer Science, vol. 13437, pp. 650–659. Springer, Cham (2022). https://doi.org/10.1007/978-3-031-16449-1_62
10. Isensee, F., Jaeger, P.F., Kohl, S.A., Petersen, J., Maier-Hein, K.H.: nnU-Net: a self-configuring method for deep learning-based biomedical image segmentation. Nat. Methods **18**(2), 203–211 (2021)
11. Lee, J., Lee, G.: Feature alignment by uncertainty and self-training for source-free unsupervised domain adaptation. Neural Netw. **161**, 682–692 (2023)
12. Liang, J., Hu, D., Feng, J.: Do we really need to access the source data? source hypothesis transfer for unsupervised domain adaptation. In: International Conference on Machine Learning, pp. 6028–6039. PMLR (2020)
13. Liu, P., Tran, C.T., Kong, B., Fang, R.: CADA: multi-scale collaborative adversarial domain adaptation for unsupervised optic disc and cup segmentation. Neurocomputing **469**, 209–220 (2022)
14. Liu, X., Yuan, Y.: A source-free domain adaptive polyp detection framework with style diversification flow. IEEE Trans. Med. Imaging **41**(7), 1897–1908 (2022)

15. Liu, Y., Zhang, W., Wang, J.: Source-free domain adaptation for semantic segmentation. In: Proceedings of the IEEE/CVF Conference on Computer Vision and Pattern Recognition, pp. 1215–1224 (2021)
16. Orlando, J., et al.: Refuge challenge: a unified framework for evaluating automated methods for glaucoma assessment from fundus photographs. Med. Image Anal. **59**, 101570 (2020)
17. Perone, C.S., Ballester, P., Barros, R.C., Cohen-Adad, J.: Unsupervised domain adaptation for medical imaging segmentation with self-ensembling. Neuroimage **194**, 1–11 (2019)
18. Sivaswamy, J., Krishnadas, S., Chakravarty, A., Joshi, G., Tabish, A.S., et al.: A comprehensive retinal image dataset for the assessment of glaucoma from the optic nerve head analysis. JSM Biomed. Imaging Data Pap. **2**(1), 1004 (2015)
19. Varsavsky, T., Orbes-Arteaga, M., Sudre, C.H., Graham, M.S., Nachev, P., Cardoso, M.J.: Test-time unsupervised domain adaptation. In: Martel, A.L., et al. (eds.) MICCAI 2020. LNCS, vol. 12261, pp. 428–436. Springer, Cham (2020). https://doi.org/10.1007/978-3-030-59710-8_42
20. Wang, S., Yu, L., Li, K., Yang, X., Fu, C.-W., Heng, P.-A.: Boundary and entropy-driven adversarial learning for fundus image segmentation. In: Shen, D., et al. (eds.) MICCAI 2019. LNCS, vol. 11764, pp. 102–110. Springer, Cham (2019). https://doi.org/10.1007/978-3-030-32239-7_12
21. Wang, S., Yu, L., Yang, X., Fu, C.W., Heng, P.A.: Patch-based output space adversarial learning for joint optic disc and cup segmentation. IEEE Trans. Med. Imaging **38**(11), 2485–2495 (2019)
22. Wilson, G., Cook, D.J.: A survey of unsupervised deep domain adaptation. ACM Trans. Intell. Syst. Technol. **11**(5), 1–46 (2020)
23. Xie, X., Niu, J., Liu, X., Chen, Z., Tang, S., Yu, S.: A survey on incorporating domain knowledge into deep learning for medical image analysis. Med. Image Anal. **69**, 101985 (2021)
24. Xu, Z., Lu, D., Wang, Y., Luo, J., Wei, D., Zheng, Y., Tong, R.K.y.: Denoising for relaxing: Unsupervised domain adaptive fundus image segmentation without source data. In: Wang, L., Dou, Q., Fletcher, P.T., Speidel, S., Li, S. (eds.) Medical Image Computing and Computer Assisted Intervention – MICCAI 2022. MICCAI 2022. Lecture Notes in Computer Science, vol. 13435, pp. 214–224. Springer, Cham (2022). https://doi.org/10.1007/978-3-031-16443-9_21
25. Yang, C., Guo, X., Chen, Z., Yuan, Y.: Source free domain adaptation for medical image segmentation with Fourier style mining. Med. Image Anal. **79**, 102457 (2022)
26. Yang, S., Wang, Y., Wang, K., Jui, S., et al.: Attracting and dispersing: a simple approach for source-free domain adaptation. In: Advances in Neural Information Processing Systems (2022)

Patch Shuffle and Pixel Contrast: Dual Consistency Learning for Semi-supervised Lung Tumor Segmentation

Chenyu Cai[1], Jianjun He[1], Manlin Zhang[1], Yanxu Hu[1], Qiong Li[2], and Andy J. Ma[1,3,4](✉)

[1] School of Computer Science and Engineering, Sun Yat-sen University, Guangzhou, China
majh8@mail.sysu.edu.cn

[2] Department of Radiology, Sun Yat-sen University Cancer Center, Guangzhou, China

[3] Guangdong Province Key Laboratory of Information Security Technology, Guangzhou, China

[4] Key Laboratory of Machine Intelligence and Advanced Computing, Ministry of Education, Guangzhou, China

Abstract. Semi-supervised learning is a promising approach for medical image segmentation with limited labeled data. Though existing consistency learning based SSL methods achieve convincing results, they neglect the finer-grained information. In this work, we propose a novel dual consistency learning (DCL) method based on characteristics of medical images for semi-supervised lung tumor segmentation. For patch shuffle consistency learning, image patches are shuffled as a strong-augmented view to improve both the student and teacher models in the mean teacher framework. For pixel contrast consistency learning, we construct a memory bank by high-quality pixel features updated with reservation and obtain anchors from wrongly classified tumor and high-confident background features, making the pixel-level feature space more discriminative. Experiments on three lung tumor datasets demonstrate the effectiveness of our method for semi-supervised medical image segmentation.

Keyword: Tumor Segmentation, Semi-supervised Learning, Patch Shuffle Consistency, Pixel Contrast Consistency

1 Introduction

Medical image segmentation aims at localizing anatomical structures of interest, such as organs, lesions and tumors, by predicting pixel-level labels, which assists

---

C. Cai and J. He are co-first authors who have contributed equally to this work.
This work was supported partially by the NSFC (No. 62276281, No. 61906218) and the Science and Technology Program of Guangzhou (No. 202002030371).

Q. Liu et al. (Eds.): PRCV 2023, LNCS 14429, pp. 504–516, 2024.
https://doi.org/10.1007/978-981-99-8469-5_40

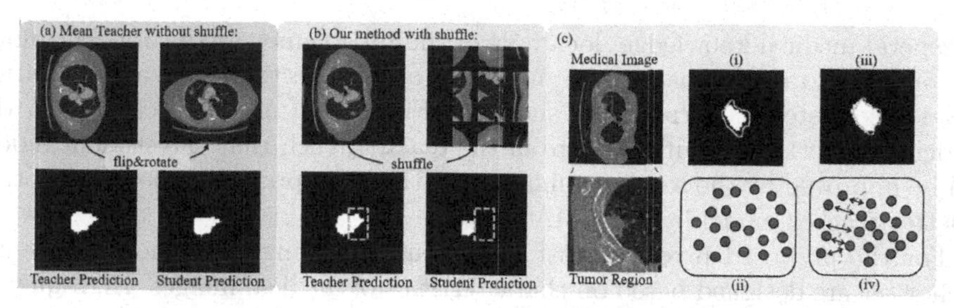

Fig. 1. Motivation of this work. (a) The predictions output are similar. (b) By shuffling patches, the student cannot predict as well as the teacher, resulting in more effective consistency learning. (c) With the existing method, some tumor pixels are misclassified (i), because the pixel-level feature space is not discriminative enough (ii). With the proposed pixel contrast consistency learning, the feature representations of tumor and background are better separated (iv), and thus improving the performance (iii).

disease analysis to improve the efficiency of diagnosis. Recently, many UNet-based encoder-decoder architectures [9,16] have been proposed and achieved encouraging results based on fully-supervised learning. However, they rely heavily on a large amount of high-quality pixel-level labeled data for training, which is expensive and time-consuming to collect. To lower the demand of the manual annotations, semi-supervised learning (SSL) is proposed, which fully utilizes the massive unlabeled data for medical image segmentation.

Among existing semi-supervised segmentation methods [4,7,17,21,27], the mean teacher [21,29] is a promising approach with effectiveness and stability to learn a teacher model by exponential moving average (EMA) of the student, and has been extended for medical image segmentation [6,12] recently. In these methods, the outputs from different views of each unlabeled image are constrained to be consistent with each other. It has been shown in existing works [10,27,29] that strong data augmentations to generate heavily-distorted views of the input images are beneficial for SSL. Nevertheless, strong augmentations like color distortion/jittering are not appropriate for medical images without color information (e.g. CT and MRI). By using only weak augmentations like flip, rotation and Gaussian noise, the outputs from student and teacher models are almost the same as shown in Fig. 1a, which leads to ineffective consistency learning.

On the other hand, most existing works on semi-supervised segmentation pay little attention to the discriminability and generalization ability of the pixel representations. To address this issue, contrastive learning initially developed for image classification [8], has been employed to solve the segmentation tasks for natural images [1,23,26,32]. Despite the success, existing methods designed for natural image segmentation do not take into account the specific characteristics of medical images. For example, a severe class-imbalance problem in lung tumor segmentation would occur compared to natural images.

To address the issues mentioned above, we propose a novel dual consistency learning (DCL) method for semi-supervised lung tumor segmentation, which consists of patch shuffle consistency and pixel contrast consistency. According to general medical knowledge, locations of the lung tumor are with uncertainty. We propose to shuffle patches in an image as an effective data augmentation way for consistency learning. As shown in Fig. 1b, by using patch shuffle, the student prediction output differs from the teacher, such that the student model can be improved by the teacher guidance. With the improved student, we obtain a better teacher model by the EMA update.

For the proposed pixel contrast consistency learning, a memory bank and anchor set are designed based on characteristics of medical images. High-quality pixel features of labeled data extracted by the teacher model are selected to construct the memory bank updated with reservation. The anchor set contains wrongly classified tumor pixel features of labeled data and high-confident background pixel features of unlabeled data extracted by the student model. By minimizing the distance between positive pairs and maximizing the distance between negative pairs from the memory bank and anchor set, the feature space become more discriminative as depicted in Fig. 1c.

Our main contributions are as follows: (1) We propose the patch shuffle consistency learning for patch position invariant tumor prediction. By EMA update, both the teacher and student models get better. (2) We propose the pixel contrast consistency learning based on characteristics of medical images. The memory bank with a update-with-reservation strategy is filled with high-quality pixel features. The anchor set is composed of wrongly classified tumor and high-confident background features, for discriminative feature learning. (3) Experiments on three lung tumor segmentation datasets demonstrate that our proposed method outperforms the state-of-the-art semi-supervised learning methods for medical image segmentation.

2 Related Work

2.1 Semi-Supervised Learning

Semi-supervised learning (SSL) has achieved great success in the field of computer vision. As stated in [1,4], there are two typical paradigms, i.e., self-training and consistency regularization.

Self-training. The self-training strategy [27] generates pseudo labels for unlabeled data, which is a form of entropy minimization. For semi-supervised image segmentation [29], both labeled data with ground truth and unlabeled data with pseudo labels are used to learn discriminative models.

Consistency Regularization. Consistency regularization aims to enforce consistent predictions among different views of an input image, which can be generated by perturbations. The Cutout [5] randomly cuts out a patch and filling it with random noise within an image. The CutMix [30] cuts and pastes patches among different images to create perturbed data. Nevertheless, they are not suitable for lung tumor segmentation, because the tumor region may be removed by using the Cutout, while the images produced by CutMix may contain pixels with confusing information from different patients. To address these issues, we propose the patch shuffle augmentation.

2.2 Semi-supervised Medical Image Segmentation

Semi-supervised learning has received much more attention in the field of medical image segmentation. One of the most important techniques is the Mean Teacher (MT) [21], which performs consistency learning via an exponential moving average (EMA) operation. In [12], an uncertainty-aware mean teacher (UA-MT) model is proposed. Mixup [2] encourages the consistency between the segmentation results of the interpolated unlabeled data and segmentation maps of the corresponding data.

Although these MT-based architectures achieve convincing results, they use only weak augmentations such as flip and rotation, which mainly focus on image-wise consistency. In contrast, our method is equipped with a stronger augmentation of patch shuffle, motivated by finer-grained patch-wise consistency, which helps utilize the uncertainty of tumor location for segmentation.

2.3 Contrastive Learning

Contrastive learning is firstly used in image classification [8] to discriminate positive image pairs from negative image pairs. In the segmentation task, it is applied on a set of pixel-wise features called anchors with their positive/negative pairs. Previous works [26,32] simply adopt the InfoNCE loss [8], resulting in only one positive sample selected for each pixel. For improvment, [1] define all the pixels of the same category as positive samples, while neglecting the importance of negative samples. To address the above issues, U^2PL [23] pays attention to false negative samples and obtain anchors with high confidence.

Despite the success, these methods lose sight of the characteristic of medical images. In contrast, our work obtains positive and negative samples from a memory bank by taking medical knowledge into account. Furthermore, we adopt a new anchor sampling criterion that differs from directly selecting pixels of the corresponding class according to the ground truth or pseudo labels.

3 Method

3.1 Overall Framework

Semi-supervised lung tumor segmentation is a pixel-wise binary classification task in which a small amount of labeled data $D_l = \{(x_i^l, y_i^l)\}_{i=1}^{N_l}$ and a large amount of unlabeled data $D_u = \{x_i^u\}_{i=1}^{N_u}$ are available for training. Here, y_i^l is the corresponding pixel-level annotations of image x_i^l, N_l and N_u are the numbers of labeled and unlabeled samples, respectively. Note that each image is made up of $H \times W$ pixels, where H denote the height and W is the width of the image. Denote the j-th pixel in image x_i^l as $x_{i,j}^l$. The label of $x_{i,j}^l$ is a one-hot vector $y_{i,j}^l \in \mathbb{R}^2$ with $y_{i,j,1}^l = 1$ for tumor or $y_{i,j,0}^l = 1$ for background pixel. Our goal is to train a lung tumor segmentation model by leveraging both labeled data D_l and unlabeled data D_u.

Innovatively, we propose a Dual Consistency Learning (DCL) method for semi-supervised lung tumor segmentation. The proposed DCL consists of Patch Shuffle Consistency Learning and Pixel Contrast Consistency Learning. The overall framework based on the teacher-student scheme is as shown in Fig. 2. Both the teacher and student models share the same architecture, which consists of a feature extractor f and a segmentation head g. For the student model, we minimize the supervised loss between the output prediction $p^l = g(f(x^l;\theta_1);\theta_2)$ and the corresponding ground truth y^l of the input labeled data x^l. Based on the proposed patch shuffle consistency elaborated in Sect. 3.2, the pseudo labels $\hat{y}^u$ of the input unlabeled data x^u output by the teacher model is used to supervise the student model with the shuffled x^u. For pixel contrast consistency learning detailed in Sect. 3.3, a memory bank $\mathcal{Q}$ is constructed saving high-quality features. Then, the pixel contrast consistency is constrained between the stored features in the memory bank and the anchors sampled from the feature extractors of both the student and the teacher models. In summary, the weights $\theta = (\theta_1, \theta_2)$ of the student model are optimized by the following loss function,

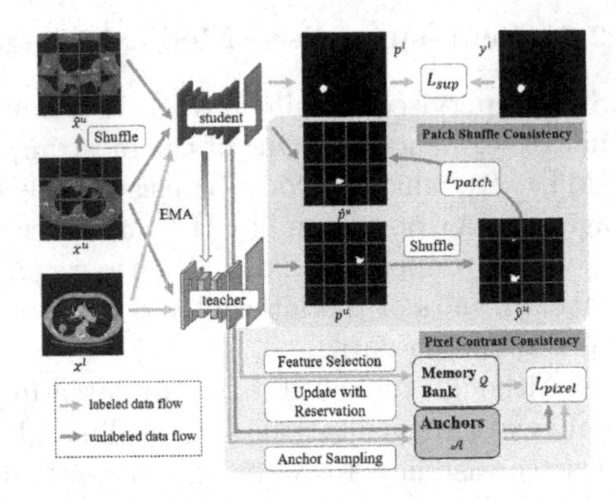

Fig. 2. Framework of our method.

$$L = L_{sup} + \lambda_{patch} L_{patch} + \lambda_{pixel} L_{pixel} \tag{1}$$

where L_{sup} is the supervised loss, L_{patch} and L_{pixel} are the proposed patch shuffle and pixel contrast consistency losses with trade-off hyper-parameters λ_{patch} and λ_{pixel} respectively. We compute the supervised loss L_{sup} by combining the Dice loss [15] and the Weighted Cross Entropy (WCE) loss to conquer the serious class-imbalance problem, i.e., $L_{sup} = L^l_{WCE} + L^l_{Dice}$.

After updating the weights θ of the student model, the weights ξ of the teacher model are computed by exponential moving average (EMA) of the student with a decay rate $\omega \in [0, 1]$, i.e., $\xi = \omega\xi + (1-\omega)\theta$.

3.2 Patch Shuffle Consistency Learning

In our method, we propose to shuffle the patches of each image for data augmentation as the input to student model. The images without the patch shuffle operation are fed to the teacher model to obtain pseudo-labels for guiding the training of the student.

Patch Shuffle Operation. We divide an unlabeled image x^u_i into $n \times n$ patches of equal size, denoted by $\{x^u_{i,k}\}^{n\times n}_{k=1}$. These patches are randomly permuted according to a shuffling order S to obtain an augmented image $\hat{x}^u_i$, which is composed of the same $n \times n$ patches as x^u_i but in a different order, i.e.,

$$\{\hat{x}^u_{i,S(k)}\}_{k=1}^{n\times n} \triangleq S(\{x^u_{i,k}\}_{k=1}^{n\times n}) \quad (2)$$

The shuffled images are fed into the student model to obtain prediction outputs $\hat{p}^u_i = g(f(\hat{x}^u_i;\theta_1);\theta_2)$. For consistency learning, the images x^u_i without shuffling are input to the teacher model to compute the prediction scores $p^u_i = g(f(x^u_i;\xi_1);\xi_2)$. Since the student outputs $\hat{p}^u_i$ are obtained through patch shuffle, the teacher predictions $p^u_i = g(f(x^u_i;\xi_1);\xi_2)$ need be shuffled by the same patch shuffle order S, i.e., $S(\{p^u_{i,k}\}_{k=1}^{n\times n})$. As shown in Fig. 2, pseudo labels are generated in every training iteration as follows:

$$\{\hat{y}^u_{i,k,1}\}_{k=1}^{n\times n} = \arg\max S(\{p^u_{i,k}\}_{k=1}^{n\times n}) \quad (3)$$

where $\hat{y}^u_{i,k}$ is the pseudo-label of the k-th patch with $\hat{y}^u_{i,k,0} = 1 - \hat{y}^u_{i,k,1}$.

Consistency Learning. We constrain the consistency between the teacher pseudo-label $\hat{y}^u_i$ and the student prediction outputs $\hat{p}^u_i$ for learning. Similar to the supervised learning process, the loss function for patch shuffle consistency learning is defined by combining the Weighted Cross Entropy loss and Dice loss between $\hat{p}^u_i$ from the student and $\hat{y}^u_i$ from the teacher model, i.e.,

$$L_{patch} = L^u_{WCE} + L^u_{Dice} \quad (4)$$

By minimizing the patch shuffle consistency loss L_{patch}, it enhances the model robustness for lung tumor segmentation. As shown in general medical knowledge, lung tumor can grow in arbitrary locations. The shuffle operation simulates such variability for model learning by detaching tumors from their original background and rearranging them in different orders. Unlike data augmentations like flip, rotation and Gaussian noise, what we propose changes the relative position of patches surrounding the tumors. Since tumors are no longer limited in their inherent neighbourhood, the semantic relationships between patches vary from the original ones. This forces the segmentation model to focus on the information within the patch containing tumors and thus improves the performance.

3.3 Pixel Contrast Consistency Learning

The flow chart of the proposed Pixel Contrast Consistency Learning is shown in Fig. 3. In our method, each pixel of the input image is projected to the high-dimensional feature space by the feature extractor f. Without loss of generality, the feature vector (or feature) of the j-th pixel in the i-th image is denoted as $z_{i,j}$. By using the teacher model, high-quality features are obtained by the *feature selection* mechanism to construct the *memory bank updated with reservation.* By using the student model, we propose an innovative *anchor sampling* strategy to select wrongly classified tumor and high-confident background features as anchors for contrastive learning. Finally, the pixel contrast loss function L_{pixel} is defined by the positive and negative pairs from the memory bank $\mathcal{Q}$ and the anchor set $\mathcal{A}$ for *optimization.* Consistency is maintained at the fine-grained pixel

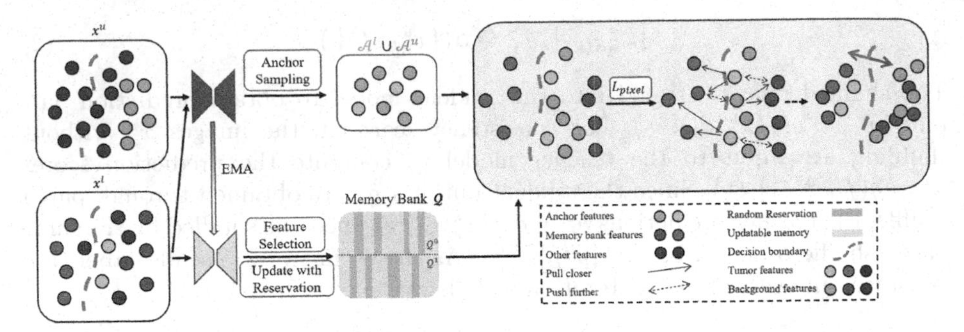

Fig. 3. Pixel contrast consistency learning. Labeled images x^l are input to the teacher model to construct the memory bank $\mathcal{Q}$ by feature selection and the update with reservation strategy. Meanwhile, anchor features are obtained by sampling from both the labeled and unlabeled data using the student model. By minimizing L_{pixel}, the feature space becomes more discriminative.

level between the memory bank constructed by the teacher model and anchor features extracted by the student model. Details are given as follows.

Feature Selection. To preserve high-quality features of both tumor and background pixels in the memory bank $\mathcal{Q}$, we devise a feature selection (FS) mechanism to filter out a large number of less representative feature vectors. Input the labeled data x_i^l to the teacher feature extractor, denote the softmax prediction probability of the j-th pixel in x_i^l as $p_{i,j}^l \in \mathbb{R}^2$. Two criteria are proposed to determine whether a pixel-level feature vector is of high quality. The first one is whether the prediction of the pixel is correct, while the second is whether the prediction confidence is higher than a fixed threshold $\tau = 0.95$ same as in [1]. The qualified feature vectors under feature selection mechanism are obtained as,

$$\mathcal{V} = \{z_{i,j}^l,\ s.t.,\ \hat{y}_{i,j}^l = y_{i,j}^l \wedge \max\{p_{i,j}^l\} > \tau\} \tag{5}$$

Memory Bank Update with Reservation. Since the supervised learning loss fits the data labels, the two criteria can be easily satisfied and numerous pixel features are selected in $\mathcal{V}$, which is likely larger than the memory bank size m. If we adopt a simple FIFO queue, all the previous feature vectors will be popped out after each iteration. This means the memory bank degrades to storing only the feature vectors in the current mini-batch without memorizing the history. To prevent from this issue, we design an update with reservation (UR) strategy for the memory bank $\mathcal{Q}$. Specifically, we set a reserved space of size $m'(m' < m)$ for the background $\mathcal{Q}^0$ and tumor memory $\mathcal{Q}^1$. In each iteration, m' randomly selected features are reserved and kept unchanged in the memory. The remaining $m - m'$ feature vectors are replaced by randomly selecting features from $\mathcal{V}$. This strategy ensures that previous high-quality feature vectors are retained in the memory bank for better contrastive learning.

Anchor Sampling. When applying contrastive learning to segmentation tasks, it will lead to unaffordable computational overhead by using all the pixel features. As discussed in [23,24], it is necessary to set a criterion to select appropriate anchor pixels for contrastive learning. In our method, the anchor sampling criterion is designed to obtain the feature vectors of anchor pixels $\mathcal{A}^l$ and $\mathcal{A}^u$ from the labeled and unlabeled data, respectively. Both the labeled data x^l and the unlabeled data x^u are fed into the student model $f(\cdot\ ;\theta_1)$ to obtain feature vectors of each pixel. Then, these features are input to the segmentation head $g(\cdot\ ;\theta_2)$ to make predictions and convert them to pseudo labels for anchor sampling.

Due to the severe class-imbalance problem, pixels are inclined to be predicted as background. A trivial solution is to predict all the pixels as background with low misclassification cost on the limited number of tumor pixels. Thus, it is more important to correct the misclassification error on the wrongly predicted tumor pixels, compared to the background pixels.

As a result, for the labeled data x^l, ground-truth tumor pixels are selected as anchors if they are wrongly predicted as background, i.e.,

$$\mathcal{A}^l = \{z^l_{i,j} \mid y^l_{i,j,1} = 1 \wedge \hat{y}^l_{i,j,1} = 0\} \tag{6}$$

On the other hand, the number of background pixels is much larger. With the larger training data, the model has stronger discrimination ability to background pixels. And the high confidence of models leads to accuracy. Consequently, for the unlabeled data x^u, pixels are selected as anchors if they are predicted as background with a confidence above the threshold τ same as in Eq. 5, i.e.,

$$\mathcal{A}^u = \{z_{i,j} \mid \hat{y}_{i,j,1} = 0 \wedge p_{i,j,0} > \tau\} \tag{7}$$

The combined anchor set is the union of $\mathcal{A}^l$ and $\mathcal{A}^u$: $\mathcal{A} = \mathcal{A}^l \cup \mathcal{A}^u$.

Optimization with L_{pixel}. Due to the relatively simple pattern of medical images, i.e., the same organs have similar shape, outline and content, the features stored in the memory bank are representative enough for contrastive learning. As a result, only the tumor and background features in the memory bank $\mathcal{Q}$ are used for optimization, without selecting specific pixels from the images as in other methods [31,32].

For anchors in $\mathcal{A}^l$, their labels are tumor, so the positive pairs are constructed by the features in the tumor memory bank, i.e., $z^{l,+}_{i,j} \in \mathcal{Q}^1$, and the negative pairs are from the background memory, i.e., $z^{l,-}_{i,j} \in \mathcal{Q}^0$. Similarly, for anchors in $\mathcal{A}^u$, since they are confidently predicted as background, the positive pairs and negative pairs are $z^{u,+}_{i,j} \in \mathcal{Q}^0$ and $z^{u,-}_{i,j} \in \mathcal{Q}^1$ respectively.

The similarity between the anchor and its positive and negative samples is measured by cosine similarity: $\mathcal{C}(z_{i,j}, q) = < z_{i,j}, q > /(\|z_{i,j}\|_2 \cdot \|q\|_2)$. Then, the cosine similarity is converted to cosine distance as: $\mathcal{D}(z_{i,j}, q) = 1 - \mathcal{C}(z_{i,j}, q)$. Finally, the pixel contrast consistency loss is formulated as follows:

$$L_{pixel} = \frac{1}{|\mathcal{A}|} \sum\nolimits_{z_{i,j} \in \mathcal{A}} \gamma\, \mathcal{D}(z_{i,j}, z^+_{i,j}) - (1-\gamma)\, \mathcal{D}(z_{i,j}, z^-_{i,j}) \tag{8}$$

where γ is the weight to balance the loss functions for the positive and negative pairs. By minimizing the pixel contrast consistency loss L_{pixel}, features of tumor and background pixels on the high-dimensional feature space become more separable, leading to a clearer classification boundary between tumor and background, and thus improving the segmentation performance.

4 Experiments

4.1 Datasets

We evaluate the proposed Dual Consistency Learning (DCL) method on three datasets including the MLT, a private dataset collected by ourselves, and two publicly available datasets MSD [19] and LIDC-IDRI [3]. More details are given in supplemental material.

Table 1. Comparison with state-of-the-art methods on MLT and MSD dataset with Dice (%), Jac. (%) and HD95.

Dataset			MLT						MSD					
Setting			5%			10%			10%			20%		
Method	Publication	MT based	Dice↑	Jac.↑	HD95↓	Dice↑	Jac.↑	HD95↓	Dice↑	Jac.↑	HD95↓	Dice↑	Jac.↑	HD95↓
CCT [17]	CVPR 2020	×	65.80	56.18	15.13	72.55	61.59	15.94	64.64	53.79	14.55	69.90	59.50	12.45
CPS [4]	CVPR 2021	×	68.68	57.94	**13.92**	73.88	63.20	15.45	65.59	55.57	13.38	71.79	62.31	10.08
URPC [25]	MICCAI 2021	×	69.18	58.78	14.13	74.15	63.79	15.75	65.76	55.63	12.19	69.63	59.85	**9.61**
MC-Net+ [28]	MEDIA 2022	×	69.41	59.15	14.42	73.04	62.95	14.62	66.01	55.33	14.02	70.27	60.06	11.41
MT [21]	ICLR 2017	✓	67.30	57.11	17.93	71.89	61.87	14.75	64.47	54.34	13.76	70.80	61.63	9.67
UAMT [12]	MICCAI 2019	✓	67.85	57.62	16.21	71.62	61.31	14.92	65.89	55.71	14.42	70.93	61.71	10.78
CMB [1]	ICCV 2021	✓	66.98	56.29	18.53	71.95	61.93	16.73	64.54	54.38	15.87	71.07	61.11	14.08
ICT [22]	NN 2022	✓	69.23	58.94	14.98	74.06	63.87	14.68	66.60	57.07	13.70	71.35	61.82	11.14
DCL (**ours**)	This paper	✓	**71.61**	**61.37**	14.54	**76.39**	**66.23**	**13.82**	**67.45**	**57.90**	**11.18**	**71.87**	**62.34**	10.56

4.2 Implementation Details

Architecture. We take the U-Net [16] as the backbone in our experiments. When compared to other methods, results are reproduced with the same network architecture (both f and g), except CCT [17] and MC-Net+ [28]. For these two methods, multiple decoders are required, so two more decoders are copied to construct a special U-Net.

Optimization. For all the experiments, we train for 300 epochs using the SGD optimizer with a momentum of 0.9 and weight decay of 1×10^{-4} following [11]. The learning rate η_0 of the feature extractor f is set to 1×10^{-2}, and 5×10^{-2} for the segmentation head g, with a poly learning rate schedule as in [23,24], $\eta = \eta_0 \times (1 - iter/max\_iter)^{0.9}$. We set $\lambda_{pixel} = 0.1$ in Eq. 1 as [1]. The remaining parameters are determined empirically as $\lambda_{patch} = 0.2$ (Eq. 1), $n = 8$ (Eq. 2), $\gamma = 0.8$ (Eq. 8). The size of memory bank is $m = 512$, with 25% reserved space. All the hyper-parameters are fixed for all the datasets and experiment settings.

Other Details. For all the datasets, we randomly apply rotate and flip augmentations to all images as pre-processing. We use several well-known metrics in medical image segmentation, i.e., Dice similarity coefficient, Jaccard (Jac.) similarity coefficient and the 95% Hausdorff Distance (HD95). Only the teacher model $g(f(\cdot;\xi_1),\xi_2)$ is used for inference following the common practice [10]. More details are given in supplemental material.

4.3 Comparison with Existing Works

This section reports experimental results of lung tumor segmentation comparing the proposed DCL with recent semi-supervised image segmentation methods [1, 4,12–14,17,18,20–22,25,28].

Results on MLT Dataset. Table 1 shows the comparison results under two partition protocols on the MLT dataset. The proposed DCL outperforms the SOTA methods for all metrics in the setting of 10% labeled data. Specifically, improvement of 4.50%, 4.36% and 2.93 are achieved compared with the mean teacher [21] baseline. Moreover, in the challenging 5% setting, our method still significantly outperforms the baseline on all three metrics by 4.31%, 4.26%, and 3.39, respectively. The proposed DCL surpasses the SOTA methods on Dice and Jac., and also achieves comparable results in terms of HD95.

Results on MSD Dataset. Table 1 compares different methods on the MSD dataset. Under the 20% labeled data setting, our performance gains over the mean teacher are 1.07% on Dice and 0.71% on Jac. When it comes to the harder setting with less annotated data (10%), the performance gains further increase to 2.98%, 3.56%, and 2.58 on Dice, Jac. and HD95. This demonstrates the great potential of DCL for working under a more challenging setting with less labeled data. The visualization results in Fig. 4 comparing to the SOTA on the MLT and MSD datasets also illustrate the advantages of our method.

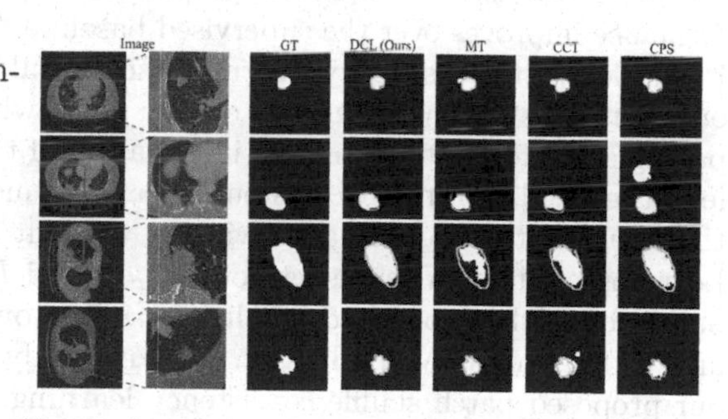

Fig. 4. Qualitative results on MLT and MSD. From left to right: Input Image, Ground truth, DCL (Ours), MT, CCT and CPS. The first and the second rows are from the MLT, the third and fourth are from the MSD. The yellow curves stand for the edge of ground truth labels. (Color figure online)

Results on LIDC-IDRI Dataset. Table 2 records the comparison results on the LIDC-IDRI. Some of the results are cited from [18]. Our DCL outperforms the UDiCT [18] designed for lung tumor segmentation by 2.29% for Dice and 1.07% for Jac. Without special design for edge segmentation, the HD95 of our method is not the best but comparable with others.

Table 2. Comparison with state-of-the-art methods on LIDC-IDRI dataset with Dice(%), Jac.(%) and HD95. The symbol † refers to results cited from [18].

Method	Publication	MT based	Dice↑	Jac.↑	HD95↓
VAT† [20]	TPAMI 2019	×	47.27	39.72	17.29
CCT† [17]	CVPR 2020	×	41.21	32.52	18.34
SASSnet† [13]	MICCAI 2020	×	45.40	37.91	19.39
DTC† [14]	AAAI 2021	×	46.95	39.98	15.48
UDiCT† [18]	TMI 2023	×	50.10	42.52	**14.67**
MC-Net+ [28]	MEDIA 2022	×	51.44	43.02	15.23
MT [21]	ICLR 2017	✓	46.81	39.05	22.70
UAMT† [12]	MICCAI 2019	✓	47.15	39.35	21.28
ICT [22]	NN 2022	✓	50.53	42.77	16.38
DCL (**ours**)	This paper	✓	**52.39**	**43.59**	16.01

4.4 Ablation Studies

Table 3. Impact of different loss functions in Eq. 1.

L_{sup}	L_{patch}	L_{pixel}	Dice (%) on MLT	Dice (%) on MSD
✓			71.70	64.27
✓	✓		74.64	66.40
✓		✓	71.92	65.62
✓	✓	✓	**76.39**	**67.45**

We investigate the effectiveness of the loss functions in Eq. 1. All the experiments are done on the MLT and MSD datasets with the setting of 10% labeled data. As shown in Table 3, on the MLT dataset, when only L_{sup} is used, Dice is 71.70%. By adding the patch shuffle consistency loss, Dice gets a 2.94% improvement. When applying the pixel contrast consistency loss, it also gets a 0.22% gain. Similarly, on the MSD, by using each of the proposed components independently, the performance improves over the supervised baseline. The improvement by using the Pixel Contrast Consistency Learning module alone is relatively slight, since it only considers the characteristics of each pixel, which is too fine-grained, without paying attention to the semantic information of the tumor as a whole. Nevertheless, the Pixel Contrast Consistency Learning module can consistently improve the segmentation performance with or without the Patch Shuffle Consistency Learning module. When using both L_{patch} and L_{pixel}, our method significantly outperforms the supervised baseline based on only L_{sup} by 4.69% on the MLT and 3.18% on the MSD dataset. This not only illustrates the effectiveness of our proposed patch shuffle consistency learning and pixel contrast consistency learning, but also validates the rationality of the dual consistency learning from the two granularities of patch and pixel. The dual consistency learning (DCL) from the two granularities of patch and pixel is reasonable. The proposed two modules complementing each other are indispensable.

5 Conclusion

In this paper, we present a novel Dual Consistency Learning (DCL) method for semi-supervised lung tumor segmentation. In patch shuffle consistency learning, patch shuffle is used to differ the student prediction output from teacher's. With pixel contrast consistency learning in the feature space, more discriminative features are learned. Extensive experiments demonstrate the effectiveness of the proposed DCL compared with the state of the art. Ablation studies verify that each component can help to boost the performance for lung tumor segmentation.

References

1. Alonso, I., et al.: Semi-supervised semantic segmentation with pixel-level contrastive learning from a class-wise memory bank. In: ICCV, pp. 8199–8208 (2021)
2. Basak, H., et al.: An exceedingly simple consistency regularization method for semi-supervised medical image segmentation. In: ISBI, pp. 1–4 (2022)
3. Charles, F., et al.: The lung image database consortium (LIDC) and image database resource initiative (IDRI): a completed reference database of lung nodules on CT scans. Med. Phys. **38**(2), 915–931 (2011)
4. Chen, X., et al.: Semi-supervised semantic segmentation with cross pseudo supervision. In: CVPR, pp. 2613–2622 (2021)
5. DeVries, T., Taylor, G.W.: Improved regularization of convolutional neural networks with cutout. arXiv preprint arXiv:1708.04552 (2017)
6. Gaurav, F., et al.: Extreme consistency: overcoming annotation scarcity and domain shifts. In: MICCAI, pp. 699–709 (2020)
7. Geoffrey, F., et al.: Semi-supervised semantic segmentation needs strong, varied perturbations. In: BMVC (2020)
8. He, K., et al.: Momentum contrast for unsupervised visual representation learning. In: CVPR, pp. 9726–9735 (2020)
9. Isensee, F., et al.: nnU-Net: a self-configuring method for deep learning-based biomedical image segmentation. Nat. Methods **18**(2), 203–211 (2021)
10. Kihyuk, S., et al.: Fixmatch: Simplifying semi-supervised learning with consistency and confidence. In: NeurIPS. pp. 596–608 (2020)
11. Lei, T., et al.: Semi-supervised medical image segmentation using adversarial consistency learning and dynamic convolution network. TMI **42**(5), 1265–1277 (2023)
12. Lequan, Y., et al.: Uncertainty-aware self-ensembling model for semi-supervised 3D left atrium segmentation. In: MICCAI, pp. 605–613 (2019)
13. Li, S., Zhang, C., He, X.: Shape-aware semi-supervised 3D semantic segmentation for medical images. In: MICCAI, pp. 552–561 (2020)
14. Luo, X., et al.: Semi-supervised medical image segmentation through dual-task consistency. In: AAAI, pp. 8801–8809 (2021)
15. Milletari, F., Navab, N., Ahmadi, S.: V-Net: fully convolutional neural networks for volumetric medical image segmentation. In: 3DV, pp. 565–571 (2016)
16. Olaf, R., et al.: U-Net: convolutional networks for biomedical image segmentation. In: MICCAI, pp. 234–241 (2015)
17. Ouali, Y., et al.: Semi-supervised semantic segmentation with cross-consistency training. In: CVPR, pp. 12671–12681 (2020)
18. Qiao, P., et al.: Semi-supervised CT lesion segmentation using uncertainty-based data pairing and SwapMix. IEEE Trans. Med. Imaging **42**(5), 1546–1562 (2023)
19. Simpson, A.L., Antonelli, M., Bakas, S., et al.: A large annotated medical image dataset for the development and evaluation of segmentation algorithms. arXiv preprint arXiv:1902.09063 (2019)
20. Takeru, M., et al.: Virtual adversarial training: a regularization method for supervised and semi-supervised learning. IEEE Trans. Pattern Anal. Mach. Intell. **41**(8), 1979–1993 (2019)

21. Tarvainen, A., Valpola, H.: Mean teachers are better role models: weight-averaged consistency targets improve semi-supervised deep learning results. In: NeurIPS, pp. 1195–1204 (2017)
22. Vikas, V., et al.: Interpolation consistency training for semi-supervised learning. Neural Netw. **145**, 90–106 (2022)
23. Wang, Y., et al.: Semi-supervised semantic segmentation using unreliable pseudo labels. In: CVPR (2022)
24. Wenguan, W., et al.: Exploring cross-image pixel contrast for semantic segmentation. In: ICCV, pp. 7283–7293 (2021)
25. Xiangde, L., et al.: Efficient semi-supervised gross target volume of nasopharyngeal carcinoma segmentation via uncertainty rectified pyramid consistency. In: MICCAI (2021)
26. Xin, L., et al.: Semi-supervised semantic segmentation with directional context-aware consistency. In: CVPR, pp. 1205–1214 (2021)
27. Yang, L., et al.: St++: make self-trainingwork better for semi-supervised semantic segmentation. In: CVPR, pp. 4258–4267 (2022)
28. Yicheng, W., et al.: Mutual consistency learning for semi-supervised medical image segmentation. Media **81**, 102530 (2022)
29. Yuan, J., et al.: A simple baseline for semi-supervised semantic segmentation with strong data augmentation. In: ICCV, pp. 8209–8218 (2021)
30. Yun, S., et al.: CutMix: regularization strategy to train strong classifiers with localizable features. In: ICCV, pp. 6022–6031 (2019)
31. Zenggui, C., Zhouhui, L.: Semi-supervised semantic segmentation via prototypical contrastive learning. In: ACMMM, pp. 6696–6705 (2022)
32. Zhong, Y., et al.: Pixel contrastive-consistent semi-supervised semantic segmentation. In: ICCV, pp. 7253–7262 (2021)

Multi-source Information Fusion for Depression Detection

Rongquan Wang[1], Huiwei Wang[1], Yan Hu[1], Lin Wei[2], and Huimin Ma[1(✉)]

[1] University of Science and Technology Beijing, Beijing 100083, China
g20208774@xs.ustb.edu.cn, {rongquanwang,huyan,mhmpub}@ustb.edu.cn
[2] Civil Aviation Flight University of China, Guanghan 618307, China
weilin@cafuc.edu.cn

Abstract. Depression is the most common psychiatric disorder. Traditional depression detection methods almost rely on structured scales and clinical opinions, which carry the risk of subjective judgment. In light of this, we investigate the potential of employing emotional images as stimuli for depression detection. Our proposed method is the first to utilize pupil dilation, blink patterns, and eye movements as features for depression detection. Notably, we introduce a comprehensive set of strategies for extracting visual cognitive features, validating the efficacy of the pupil emotion response theory and blink emotion response theory. Finally, we train a Support Vector Machine (SVM) classifier to differentiate between depressed and normal subjects, achieving an impressive accuracy of 89.5%, which is higher than other state-of-the-art methods in automatic depression detection.

Keywords: Depression detection · Multi-source information fusion · Visual cognition · Eye movements · Pupil dilation · blinking patterns

1 Introduction

According to the Global Mental Health Report, the prevalence of common mental disorders such as depression increased by 25% in the first year of the COVID-19 pandemic, with 251 to 310 million people suffering from depression worldwide [12]. Most of traditional depression detection approaches heavily rely on subjective structured scales and clinical diagnoses, resulting in subjectivity, time consumption, and substantial resource requirements. Therefore, traditional methods are likely to cause many patients and even severe patients to fail to diagnose and treat in time. According to the American Psychiatric Association, more than 50% of patients do not obtain effective treatment [16].

In recent years, researchers in the fields of psychology and artificial intelligence have attempted to develop more objective and efficient methods for depression detection. Physiological signals such as reaction time [8], electroencephalography (EEG) [11], and functional magnetic resonance imaging (fMRI) [1] have been shown to be useful in identifying depression. With the development of attentional biases theory, eye-tracking becomes an important feature for the study of

Q. Liu et al. (Eds.): PRCV 2023, LNCS 14429, pp. 517–528, 2024.
https://doi.org/10.1007/978-981-99-8469-5_41

depression, with traditional methods including the Stroop task, spatial cueing task, visual search task, and point detection task [4,6,14].

Many image-based methods for depression detection have emerged recently. Eizenman et al. [2], Kellough et al. [7], and Lu et al. [9] found that depressed subjects spent more time gazing at negative images than normal subjects. Zeng et al. [18] used a free viewing paradigm to extract eye-movement features to classify 36 subjects (18D (Depressed subjects) +18N(Normal subjects)) and obtained a maximum of 76.04% of accuracy. Zhu et al. [19] also used a free viewing paradigm with four emotional faces to classify 36 subjects (18D+18N), and they obtained a maximum accuracy of 82.5% using a content-based integration method. Shen et al. [13] used a free viewing and task-driven paradigm to extract eye-movement features to classify 56 subjects (29D+27N) and achieved 77.0% of accuracy.

However, the limited diversity of feature extraction methods employed in the aforementioned approaches constrains their detection accuracy. Existing studies have demonstrated that pupil dilation and constriction serve as outward indicators of cognitive and emotional processing. Compared with normal individuals, the pupil dilation of depressed patients gazing at negative emotional stimuli is greater and longer lasting [15,17]. Meanwhile, depressed patients' blink more frequently than normal individuals [10]. Furthermore, people blink more frequently in stressful situations compared to neutral or relaxed situations [3]. Therefore, according to the above theories, we propose a depression detection method based on pupil dilation, blink patterns, and eye movement features. The main contributions of this paper are as follows:

1. Based on the theories of pupil emotion response, blink emotion response, and attentional bias, this paper introduces a novel experimental paradigm. This paradigm effectively elicits emotional responses from participants, facilitating the extraction of various visual cognitive features. Additionally, we construct a multidimensional semantic image dataset, utilizing the images from the dataset as emotional stimuli for the subjects.
2. This paper presents a depression detection method based on multi-source information features. Here, we first use pupil features and blink features to detect depression, novel approaches for extracting pupil features and blink features are proposed. The incorporation of these features increases the dimensionality of the method, leading to improved accuracy and stability.
3. We trained a Support Vector Machine (SVM) classifier to detect depressed and normal subjects for depression classification. Our method achieves a detection accuracy of 89.5%, outperforming state-of-the-art comparative algorithms. Furthermore, the effectiveness of the pupil emotion response theory and blink emotion response theory for depression detection is validated through significance testing.

2 Method

2.1 Overview

In this section, Sect. 2.2 describes the experimental paradigm for depression detection, Sect. 2.3 introduces the preprocessing method of the pupil data and the pupil features extraction process, Sect. 2.4 introduces the preprocessing method of the blink and eye-movement data and the blink and eye-movement features extraction process, Sect. 2.5 introduces feature selection, and Sect. 2.6 introduces depression classification model selection. Figure 1 shows the fundamental framework of our method.

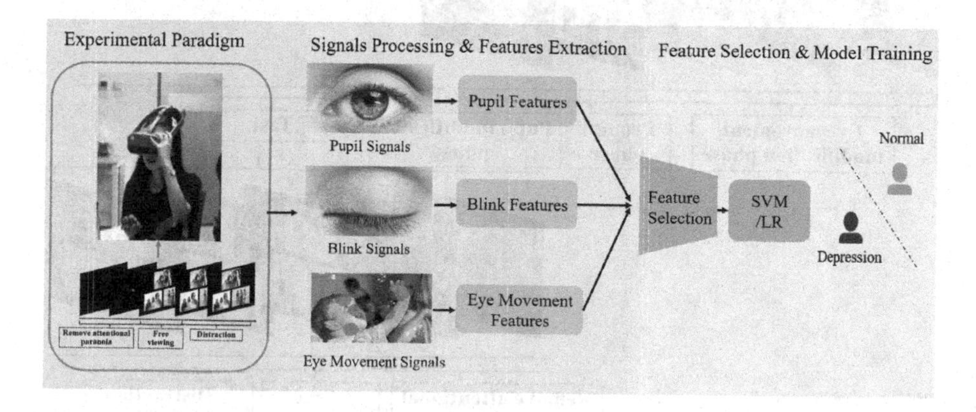

Fig. 1. Fundamental framework of the research.

2.2 Experimental Paradigm for Depression Detection

The emotional images used in the experiments come from the multi-dimensional semantic image set we built. The image set is based on the MMPI scale [5], and the semantics expressed by each group of images can find corresponding entries in the MMPI scale. This experiment contains a total of 30 sets of tasks, each of which presents one positive, one neutral, and one negative image at the same time. In total, there are 90 selected affective images.

The experimental paradigm consists of an eye-movement modification phase, a prompt phase, a pupil modification phase, and 30 groups of tasks. During the eye-movement modification phase, the subjects need to follow the prompts to gaze at the blue dots that appear on the screen in turn. The eye-tracking device tracks the gaze point of the subject and determines that after the subject fixates on the blue dot, the next blue dot will appear. The blue dots appear 5 times in sequence. If the eye tracking succeeds, it means that the eye-movement modification is completed. Before the start of the task, there is a prompt stage, in which the experimental process and the tasks that the subjects need to do are explained to the subjects in the form of pictures and text. During the pupil modification phase, a white pattern appeared in the center of a black screen, and the subjects were asked to fixate on the cross for 10 s. During this process, the pupils are fully dilated. Baseline pupil diameters were therefore obtained to eliminate individual differences. In the 30 groups of tasks, the subjects only need to complete the fixation according to the prompts. Figure 2 shows the overall flow of the experimental paradigm.

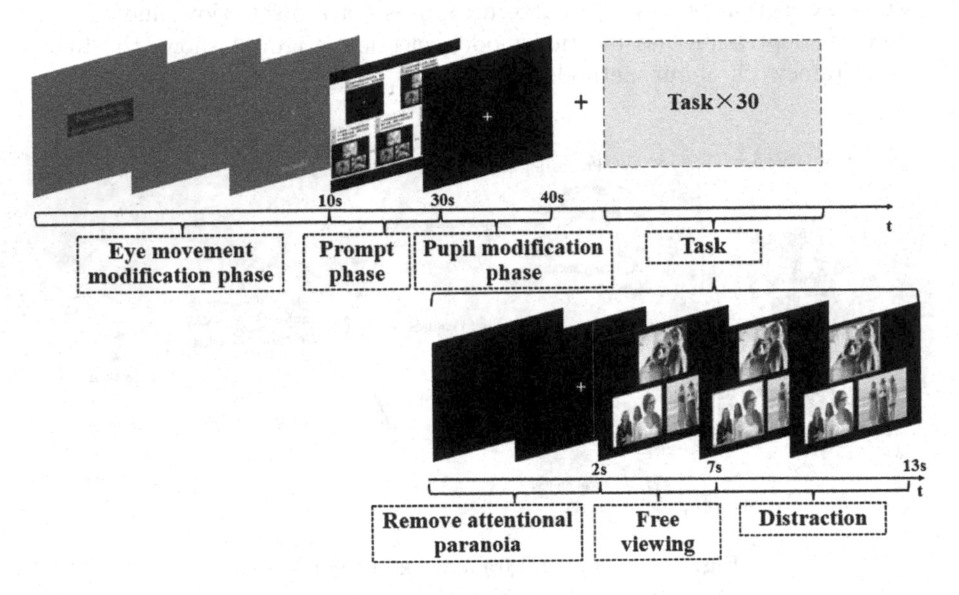

Fig. 2. Test procedure for a set of tasks in the experimental paradigm. (Color figure online)

Pupil, blink, and eye-movement data are acquired by a Tobii eye-tracking device throughout the experiment. We collected these data from a total of 82 participants from schools and hospitals. It included 39 depressed subjects (9 males and 30 females, age mean = 23, standard deviation = 1.34) and 43 normal controls (17 males and 26 females, age mean = 27, standard deviation = 1.41). All data were obtained after getting informed consent from the participants.

In order to eliminate the bias or variation inherent in the emotional images used in the experiment may affect the responses of the participants and, in turn, the generalization ability and reliability of the model. For the dataset, we have fully done the validity verification for the emotional attributes of the image (positive/neutral/negative). The verification method adopted is questionnaire scoring. In the questionnaire, 30 groups of images were set as 30 questions, and the three images in each question were scored by the respondents according to the emotional attributes of the images. The results showed that almost all images show their corresponding emotions.

2.3 Pupil Data Processing and Features Extraction

Due to objective factors such as the accuracy of the eye-tracking device and the individual differences of the subjects, there are some problems in the original pupil data, so the original pupil data need to be preprocessed. Here, we proceed as follows:

(1) Problem of uneven sampling points: Resampling of data.
(2) Missing data: Sect. 2.4 introduces the method of identifying the blinking point from the pupil diameter, and sets the pupil diameter value at the blinking point to 0. In order to study the cognitive and psychological characteristics of the pupil, it is necessary to process the blinking point. Since the subject blinked very little time, from the perspective of time, directly deleting the pupil data during the blink will not have a significant impact on the entire data. Secondly, we analyze the characteristics of eye blinking separately. Therefore, based on the above considerations, the missing pupil diameter data caused by the subject blinking was deleted. For the problem of missing data caused by non-blinking, we use the nearest neighbor interpolation method for interpolation.
(3) Binocular data inconsistency: For the problem of inconsistent pupil aperture values in the left and right eyes, the averaging process is performed to average the binocular data into monocular data.
(4) Individual differences: Due to the influence of individual physiological factors, the pupil diameter values of different individuals in the same environment are different, so it is unreasonable to directly use pupil diameter values to judge the emotional response of the subject when gazing at images. Therefore, we obtain the pupil diameter values of the subjects during the pupil modification phase to eliminate the factors of individual differences. Firstly, calculate the average pupil diameter of the subject in the pupil modification phase as the baseline, as Eq. 1. D_i is the baseline of subject i in the pupil modification phase, and n is the number of data collected in this phase. C is the pupil modification phase, and d_i^t is the pupil diameter of subject i at time t. Then calculate the calibration value based on the baseline, as Eq. 2. D_c is the calibrated diameter, d is the original diameter value, and D_i is the diameter baseline.
(5) Initial light response: The pupil diameter decreases significantly within the first one to two seconds of gazing at an emotional image, a phenomenon that has little to do with the emotional content of the image and is mainly caused by the image's brightness, called the initial light response. The features of the initial light response are extracted below.

$$D_i = \frac{1}{n}\sum_{t \in C} d_i^t \quad (1)$$

$$D_c = \frac{d}{D_i} \quad (2)$$

As in Fig. 3, based on the pupil response properties, we extract the processed pupil data into cognitive features. The pupillary base diameter indicates the mean value of the pupil diameter during the pupil modification phase. The pupillary constriction latency indicates the duration of the initial light response. The pupillary constriction velocity indicates the average pupil contraction velocity during the initial light response. The rank of diameter is the characteristic amount of pupil diameter after equalizing image brightness, arousal, and image position. The pupillary relative value indicates the relative value of pupil diameter for each type of image or each location. More features are extracted based on the base features, including the standard deviation of 30 sets of base features for all base features except the pupil reference value, for a total of 23 pupil features.

The pupil diameter of a person is affected by the brightness factor, and the pupil diameter decreases as the brightness increases. The brightness of each group of images viewed by the subjects was different, which led to the change of pupil diameter not being affected by the single factor of emotional response. Therefore, we define a "Rank of diameter for three types(ROD)" feature to eliminate the influence of brightness on pupil diameter. We calculate the average pupil diameter of subjects gazing at positive/negative/neutral images in each group of experiments, sort them from low to high, and record ordinal values (rank = 1,2,3), and accumulate 30 sets of experiments, and finally perform normalization. As Eq. 3.

$$ROD = \frac{\sum_{i=1}^{30} rank\,(i, type)}{\sum_{type} \sum_{i=1}^{30} rank\,(i, type)} type \in \{positive, neutral, negative\} \quad (3)$$

2.4 Blink and Eye-Movement Data Processing and Features Extraction

The eyelids partially or completely cover the pupil during blinking, so the pupil diameter collected by the eye-tracking device decreases rapidly or even becomes zero during blinking (when the eye-tracking device does not collect the pupil diameter, the pupil diameter is recorded as 0 or -1). Therefore, a blink is defined as an intermittent point that is continuously below the normal pupil diameter threshold (calculated as three times the standard deviation from the mean diameter) and lasts for 100 ms, and the number of consecutive groups of blink points is recorded as the number of blinks. After identifying the blink data and calculating the number of blinks, the intermittent points are then treated according to the treatment of the blink-induced data loss problem in Sect. 2.3.

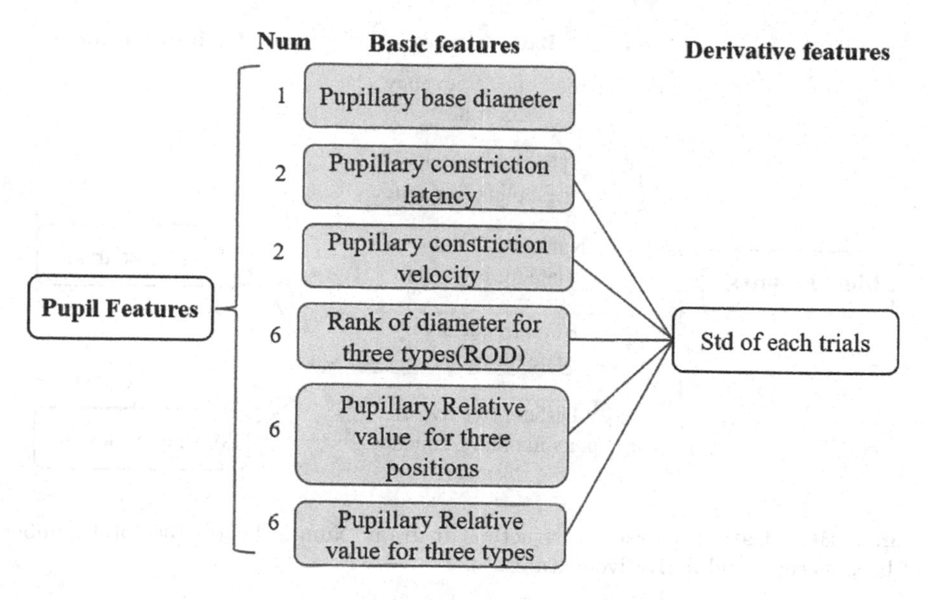

Fig. 3. Pupil features extraction structure diagram. Num indicates the total number of base features and derivative features.

Based on the blink emotional response properties, we extract blink-related features, which are divided into base features and derived features, as shown in Fig. 4. The states mentioned in the basic features represent the three states of the remove attentional paranoia phase, free viewing phase, and distraction phase during the experimental paradigm. We calculate the mean and standard deviation of the "Number of blinks by state" and "Blink frequency by state" in 30 sets of tasks; we also calculate the difference between the two blink frequencies in each state, totaling 32 blink features.

In addition to pupil features and blink features, we also extract eye-movement features based on attentional biases theory. Among them, 51 eye-movement features are studied for attentional biases theory, and 16 eye-movement features were studied for attentional orienting and attentional shifting characteristics, totaling 67 eye-movement features.

2.5 Feature Selection

Feature selection is a way to reduce the dimension of the data. In the case of high feature dimension, feature selection can improve classification efficiency and classification accuracy. The candidate features at this point have a high correlation, so we first use pearson correlation coefficient to calculate the features with correlation coefficients exceeding 0.8 in absolute value are filtered out, and only one of them is retained. Then, the embedding of random forest method are used to filter out the top 80% of features with cumulative importance. Finally, the features with p-value greater than 0.05 were filtered out using the correlation

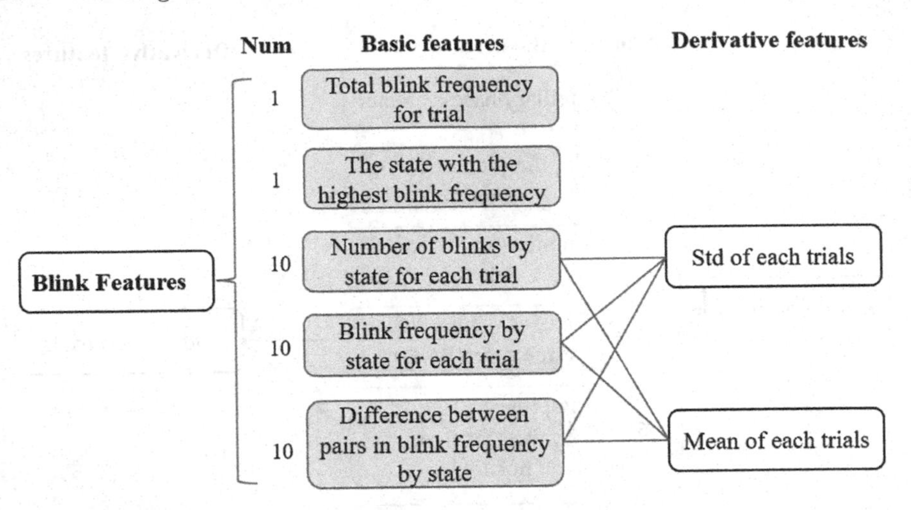

Fig. 4. Blink features extraction structure diagram. Num indicates the total number of base features and derivative features.

test. P-value less than 0.05 means that there is a significant difference between the depressed population and the normal population on the feature. There were 122 original candidate features, and after feature selection, 40 features were finally retained.

2.6 Depression Classification Model Selection

SVM models are suitable for classification tasks, and it can deal with high-dimensional data, has strong generalization ability, is suitable for small sample data, can deal with nonlinear problems, and has good robustness and interpretability. Logistic Regression (LR) is a linear model, and its advantages are its simple form and good interpretability. Therefore, it is the most widely used model in practical applications. Since the number of samples is limited, we train SVM and LR on a dataset of size 82 (39D+43N), respectively, and use a five-fold cross-validation method to obtain detection performance.

3 Experiment

3.1 Analysis of Feature Selection Results

Table 1 shows the results of the correlation test. S indicates that the p-value for the feature is less than 0.05, and depressed subjects are significantly different from normal subjects on this feature. Table 1 shows that there are significant differences in pupil diameter features between depressed subjects and normal subjects on both negative images and positive images. As shown in Fig. 5, when gazing at negative images, the pupil diameters of depressed subjects were generally larger than those of normal subjects. This verifies the previous conclusion that the depressed subjects had a greater emotional response when gazing at negative images, so they had greater pupil diameter dilation. Also, the blink frequency is significantly different in the remove attentional paranoia phase, free viewing phase, and distraction phase, verifying the previous conclusion that depressed patients blink more frequently than normal individuals; and the p-value for blink frequency is smaller in the free viewing phase and the distraction phase than in the remove attentional paranoia phase, because the remove attentional paranoia phase is not presented with emotional images, which also verify the previous conclusion that depressed patients are more likely to have a tense mental state than normal individuals when there is an emotional task, and so blink frequency increases.

Table 1. The significant analysis of significant features based on p-values.

Feature	P-value	5% level of significant
pupil_rela_negative	0.001	S
pupil_rela_positive	0.001	S
state_2_blink_v	0.031	S
state_3_blink_v	0.009	S
state_5_blink_v	<.001	S

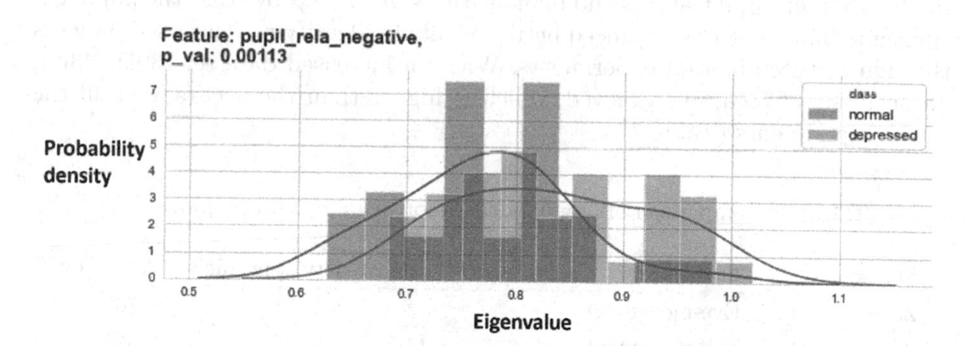

Fig. 5. Pupil diameter characteristic distribution.

3.2 Model Training and Classification Performance

We perform feature ablation experiments. We use sensitivity, specificity, and accuracy as the final evaluation indicators. As shown in Table 2, based on eye-movement, pupil, and blink features, the sensitivity and accuracy obtained by training the SVM classifier are the highest; the specificity obtained by training the LR classifier is the highest. However, since our goal is depression detection, higher sensitivity is more in line with the purpose of our study, so SVM is finally selected as the model we use, we obtain the best accuracy of 89.5%. Here, the training parameters of the SVM classifier are determined by a grid search method.

Table 2. The performance of different classifiers and features.

Classifier	Features	Sensitivity	Specificity	Accuracy
LR	Eye-movement	0.757	0.794	0.775
LR	Eye-movement&Pupil&Blink	0.818	**0.853**	0.837
SVM	Eye-movement	0.878	0.765	0.819
SVM	Eye-movement&Pupil&Blink	**0.969**	0.824	**0.895**

3.3 Comparison with Existing Work

Currently, most of the experiments to classify depressed and normal people based on eye-movement use a free-viewing experimental paradigm and use emotional face images. Compare with the work of Zeng et al. [18], Zhu et al. [19], and Shen et al. [13], we propose a different experimental paradigm to study pupillary and blink properties. Based on the extraction of eye-movement features, we further extract pupil features and blink features. And we prove that the pupil features and blink features produced better results in identifying depressed patients through feature ablation experiments. With an increased data set (39D+43N), an accuracy of 89.5% is achieved, which is higher than the accuracy of all the above experiments (Table 3).

Table 3. Comparison of original depression classification performance.

Method	Features	Sensitivity	Specificity	Accuracy
Zeng et al. [18]	Eye-movement	*	*	0.760
Zhu et al. [19]	Eye-movement	*	*	0.825
Shen et al. [13]	Eye-movement	0.690	**0.851**	0.770
Ours	Eye-movement&Pupil&Blink	**0.969**	0.824	**0.895**

*indicates that the value cannot be obtained

4 Conclusion

Our goal is to achieve an objective, efficient, and accurate method of Automated detecting depression. In this paper, we propose an experimental paradigm to elicit emotions from subjects. Based on the study of eye-movement features, we fuse pupil and blink features, and this multivariate feature fusion method for depression detection achieves better result. Meanwhile, we process and select the original features, and finally obtain 40 visual cognitive features. We perform a significant analysis of the features and validate the pupil response theory and blink theory. We conduct feature ablation experiments to verify that multi-feature fusion has a better effect on depression detection. Finally, we train the SVM classifier to classify depressed and normal subjects. With an expanded dataset we achieved 89.5% accuracy, which achieves a competitive level compared with state-of-the-art algorithms.

Acknowledgment. This work was supported by the National Nature Science Foundation of China (No. U20B2062 and No. 62227801), Central Government Guided Local Science and Technology Development Project (22-1-3-11-zyyd-nsh), the R&D Program of CAAC Key Laboratory of Flight Techniques and Flight Safety (NO. FZ2021ZZ05), and the Interdisciplinary Research Project for Young Teachers of USTB (Fundamental Research Funds for the Central Universities) (No. FRF-IDRY-21-001).

References

1. Drysdale, A.T., et al.: Resting-state connectivity biomarkers define neurophysiological subtypes of depression. Nat. Med. **23**(1), 28–38 (2017)
2. Eizenman, M., et al.: A naturalistic visual scanning approach to assess selective attention in major depressive disorder. Psychiatry Res. **118**(2), 117–128 (2003)
3. Giannakakis, G., et al.: Stress and anxiety detection using facial cues from videos. Biomed. Signal Process. Control **31**, 89–101 (2017)
4. Gilboa, E., Gotlib, I.H.: Cognitive biases and affect persistence in previously dysphoric and never-dysphoric individuals. Cogn. Emotion **11**(5–6), 517–538 (1997)
5. Graham, J.R.: MMPI-2: Assessing Personality and Psychopathology. Oxford University Press (1990)
6. Hedlund, S., Rude, S.S.: Evidence of latent depressive schemas in formerly depressed individuals. J. Abnorm. Psychol. **104**(3), 517 (1995)
7. Kellough, J.L., Beevers, C.G., Ellis, A.J., Wells, T.T.: Time course of selective attention in clinically depressed young adults: an eye tracking study. Behav. Res. Ther. **46**(11), 1238–1243 (2008)
8. Li, W., Ma, H., Wang, X., Shi, D.: Features derived from behavioral experiments to distinguish mental healthy people from depressed people. In: The 11th IASTED International Conference on Biomedical Engineering. ACTAPRESS (2014)
9. Lu, S., et al.: Attentional bias scores in patients with depression and effects of age: a controlled, eye-tracking study. J. Int. Med. Res. **45**(5), 1518–1527 (2017)
10. Mackintosh, J., Kumar, R., Kitamura, T.: Blink rate in psychiatric illness. Br. J. Psychiatry **143**(1), 55–57 (1983)
11. Newson, J.J., Thiagarajan, T.C.: EEG frequency bands in psychiatric disorders: a review of resting state studies. Front. Hum. Neurosci. **12**, 521 (2019)
12. WHO Organization: World mental health report: transforming mental health for all. World mental health report: transforming mental health for all (2022)
13. Shen, R., Zhan, Q., Wang, Y., Ma, H.: Depression detection by analysing eye movements on emotional images. In: ICASSP 2021–2021 IEEE International Conference on Acoustics, Speech and Signal Processing (ICASSP), pp. 7973–7977. IEEE (2021)

14. Li, R., Ma, H., Wang, R., Ding, J.: Device-adaptive 2D gaze estimation: a multi-point differential framework. In: Peng, Y., Hu, S.-M., Gabbouj, M., Zhou, K., Elad, M., Xu, K. (eds.) ICIG 2021. LNCS, vol. 12889, pp. 485–497. Springer, Cham (2021). https://doi.org/10.1007/978-3-030-87358-5_39
15. Siegle, G.J., Steinhauer, S.R., Carter, C.S., Ramel, W., Thase, M.E.: Do the seconds turn into hours? Relationships between sustained pupil dilation in response to emotional information and self-reported rumination. Cogn. Ther. Res. **27**(3), 365–382 (2003)
16. Skowron, K., et al.: The role of psychobiotics in supporting the treatment of disturbances in the functioning of the nervous system-a systematic review. Int. J. Mol. Sci. **23**(14), 7820 (2022)
17. Steidtmann, D., Ingram, R.E., Siegle, G.J.: Pupil response to negative emotional information in individuals at risk for depression. Cogn. Emot. **24**(3), 480–496 (2010)
18. Zeng, S., Niu, J., Zhu, J., Li, X.: A study on depression detection using eye tracking. In: Tang, Y., Zu, Q., Rodríguez García, J.G. (eds.) HCC 2018. LNCS, vol. 11354, pp. 516–523. Springer, Cham (2019). https://doi.org/10.1007/978-3-030-15127-0_52
19. Zhu, J., et al.: An improved classification model for depression detection using EEG and eye tracking data. IEEE Trans. Nanobiosci. **19**(3), 527–537 (2020)

Author Index

Q. Liu et al. (Eds.): PRCV 2023, LNCS 14429, pp. 529–531, 2024.
https://doi.org/10.1007/978-981-99-8469-5

Printed in the United States
by Baker & Taylor Publisher Services